Contact Geometry and Nonlinear Differential Equations

Methods from contact and symplectic geometry can be used to solve highly non-trivial non-linear partial and ordinary differential equations without resorting to approximate numerical methods or algebraic computing software. This book explains how it's done. It combines the clarity and accessibility of an advanced textbook with the completeness of an encyclopedia. The basic ideas that Lie and Cartan developed at the end of the nineteenth century to transform solving a differential equation into a problem in geometry or algebra are here reworked in a novel and modern way. Differential equations are considered as a part of contact and symplectic geometry, so that all the machinery of Hodge–de Rham calculus can be applied. In this way a wide class of equations can be tackled, including quasi-linear equations, Monge–Ampère equations (which play an important role in modern theoretical physics and meteorology).

The main features of the book are geometric transparency, clear and almost immediate applications to interesting problems, and exact solutions clarifying how approximate numerical solutions can be better obtained. The types of problem considered range from the classical (e.g., Lie's classification probelm) to the analysis of laser beams or the dynamics of cyclones. The authors balance rigor with the need to solve problems, so it will serve as a reference and as a user's guide.

ENCYCLOPEDIA OF MATHEMATICS AND ITS APPLICATIONS

Volume 101

Contact Geometry and Nonlinear Differential Equations

All the titles listed below can be obtained from good booksellers or from Cambridge University Press.

For a complete series listing visit http://www.cambridge.org/uk/series/

60 J. Krajicek *Bounded Arithmetic, Propositional Logic, and Complexity Theory*
61 H. Groemer *Geometric Applications of Fourier Series and Spherical Harmonics*
62 H. O. Fattorini *Infinite Dimensional Optimization and Control Theory*
63 A. C. Thompson *Minkowski Geometry*
64 R. B. Bapat and T. E. S. Raghavan *Nonnegative Matrices with Applications*
65 K. *Engel Sperner Theory*
66 D. Cvetkovic, P. Rowlinson and S. Simic *Eigenspaces of Graphs*
67 F. Bergeron, G. Labelle and P. Leroux *Combinatorial Species and Tree-Like Structures*
68 R. Goodman and N. Wallach *Representations and Invariants of the Classical Groups*
69 T. Beth, D. Jungnickel, and H. Lenz *Design Theory I, 2nd edn*
70 A. Pietsch and J. Wenzel *Orthonormal Systems for Banach Space Geometry*
71 G. E. Andrews, R. Askey and R. Roy *Special Functions*
72 R. Ticciati *Quantum Field Theory for Mathematicians*
73 M. Stern *Semimodular Lattices*
74 I. Lasiecka and R. Triggiani *Control Theory for Partial Differential Equations I*
75 I. Lasiecka and R. Triggiani *Control Theory for Partial Differential Equations II*
76 A. A. Ivanov *Geometry of Sporadic Groups I*
77 A. Schinzel *Polymomials with Special Regard to Reducibility*
78 H. Lenz, T. Beth, and D. Jungnickel *Design Theory II, 2nd edn*
79 T. Palmer *Banach Algebras and the General Theory of *-Algebras II*
80 O. Stormark *Lie's Structural Approach to PDE Systems*
81 C. F. Dunkl and Y. Xu *Orthogonal Polynomials of Several Variables*
82 J. P. Mayberry *The Foundations of Mathematics in the Theory of Sets*
83 C. Foias, O. Manley, R. Rosa and R. Temam *Navier–Stokes Equations and Turbulence*
84 B. Polster and G. Steinke *Geometries on Surfaces*
85 R. B. Paris and D. Kaminski *Asymptotics and Mellin–Barnes Integrals*
86 R. McEliece *The Theory of Information and Coding, 2nd edn*
87 B. Magurn *Algebraic Introduction to K-Theory*
88 T. Mora *Solving Polynomial Equation Systems I*
89 K. Bichteler *Stochastic Integration with Jumps*
90 M. Lothaire *Algebraic Combinatorics on Words*
91 A. A. Ivanov and S. V. Shpectorov *Geometry of Sporadic Groups II*
92 P. McMullen and E. Schulte *Abstract Regular Polytopes*
93 G. Gierz et al. *Continuous Lattices and Domains*
94 S. Finch *Mathematical Constants*
95 Y. Jabri *The Mountain Pass Theorem*
96 G. Gasper and M. Rahman *Basic Hypergeometric Series, 2nd edn*
97 M. C. Pedicchio and W. Tholen (eds.) *Categorical Foundations*
98 M. E. H. Ismail *Classical and Quantum Orthogonal Polynomials in One Variable*
99 T. Mora *Solving Polynomial Equation Systems II*
100 E. Olivieri and M. Eulália Vares *Large Deviations and Metastability*
101 A. Kushner, V. Lychagin and V. Rubtsov *Contact Geometry and Nonlinear Differential Equations*
102 L.W. Beineke, R. J. Wilson, P. J. Cameron. (eds.) *Topics in Algebraic Graph Theory*
103 O. Staffans *Well-Posed Linear Systems*
104 J. M. Lewis, S. Lakshmivarahan and S. Dhall *Dynamic Data Assimilation*
105 M. Lothaire *Applied Combinatorics on Words*

Contact Geometry and Non-linear Differential Equations

ALEXEI KUSHNER, VALENTIN LYCHAGIN AND
VLADIMIR RUBTSOV

CAMBRIDGE
UNIVERSITY PRESS

CAMBRIDGE
UNIVERSITY PRESS

University Printing House, Cambridge CB2 8BS, United Kingdom

One Liberty Plaza, 20th Floor, New York, NY 10006, USA

477 Williamstown Road, Port Melbourne, VIC 3207, Australia

314-321, 3rd Floor, Plot 3, Splendor Forum, Jasola District Centre, New Delhi - 110025, India

79 Anson Road, #06-04/06, Singapore 079906

Cambridge University Press is part of the University of Cambridge.

It furthers the University's mission by disseminating knowledge in the pursuit of education, learning and research at the highest international levels of excellence.

www.cambridge.org
Information on this title: www.cambridge.org/9780521824767

First published 2007

A catalogue record for this publication is available from the British Library

ISBN 978-0-521-82476-7 Hardback

Contents

Preface

The aim of this book is to introduce the reader to a geometric study of partial differential equations of second order.

We begin the book with the most classical subject: the geometry of ordinary differential equations, or more general, differential equations of finite type. The main item here is the various notions of symmetry and their use in solving a given differential equation. In Chapter 1 we discuss the distributions, integrability and symmetries. In a form appropriate to our aims, we remind the reader of the main notions of the geometry of distributions: complete integrability, curvature, integral manifolds and symmetries. The Frobenius integrability theorem is presented in its geometric form: as a flatness condition for the distribution.

The main result of this chapter is the famous Lie–Bianchi theorem which gives a condition and an constructive algorithm for integrability in quadratures of a distribution in terms of a Lie algebra of the shuffling symmetries. The theorem clearly explains the etymology of the expression "solvable Lie algebra."

In Chapter 2 we apply these results to explicit integration of scalar ordinary differential equations. We consider some standard examples of differential equations integrable in quadratures but treat them in quite non-standard geometric way to demonstrate the advantage of the language and the method of symmetries. Even in the case of linear differential equations one is able to find some new and interesting results by systematically exploiting the notion of symmetries. The most instructive illustration of this methodology is the application of the linear symmetries of (skew) self-adjoint linear operators. The space of linear symmetries admits in this case the structure of a Lie superalgebra. For example, the even part of the linear symmetries for the Schrödinger operator $L = \partial^2 + W$ is isomorphic to the Lie algebra sl_2, and the generating functions of the linear symmetries satisfy the third-order differential equation. The corresponding third-order operator is the second symmetric power of the Schrödinger one. This operator is also known as a *second Gelfand–Dikii Hamiltonian operator*, which transforms the functional space of the potentials W under appropriate boundary conditions into the infinite-dimensional Poisson algebra known as the *Virasoro algebra*. We use this operator to obtain a description of *integrable* potentials W such that the solutions of the Schrödinger equation $Lu = 0$ can be obtained in quadratures. The operator is also used to find symmetries of the eigenvalue problem for the Schrödinger operator. We show that if the potential W satisfies the KdV equation, or one of their higher analogs, then the eigenvalues and eigenfunctions can be found by quadratures.

In Chapter 3 we illustrate the potency of the geometric approach to the symmetries developing two constructions: a symmetry reduction and Lie's superposition principle.

The first construction is very natural: given an ideal τ of the Lie algebra $\mathfrak{g}$ of shuffling symmetries of a completely integrable distribution, we decompose the integrability problem into two steps: integrability of a completely integrable with symmetry algebra τ, and then the new one with the symmetry factor algebra $\mathfrak{g}/\tau$.

Taking τ to be the radical of $\mathfrak{g}$ we reduce by quadratures the integration of the initial distribution to a distribution with a semi-simple or simple symmetry algebra. The last distributions correspond to ordinary differential equations which we call *model equations.* We give a description of the model equations which correspond to the classical simple Lie algebras. One can see the advantage of using the model system from Lie's *superposition principle.* The principle provides us with *all* the solutions of the model differential equation when we know some finite number of solutions (*a fundamental system of solutions*) and a (in general-non-linear) superposition law. Thus, for the case of the three-dimensional Lie algebra sl_2, the model is the Riccati equation and the superposition rule is given by the *cross-ratio* of four points.

Part II of the book is devoted to *symplectic algebra*. Here we collect necessary information associated with the existence of a symplectic structure on a basic vector space. We had decided to gather together here all the main results not only for the sake of completeness and to make the book self-contained but also because of the conceptual importance of the symplectic structure for Monge–Ampère differential equations.

Historically, the appearance of a symplectic structure in the geometric studies of differential equation has traditionally been attributed to Huygens papers in geometric optics (though, strictly speaking, he had used merely a *contact structure* – an odd-dimensional cousin of the symplectic structure). The importance of symplectic geometry was recognized by J. Lagrange, G. Monge, A. Legendre and especially by Sophus Lie. E. Cartan and his Belgian student T. Lepage had used the symplectic machinery to study the geometry of Monge–Ampère equations at the beginning of the twentieth century. It is curious to note that T. Lepage had introduced a symplectic analog of Hodge theory before the appearance the "very" Hodge decomposition theorem on Riemannian manifolds.

The necessity of symplectic and contact geometry in mechanics is well known. For Monge–Ampère differential equations one should go further and use differential forms in the middle dimension.

The algebra of exterior forms on a symplectic vector spaces has some interesting specific features. In Chapter 5 we study as sl_2- structure given by a couple of "rising" $\top$ and "lowering" $\perp$ operators on the exterior forms:

$$\top\omega = \omega \wedge \Omega, \qquad \perp \omega = \iota_{X_\Omega}\omega$$

and by its commutator.

Here Ω is the given symplectic 2-form and ι_{X_Ω} is the contraction (the "inner product") with the symplectically dual bivector X_Ω.

Their commutator acts on k–forms by a multiplication:

$$\omega \to (n-k)\omega.$$

The form ω **annihilated** by $\perp$ is called a "primitive" or *effective k-form*. These forms are extremely important – they correspond to Monge–Ampère operators. The arguments of the sl_2-representation theory give the *Hodge – Lepage* decomposition theorem – the main result of this chapter – which states that any exterior k-form ω on the symplectic vector space V is a sum of the forms $\omega_i \wedge \Omega^i, i = 0, \ldots,$ where ω_i are effective forms uniquely determined by ω.

The classification problems for differential equations and operators have their trace in linear algebra – this is a classification of effective forms with respect to the symplectic group. Chapter 6 deals with the easiest classifications problems in dimension 4. In the next chapter we give a symplectic classification of exterior 2-forms in arbitrary dimensions.

In Chapter 8 we classify effective 3-forms in six-dimensional symplectic space with respect to a natural action of the symplectic group Sp_3 . The problem has a long history. Being in the spirit of the classical questions of the theory of geometric invariants, this problem was well known within a classification of spinors of dimension 12 and 14 (see [40], [90]) for the case when the base field is algebraically closed. Their methods do not work for the *real* classification. The first classification was obtained in our papers [74, 77] where the list of normal forms had a gap that was later filled by B. Banos [4, 5].

Part IV is devoted to the Monge–Ampère equations and to the related objects: Monge–Ampère operators and partial differential equation (PDE) systems on two-dimensional manifolds.

Chapters 9 and 10 contain some necessary information about symplectic and contact manifolds.

The application of the algebraic machinery of Chapter 5 gives a description of the Monge–Ampère equations and Monge – Ampère operators. The initial point of our approach is the following observation: to any differential k-form $\omega \in \Omega^k(J^1M)$, where J^1M is the space of 1-jet functions on a manifold M, we attach a non-linear second-order differential operator $\Delta_\omega : C^\infty(M) \to \Omega^k(M)$, acting as

$$\Delta_\omega(h) = j_1(h)^*(\omega),$$

where $j_1(h) : M \to J^1M$ is the 1-jet prolongation of a function $h \in C^\infty(M)$.

We see that the first advantage of this approach is a reduction of the order of the jet spaces: we use a simpler space J^1M instead of the space J^2M where the Monge–Ampère equations should lie *ad hoc*, being second-order partial differential equations. The space J^1M has the Cartan distribution which in this case is nothing but the aforementioned contact structure which impacts fascinatingly on the treatment of second-order differential operators and equations.

We should stress that our definition does not cover *all* non-linear second-order differential equations but only a certain subclass of them. This subclass is rather wide and contains all linear, quasi-linear and Monge–Ampère equations. We call the operators Δ_ω with $\omega \in \Omega^n(M)$, where $n = \dim M$, *Monge–Ampère operators*. The following observation justifies this definition: being written in a local canonical contact coordinates on J^1M the operators Δ_ω have the same type of non-linearity as the Monge – Ampère operators. Namely, the non-linearity involves the determinant of the Hesse matrix and its minors.

The correspondence $\omega \to \Delta_\omega$ is not one-to-one: this map has a huge kernel. If we denote the canonical contact 1-form on J^1M by ω_0, then the kernel is generated by the forms $\alpha \wedge \omega_0 + \beta \wedge d\omega_0$.

It is not hard to check that these forms produce an ideal $\mathcal{C}$ in the exterior algebra $\Omega^*(J^1M)$ which we call Cartan ideal, and the quotient $\Omega^*(J^1M)/\mathcal{C}$ by this ideal is isomorphic to the *effective* exterior forms $\Omega^*_\epsilon(J^1M)$ which we had discussed above. Hence, the effective exterior forms uniquely define Monge – Ampère operators and we can apply all of the machinery of contact/symplectic geometry to a study of these operators and the related non-linear differential equations. For example, from the geometrical point of view, solutions of differential equations corresponding to Δ_ω are nothing but the Legendre submanifolds L in J^1M which are *integral* with respect to the form ω, that is, $\omega|_L = 0$. It is also much easier to apply the contact transformations to differential forms than to the differential operators, so one can define (infinitesimal) symmetries of the Monge–Ampère operators and Monge–Ampère equations by using the induced action of the contact diffeomorphisms (respectively contact vector fields) on the effective differential forms.

In Chapter 11 we introduce and discuss some operators acting on the effective forms and (by correspondence) on the Monge – Ampère operators. First of all the de Rham operator induces a complex on the algebra of effective forms. The cohomology of the complex coincides with the de Rham cohomology of the base M up to dimension $n - 1$, where $n = \dim M$. They are trivial in dimensions greater than n, and only in dimension n do we have an essential difference with the cohomology of M. This relates to conservation laws and an Euler operator $\mathcal{E}$. By a conservation law we mean an $(n - 1)$-differential form θ on J^1M, such that $d\theta|_L = 0$ for all solutions L. One can check that this is possible if and only if $d\theta = g\omega \bmod C$ for some function $g \in C^\infty(J^1M)$. We call such a function a *generating function* of the conservation law. There is one-to-one correspondence between generating functions and conservation laws considered up to the trivial ones, and a function g is a generating function if and only if $\mathcal{E}(g\omega) = 0$. We study conservation laws in Chapter 13. We show a relation between contact symmetries and conservation laws for Monge–Ampère equations of divergent type that generalize the classical Noeter theorem in variational calculus. Conservation laws can be used in different directions and here we discuss their application to the classical problem of "sewing" of two solutions by a border of codimension 1. This leads us to the contact analog of the classical Hugoniot–Rankin conditions which are used further for description of shock waves and discontinuous solutions. The end of this chapter is devoted to an application of the developed approach to variational problems. We show that the Euler operator is exactly the operator in the corresponding Euler–Lagrange equations. The chapter closes with a description of non-holonomic filtration in the exterior algebra of J^1M and with an interpretation of the Euler operator as a connecting differential in the spectral sequence associated with this filtration.

Chapter 14 deals with the special case when the base manifold M is two-dimensional. This situation has a lot of complementary geometric and algebraic structures which come into play. The structures resulte from the naturally defined non-holonomic field of endomorphism on J^1M, that is, a field of operators defined on the Cartan distribution only. This leads to a *non-holonomic almost complex structure,* for elliptic differential equations, and to a *non-holonomic almost product structure* for the case of hyperbolic equations. For the parabolic one we get a *non-holonomic almost tangent structure*.

The theory becomes more enlightened in the case when J^1M is replaced by the cotangent bundle T^*M. We call the corresponding Monge–Ampère equation *symplectic* and their geometry is defined by the corresponding structure on the phase space. Thus, for example, elliptic equations define an almost complex structure on T^*M. The liaison between the geometric structures and equations will be used profoundly in subsequent chapters to establish and to clarify the classification and equivalence problems.

Part IV of the book has a somewhat specific feature – on one hand it is so important and voluminous that it could be chosen as the foundation for a separate book. On the other hand, we had decided to include this part as an illustration how the proposed approach can be effectively used in practice for many different type of equations, coming from practically all branches of the natural sciences. We were clearly unable to cover all possible applications but we had focused on the examples which were elaborated by ourselves during long-time contacts with physicists, engineers, biologists, etc. Most of the examples were developed up to numerical results and had figured as a useful part of some joint technical and scientific projects which were undertaken at the Applied Mathematical Department of Moscow Technical University of Civil Constructions during the period 1978–1990 under the direction of one of the authors (VL) and with a strong participation of two others. Some of the examples in this part were developed at Astrakhan State University under the guidance of one of the authors (AK) in a collaboration with the Biology Department of Moscow State University between 1999 and 2003. Another important collaboration which we acknowledge in this part is the long-time cooperation of third author (VR) with applied mathematicians and meteorologists from the Meteorological Office (Bracknell, UK) and Reading University.

Chapter 15 is devoted to a study of the Khokhlov–Zabolotskaya (KZ) equation. We knew about this equation in the mid-1970s from contacts with the theoretical acoustics group of R. Khokhlov (Physics Department of Moscow State University). The equation describes the propagation of three-dimensional sound beams in a non-linear medium. We treat this equation in its full three-dimensional version. It worth mentioning that the two-dimensional version of this equation is a hierarchy member of the famous dispersionless Kadomtsev–Petviashvili integrable system. The numerous applications and physically relevant versions of this dispersionless hierarchy are beyond the scope of our book (for statements and references see [78]).

We describe symmetries, conservation laws and exact solutions of the KZ equation. We discuss singularities of the solutions, Hugoniot–Rankin conditions and shock acoustic waves. Using this information we give a mathematical explanation to an experimentally verified phenomena of self-diffraction and periodic oscillation of sound beams (which is completely similar to the behavior of beams) and give some explicit formulas for the parameters of this behavior.

A version of the Kolmogorov–Petrovsky–Piskunov equation with a non-linear diffusion coefficient is the subject of our study in Chapter 16. This equation has a lot of interest in biology, ecology, and heat and mass transfer theory. We compute the Lie algebra of its symmetries and show how to use them to construct invariant solutions.

In Chapter 17 we collect all the applications of geometric studies of Monge–Ampère equations in theoretical meteorology where Monge–Ampère like operators had appeared some time ago in so-called *semi-geostrophic* models. They constitute an important class of models which are very useful in numerical weather prediction. This chapter is based on some on-going research by one of the authors with I. Roulstone (Surrey University, UK). We give a short account of the geometric study of balanced rotational models, which mathematically means a very special case of the Navier–Stokes system with the presence of Coriolis-like forces. An important aero- and hydrodynamical notion associated with these models is a *potential vorticity*. The conservation law of this quantity (under some mild restrictions) gives a non-linear differential equation which is easily represented in a form of the symplectic Monge–Ampère equation. The geometric structures related to this symplectic Monge–Ampère equation define and sometimes (and in turn are defined by) some nearly balanced two-dimensional model.

The second main motive of this book is a contact equivalence problem for differential equations. The first and important algebraic step to this problem was made in the previous parts of the book when we had discussed the classification and equivalence for (effective) exterior forms.

Part V of the book contains the contact classification results on Monge–Ampère equations (in analytic and smooth categories) which can be obtained by our geometric approach and which have the most complete form when the base manifolds are two- or three-dimensional.

The first case ($\dim M = 2$) can be attributed as a *classical Sophus Lie problem*. This problem was raised in the S. Lie article [66] and in our language it may be reformulated in the following way: *to find the equivalence classes of second-order (non-linear) differential equations with respect to the (local) group of contact diffeomorphisms.* Lie himself had stated some theorems (or without proofs or with some indications/hints of them) which can be considered as an attempt to give answers to the problem in some special cases. One of his main results is a statement about (quasi-)linearizability of any *analytic* Monge–Ampère equation. He had also considered the Monge–Ampère equation with constant coefficients and the Monge–Ampère equation in the presence of so-called *intermediate integrals*.

The essential inroad to the classical Lie problem was given by the French School and mainly by G. Darboux and E. Goursat who gave a classification of the two-dimensional hyperbolic Monge–Ampère equation under some restrictions. We should mention that the geometric version of Goursat's results was given by T. Morimoto who had used the language of G-structures and whose approach differs cardinally from ours (see [83]).

Chapters 18–21 contain a modern version of S. Lie's results (which to our knowledge were never completely proven until our papers in the early 1980s) and more general results on the symplectic and contact classification (A. Kushner, B. Kruglikov and D. Tunitskii), based on e-structures naturally defined by Monge–Ampère equations of the general type. Concerning the general equivalence problem we outline classification results for Monge–Ampère equations in the general case. We apply these results for classification and normal forms of symplectic Monge–Ampère equations on three-dimensional manifolds.

You will discover a family of sympathic cats which decorate the main text. Each cat has his own personal name and we hope that it will not be an enigmatic problem to our cleverminded and brilliant readers to understand why one or another cat appears in its proper place place in the text. A list of pictures of these cats with their names appear below.

Welcome Eureka Mentor Thinking Lazy Terminator

In conclusion we wish to thank our friends, colleagues and students for their help and support during the preparation of this book. In particular, we wish to thank Marat Djamaletdinov for his beautiful pictures which, we believe, should captivate readers.

PART I

Symmetries and Integrals

1

Distributions

1.1 Distributions and integral manifolds

1.1.1 Distributions

Let M be an $(n + m)$-dimensional smooth manifold, and let $\tau : TM \to M$ be the tangent bundle. By a *distribution* P on M one means a *smooth* field $P : a \in M \longmapsto P_a = P(a) \subset T_aM$ of m-dimensional subspaces of the tangent spaces. The number m is called a *dimension* of the distribution, $m = \dim P$, and n is called a *codimension* of $P, n = \operatorname{codim} P$.

There are different ways to say that P is a smooth field, and to define a distribution. We give the more important ones.

1. **As subbundles**
   Let $E_P = \bigcup_{a \in M} P(a) \subset TM$. Then P is a distribution if and only if $\tau_P : E_P \to M$ is a subbundle of the tangent bundle.
2. **By local bases**
   We say that a vector field $X \in D(M)$ *belongs* (or is *tangent*) to P on a subset $N \subset M$, if $X_a \in P(a)$ for all $a \in N$. Then smoothness of P means that there are local bases for P consisting of vector fields that belong to P.

   In other words, for any point $a \in M$ there exists a neighborhood $\mathcal{O}$ of a and m vector fields $X_1, \ldots, X_m$ that belong to P on $\mathcal{O}$ and such that vectors $X_{1,b}, \ldots, X_{m,b}$ form a basis of P_b at any $b \in \mathcal{O}$. Note, that the condition X *belongs* to P means that the vector field X is a section of the bundle $\tau_P : E_P \to M$.

   We denote by $D(P)$ the set of all vector fields that belong to P. It is clear that $D(P)$ is a module over the algebra $C^\infty(M)$ of smooth functions on M:

$$X, Y \in D(P) \Longrightarrow X + Y \in D(P),$$
$$f \in C^\infty(M), \quad X \in D(P) \Longrightarrow f X \in D(P).$$

In the case when $D(P)$ admits a global basis, consisting, say, of vector fields $X_1, \ldots, X_m$ we write $P = \mathcal{F}\langle X_1, \ldots, X_m\rangle$.

3. **By equations**
Let $\mathrm{Ann}(P(a)) \subset T_a^*M$ be the *annihilator* of $P(a) \subset T_aM$, that is

$$\mathrm{Ann}(P(a)) = \{\omega_a \in T_a^*M | \omega_a = 0 \text{ on } P(a)\}.$$

In other words, $\mathrm{Ann}(P(a))$ contains all linear equations for $P(a)$, that is all covectors ω_a vanishing on vectors from $P(a)$. Note that $\dim \mathrm{Ann}(P(a)) = n$.

We say that a differential 1-form $\omega \in \Omega^1(M)$ *annihilates* P on a subset $N \subset M$ if and only if $\omega_a \in \mathrm{Ann}(P(a))$ for all $a \in N$. We denote by $\mathrm{Ann}(P)$ the set of all differential forms on M that annihilate P and by $\Omega^1(M)$ the $C^\infty(M)$-module of differential 1-forms on M. Obviously, $\mathrm{Ann}(P)$ is a module over $C^\infty(M)$:

$$\alpha, \beta \in \mathrm{Ann}(P) \Longrightarrow \alpha + \beta \in \mathrm{Ann}(P),$$
$$f \in C^\infty(M), \quad \alpha \in \mathrm{Ann}(P) \Longrightarrow f\alpha \in \mathrm{Ann}(P).$$

In this terms the smoothness of P means that locally P can be defined by n differential 1-forms; that is, for any point $a \in M$ there exists a neighborhood $\mathcal{O}$ of a and n differential 1-forms $\omega_1, \ldots, \omega_n$ that annihilate P on $\mathcal{O}$ and such that covectors $\omega_{1,b}, \ldots, \omega_{n,b}$ form a basis of $\mathrm{Ann}(P(b))$ at any $b \in \mathcal{O}$. In the case when $\mathrm{Ann}(P)$ admits a global basis, say, $\omega_1, \ldots, \omega_n$ we write

$$P = \mathcal{F}\langle \omega_1, \ldots, \omega_n\rangle.$$

For the distribution $P = \mathcal{F}\langle X_1, \ldots, X_m\rangle$ one can define its derivatives. The distribution $P^{(1)}$ on M which is generated by the vector fields $X_1, \ldots, X_m$ and by all possible sorts of commutators $[X_i, X_j]$ $(i < j; i,j = 1, \ldots, m)$ is called the *first derivative* of P, i.e.,

$$P^{(1)} = \mathcal{F}\langle X_1, \ldots, X_m, [X_1, X_2], \ldots, [X_1, X_m], \ldots, [X_{m-1}, X_m]\rangle.$$

Analogously one can define the higher derivatives: $P^{(k+1)} \stackrel{\text{def}}{=} \left(P^{(k)}\right)^{(1)}$ $(k > 1)$.

1.1.2 Morphisms of distributions

Let $F : N \to M$ be a smooth map and let P be a distribution on M. Then differentials

$$F_{*,b} : T_bN \to T_aM$$

where $a = F(b)$, define a family $F^*(P)$ of vector spaces on N by

$$F^*(P)(b) = F_*^{-1}(P(a))$$

for all $b \in N$.

Dimension $\dim F^*(P)(b)$ is equal to $\dim \ker F_{*,b} + \dim(\operatorname{Im} F_{*,b} \cap P(a))$ and therefore varies in the general case.

Note that if $\omega_{1,a}, \ldots, \omega_{n,a}$ is a basis in $\operatorname{Ann}(P(a))$ then $F^*(\omega_{1,a}), \ldots, F^*(\omega_{n,a})$ generates $\operatorname{Ann}(F^*(P)(b))$.

We say that a map F is *P-regular* if the dimension function $b \in N \longmapsto \dim F^*(P)(b)$ is locally constant.

For the case of P-regular maps $F^*(P)$ is a distribution on N. We call this distribution the *image of P under F*.

The following three cases have great importance in applications.

1. F is a diffeomorphism, then the image $F^*(P)$ is well defined for any distribution P.
2. F is a surjection, or a smooth bundle. Then the image $F^*(P)$ is well defined for any distribution P also.
3. F is an embedding. Thus N is a submanifold of M. Then regularity F means that the intersection $T_bN \cap P(b)$ has a constant dimension for all $b \in N$. In this case we call $F^*(P)$ the *restriction of* P on N.

1.1.3 Integral manifolds

Let P be a distribution on M. A submanifold $i : N \hookrightarrow M$ is said to be *integral* for P if the restriction of P to N is equal to the tangent bundle, that is,

$$T_aN \subset P(a) \tag{1.1}$$

for any point $a \in N$.

This definition implies that the dimension of an integral manifold cannot exceed the dimension of a distribution.

If we define distributions in terms of differential 1-forms, say, locally $P = \mathcal{F}\langle\omega_1, \ldots, \omega_n\rangle$ then condition (1.1) takes the form

$$\omega_1|_N = 0, \ldots, \omega_n|_N = 0$$

of the so-called *Pfaff system*.

An integral manifold N is called a *maximal integral manifold* if for any point $a \in N$ one can find a neighborhood $\mathcal{O}$ of a such that there is no integral manifold N' such that $N' \supset N \cap \mathcal{O}$.

From this it is clear that the dimension of maximal integral manifolds does not exceed the dimension of the distribution. The distributions which have integral manifolds of the dimension equal to the dimension of the distribution are the simpler ones. They are called *completely integrable distributions*, or *CIDs*.

A smooth function $H \in C^\infty(M)$ is called *first integral* for P if $dH \in \mathrm{Ann}(P)$.

As we will see later on a distribution P, is completely integrable if and only if it has (locally) $n = \operatorname{codim} P$ first functional independent integrals $H_1, \ldots, H_n$; that is, (locally) $P = \mathcal{F}\langle dH_1, \ldots, dH_n \rangle$.

Note that any distribution has at least one-dimensional integral manifolds (integral curves). Indeed, any integral curve of a vector field $X \in D(P)$ is an integral curve of the distribution P. To find this curve one should solve some ordinary differential equations (ODEs). This observation has a general nature; namely, the problem of finding integral manifolds of a distribution implies a solution of some differential equation, and vice versa, the problem of finding solutions of differential equations is equivalent to finding integral manifolds of some distributions.

Let us look at a few examples.

Example 1.1.1 *The simplest non-trivial distribution is a one-dimensional distribution on the plane $M = \mathbb{R}^2$. Let x, y be coordinates on M, and let $P = \mathcal{F}\langle \omega \rangle$ where $\omega = a(x, y)\, dx + b(x, y)\, dy$, and $a^2 + b^2 \neq 0$. Let $N \subset M$ be an integral curve. Assume, for example, that x can be chosen as a (local) coordinate on N. Then N is a graph of a function $h(x)$,*

$$N = \{(x, h(x)), x \in \mathbb{R}\}$$

and the Pfaff equation $\omega|_N = 0$ takes the form of the first-order differential equation

$$a(x, h(x)) + b(x, y(x))y'(x) = 0.$$

Note also that $P = \mathcal{F}\langle X \rangle$ where

$$X = b(x, y)\frac{\partial}{\partial x} - a(x, y)\frac{\partial}{\partial y}$$

and therefore to find integral curves of the distribution one should solve the system of differential equations:

$$\dot{x} = b(x, y), \quad \dot{y} = -a(x, y).$$

The existence theorem shows that these equations have smooth solutions and therefore the distribution has integral manifolds of dimension 1. Thus P is a CID.

We observe from this example that integration of a first-order ODE, say,

$$y' = F(x, y)$$

is equivalent to finding integral curves of the distribution $P = \mathcal{F}\langle\omega\rangle$ where

$$\omega = dy - F(x, y)dx.$$

In this case we also have

$$P = \mathcal{F}\langle D\rangle$$

where

$$D = \frac{\partial}{\partial x} + F(x, y)\frac{\partial}{\partial y}.$$

The next example generalizes this observation for ODEs of arbitrary order.

Example 1.1.2 (Cartan distribution) *Let $M = \mathbb{R}^{k+1}$. Denote the coordinates in M by $x, p_0, p_1, \ldots, p_k$ and given a function $F(x, p_0, \ldots, p_{k-1})$ consider the following differential 1-forms:*

$$\begin{aligned}
\omega_0 &= dp_0 - p_1\,dx,\\
\omega_1 &= dp_1 - p_2\,dx,\\
\vdots\quad & \quad\vdots \quad\quad \vdots \quad\quad \vdots\\
\omega_{k-2} &= dp_{k-2} - p_{k-1}\,dx,\\
\omega_{k-1} &= dp_{k-1} - F(x, p_0, \ldots, p_{k-1})\,dx
\end{aligned}$$

and the distribution $P = \mathcal{F}\langle\omega_0, \ldots, \omega_{k-1}\rangle$. This is the one-dimensional distribution, called the Cartan distribution.

This distribution can also be described by a single vector field D, $P = \mathcal{F}\langle D\rangle$, where

$$D = \frac{\partial}{\partial x} + p_1\frac{\partial}{\partial p_0} + p_2\frac{\partial}{\partial p_1} + \cdots + p_{k-1}\frac{\partial}{\partial p_{k-2}} + F(x, p_0, \ldots, p_{k-1})\frac{\partial}{\partial p_{k-1}}.$$

If N is an integral curve of the distribution then x can be chosen as a coordinate on N, and therefore

$$N = \{(x, h_0(x), h_1(x), \dots, h_{k-1}(x)),\ x \in \mathbb{R}\}.$$

Conditions

$$\omega_0|_N = 0, \quad \dots, \quad \omega_{k-2}|_N = 0$$

imply that

$$h_1 = h_0', \quad h_2 = h_1', \quad \dots, \quad h_{k-1} = h_{k-2}'$$

or that

$$N = \left\{\left(x, h(x), h'(x), \dots, h^{(k-1)}(x)\right), x \in \mathbb{R}\right\}$$

for some function $h(x)$.

The last equation $\omega_{k-1}|_N = 0$ gives us an ordinary differential equation

$$h^{(k)}(x) = F\left(x, h(x), h'(x), \dots, h^{(k-1)}(x)\right).$$

The existence theorem shows us once more that the integral curves do exist, and therefore the Cartan distribution is a CID.

Note that three is the lowest number of dimensions where one can encounter a non-CID.

Example 1.1.3 (Contact distribution, see Figure 1.1) *Let $M = \mathbb{R}^{2n+1}$ and let $P = \mathcal{F}\langle\omega\rangle$ where*

$$\omega = du - \sum_{i=1}^{n} p_i\, dq_i$$

in the coordinates $(q_1, \dots, q_n, u, p_1, \dots, p_n)$ on $\mathbb{R}^{2n+1}$.

Then P is a $2n$-dimensional distribution, but there are no $2n$-dimensional integral manifolds of P. Let us assume, for example, that $n = 1$, and that N is a two-dimensional integral manifold such that q and p, say, are (local) coordinates on N. Then N given by $u = h(q, p)$ for some function $h(q, p)$ and $\omega|_N = 0$ imply contradictory differential equations for h:

$$h_q = p, \quad h_p = 0.$$

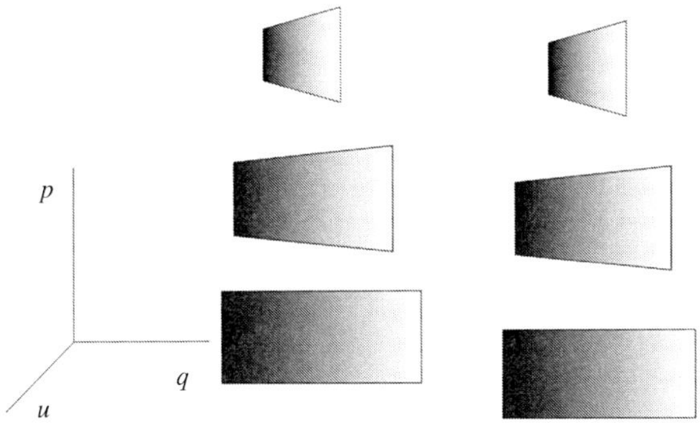

Figure 1.1. The contact distribution in $\mathbb{R}^3$.

On the other hand, every smooth function $f(q_1, \ldots, q_n)$ determines an n-dimensional submanifold

$$L_f = \left\{ u = f(q), p_1 = \frac{\partial f}{\partial q_1}, \ldots, p_n = \frac{\partial f}{\partial q_n} \right\}$$

which is integral because of

$$\omega|_{L_f} = 0.$$

Note that this distribution can also be defined by $2n$ vector fields

$$X_1 = \frac{\partial}{\partial q_1} + p_1 \frac{\partial}{\partial u}, \quad \ldots, \quad X_n = \frac{\partial}{\partial q_n} + p_n \frac{\partial}{\partial u},$$
$$Y_1 = \frac{\partial}{\partial p_1}, \quad \ldots, \quad Y_n = \frac{\partial}{\partial p_n}.$$

This example introduces the special case of a *contact distribution*, which is of great importance throughout this book.

Example 1.1.4 (Oricycle distribution) *Let $M = \mathbb{R} \times \mathbb{R}^+ \times S^1$ be a three-dimensional manifold. The following differential form*

$$\omega = (1 - \cos\phi)\, dx + \sin\phi\, dy - y\, d\phi$$

determines the so-called oricycle distribution on M.

Here $x, y > 0$ are coordinates on $\mathbb{R} \times \mathbb{R}^+$ and ϕ is the angle on S^1.

Integral curves of this distribution are oricycles.

Example 1.1.5 *Let M be a Möbius strip (see Figure 1.2). Define a one-dimensional distribution P on M where $P(a)$ is the line perpendicular to the*

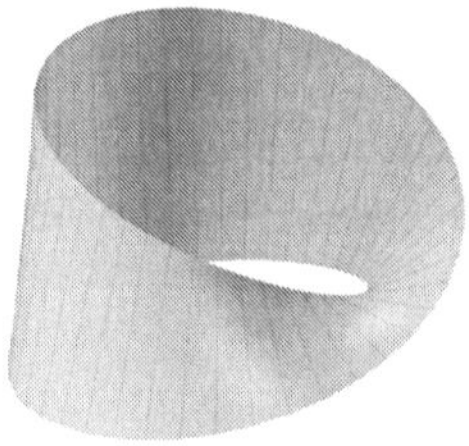

Figure 1.2. The distribution on the Möbius strip.

central circle of the strip. Both modules $D(P)$ and $\mathrm{Ann}(P)$ associated with the distribution are not free because the fibering of the Möbius strip over its central circle is non-trivial. But both modules become free as soon we cut the bundle in such a way that it becomes trivial. Integral curves of this distribution are the fibres of the bundle $M \to S^1$.

Example 1.1.6 (Overdetermined system of PDEs) *Consider the following overdetermined system of partial differential equations (PDEs):*

$$\phi_x = A(x, y, \phi, \psi),$$

$$\phi_y = B(x, y, \phi, \psi),$$

$$\psi_x = C(x, y, \phi, \psi),$$

$$\psi_y = D(x, y, \phi, \psi),$$

with respect to functions $\phi(x, y)$ and $\psi(x, y)$.

Define two differential 1-forms

$$\omega_1 = du - A(x, y, u, v)\, dx - B(x, y, u, v)\, dy$$

and

$$\omega_2 = dv - C(x, y, u, v)\, dx - D(x, y, u, v)\, dy$$

on the space $\mathbb{R}^4$ with coordinates x, y, u, v.

Then the pair of functions $(\phi(x, y), \psi(x, y))$ is a solution of the system if and only if the surface

$$\Gamma = \{u = \phi(x, y),\ v = \psi(x, y)\} \subset \mathbb{R}^4$$

is an integral manifold of the distribution $\mathcal{F}\langle\omega_1, \omega_2\rangle$.

1.2 Symmetries of distributions

A symmetry of a distribution is a transformation of the manifold M that maps the distribution into itself.

In other words, a diffeomorphism $F : M \to M$ is a *symmetry of a distribution P* if

$$F_*(P_a) = P_{F(a)} \tag{1.2}$$

for all $a \in M$.

Assume that $P = \mathcal{F}\langle\omega_1, \dots, \omega_n\rangle$, then (1.2) means that the differential 1-forms $F^*(\omega_1), \dots, F^*(\omega_n)$ determine the same distribution P and therefore can be expressed in terms of basis forms $\omega_1, \dots, \omega_n$.

That is

$$\begin{aligned} F^*(\omega_1) &= a_{11}\omega_1 + \cdots + a_{1n}\omega_n, \\ &\vdots \quad \vdots \quad \vdots \quad \vdots \\ F^*(\omega_n) &= a_{n1}\omega_1 + \cdots + a_{nn}\omega_n \end{aligned} \tag{1.3}$$

for some non-degenerate matrix $\|a_{ij}\|$ with $a_{ij} \in C^\infty(M)$.

On the other hand, the last conditions could be written without additional functions a_{ij}:

$$F^*(\omega_1) \wedge \omega_1 \wedge \cdots \wedge \omega_n = 0, \quad \dots, \quad F^*(\omega_n) \wedge \omega_1 \wedge \cdots \wedge \omega_n = 0. \tag{1.4}$$

Example 1.2.1 *The Legendre transformation*

$$F : (q, u, p) \mapsto (p, u - qp, -q)$$

is a symmetry of the contact distribution.

Indeed, in this case

$$F^*(\omega) = d(u - pq) + q\,dp = \omega.$$

Example 1.2.2 *The transformation*

$$G : (q, u, p) \mapsto (p, \lambda qp - \lambda u, \lambda q),$$

where $\lambda \neq 0$ is some constant, is a symmetry of the contact distribution. Really, $G^*(\omega) = -\lambda\omega$.

A vector field X on the manifold M is called an *infinitesimal symmetry*, or simply a *symmetry*, of the distribution P if the flow A_t along the vector field X

consist of symmetries of P, i.e.,

$$(A_t)_*(P(a)) = P(A_t a)$$

for all points $a \in M$ and t.

Denote by $\mathrm{Sym}(P)$ the set of all infinitesimal symmetries of the distribution P.

Theorem 1.2.1 *The following conditions are equivalent:*

(*a*) $X \in \mathrm{Sym}(P)$,

(*b*) $\forall Y \in D(P) \Longrightarrow [X, Y] \in D(P)$,

(*c*) $\forall \omega \in \mathrm{Ann}(P) \Longrightarrow L_X(\omega) \in \mathrm{Ann}(P)$.

Proof

(a)⇒(b) Suppose that P is an m-dimensional distribution generated by vector fields $X_1, \ldots, X_m$ or by n differential 1-forms $\omega_1, \ldots, \omega_n$. Let $X \in \mathrm{Sym}(P)$ and let $\{A_t\}_{t \in I}$ be the one-parametric transformation group of X. Then the transformations A_t preserve the distribution P. Therefore,

$$(A_t)_*(X_i) = \sum_{j=1}^{m} A_{ij}(t) X_j, \tag{1.5}$$

where $A_{ij}(t) \in C^\infty(M)$ are functions that depend smoothly on the parameter t. Differentiating at $t = 0$ and using the definition of the Lie derivative we obtain:

$$[X_i, X] = L_X(X_i) = \sum_{j=1}^{m} a_{ij} X_j$$

with $a_{ij} \overset{\text{def}}{=} (d/dt) A_{ij}(t)|_{t=0}$. Hence, $[D(P), X] \subset D(P)$.

(b)⇒(c) Let $\omega \in \mathrm{Ann}(P)$. Then using $L_X(\omega) = \iota_X(d\omega) + d(\iota_X \omega)$ and the infinitesimal Stokes' formula one obtains:

$$L_X(\omega)(X_i) = d\omega(X, X_i) + X_i(\omega(X)) = X(\omega(X_i)) - \omega([X, X_i]) = 0$$

for all $i = 1, \ldots, m$. Therefore, $L_X(\omega) \in \mathrm{Ann}(P)$. Here ι_X is an operation of interior multiplication by the vector field X:

$$(\iota_X \theta)(Y_1, \ldots, Y_{k-1}) \overset{\text{def}}{=} \theta(X, Y_1, \ldots, Y_{k-1})$$

for any k-form θ. We also denote this operation by $X \rfloor$.

(c)⇒(a) Let us consider the following differential $(n+1)$-forms:

$$\Omega_j(t) \stackrel{\text{def}}{=} A_t^*(\omega_j) \wedge \omega_1 \wedge \cdots \wedge \omega_n, \tag{1.6}$$

which depend smoothly on the parameter t. We have $\Omega_j(0) = 0$, since $A_0 = \text{id}$. Now we are going to prove that $\Omega_j(t) \equiv 0$ for all $t \in I$. This will imply that $A_t^*(\omega_j)$ are linear combination of $\omega_1, \ldots, \omega_n$ for all t and that X is a symmetry of the distribution P. We find

$$\frac{d}{dt}\Omega_j(t) = A_t^*(L_X(\omega_j)) \wedge \omega_1 \wedge \cdots \wedge \omega_n.$$

From (b) it follows that

$$L_X(\omega_j) = \sum_{i=1}^n b_{ji}\omega_i$$

$(j = 1, \ldots, n)$ for some functions $b_{ji} \in C^\infty(M)$. Moreover,

$$A_t^*(L_X(\omega_j)) = \sum_{i=1}^n A_t^*(b_{ji}) A_t^*(\omega_i)$$

and, therefore,

$$\frac{d}{dt}\Omega_j(t) = \sum_{i=1}^n A_t^*(b_{ji})\Omega_i(t). \tag{1.7}$$

Equation (1.7) shows that the vector consisting of $(n+1)$-forms $\Omega_1(t), \ldots, \Omega_n(t)$ is a solution of the linear homogeneous system of ordinary differential equations with zero initial conditions. Hence, it must be equal to zero identically. ■

Corollary 1.2.1 Sym(P) *is a Lie* $\mathbb{R}$*-algebra with respect to the commutator of the vector fields.*

Proof We shall prove the following statements:

1. $X, Y \in \text{Sym}(P) \Longrightarrow X + Y \in \text{Sym}(P)$;
2. $X \in \text{Sym}(P), \lambda \in \mathbb{R} \Longrightarrow \lambda X \in \text{Sym}(P)$;
3. $X, Y \in \text{Sym}(P) \Longrightarrow [X, Y] \in \text{Sym}(P)$.

All of these statements follow from (c) and the properties of the Lie derivative: $L_{X+Y} = L_X + L_Y, L_{\lambda X} = \lambda L_X, L_{[X,Y]} = [L_X, L_Y]$. ■

Example 1.2.3 *The vector field*

$$X = a\frac{\partial}{\partial q} + b\frac{\partial}{\partial u},$$

where $a, b \in \mathbb{R}$, is an infinitesimal symmetry of the contact distribution on $\mathbb{R}^3$. Indeed, the one-parameter group corresponding to the vector field is

$$A_t : (q, u, p) \longmapsto (q + at,\ u + bt,\ p).$$

Therefore, for $\omega = du - p\,dq$, we have

$$(A_t)^*(\omega) = \omega.$$

Example 1.2.4 *Consider a distribution P on $\mathbb{R}^n$ generated by differential 1-forms*

$$\omega_j = \omega_{j1}(x)\,dx_1 + \cdots + \omega_{jn}(x)\,dx_n$$

where $j = 1, \ldots, n$ and $\omega_{j1}, \ldots, \omega_{jn}$ are smooth functions.

If all of these functions do not depend on the variable x_1, then the vector field $\partial/\partial x_1$ is a symmetry of the distribution P because

$$L_{\partial/\partial x_1}(\omega_j) = 0.$$

Example 1.2.5 *Let P be a one-codimensional distribution generated by a differential 1-form ω. A vector field X is an infinitesimal symmetry of this distribution if and only if the Lie derivative along X of the form ω is proportional to ω. It means that $L_X(\omega) = f\omega$ for some smooth function f, or $L_X(\omega) \wedge \omega = 0$.*

Example 1.2.6 *The vector fields*

$$X = \frac{\partial}{\partial x}, \quad Y = y\frac{\partial}{\partial y} + \sin\phi\frac{\partial}{\partial \varphi}$$

are symmetries of the oricycle distribution.

Example 1.2.7 *Let us find infinitesimal symmetries of the contact distribution in $\mathbb{R}^3$. We shall consider the more general case later on.*

Let

$$X = a(q,u,p)\frac{\partial}{\partial q} + b(q,u,p)\frac{\partial}{\partial u} + c(q,u,p)\frac{\partial}{\partial p}$$

be an infinitesimal symmetry of the distribution. Then the Lie derivative

$$L_X(du - p\,dq) = db - c\,dq - p\,da$$

should vanish when $du = p\,dq \,\mathrm{mod}(\omega)$.

This gives two differential equations on the components of X*:*

$$p\frac{\partial a}{\partial p} - \frac{\partial b}{\partial p} = 0,$$

$$\frac{\partial b}{\partial q} - p\frac{\partial a}{\partial q} - c + p\left(\frac{\partial b}{\partial u} - p\frac{\partial a}{\partial u}\right) = 0.$$

Let us put

$$f(q,u,p) = b(q,u,p) - p\,a(q,u,p).$$

Then,

$$a = -\frac{\partial f}{\partial p}, \quad b = f - p\frac{\partial f}{\partial p}, \quad c = \frac{\partial f}{\partial q} + p\frac{\partial f}{\partial u},$$

and any infinitesimal symmetry of the contact distribution in $\mathbb{R}^3$ *is determined by a smooth function* f *and has the following form:*

$$X_f = -\frac{\partial f}{\partial p}\frac{\partial}{\partial q} + \left(f - p\frac{\partial f}{\partial p}\right)\frac{\partial}{\partial u} + \left(\frac{\partial f}{\partial q} + p\frac{\partial f}{\partial u}\right)\frac{\partial}{\partial p}.$$

Note that $\omega(X_f) = f$.

1.3 Characteristic and shuffling symmetries

Let X be a symmetry of a distribution P and let A_t be the flow along X. Then $A_t(N)$ is an integral manifold of P for any integral manifold N. This reflects the main property of symmetries: one-parameter transformation groups generated by symmetries act on the set of integral manifolds.

There is, however, a very distinguished class of symmetries that transform each maximal integral manifold into itself. They are called *characteristic symmetries*.

Namely, if an (infinitesimal) symmetry X belongs to the distribution P, then it is called a *characteristic symmetry*.

Denote by $\mathrm{Char}(P)$ the set of all characteristic symmetries

$$\mathrm{Char}(P) \stackrel{\text{def}}{=} \mathrm{Sym}(P) \cap D(P).$$

Lemma 1.3.1 *Let X be a characteristic symmetry of a distribution P and let N be a maximal integral manifold of P. Then X is tangent to N.*

Proof Denote by A_t the corresponding one-parameter transformation group generated by X, and assume that X is not tangent to X at some point $a \in N$. Then there is a neighborhood $\mathcal{O}$ of a such that

$$N' = \bigcup_{-\varepsilon<t<\varepsilon} A_t(N \cap \mathcal{O})$$

is a smooth submanifold of M (for sufficiently small ε).

Moreover, $N' \supset N \cap \mathcal{O}$ and N' is an integral manifold. To see this note that tangent spaces to N' are sums of tangent spaces to $A_t(N \cap \mathcal{O})$ and the one-dimensional subspace generated by X, but both of them belong to P. Hence with N' integral, N' should belong to N and X should be tangent to N. ∎

Theorem 1.3.1

1. $\mathrm{Char}(P)$ *is an ideal of the Lie algebra* $\mathrm{Sym}(P)$. *That is,* $\mathrm{Char}(P)$ *is a linear subspace of* $\mathrm{Sym}(P)$ *and*

$$X \in \mathrm{Sym}(P), \quad Y \in \mathrm{Char}(P) \Longrightarrow [X, Y] \in \mathrm{Char}(P).$$

2. $\mathrm{Char}(P)$ *is a* $C^\infty(M)$*-module:*

$$f \in C^\infty(M), \quad Y \in \mathrm{Char}(P) \Longrightarrow f\,Y \in \mathrm{Char}(P).$$

Proof

1. It is clear that $[X, Y] \in \mathrm{Sym}(P)$. Applying the formula $[L_X, \iota_Y] = \iota_{[X,Y]}$ to a differential 1-form $\omega \in \mathrm{Ann}(P)$ we obtain
$$\omega([X, Y]) = \iota_{[X,Y]}(\omega) = L_X(\iota_Y\omega) - \iota_Y(L_X(\omega)) = 0.$$
Because $\iota_Y\omega = 0$ and $L_X(\omega) \in \mathrm{Ann}(P)$.
2. We have $[f\,Y, X] = f[Y, X] - X(f)Y \in D(P)$ for any vector field $X \in D(P)$. Thus, $f\,Y \in \mathrm{Sym}(P)$, but $f\,Y \in D(P)$ too. Therefore, $f\,X \in \mathrm{Char}(P)$.

■

Since $\mathrm{Char}(P)$ is an ideal of the Lie algebra $\mathrm{Sym}(P)$, we can define the quotient Lie algebra
$$\mathrm{Shuf}(P) \overset{\mathrm{def}}{=} \mathrm{Sym}(P)/\,\mathrm{Char}(P).$$

Elements of $\mathrm{Shuf}(P)$ are called *shuffling symmetries* of P.

The geometrical sense of shuffling symmetries can be explained as follows.

Let us denote by $\mathrm{Sol}(P)$ the set of all maximal integral manifolds of P. Then any symmetry $X \in \mathrm{Sym}(P)$ generates a flow on $\mathrm{Sol}(P)$, and, as we have seen, the characteristic symmetries generate trivial flows. Moreover, if for some symmetries $X, Y \in \mathrm{Sym}(P)$ the difference $X - Y \in \mathrm{Char}(P)$, then the corresponding flows on $\mathrm{Sol}(P)$ are the same.

In other words, classes $X \bmod \mathrm{Char}(P)$ mix or "shuffle" the set of all maximal manifolds like a player shuffles a pack of cards.

Example 1.3.1 *Let the distribution P on the puncturing plane $\mathbb{R}^2\backslash\{0\}$ with coordinates x, y be given by $\omega = x\,dx + y\,dy$. Maximal integral manifolds of this distribution are circles $x^2 + y^2 = \mathrm{const}$. The vector field*
$$X = -y\frac{\partial}{\partial x} + x\frac{\partial}{\partial y}$$
generates characteristic symmetries as a module over $C^\infty(\mathbb{R}^2 \setminus \{0\})$.

The class of symmetry
$$Y = x\frac{\partial}{\partial x} + y\frac{\partial}{\partial y}$$
is a non-trivial shuffling symmetry of P (see Figure 1.3).

Example 1.3.2 *The contact distribution in $\mathbb{R}^3$ has no characteristic symmetries. We shall see later on that this is true for any contact distribution. Indeed, let $X = X_f$ be a characteristic symmetry. Then $\omega(X) = 0$. But $\omega\left(X_f\right) = f$, and therefore $f = 0$ and $X = 0$.*

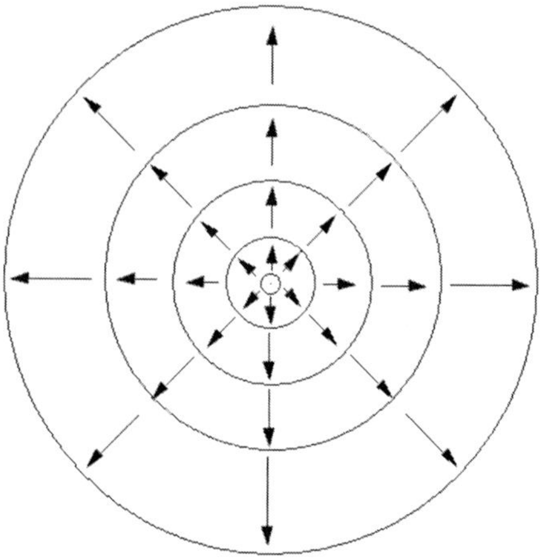

Figure 1.3. The radial symmetry.

1.4 Curvature of a distribution

The commutator of vector fields

$$X, Y \longmapsto [X, Y]$$

is the skew-symmetric and $\mathbb{R}$-bilinear map.

This is not $C^{\infty}(M)$-bilinear because

$$[fX, gY] = fg[X, Y] + f\,X(g)Y - gY(f)X.$$

Let us introduce a *bundle* ν_P *normal to* a distribution P as a factor-bundle TM/P, that is

$$\nu_{P,x} = T_xM/P_x$$

for any $x \in M$, and restrict the commutator map on $D(P)$.

The above formula shows that map

$$\Omega_P : D(P) \times D(P) \to \Gamma(\nu_P),$$
$$\Omega_P : X \in D(P), \quad Y \in D(P) \longmapsto [X, Y] \bmod D(P)$$

is $C^{\infty}(M)$-bilinear, and therefore gives a skew-symmetric 2-form on the distribution with values in the normal bundle.

We call Ω_P a *curvature of the distribution P*.

The value of Ω_P at a point $x \in M$ can be computed as follows:

$$\Omega_P(X_x, Y_x) = [X, Y] \operatorname{mod} P_x$$

where vector fields X, Y are any extensions of vectors $X_x, Y_x \in T_x M$ on M.

Note that $\operatorname{Ann}(P_x)$ annihilates P_x and

$$\nu_x^* = \operatorname{Ann}(P_x), \quad P_x^* = T^*M / \operatorname{Ann}(P_x).$$

Therefore, if $\omega_1, \ldots, \omega_n$ is a (local) free basis for $\operatorname{Ann}(P)$ and $X_{\omega_1}, \ldots, X_{\omega_n}$ is the dual basis in ν_P then we obtain for the curvature form

$$\Omega_P(X, Y) = -\sum_{i=1}^{n} d\omega_i(X, Y)\, X_{\omega_i}.$$

Indeed, if $X, Y \in D(P)$ and $\theta \in \operatorname{Ann}(P)$, then by definition

$$\langle \theta, \Omega_P(X, Y)\rangle = \langle \theta, [X, Y]\rangle = X(\theta(Y)) - Y(\theta(X)) - d\theta(X, Y) = -d\theta(X, Y),$$

and therefore the ith coordinate of $\Omega_P(X, Y)$ with respect to the basis $X_{\omega_1}, \ldots, X_{\omega_n}$ is equal to $d\omega_i(X, Y)$.

We shall write

$$\Omega_P = -\sum_{i=1}^{n} d\omega_i \otimes X_{\omega_i}. \tag{1.8}$$

We say that a distribution P is *flat*, or, as we shall see later on, *completely integrable*, if $\Omega_P = 0$.

Example 1.4.1 *All one-dimensional distributions are flat.*

Example 1.4.2 *If* $\operatorname{Ann}(P)$ *admits a (local) basis of closed 1-forms then (1.8) shows that P is flat.*

Example 1.4.3 *The contact distribution on* $\mathbb{R}^3$ *given by the 1-form* $\omega = du - p\,dq$ *has the following curvature:*

$$\Omega = dp \wedge dq \otimes \frac{\partial}{\partial u}.$$

Example 1.4.4 *The proposed construction of the curvature forms for distributions is very similar to the construction of the curvature of connections. Namely, let* $\pi : M \to B$ *be a bundle over a base manifold B. Recall that vectors tangent to the fibres of the projection* π *are called* vertical, *and the subspace of*

T_xM consisting of all vertical vectors at the point is called vertical *and denoted by T_x^v:*

$$T_x^v = T_x(\pi^{-1}(b))$$

where $b = \pi(x)$.

A connection in the bundle *is given by a distribution P,* $\dim P = \dim B$, *such that subspaces P_x and T_x^v are transversal at each point $x \in M$. In other words,*

$$T_xM = T_x^v \oplus P_x.$$

Vectors of P_x are called horizontal *with respect to connection.*

In this case the normal bundle ν_P can be identified with the vertical vectors bundle and the curvature form Ω_P defines a 2-form

$$\Omega_P : P_x \wedge P_x \to T_x^v$$

or, due to isomorphism $\pi_* : P_x \to T_bB$, where $b = \pi(x) \in B$, a 2-form

$$\Omega_P : T_bB \wedge T_bB \to T_x^v.$$

which is a curvature form of the connection.

1.5 Flat distributions and the Frobenius theorem

A connection in Example (1.4.4) is called *flat* if the curvature 2-form vanishes. In a similar way, we say that a distribution P on a manifold is *flat* if the curvature 2-form vanishes.

The condition of flatness can be reformulated in terms of vector fields or differential 1-forms generating the distribution.

Lemma 1.5.1 *The following conditions are equivalent:*

(*a*) *$P = \mathcal{F}\langle X_1, \ldots, X_m\rangle = \mathcal{F}\langle \omega_1, \ldots, \omega_n\rangle$ is flat.*

(*b*) *There are smooth functions c_{ij}^l such that*

$$[X_i, X_j] = \sum_l c_{ij}^l X_l$$

for all $i, j = 1, \ldots, m$.

(c) *There are differential 1-forms α_{ij} such that*

$$d\omega_i = \sum_j \alpha_{ij} \wedge \omega_j$$

for all $i = 1, \ldots, n$.

(d)

$$d\omega_i \wedge \omega_1 \wedge \cdots \wedge \omega_n = 0$$

for all $i = 1, \ldots, n$.

Proof

(a)$\Longleftrightarrow$(b) Indeed, $\Omega_P = 0$ if and only if the commutator $[X, Y]$ of any two vector fields $X, Y \in D(P)$ belongs to $D(P)$.

(a)$\Longleftrightarrow$(c) Condition (c) means that all $d\omega_i$ vanish on P, and due to (1.8) this is equivalent to (a).

(c)$\Longleftrightarrow$(d) Both conditions state that $d\omega_i$ equals zero on $\omega_1 = \cdots = \omega_n = 0$. ∎

Lemma 1.5.2 *Let P be a distribution satisfying one of the above conditions. Suppose that a vector field X belongs to P and A_t is the one-parameter group of shifts along X. Then the differential 1-forms $A_t^*(\omega_1), \ldots, A_t^*(\omega_n)$ vanish on P for all t and therefore they are linear combinations of the forms $\omega_1, \ldots, \omega_n$.*

Proof Consider the differential $(n+1)$-forms

$$\Theta_j(t) \stackrel{\text{def}}{=} A_t^*(\omega_j) \wedge \omega_1 \wedge \cdots \wedge \omega_n,$$

where $j = 1, \ldots, n$.

Then

$$\frac{d\Theta_j}{dt} = \frac{d}{dt} A_t^*(\omega_j) \wedge \omega_1 \wedge \cdots \wedge \omega_n = A_t^*(L_X(\omega_j)) \wedge \omega_1 \wedge \cdots \wedge \omega_n$$

and

$$L_X(\omega_j) = \iota_X(d\omega_j) + d(\omega_j(X)) = \iota_X(d\omega_j) = \alpha_{j1}(X)\omega_1 + \cdots + \alpha_{jn}(X)\omega_n.$$

Therefore,

$$\frac{d\Theta_j}{dt} = \Phi_{j1}(t)\Theta_1 + \cdots + \Phi_{jn}(t)\Theta_n, \tag{1.9}$$

where $\Phi_{js}(t) = A_t^*(\alpha_{js}(X))$.

In other words, the forms $\Theta_1(t), \ldots, \Theta_n(t)$ satisfy the linear ordinary differential equations (1.9). Since $\Theta_j(0) = 0$ for all $j = 1, \ldots, n$, the uniqueness theorem implies that $\Theta^j(t) \equiv 0$ for all $j = 1, \ldots, n$ and all t. ∎

Lemma 1.5.3 *Let P be a distribution satisfying one of the above conditions, and let L be an integral manifold of the distribution P. Let X be a vector field such that X belongs to P and X is not tangent to L. Then for sufficiently small t*

$$\widehat{L} \stackrel{\text{def}}{=} \cup_t A_t(L)$$

is an integral manifold of P too.

Proof From the previous lemma it follows that

$$\omega_j|_{A_t(L)} = A_t^*(\omega_j)|_L = 0.$$

Therefore the submanifolds $A_t(L)$ are also integral manifolds of the distribution P. For sufficiently small t the set $\widehat{L} = \cup_t A_t(L)$ is a smooth submanifold of M and tangent vectors to $\widehat{L}$ are sums of vectors tangent to $A_t(L)$ for some t and vectors proportional to X. Both of them belong to P. ∎

We say, that a distribution P is *completely integrable* if for any point $a \in M$ there exists an integral manifold N of P such that $a \in N$ and $\dim N = \dim P$.

Theorem 1.5.1 (Frobenius) *A distribution P is completely integrable if and only if P is flat. Moreover, if P is completely integrable, and if N_1 and N_2 are maximal integral submanifolds of P passing through a point $a \in N_1 \cap N_2$. Then $N_1 = N_2$ in some neighborhood of a.*

Proof The previous lemma shows how to construct locally the submanifold N by integrating vector fields $X \in D(P)$, and this N is a locally unique if $\dim N = \dim P$.

It is also shows that any vector field $X \in D(P)$ should be tangent to N, and therefore $[X, Y] \in D(P)$ if $X, Y \in D(P)$. ∎

Corollary 1.5.1 *The distribution P is completely integrable if and only if P and its first derivative coincide: $P^{(1)} = P$.*

Example 1.5.1 (Distributions of codimension one) *A distribution of codimension one can be defined locally by one differential 1-form ω. This distribution is completely integrable if and only if*

$$\omega \wedge d\omega = 0.$$

For example, for the distribution $\mathcal{F}\langle\omega\rangle$ in $\mathbb{R}^3$, where

$$\omega = A(x, y, z)dx + B(x, y, z,)dy + C(x, y, z)dz$$

we get

$$\omega \wedge d\omega = \left[A\left(\frac{\partial B}{\partial z} - \frac{\partial C}{\partial y}\right) - B\left(\frac{\partial A}{\partial z} - \frac{\partial C}{\partial x}\right) + C\left(\frac{\partial B}{\partial x} - \frac{\partial A}{\partial y}\right)\right] dx \wedge dy \wedge dz$$

and therefore, this distribution is completely integrable if and only if

$$A\left(\frac{\partial B}{\partial z} - \frac{\partial C}{\partial y}\right) - B\left(\frac{\partial A}{\partial z} - \frac{\partial C}{\partial x}\right) + C\left(\frac{\partial B}{\partial x} - \frac{\partial A}{\partial y}\right) = 0.$$

Example 1.5.2 *The oricycle distribution is completely integrable because of*

$$d\omega = \sin\phi \, d\phi \wedge dx + (1 + \cos\phi) \, d\phi \wedge dy$$

and $\omega \wedge d\omega = 0$.

Example 1.5.3 *For distribution from Example 1.1.6 conditions of complete integrability are simply conditions of integrability of the corresponding system of differential equations. The Frobenius theorem requires that $d\omega_i \wedge \omega_1 \wedge \omega_2 = 0$ for $i = 1, 2$, or*

$$\frac{\partial A}{\partial y} - \frac{\partial B}{\partial x} + B\frac{\partial A}{\partial u} + D\frac{\partial A}{\partial v} - A\frac{\partial B}{\partial u} - C\frac{\partial B}{\partial v} = 0,$$

$$\frac{\partial C}{\partial y} - \frac{\partial D}{\partial x} - B\frac{\partial C}{\partial u} - D\frac{\partial C}{\partial v} + A\frac{\partial D}{\partial u} + C\frac{\partial D}{\partial v} = 0.$$

1.6 Complex distributions on real manifolds

Let M be a real $(n+m)$-dimensional smooth manifold, and let $\tau : TM^{\mathbb{C}} \to M$ be the complexification of a tangent bundle. By a *complex distribution* P on M one means a *smooth* field $P : a \in M \longmapsto P_a = P(a) \subset T_a M^{\mathbb{C}}$ of complex subspaces ($\dim_{\mathbb{C}} = m$) of the complexification $T_a M^{\mathbb{C}}$.

The number m is called a (complex) *dimension* of the distribution, $m = \dim_{\mathbb{C}} P$, and n is called a (complex) *codimension* of P, $n = \operatorname{codim}_{\mathbb{C}} P$.

Similar to the real case, complex distributions can be described by complex vector fields or by complex-valued differential 1-forms: $P = \mathcal{F}\langle X_1, \ldots, X_m\rangle$, where $X_i \in D(M)^{\mathbb{C}}$, or $P = \mathcal{F}\langle\omega_1, \ldots, \omega_n\rangle$, where $\omega_i \in \Omega^1(M)^{\mathbb{C}}$, and $D(M)^{\mathbb{C}}$ and $\Omega^1(M)^{\mathbb{C}}$ are complexification of the modules $D(M)$ and $\Omega^1(M)$ respectively:

$$D(M)^{\mathbb{C}} \stackrel{\text{def}}{=} \{X + \iota Y \,|\, X, Y \in D(M)\},$$

$$\Omega^1(M)^{\mathbb{C}} \stackrel{\text{def}}{=} \{\alpha + \iota\beta \,|\, \alpha, \beta \in \Omega^1(M)\}.$$

They are modules over the algebra $C^\infty(M, \mathbb{C})$ of smooth complex-valued functions on M.

We denote by $D(P)$ the $C^\infty(M, \mathbb{C})$-module of complex vector fields on M which belong to P and by Ann P the *annihilator* of P, that is

$$\operatorname{Ann}(P(a)) \stackrel{\text{def}}{=} \{\omega_a \in T_a^* M^{\mathbb{C}} \,|\, \omega_a = 0 \text{ on } P(a)\}.$$

Let P be a complex distribution on M. A submanifold $i : N \hookrightarrow M$ is said to be *integral* for P if

$$T_a N^{\mathbb{C}} \subset P(a)$$

for any point $a \in N$.

If distribution P is defined by complex-valued differential 1-forms, say, $P = \mathcal{F}\langle\omega_1, \ldots, \omega_n\rangle$, then $i : N \hookrightarrow M$ is integral if and only if

$$\omega_1|_N = 0, \quad \ldots, \quad \omega_n|_N = 0.$$

A smooth function H is called *first integral* for P if $dH \in \operatorname{Ann}(P)$.

A complex distribution P, is called *completely integrable* if it has (locally) $n = \operatorname{codim}_{\mathbb{C}} P$ functional independent first integrals $H_1, \ldots, H_n$; that is, (locally) $P = \langle dH_1, \ldots, dH_n\rangle$.

A complex distribution P is *involutive* if $[X, Y] \in D(P)$ for any $X, Y \in D(P)$.

Theorem 1.6.1 (Frobenius–Nirenberg [35]) *Let P be a complex involutive distribution such that the distribution $P + \overline{P}$ is an involutive distribution and* $\dim_{\mathbb{C}}\left(P \cap \overline{P}\right) = \text{const}$. *Then P is a completely integrable distribution.*

1.7 The Lie–Bianchi theorem

1.7.1 The Maurer–Cartan equations

Let P be an n-codimensional completely integrable distribution generated by the differential 1-forms $\omega_1, \ldots, \omega_n$ and let $\mathfrak{g}$ be a Lie subalgebra of $\operatorname{Shuf}(P)$.

We say that $\mathfrak{g}$ is *transversal* to the distribution P if $\dim \mathfrak{g} = n$, and values of symmetries generate the factor $T_a M / P_a$ at any point $a \in M$.

Denote by $\overline{X} \in \mathrm{Sym}(P)$ a representative of a shuffling symmetry $X \in \mathrm{Shuf}(P)$. It is clear that the value $\theta(X) \overset{\text{def}}{=} \theta(\overline{X})$ is correctly defined for any $\theta \in \mathrm{Ann}(P)$.

Let $X_1, \ldots, X_n$ be a basis of the Lie algebra $\mathfrak{g}$. Then the transversality condition for the algebra $\mathfrak{g}$ means that the matrix

$$W \overset{\text{def}}{=} \left|\left|\omega_i(X_j)\right|\right|_{i,j=1,\ldots,n} \tag{1.10}$$

is non-degenerate at any point.

Let us choose another basis $\overline{\omega}_1, \ldots, \overline{\omega}_n$ in the module $\mathrm{Ann}(P)$ such that

$$\overline{\omega}_i\left(\overline{X}_j\right) = \delta_{ij} \qquad (i,j = 1, \ldots, n)\,. \tag{1.11}$$

To find this basis, it is enough to set

$$\left\|\begin{matrix} \overline{\omega}_1 \\ \vdots \\ \overline{\omega}_n \end{matrix}\right\| \overset{\text{def}}{=} W^{-1} \left\|\begin{matrix} \omega_1 \\ \vdots \\ \omega_n \end{matrix}\right\|. \tag{1.12}$$

Suppose now that for the forms $\omega_1, \ldots, \omega_n$ condition (1.11) holds, and let $c_{ij}^l \in \mathbb{R}$ be structure constants of the Lie algebra $\mathfrak{g}$ $(i, j, l = 1, \ldots, m)$, i.e.,

$$\left[X_i, X_j\right] = \sum_{l=1}^{n} c_{ij}^l X_l \tag{1.13}$$

for all $i, j = 1, \ldots, n$.

Theorem 1.7.1 *Let P be a completely integrable distribution which is defined by differential 1-forms $\omega_1, \ldots, \omega_n$ that satisfy (1.11) and (1.13). Then the following formulae*

$$d\omega_l + \sum_{i<j} c_{ij}^l \omega_i \wedge \omega_j = 0, \tag{1.14}$$

hold for all $l = 1, \ldots, n$.

We say that (1.14) is the *Maurer–Cartan equation* by analogy with the Maurer–Cartan equation in Lie group theory.

Proof From the Frobenius theorem it follows that

$$d\omega_l = \sum_{s=1}^{n} \gamma_{ls} \wedge \omega_s,$$

where γ_{ls} are some differential 1-forms.

Under the conditions specified, these 1-forms belong to $\mathrm{Ann}(P)$, i.e., γ_{ls} vanish on the vectors of the distribution P.

To prove this, we use the formula:

$$L_{\overline{X}_j}(\omega_l) = d(\omega_l(\overline{X}_j)) + \overline{X}_j \lrcorner\, d\omega_l = \overline{X}_j \lrcorner\, d\omega_l.$$

Since $X_j \in \mathrm{Sym}(P)$, then $L_{\overline{X}_j}(\omega_l) \in \mathrm{Ann}(P)$ for all $j, l = 1, \ldots, n$, and

$$L_{\overline{X}_j}(\omega_l) = \overline{X}_j \lrcorner\, d\omega_l = \sum_{s=1}^{n} \gamma_{ls}(\overline{X}_j)\omega_j - \gamma_{lj} \in \mathrm{Ann}(P),$$

which implies that $\gamma_{lj} \in \mathrm{Ann}(P)$.

Therefore,

$$\gamma_{lj} = \sum_{i=1}^{n} a_{ij}^{l} \omega_i$$

for some smooth functions $a_{ij}^{l} \in C^{\infty}(M)$, and

$$d\omega_l = 2 \sum_{i<j} a_{ij}^{l} \omega_i \wedge \omega_j,$$

and

$$d\omega^l(\overline{X}_i, \overline{X}_j) = 2a_{ij}^{l}.$$

On the other hand,

$$\begin{aligned} d\omega_l(\overline{X}_i, \overline{X}_j) &= \overline{X}_i(\omega_l(\overline{X}_j)) - \overline{X}_j(\omega_l(\overline{X}_j)) - \omega_l([\overline{X}_i, \overline{X}_j]) \\ &= -\omega_l([\overline{X}_i, \overline{X}_j]) = -\omega_l\left(\sum_{k=i}^{n} c_{ij}^{k} \overline{X}_k\right) = -c_{ij}^{l}. \end{aligned}$$

Therefore,

$$2a_{ij}^{l} = -c_{ij}^{l}$$

and

$$d\omega_l = -\sum_{i<j} c_{ij}^l \omega_i \wedge \omega_j. \qquad \blacksquare$$

1.7.2 Distributions with a commutative symmetry algebra

Now let $\mathfrak{g}$ be a commutative symmetry Lie algebra which is transversal to a completely integrable distribution P. Then the above discussion gives the following algorithm for finding first integrals.

Namely, let $X_1, \ldots, X_n$ be a basis of $\mathfrak{g}$ and let the distribution P be generated by the differential 1-forms $\omega_1, \ldots, \omega_n$. Then,

$$\Omega_P(X, Y) = -\sum_{i=1}^{n} d\omega_i(X, Y)\, X_{\omega_i}. \tag{1.15}$$

Step 1. Form the matrix

$$W = \begin{Vmatrix} \omega_1(\overline{X}_1) & \cdots & \omega_1(\overline{X}_n) \\ \vdots & \vdots & \vdots \\ \omega_n(\overline{X}_1) & \cdots & \omega_n(\overline{X}_n) \end{Vmatrix}$$

and find W^{-1}.

Step 2. Change the basis in $\mathrm{Ann}(P)$ for

$$\begin{Vmatrix} \overline{\omega}_1 \\ \vdots \\ \overline{\omega}_n \end{Vmatrix} = W^{-1} \begin{Vmatrix} \omega_1 \\ \vdots \\ \omega_n \end{Vmatrix}.$$

Step 3. Differential 1-forms $\overline{\omega}_1, \ldots, \overline{\omega}_n$ are closed due to the Maurer–Cartan equations. Therefore, a complete set of first integrals of the distribution can be found by quadratures:

$$H_j(a) = \int_\Gamma \overline{\omega}_j, \tag{1.16}$$

if the base manifold M is, for example, connected and simply connected.

Here Γ is a path from the fixed point a_0 to a point $a \in M$.

We shall use formula (1.16) for all manifolds and consider $H_1, \dots, H_n$ as multivalued first integrals for P.

Example 1.7.1 (Distributions of codimension one) *Let $P = \mathcal{F}\langle\omega\rangle$ be a completely integrable codimension one distribution and X be a transversal symmetry of P. Then the differential 1-form*

$$\overline{\omega} = \frac{1}{\omega(\overline{X})}\omega$$

is closed and

$$H = \int_\Gamma \frac{\omega}{\omega(\overline{X})}$$

is a first integral of P.

Example 1.7.2 *Consider the oricycle distribution. It has the infinitesimal symmetry $X = \partial/\partial x$ and $W = \omega(X) = 1 - \cos\phi$. Therefore,*

$$\overline{\omega} = dx + \frac{\sin\phi}{1-\cos\phi}dy - \frac{y}{1-\cos\phi}d\phi$$

is closed.

Then,

$$\begin{aligned} H(x, y, z) &= \int_0^x a\left(t, 0, \frac{\pi}{2}\right) dt + \int_0^y b\left(x, t, \frac{\pi}{2}\right) dt + \int_{\pi/2}^{\phi} c(x, y, t)\, dt \\ &= x + y\cot\frac{\phi}{2} \end{aligned}$$

for the following path Γ in the space (x, y, ϕ):

$$\left(0, 0, \frac{\pi}{2}\right) \to \left(x, 0, \frac{\pi}{2}\right) \to \left(x, y, \frac{\pi}{2}\right) \to (x, y, \phi).$$

Therefore,

$$H(x, y, \phi) = x + y\cot\frac{\phi}{2}$$

is a first integral for the oricycle distribution.

Example 1.7.3 (Integrating factor) *Consider a first-order differential equation*

$$y' = F(x, y). \tag{1.17}$$

The corresponding 1-form is

$$\omega = dy - F(x, y)\, dx.$$

Suppose that $X = a(x, y)\frac{\partial}{\partial x} + b(x, y)\frac{\partial}{\partial y}$ *is a transversal symmetry of the distribution, that is the symmetry of the equation. Then,* $W = b(x, y) - F(x, y)a(x, y)$ *and the differential 1-form*

$$\overline{\omega} = \frac{dy - F\, dx}{b(x, y) - F(x, y)a(x, y)}$$

is closed, and the function

$$H = \int_\Gamma \overline{\omega}$$

is a first integral of the equation. The function W^{-1} *is called an* integrating factor *for the equation.*

Example 1.7.4 (Homogeneous equations) *Consider a homogeneous differential equation*

$$y' = \Phi(x, y) \tag{1.18}$$

where $\Phi(\lambda x, \lambda y) = \Phi(x, y)$ *for all* x, y *and* $\lambda \neq 0$.

This equation is invariant under the scale transformation $x \longmapsto \lambda x$, $y \longmapsto \lambda y$, *and therefore the vector field*

$$X = x\frac{\partial}{\partial x} + y\frac{\partial}{\partial y}$$

is a symmetry. In this case $W = \omega(X) = y - x\Phi(x, y)$ *and the form*

$$\overline{\omega} = \frac{dy - \Phi(x, y)\, dx}{y - x\, \Phi(x, y)}$$

is closed and the function $H = \int_\Gamma \overline{\omega}$ *is a first integral of the equation.*

Example 1.7.5 (Generalized homogeneous equations) *Let us consider a generalization of (1.18):*

$$y' = \Phi(x, y) \tag{1.19}$$

where $\Phi\left(\lambda^a x, \lambda^b y\right) = \lambda^{b-a}\Phi(x, y)$ *for some real numbers* a, b. *This equation has the scale transformations symmetry group* $x \longmapsto \lambda^a x$, $y \longmapsto \lambda^b y$. *Therefore,*

$$X = ax\frac{\partial}{\partial x} + by\frac{\partial}{\partial y}$$

is a symmetry and

$$W^{-1} = \frac{1}{by - ax\ \Phi(x, y)}$$

is an integrating factor.

1.7.3 Lie–Bianchi theorem

Let us now analyze the case of general symmetry Lie algebra.

Denote by $\mathfrak{g}^{(1)} = [\mathfrak{g}, \mathfrak{g}]$ the commutator of $\mathfrak{g}$; that is, the ideal of $\mathfrak{g}$ generated by all commutators $[X, Y]$ for $X, Y \in \mathfrak{g}$. Suppose that $\mathfrak{g}^{(1)} \neq \mathfrak{g}$ and $l = \operatorname{codim}_{\mathfrak{g}} \mathfrak{g}^{(1)}$.

Let us choose a basis $X_1, \ldots, X_l, \ldots, X_n$ in $\mathfrak{g}$ such that

$$X_1, \ldots, X_l \notin \mathfrak{g}^{(1)} \quad \text{but} \quad X_{l+1}, \ldots, X_n \in \mathfrak{g}^{(1)}.$$

Then the Maurer–Cartan equations imply that

$$d\overline{\omega}_i = 0, \quad i = 1, \ldots, l.$$

Let

$$H_1 = \int_\Gamma \overline{\omega}_1, \ldots, H_l = \int_\Gamma \overline{\omega}_l$$

be the corresponding first integrals.

Then submanifolds

$$M_c = \{H_1 = c_1, \ldots, H_l = c_l\},$$

where $c = (c_1, \ldots, c_l) \in \mathbb{R}^l$, are invariant with respect to the commutator $\mathfrak{g}^{(1)}$ because

$$X_i(H_j) = dH_j(X_i) = \omega_j(X_i) = 0$$

for $i = l + 1, \ldots, n$ and $j = 1, \ldots, l$.

Denote by P_c the restriction of distribution P on M_c. Since H_i are first integrals, P_c is a completely integrable distribution of the same dimension n and codimension $\dim \mathfrak{g}^{(1)}$.

Therefore, $\mathfrak{g}^{(1)}$ is a transversal symmetry algebra for P_c for any c.

We can apply the same procedure for the distribution P_c if the next commutator

$$\mathfrak{g}^{(2)} = [\mathfrak{g}^{(1)}, \mathfrak{g}^{(1)}]$$

does not coincide with $\mathfrak{g}^{(1)}$, etc.

Define by induction the *derived subalgebras* $\mathfrak{g}^{(s+1)} = [\mathfrak{g}^{(s)}, \mathfrak{g}^{(s)}]$ and recall that a Lie algebra $\mathfrak{g}$ is said to be *solvable* if there is a natural number r such that

$$\mathfrak{g}^{(r)} = 0.$$

For a solvable Lie algebra $\mathfrak{g}$ one has the descending sequence of Lie subalgebras

$$\mathfrak{g} = \mathfrak{g}^{(0)} \supset \mathfrak{g}^{(1)} \supset \cdots \supset \mathfrak{g}^{(r-1)} \supset \mathfrak{g}^{(r)} = 0,$$

and the above procedure shows that we can find the complete sequence of first integrals by integration of closed 1-forms only, or *quadratures*.

In other words we get the following result.

Theorem 1.7.2 (Lie–Bianchi) *Let P be a completely integrable distribution and let $\mathfrak{g} \subset \mathrm{Shuf}(P)$ be a solvable symmetry Lie algebra transversal to P, $\dim \mathfrak{g} = \operatorname{codim} P$. Then P is integrable by quadratures.*

Note that the above proof of this theorem provides us with an algorithm for finding first integrals by quadratures.

2
Ordinary differential equations

2.1 Symmetries of ODEs

2.1.1 Generating functions

Let us consider an ordinary differential equation (ODE) of $(k+1)$-th order which is resolved with respect to the highest derivative:

$$y^{(k+1)} = F(x,\ y,\ y',\ldots,\ y^{(k)}). \tag{2.1}$$

As we have seen this equation determines a one-dimensional distribution P on the *manifold of k-jets* $J^k = J^k\mathbb{R}$ with coordinates $x, p_0 \stackrel{\text{def}}{=} y, p_1, \ldots, p_k$. The distribution is generated by the vector field

$$\mathcal{D} = \frac{\partial}{\partial x} + p_1\frac{\partial}{\partial p_0} + \cdots + p_k\frac{\partial}{\partial p_{k-1}} + F\frac{\partial}{\partial p_k},$$

or by the Cartan differential 1-forms

$$\omega_1 = dp_0 - p_1\,dx, \quad \ldots, \quad \omega_k = dp_{k-1} - p_k\,dx, \quad \omega_{k+1} = dp_k - F\,dx.$$

Moreover, the vector field $\mathcal{D}$ is a characteristic symmetry of the distribution P.

In order to describe shuffling symmetries we note that in the module $\mathrm{Char}(P)$ the following relation

$$\frac{\partial}{\partial x} \equiv -p_1\frac{\partial}{\partial p_0} - \cdots - p_k\frac{\partial}{\partial p_{k-1}} - F\frac{\partial}{\partial p_k} \mod \mathrm{Char}(P)$$

holds.

Therefore, any shuffling symmetry $S \in \mathrm{Shuf}(P)$ has a unique representative of the form

$$S = \alpha_0(x, p_0, p_1, \dots, p_k)\frac{\partial}{\partial p_0} + \cdots + \alpha_k(x, p_0, p_1, \dots, p_k)\frac{\partial}{\partial p_k}.$$

Moreover,

$$dp_i \equiv p_{i+1}\, dx \bmod \omega_{i+1},$$

and therefore,

$$df \equiv \mathcal{D}(f)\, dx \bmod \mathrm{Ann}(P),$$

for any function $f = f(x, p_0, p_1, \dots, p_k)$.

For Lie derivatives of the Cartan forms one obtains

$$L_S(\omega_i) = (\mathcal{D}(\alpha_i) - \alpha_{i+1})dx \bmod \mathrm{Ann}(P)$$

if $i = 1, \dots, k$.

Therefore, $S \in \mathrm{Sym}(P)$ if

$$\alpha_{i+1} = \mathcal{D}(\alpha_i)$$

for all $i = 0, \dots, k-1$.

In other words, any shuffling symmetry can be represented in the form

$$S = \phi\frac{\partial}{\partial p_0} + \mathcal{D}(\phi)\frac{\partial}{\partial p_1} + \cdots + \mathcal{D}^k(\phi)\frac{\partial}{\partial p_k}, \tag{2.2}$$

where $\phi \overset{\text{def}}{=} \alpha_0$ and $\mathcal{D}^i \overset{\text{def}}{=} \mathcal{D}(\mathcal{D}^{i-1})$, $\mathcal{D}^0(\phi) \overset{\text{def}}{=} \phi$.

The function ϕ is called a *generating function* of the symmetry S and we write S_ϕ instead of S.

Note also that

$$\phi = \omega_1(S_\phi).$$

For the last Cartan form ω_{k+1} we have

$$\begin{aligned} L_{S_\phi}(\omega_{k+1}) &= (\mathcal{D}(\alpha_k) - S_\phi(F))dx \bmod \mathrm{Ann}(P) \\ &= (\mathcal{D}^{k+1}(\phi) - S_\phi(F))dx \bmod \mathrm{Ann}(P). \end{aligned}$$

Therefore, S_ϕ is a shuffling symmetry of the ordinary differential equation if and only if the generating function ϕ satisfies the following *Lie equation*:

$$\mathcal{D}^{k+1}(\phi) - \sum_{i=0}^{k} \frac{\partial F}{\partial p_i} \mathcal{D}^i(\phi) = 0. \tag{2.3}$$

Let us denote by

$$\Delta_F : C^\infty(\mathbb{R}^{k+2}) \longrightarrow C^\infty(\mathbb{R}^{k+2})$$

the following linear k-th order scalar differential operator:

$$\Delta_F \overset{\text{def}}{=} \mathcal{D}^{k+1} - \sum_{i=0}^{k} \frac{\partial F}{\partial p_i} \mathcal{D}^i$$

and call it *linearization of* (2.1).

Theorem 2.1.1 *There exists the isomorphism*

$$\text{Shuf}(P) \cong \ker \Delta_F$$

between solutions of the Lie equation and shuffling symmetries given by

$$\ker \Delta_F \ni \phi \longleftrightarrow S_\phi \in \text{Shuf}(P).$$

Remark 2.1.1 *To understand the geometrical meaning of the generating functions, let us consider a symmetry S_ϕ and let $h(x)$ be a solution of the equation. Then the corresponding curve $L_h = (p_0 = h(x), p_1 = h'(x), \ldots, p_k = h^{(k)}(x))$ is an integral curve of the Cartan distribution. Let A_t be a one-parametric group of shifts along the trajectories of S_ϕ. Then locally and for small t we have $A_t(L_h) = L_{h_t}$ and $h_t = h + t\,\phi|_{L_h} + o(t), h_t^{(i)} = h + \mathcal{D}^i(\phi)\big|_{L_h} t + o(t)$ for $i = 1, \ldots, k$. In other words, the action of symmetry S_ϕ on a solution h corresponds to a transformation of the form $h_t = h + t\,\phi|_{L_h} + o(t)$ on functions, or solutions of the evolutionary equation*

$$\frac{\partial h}{\partial t} = \phi\left(x, h, \ldots, \frac{\partial^k h}{\partial x^k}\right).$$

Example 2.1.1 (The translation in x, see Figure 2.1) *This symmetry corresponds to transformations $T_t : (x, p_0, \ldots, p_k) \longmapsto (x + t, p_0, \ldots, p_k)$ and*

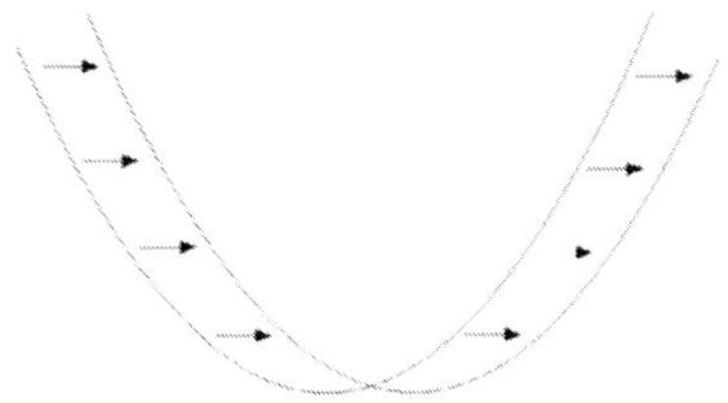

Figure 2.1. The translation in x.

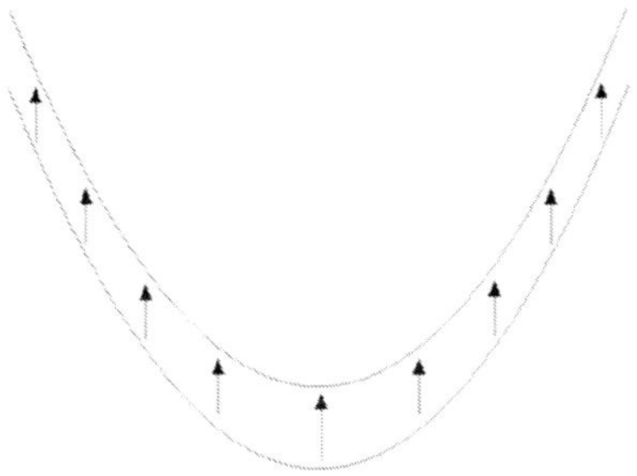

Figure 2.2. The translation in p_0.

has generating function $\phi = -p_1$, *and*

$$S_\phi = -p_1 \frac{\partial}{\partial p_0} - \cdots - -p_k \frac{\partial}{\partial p_{k-1}} - F \frac{\partial}{\partial p_k} = \frac{\partial}{\partial x} - \mathcal{D}.$$

Equation (2.1) admits the symmetry if and only if

$$\frac{\partial F}{\partial x} = 0.$$

Example 2.1.2 (The translation in p_0, see Figure 2.2) *This symmetry corresponds to transformations*

$$T_t : (x, p_0, \ldots, p_k) \longmapsto (x, p_0 + t, \ldots, p_k)$$

and has the generating function $\phi = 1$ *and*

$$S_\phi = \frac{\partial}{\partial p_0}.$$

Equation (2.1) admits the symmetry if and only if

$$\frac{\partial F}{\partial p_0} = 0.$$

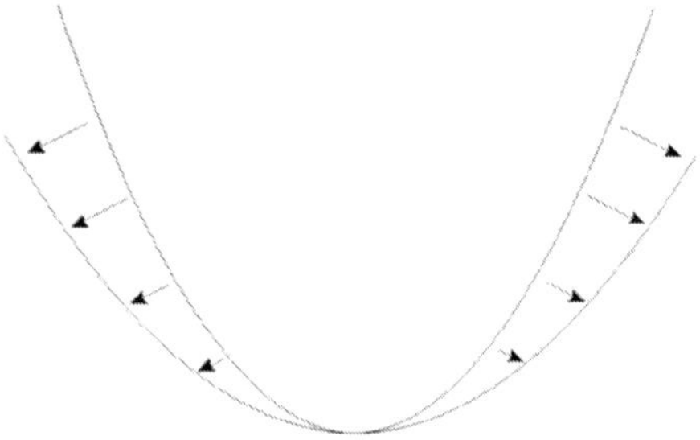

Figure 2.3. The scale symmetry.

Example 2.1.3 (Scale transformations, see Figure 2.3) *This symmetry corresponds to transformations*

$$T_t : (x, p_0, \ldots, p_i, \ldots p_k) \longmapsto (e^{\alpha t}x, e^{\beta t}p_0, \ldots, e^{(\beta - i\alpha)t}p_i, \ldots, e^{(\beta - k\alpha)t}p_k)$$

and has the generating function

$$\phi = \alpha x p - \beta p_0.$$

Example 2.1.4 (Translations in $h(x)$) *Functions $\phi = h(x)$ generate transformations*

$$T_t : (x, p_0, \ldots, p_i, \ldots p_k) \longmapsto (x, p_0 + h(x), \ldots, p_i + h^{(i)}(x), \ldots, p_k + h^{(k)}(x)).$$

Example 2.1.5 (Point transformations) *Functions $\phi = a(x, p_0)p + b(x, p_0)$ generate so-called* point transformations. *They correspond to vector fields on $\mathbb{R}^2$ of the form*

$$b(x, p_0)\frac{\partial}{\partial p_0} - a(x, p_0)\frac{\partial}{\partial x}.$$

Example 2.1.6 (Contact transformations) *They correspond to the contact vector fields X_f on $\mathbb{R}^3$ and have generating functions of the form*

$$\phi = f(x, p_0, p_1).$$

Example 2.1.7 (KdV symmetries, see Figure 2.4) *The picture below shows symmetries for the elliptic function differential equation*

$$(y')^2 - 4y^3 + a_1 y + a_0$$

generating by the Korteweg–de Vries (KdV) equation (see [68]).

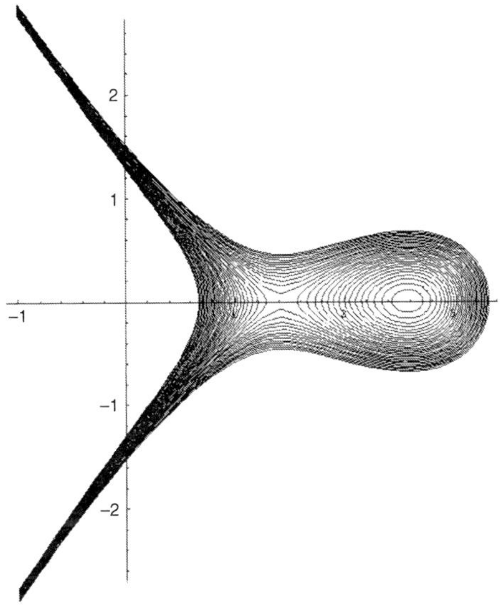

Figure 2.4. KdV symmetries.

2.1.2 Lie algebra structure on generating functions

We know that Shuf(P) is a Lie algebra for any distribution P. Because of this fact and the isomorphism between solutions of the Lie equation (i.e., generating functions) and symmetries we are able to introduce a Lie algebra structure on a space of solutions $\ker \triangle_F$.

Namely, we define a bracket $[\phi, \psi]$ between two generating functions ϕ and ψ as the generating function of the commutator $\left[S_\phi, S_\psi\right]$, that is

$$S_{[\phi,\psi]} \stackrel{\text{def}}{=} \left[S_\phi, S_\psi\right].$$

We call this a *Poisson–Lie bracket* of functions $\phi, \psi \in \ker \triangle_F$. Obviously, this bracket defines a Lie algebra structure on $\ker \triangle_F$.

To find the bracket we remark that $\phi = \omega_1(S_\phi)$, and therefore

$$[\phi, \psi] = \omega_1(S_{[\phi,\psi]}) = S_{[\phi,\psi]}(p_0) = [S_\phi, S_\psi](p_0) = S_\phi(\psi) - S_\psi(\phi)$$

or

$$[\phi, \psi] = S_\phi(\psi) - S_\psi(\phi). \tag{2.4}$$

In coordinates one finds

$$[\phi, \psi] = \sum_{i=0}^{k} \left(\mathcal{D}^i(\phi) \frac{\partial \psi}{\partial p_i} - \mathcal{D}^i(\psi) \frac{\partial \phi}{\partial p_i} \right).$$

Theorem 2.1.2 *The generating functions of symmetries (2.1) form a Lie algebra with respect to the Poisson–Lie bracket.*

2.1.3 Commutative symmetry algebra

We begin with the following observation. Let $v_1, \ldots, v_n$ be linear independent vector fields in a domain $D \subset \mathbb{R}^n$, and let

$$[v_i, v_j] = 0$$

for all $i, j = 1, \ldots, n$.

Take independent functions $f_1, \ldots, f_n$ on D and define differential 1-forms $\theta_1, \ldots, \theta_n$ as

$$\theta = W^{-1}\, df,$$

where

$$\theta = \begin{Vmatrix} \theta_1 \\ \theta_2 \\ \vdots \\ \theta_n \end{Vmatrix}, \quad df = \begin{Vmatrix} df_1 \\ df_2 \\ \vdots \\ df_n \end{Vmatrix},$$

and

$$W = \begin{Vmatrix} v_1(f_1) & v_2(f_1) & \cdots & v_n(f_1) \\ v_1(f_2) & v_2(f_2) & \cdots & v_n(f_2) \\ \vdots & \vdots & \cdots & \vdots \\ v_1(f_n) & v_2(f_n) & \cdots & v_n(f_n) \end{Vmatrix}.$$

Then $\theta_1, \ldots, \theta_n$ constitute the dual basis to $v_1, \ldots, v_n$.

Lemma 2.1.1 $d\theta_i = 0$ *for all* $i = 1, \ldots, n$.

Proof We have $\theta_i(v_j) = \delta_{ij}$, and therefore

$$d\theta_i(v_a, v_b) = v_a(\theta_i(v_b)) - v_b(\theta_i(v_a)) - \theta_i([v_a, v_b]) = 0. \qquad \blacksquare$$

We apply this result for integration of ordinary differential equations.

Assume that one has $(k+1)$ commuting linear independent shuffling symmetries $\phi_1, \ldots, \phi_{k+1}$ in a domain $D = \{p_{k+1} = F(x, p_0, p_1, \ldots, p_k)\} \subset J^{k+1}$. Then $[\mathcal{D}, X_{\phi_i}] = 0$. Indeed, by the definition of a symmetry $[\mathcal{D}, X_{\phi_i}] = \lambda \mathcal{D}$ for some function λ. Applying both sides to the coordinate function x we obtain $\lambda = 0$.

Therefore, the vector fields $\mathcal{D}, X_{\phi_1}, \ldots, X_{\phi_{k+1}}$ commute and are linear independent at each point. To obtain first integrals we need the following construction. Let us define a *Cartan form* ω_f *corresponding to a function* $f \in C^\infty(J^{k+1})$ as

$$\omega_f = \sum_{i=0}^{k+1} \frac{\partial f}{\partial p_i} \omega_i,$$

where

$$\omega_i = dp_i - p_{i+1}\, dx \quad (0 \leq i \leq k)$$

and

$$\omega_{k+1} = dp_{k+1} - \mathcal{D}(F)\, dx.$$

Then,

$$\omega_f(X_\phi) = X_\phi(f) \quad \text{and} \quad \omega_f(\mathcal{D}) = 0.$$

Theorem 2.1.3 *Let $\phi_1, \ldots, \phi_{k+1}$ be commuting shuffling symmetries for ordinary differential equation $\mathcal{E} = \{p_{k+1} = F\} \subset J^{k+1}$, and let $D \subset \mathcal{E}$ be a domain where vector fields $X_{\phi_1}, \ldots, X_{\phi_{k+1}}$ are linear independent. Let $f_1, \ldots, f_{k+1}$ be functions such that the functions $x, f_1, \ldots, f_{k+1}$ are independent in D. Then differential 1-forms $\theta_1, \ldots, \theta_{k+1}$ defined by*

$$\theta = W^{-1} \omega_f,$$

where

$$\theta = \left\| \begin{matrix} \theta_1 \\ \theta_2 \\ \vdots \\ \theta_{k+1} \end{matrix} \right\|, \quad \omega_f = \left\| \begin{matrix} \omega_{f_1} \\ \omega_{f_2} \\ \vdots \\ \omega_{f_{k+1}} \end{matrix} \right\|,$$

and

$$W = \left\| \begin{matrix} X_{\phi_1}(f_1) & X_{\phi_2}(f_1) & \cdots & X_{\phi_{k+1}}(f_1) \\ X_{\phi_1}(f_2) & X_{\phi_2}(f_2) & \cdots & X_{\phi_{k+1}}(f_2) \\ \vdots & \vdots & \cdots & \vdots \\ X_{\phi_1}(f_{k+1}) & X_{\phi_2}(f_{k+1}) & \cdots & X_{\phi_{k+1}}(f_{k+1}) \end{matrix} \right\|,$$

are closed in D and $\theta_i(\mathcal{D}) = 0$ for all $i = 1, \ldots, k+1$.

Proof We have

$$dx(\mathcal{D}) = 1, \quad \omega_{f_i}(\mathcal{D}) = 0,$$

and

$$dx(X_{\phi_i}) = 0, \quad \omega_{f_i}(X_{\phi_j}) = X_{\phi_j}(f_i)$$

for all $i, j = 1, \ldots, k+1$.

Therefore, the differential 1-forms

$$dx,\ \theta_1,\ \ldots,\ \theta_{k+1}$$

constitute a dual basis for

$$\mathcal{D},\ X_{\phi_1}, \ldots, X_{\phi_{k+1}}$$

and the theorem follows from the above lemma. ■

Theorem 2.1.4 *Let $\phi, \phi_1, \ldots, \phi_r$ be shuffling symmetries for an ordinary differential equation $\mathcal{E} \subset J^m$, and let $D \subset \mathcal{E}$ be a domain where vector fields $X_{\phi_1}, \ldots, X_{\phi_r}$ are linear independent. If a vector field X_ϕ is a linear combination of the fields $X_{\phi_1}, \ldots, X_{\phi_r}$ then the shuffling symmetry ϕ is a linear combination of $\phi_1, \ldots, \phi_r$ in the domain D:*

$$\phi = \lambda_1 \phi_1 + \cdots + \lambda_r \phi_r$$

where coefficients $\lambda_1, \ldots, \lambda_r$ are first integrals for $\mathcal{E}$.

Proof In the domain D we have

$$X_\phi = \lambda_1 X_{\phi_1} + \cdots + \lambda_m X_{\phi_m}$$

for some functions $\lambda_1, \ldots, \lambda_m$. Then, $\phi = X_\phi(p_0) = \lambda_1 \phi_1 + \cdots + \lambda_r \phi_r$.

As we have seen $[\mathcal{D}, X_\psi] = 0$, for all shuffling symmetries, therefore

$$[\mathcal{D}, X_\phi] = \mathcal{D}(\lambda_1) X_{\phi_1} + \cdots + \mathcal{D}(\lambda_m) X_{\phi_m} = 0$$

and $\mathcal{D}(\lambda_i) = 0$ for all $i = 1, \ldots, m$. ■

2.2 Non-linear second-order ODEs

In this section we shall apply the Lie–Bianchi theorem and the above constructions for ODEs of second order. Let

$$y'' = F(x, y, y')$$

be such an equation, and let $\phi(x, p_0, p_1)$ and $\psi(x, p_0, p_1)$ be generating functions of symmetries of the equation.

Here

$$\mathcal{D} = \frac{\partial}{\partial x} + p_1 \frac{\partial}{\partial p_0} + F \frac{\partial}{\partial p_1},$$

$$\omega_1 = dp_0 - p_1\, dx, \quad \omega_2 = dp_1 - F\, dx,$$

and the shuffling symmetry corresponding to the generating function ϕ has the form

$$S_\phi = \phi \frac{\partial}{\partial p_0} + \mathcal{D}(\phi) \frac{\partial}{\partial p_1}.$$

The Poisson–Lie bracket has the form

$$[\phi, \psi] = \phi \frac{\partial \psi}{\partial p_0} - \psi \frac{\partial \phi}{\partial p_0} + \mathcal{D}(\phi) \frac{\partial \psi}{\partial p_1} - \mathcal{D}(\psi) \frac{\partial \phi}{\partial p_1}.$$

Assume that ϕ and ψ generates a two-dimensional Lie algebra. It is known that two-dimensional Lie algebras are either commutative or solvable and one can find a new basis (we will use the same notation ϕ and ψ for the basis) in the Lie algebra, such that $[\phi, \psi] = c\psi$, where $c = 0$ for a commutative Lie algebra and $c = 1$ for a solvable Lie algebra.

Note also that

$$\omega_1(S_\phi) = \phi, \quad \omega_2(S_\phi) = \mathcal{D}(\phi).$$

Therefore matrix (20.14) takes the form of the Wandermond matrix

$$W = \begin{Vmatrix} \phi & \psi \\ \mathcal{D}(\phi) & \mathcal{D}(\psi) \end{Vmatrix}$$

and

$$\begin{Vmatrix} \overline{\omega}_1 \\ \overline{\omega}_2 \end{Vmatrix} = W^{-1} \begin{Vmatrix} \omega_1 \\ \omega_2 \end{Vmatrix} = \frac{1}{|W|} \begin{Vmatrix} \mathcal{D}(\psi)\omega_1 - \psi\omega_2 \\ -\mathcal{D}(\phi)\omega_1 + \phi\omega_2 \end{Vmatrix}$$

or

$$\overline{\omega}_1 = \frac{\mathcal{D}(\psi)\omega_1 - \psi\omega_2}{\phi\mathcal{D}(\psi) - \psi\mathcal{D}(\phi)},$$

$$\overline{\omega}_2 = -\frac{\mathcal{D}(\phi)\omega_1 - \phi\omega_2}{\phi\mathcal{D}(\psi) - \psi\mathcal{D}(\phi)}.$$

Here $|W|$ is the determinant of W.

It follows from the Maurer–Cartan equations that $\overline{\omega}_1$ is closed and the restrictions of $\overline{\omega}_2$ on first integrals of $\overline{\omega}_1$ are closed also.

Example 2.2.1 *Let us consider the following differential equation* $y'' = x(y - xy')^4$. *This equation shall appear later on when we will find solutions of the Monge–Ampère equation (see Section 12.5). The vector field* $\mathcal{D}$ *has the form*

$$\mathcal{D} = \frac{\partial}{\partial x} + p_1 \frac{\partial}{\partial p_0} + x(p_0 - p_1 x)^4 \frac{\partial}{\partial p_1}.$$

Let us find partial solutions of the Lie equation which are linear in p_1. *These are so-called point symmetries of the equation. Substituting a function* $\phi(x, y, p) = \alpha(x, p_0)p_1 + \beta(x, p_0)$ *in the Lie equation we find two linear independent solutions:*

$$\phi = p_0 + xp_1, \quad \psi = x$$

and

$$[\phi, \psi] = -2\psi.$$

In the new basis

$$\overline{\omega}_1 = \frac{1}{|W|}\left(\left(2p_1^2 - xp_0(p_0 - xp_1)^4\right) dx - \left(2p_1 + x^2(p_0 - xp_1)^4\right) dp_0 + (p_0 + xp_1)\, dp\right.$$

$$\overline{\omega}_2 = \frac{1}{|W|}\left(\frac{x^2 - p_1(p_0 - xp_1)^2}{(p_0 - xp_1)^2} dx + dp_0 - x\, dp_1\right),$$

where $|W| = (p_0 - xp_1)(1 - x^3(p_0 - xp_1)^3)$, *the first differential 1-form is closed, and indeed*

$$\overline{\omega}_1 = d\left(\ln \frac{|p_0 - xp_1|}{\sqrt[3]{|1 - x^3(p_0 - xp_1)^3|}}\right).$$

Therefore we get the first integral

$$H(x, p_0, p_1) = \frac{|p_0 - xp_1|}{\sqrt[3]{|1 - x^3(p_0 - xp_1)^3|}}.$$

The restrictions of the 1-form $\overline{\omega}_1$ *on* $M_a = \{H = a\}$ *are*

$$\overline{\omega}_2|_{M_a} = \left(\frac{1}{x^2\sqrt[3]{x^3 + a^3}} - x^2 p_0\right) dx + \frac{1}{x}\, dp_0.$$

Integrating these forms, we find general solutions of the equation:

$$y = -x \int \frac{dx}{x^2 \sqrt[3]{x^3 + a^3}} + bx$$
$$= \frac{\sqrt[3]{(x^3 + a^3)^2}}{a^3} - \frac{x_2^3 F_1(\frac{2}{3}, \frac{1}{3}; \frac{5}{3}; -x^3/a^3)}{2a^4} + bx$$

where a and b are constants and ${}_2F_1$ is the hypergeometric function.

2.2.1 Equation $y'' = y' + F(y)$

In this section we find (see [22]) all differential equations of the form $y'' = y' + F(y)$, which have a 2-dimensional Lie algebra of point symmetries and then solve these equations by quadratures.

First of all, the independent variable x does not enter explicitly into the equation, hence translation along the x axis is a symmetry. This symmetry has the generating function $\phi = p_1$.

The problem is to determine when the equation has an additional symmetry with a generating function of the form $\psi = \alpha(x, p_0)p_1 + \beta(x, p_0)$.

In what follows, we exclude the particular case of linear functions $F(y)$.

The Lie equation for the generating function ψ takes the following form:

$$\frac{\partial^2\alpha}{\partial p_0^2} p_1^3 + \left(2\frac{\partial^2\alpha}{\partial x \partial p_0} + 2\frac{\partial\alpha}{\partial p_0} + \frac{\partial^2\beta}{\partial p_0^2}\right) p_1^2 + \left(3\frac{\partial\alpha}{\partial p_0} F + \frac{\partial\alpha}{\partial x} + \frac{\partial^2\alpha}{\partial x^2} + 2\frac{\partial^2\beta}{\partial x \partial p_0}\right) p_1$$
$$+ \left(2F\frac{\partial\alpha}{\partial x} + F\frac{\partial\beta}{\partial p_0} + \frac{\partial^2\beta}{\partial x^2} - \frac{\partial\beta}{\partial x} - F'\beta\right) = 0.$$

Taking coefficients in p_1s we obtain the following system of equations:

$$\frac{\partial^2\alpha}{\partial p_0^2} = 0,$$

$$2\frac{\partial^2\alpha}{\partial x \partial p_0} + 2\frac{\partial\alpha}{\partial p_0} + \frac{\partial^2\beta}{\partial p_0^2} = 0,$$

$$3F\frac{\partial\alpha}{\partial p_0} + \frac{\partial\alpha}{\partial x} + \frac{\partial^2\alpha}{\partial x^2} + 2\frac{\partial^2\beta}{\partial x \partial p_0} = 0,$$

$$2F\frac{\partial\alpha}{\partial x} + F\frac{\partial\beta}{\partial p_0} + \frac{\partial^2\beta}{\partial x^2} - \frac{\partial\beta}{\partial x} - F'\beta = 0.$$

The first equation yields

$$\alpha = \gamma p_0 + \delta,$$

where γ and δ are some functions of x.

Substituting this into the second equation, we obtain

$$\beta = -(\gamma + \gamma')p_0^2 + \varepsilon p_0 + \zeta,$$

where ε and ζ are again functions of x.

Then the third equation is reduced to

$$3\gamma F = 3(\gamma + \gamma')p_0 - \delta' - \delta'' - 2\varepsilon'.$$

Since the function $F(y)$ was supposed to be non-linear, it follows that

$$\gamma = 0,$$

and

$$\varepsilon = (d - \delta - \delta')/2$$

where $d = \text{const.}$

Finally, we find

$$(\varepsilon p_0 + \zeta)F' - \eta F = \theta p_0 + \lambda$$

with

$$\eta = 2\delta' + \varepsilon, \quad \theta = \varepsilon'' - \varepsilon'.$$

This is an ordinary differential equation with respect to $F(p_0)$, where the variable x enters as a parameter.

Its general solution, under the assumption

$$\varepsilon \neq 0, \quad \eta \neq 0, \quad \varepsilon \neq \eta,$$

is given by the following formula:

$$F = \mu\left(p_0 + \frac{\zeta}{\varepsilon}\right)^{\eta/\varepsilon} + \frac{\theta}{\varepsilon - \eta}\left(p_0 + \frac{\zeta}{\varepsilon}\right) + \frac{\theta\zeta - \varepsilon\lambda}{\varepsilon\eta}$$

where μ is an arbitrary function of x.

From all such functions, we have to choose those which depend only on p_0 and are non-linear in p_0. The former requirement holds if and only if all the functions

$$\mu, \quad \frac{\zeta}{\varepsilon}, \quad \frac{\eta}{\varepsilon}, \quad \frac{\theta}{\varepsilon - \eta}, \quad \frac{\theta\zeta - \varepsilon\lambda}{\varepsilon\eta}$$

are constants.

Denoting the constants μ, ζ/ε, η/ε by a, b and c respectively, and taking into account all the relations among functions under consideration, we arrive at the following expressions:

$$\varepsilon = -\frac{k+1}{2} a \exp(kx), \quad \lambda = (k^2 - k)b\varepsilon,$$
$$\eta = c\varepsilon, \quad \theta = (k^2 - k)\varepsilon, \quad \zeta = b\varepsilon,$$

where

$$k = \frac{1-c}{3+c}.$$

Hence,

$$F(p_0) = a(p_0 + b)^c - \frac{2c+2}{(c+3)^2} p_0.$$

Now we consider the previously excluded cases. Either assumption $\eta = 0$ or $\eta = \varepsilon$ results in linearity of the function $F(p_0)$.

In the case where $\varepsilon = 0$, we obtain a new series of functions

$$F(p_0) = a \exp(bp_0) - \frac{2}{b}, \qquad a, b \in \mathbb{R}.$$

These computations can be summarized as follows.

Theorem 2.2.1 *Non-linear differential equations of the form*

$$y'' = y' + F(y)$$

that have a two-dimensional Lie algebra of point symmetries split in two classes:

1.

$$y'' = y' + a(y+b)^c - \frac{2c+2}{(c+3)^2}(y+b),$$

where $a, b, c \in \mathbb{R}, c \neq -3, 0, 1$ *and*

2.

$$y'' = y' + a \exp(by) - \frac{2}{b},$$

where $a, b \in \mathbb{R}, b \neq 0$.

In the first case the symmetry Lie algebra is generated by

$$\phi = p_1, \quad \psi = \left(p_1 - \frac{k+1}{2}(p_0 + b)\right)\exp(kx)$$

where

$$k = \frac{1-c}{3+c},$$

and

$$[\phi, \psi] = -k\psi.$$

In the second case

$$\phi = p_1, \quad \psi = \left(p_1 - \frac{2}{b}p_0\right)\exp(-x)$$

and

$$[\phi, \psi] = \psi.$$

2.2.2 Integration

We will carry out the integration procedure for the equation of the first series (see Theorem 2.2.1) with $a = 1$, $b = 0$, $c = -2$. In this case we have the following equation:

$$y'' = y' + y^{-2} + 2y. \tag{2.5}$$

The corresponding Lie algebra of point symmetries is generated by two functions

$$\phi = p_1 \quad \text{and} \quad \psi = (p_1 - 2p_0)\exp(3x)$$

and we have

$$[\phi, \psi] = -3\psi.$$

We see that the first commutator subalgebra $\mathfrak{g}^{(1)}$ is generated by the function ψ. Hence, the basis ϕ, ψ is written in the order prescribed by the algorithm.

The Wronski matrix W is

$$W = \begin{Vmatrix} \phi & \psi \\ \mathcal{D}(\phi) & \mathcal{D}(\psi) \end{Vmatrix} = \begin{Vmatrix} p_1 & (p_1 - 2p_0)\exp(3x) \\ p_1 + p_0^{-2} + 2p_0 & (p_1 + p_0^{-2} - 4p_0)\exp(3x) \end{Vmatrix}$$

Its determinant $|W|$ is equal to $T\exp(3x)$, where

$$T = p_1^2 - 4p_0p_1 + 2p_0^{-1} + 4p_0^2.$$

The new basis of differential 1-forms is as follows:

$$\overline{\omega}_1 = \frac{1}{T}[(2p_1 + p_0^{-2} - 4p_0)\,dp_0 - (2p_0 - p_1)\,dp_1] - dx,$$

$$\overline{\omega}_2 = \frac{1}{T}[-\exp(-3x)(p_1 + p_0^{-2} + 2p_0)\,dp_0 + \exp(-3x)p_1\,dp_1].$$

The form $\overline{\omega}_1$ is closed and its integral is

$$-\frac{1}{2}\ln|T| - x.$$

Instead of this function it is more convenient to take

$$H_1 = \exp(2x)(p_1^2 - 4p_0p_1 + 2p_0^{-1} + 4p_0^2)$$

which is also an integral of the equation.

Now we integrate the differential 1-forms $\overline{\omega}_2$ on the surfaces $M_{C_1} \stackrel{\text{def}}{=} \{H_1 = C_1 = \text{ const}\}$.

On these surfaces

$$p_1 = 2p_0 \pm \sqrt{C_1\exp(-2x) - 2p_0^{-1}}$$

and

$$\overline{\omega}_2 = \left(-\exp(-3x) \mp \frac{2p_0\exp(-3x)}{\sqrt{C_1\exp(-2x) - 2p_0^{-1}}}\right)dx \pm \frac{\exp(-3x)}{\sqrt{C_1\exp(-2x) - 2p_0^{-1}}}\,dp_0.$$

The integral of this differential form depends on the sign of the constant C_1. For $C_1 > 0$ we have

$$\begin{aligned} H_2 =& \frac{1}{3}\exp(-3x) \pm \frac{y\exp(-2x)}{C_1}\sqrt{C_1 - 2y^{-1}\exp(2x)} \\ &\pm C_1^{-3/2}\ln\left|\frac{\sqrt{C_1 - 2y^{-1}\exp(2x)} + \sqrt{C_1}}{\sqrt{C_1 - 2y^{-1}\exp(2x)} - \sqrt{C_1}}\right| + C_2, \end{aligned}$$

and for $C_1 < 0$

$$\begin{aligned} H_2 =& \frac{1}{3}\exp(-3x) \pm \frac{y\exp(-2x)}{C_1}\sqrt{C_1 - 2y^{-1}\exp(2x)} \\ &\mp(-C_1)^{-3/2}\arctan\sqrt{\frac{2}{C_1}y^{-1}\exp(2x) - 1} + C_2. \end{aligned}$$

2.2.3 Non-linear third-order equations

Here we consider ordinary differential equations of the following type:

$$y^2y''' = ayy'y'' + b(y')^3$$

where $a, b \in \mathbb{R}$.

Equations of this type have a three-dimensional solvable Lie algebra of symmetries:

$$\phi_1 = p_0, \quad \phi_2 = p_0 - xp_1, \quad \phi_3 = p_1$$

with commutation relations

$$[\phi_1, \phi_2] = 0, \quad [\phi_1, \phi_3] = 0, \quad [\phi_2, \phi_3] = -\phi_3.$$

The corresponding transformations for these symmetries are

$$\begin{aligned} A_t^1 &: (x, p_0, p_1, p_2) \longmapsto (x,\ e^t p_0, e^t p_1, e^t p_2), \\ A_t^2 &: (x, p_0, p_1, p_2) \longmapsto (e^t x,\ e^t p_0, p_1, e^{-t} p_2), \\ A_t^3 &: (x, p_0, p_1, p_2) \longmapsto (x + t,\ p_0, p_1, p_2) \end{aligned}$$

respectively.

Hence, if we take the symmetry Lie algebra $\mathfrak{g} = \langle S_{\phi_1}, S_{\phi_2}, S_{\phi_3}\rangle$ generated by these symmetries, then the commutator subalgebra $\mathfrak{g}^{(1)}$ is generated by S_{ϕ_3}, and $\mathfrak{g}^{(2)} = 0$.

Therefore the equation can be solved in two steps.
First of all, we find the Wronski matrix

$$W = \left\| \begin{matrix} p_0 & p_0 - xp_1 & p_1 \\ p_1 & -xp_2 & p_2 \\ p_2 & -p_2 - xF & F \end{matrix} \right\|$$

and the inverse matrix

$$W^{-1} = \frac{1}{|W|} \left\| \begin{matrix} p_2^2 & -p_0F - p_1p_2 & p_0p_2 \\ -p_1F + p_2^2 & p_0F - p_1p_2 & p_1^2 - p_0p_2 \\ xp_1F + p_2(xp_2 - p_1) & xp_0F + 2p_0p_2 - xp_1p_2 & xp_1^2 - p_0(p_1 + xp_2) \end{matrix} \right\|$$

where $|W|$ is the determinant of the matrix W and

$$F(x, p_0, p_1, p_2) = \frac{ap_0p_1p_2 + bp_1^3}{p_0^2}, \qquad |W| = 2p_0p_2^2 - p_1^2p_2 - p_1p_0F.$$

Thus, the differential 1-forms $\overline{\omega}_1, \overline{\omega}_2, \overline{\omega}_3$ are

$$\begin{aligned} |W|\overline{\omega}_1 &= p_2^2\, dp_0 - (p_0F + p_1p_2)dp_1 + p_0p_2\, dp_2, \\ |W|\overline{\omega}_2 &= (p_2^2 - p_1F)\, dp_0 + (p_0F - p_1p_2)\, dp_1 + (p_1^2 - p_0p_2)\, dp_2, \\ |W|\overline{\omega}_3 &= (p_0p_1F + p_1^2p_2 - 2p_0p_2^2)\, dx + (xp_2^2 - p_1p_2 - xp_1F)\, dp_0 \\ &\quad + (xp_0F + 2p_0p_2 - xp_1p_2)\, dp_1 + (xp_1^2 - xp_0p_2 - p_0p_1)\, dp_2. \end{aligned}$$

It follows from the Maurer–Cartan equations that

$$\begin{aligned} d\overline{\omega}_1 &= 0, \\ d\overline{\omega}_2 &= 0, \\ d\overline{\omega}_3 &= \overline{\omega}_2 \wedge \overline{\omega}_3. \end{aligned}$$

Hence, we find two integrals H_1 and H_2 such that $dH_1 = \overline{\omega}_1,\ dH_2 = \overline{\omega}_2$ and then on the level surface

$$M_C = \{H_1 = c_1,\ H_2 = c_2\}$$

we find the third integral.

2.3 Linear differential equations and linear symmetries

2.3.1 The variation of constants method

Let us consider a linear ordinary differential equation

$$y^{(k+1)} + A_k(x)y^{(k)} + \cdots + A_0(x)y = g(x), \tag{2.6}$$

where $A_0(x), \ldots, A_0(x),\ g(x)$ are given smooth functions of x.

Note that a function $f(x)$ is a generating function for some symmetry of the equation if and only if

$$\Delta_F(f) = f^{(k+1)} + A_k(x)f^{(k)}(x) + \cdots + A_0(x)f(x) = 0. \tag{2.7}$$

Hence, an arbitrary solution of the corresponding homogeneous equation is a symmetry (a generating function of a symmetry, to be more precise) of the initial equation.

The space of all solutions forms a $(k+1)$-dimensional commutative algebra, and if one knows its basis (i.e., a fundamental system of solutions of the homogeneous equation), then the non-homogeneous equation can be integrated by means of quadratures.

Let $f_1(x), \ldots, f_{k+1}(x)$ be a fundamental system of solutions of the homogeneous equation (2.7). Then,

$$S_f = f(x)\frac{\partial}{\partial p_0} + f'(x)\frac{\partial}{\partial p_1} + \cdots + f^{(k)}(x)\frac{\partial}{\partial p_k},$$

and

$$[S_{f_i}, S_{f_j}] = 0.$$

We have

$$\omega_i(S_f) = f^{(i-1)},$$

and

$$W = \begin{Vmatrix} f_1 & \cdots & f_{k+1} \\ f_1' & \cdots & f_{k+1}' \\ \vdots & \vdots & \vdots \\ f_1^{(k)} & \cdots & f_{k+1}^{(k)} \end{Vmatrix}$$

is the ordinary *Wronski matrix*.

The corresponding forms $\overline{\omega}_1, \ldots, \overline{\omega}_k$ are closed and define a complete set of integrals,

$$H_j = \int \overline{\omega}_i.$$

2.3.2 Linear symmetries

A shuffling symmetry S_ϕ is called a *linear symmetry* if the generating function ϕ is linear in $p_0, \ldots, p_k$:

$$\phi = b_0(x)p_0 + \cdots + b_k(x)p_k.$$

With any linear symmetry we associate a linear operator

$$\Delta_\phi = b_0 + \cdots + b_k \partial^k$$

where $\partial = d/dx$, and we rewrite the Lie equation for linear symmetries in terms of the algebra of linear differential operators. To this end we need the following lemma.

Lemma 2.3.1 *For any linear differential operators* $A = a_0 + \cdots + a_n \partial^n$ *and*

$$L = l_0 + \cdots + l_k \partial^k + \partial^{k+1}$$

there are differential operators C_A *and* R_A *of order* $\leq n - k - 1$ *and* $\leq k$ *respectively such that*

$$A = C_A \circ L + R_A.$$

Moreover, these operators C_A *and* R_A *are uniquely determined by* A.

Proof The result follows from the observation that

$$A = a_n \partial^{l-n-1} \circ L + A_1$$

where the order of A_1 is not greater than $n - 1$ and induction on n.

To prove the uniqueness assume that $A = C_1 \circ L + R_1 = C_2 \circ B + R_2$. Then $(C_2 - C_1) \circ L = R_1 - R_2$. If $C_1 \neq C_2$ then the order of $(C_2 - C_1) \circ L$ is not less then $k + 1$ and the order of $R_1 - R_2$ is not greater than $k - 1$. Therefore, $C_1 = C_2$ and $R_1 = R_2$. ∎

Remark 2.3.1 *In a similar way one proves that there are operators c_A and r_A such that*

$$A = L \circ c_A + r_A$$

and these operators are uniquely determined by A.

Lemma 2.3.2 *Let A be a differential operator such that $A(h) = 0$ for any solution h of equation $L(h) = 0$. Then $A = C \circ L$ for some differential operator C.*

Proof Using the above lemma we obtain $A = C \circ L + R$ for some operator R of order $\leq k$. Moreover, $\ker R \supset \ker L$ by assumption. If $R \neq 0$ then $\dim \ker R \leq k$ and $\dim \ker L = k + 1$. Therefore, $R = 0$. ■

Theorem 2.3.1 *A differential operator $\Delta_\phi = b_0 + \cdots + b_k \partial^k$ corresponds to a shuffling symmetry $\phi = b_0 p_0 + \cdots + b_k p_k$ of the linear differential equation $L(h) = 0$, where*

$$L = A_0 + \cdots + A_k \partial^k + \partial^{k+1}$$

if and only if there is a differential operator ∇_ϕ of order k and such that

$$L \circ \Delta_\phi = \nabla_\phi \circ L.$$

Moreover, the commutator $[\phi, \psi]$ of linear symmetries corresponds to the remainder R of division $[\Delta_\phi, \Delta_\psi]$ by L; that is,

$$R_{[\Delta_\phi, \Delta_\psi]} = \Delta_{[\phi,\psi]}.$$

Proof Consider a solution h and let h_t be the 1-parameter family of solutions, obtained by shifting h along symmetry S_ϕ. Then, (see Remark 2.1.1), $h_t = h + t\Delta_\phi(h) + o(t)$ and $L(h_t) = tL(\Delta_\phi(h)) + o(t)$ and therefore, $L \circ \Delta_\phi(h) = 0$ for all solutions $L(h) = 0$. The existence of ∇_ϕ now follows from the above lemma.

To prove the last statement of the theorem we denote the above transformation by $\Phi_t : h \longmapsto h_t = h + t\Delta_\phi(h) + o(t)$.

Let Ψ_s be the transformation corresponding to a linear symmetry ψ. Then $\Psi_s : h \longmapsto h_s = h + s\Delta_\psi(h) + o(s)$, and the commutator $[\phi, \psi]$ corresponds to the st-term in the difference $\Phi_t \circ \Psi_s - \Psi_s \circ \Phi_t$. We have

$$\begin{aligned} \Phi_t \circ \Psi_s : h &\overset{\Psi_s}{\longmapsto} h + s\Delta_\psi(h) + o(s) \\ &\overset{\Phi_t}{\longmapsto} h + s\Delta_\psi(h) + t\Delta_\phi(h) + st\Delta_\phi \circ \Delta_\psi(h) + o(t) + o(s) \end{aligned}$$

and, on the other hand,

$$\Psi_s \circ \Phi_t : h \overset{\Phi_t}{\longmapsto} h + t\Delta_\phi(h) + o(t)$$
$$\overset{\Psi_s}{\longmapsto} h + s\Delta_\psi(h) + t\Delta_\phi(h) + st\Delta_\psi \circ \Delta_\phi(h) + o(t) + o(s).$$

Therefore,

$$\Phi_t \circ \Psi_s - \Psi_s \circ \Phi_t : h \longmapsto st[\Delta_\phi, \Delta_\psi](h) + o(t) + o(s)$$
$$= stR_{[\Delta_\phi, \Delta_\psi]}(h) + o(t) + o(s).$$

■

Let us denote by $\mathfrak{G}(L)$ a vector space of all differential operators Δ such that

$$L \circ \Delta = \nabla \circ L$$

for some (uniquely determined) differential operator ∇.

First of all $\mathfrak{G}(L)$ is an algebra with respect to the composition of operators, and a Lie algebra with respect to the commutator of operators.

Denote by $\mathrm{Sym}(L)$ the Lie algebra of linear symmetries of L.

Theorem 2.3.2 *1. If $\Delta \in \mathfrak{G}(L)$ then $R_\Delta \in \mathrm{Sym}(L)$.*
2. The residue map

$$R : \mathfrak{G}(L) \to \mathrm{Sym}(L),$$
$$R : \Delta \in \mathfrak{G}(L) \longmapsto R_\Delta \in \mathrm{Sym}(L)$$

is a Lie algebra homomorphism.

Proof Indeed, from $L \circ \Delta = \nabla \circ L$ we find

$$L \circ (C_\Delta - c_\nabla) \circ L = r_\nabla \circ L - L \circ R_\Delta.$$

If $C_\Delta \neq c_\nabla$ then the operator $L \circ (C_\Delta - c_\nabla) \circ L$ has order greater than $2k + 2$ but the order of $r_\nabla \circ L - L \circ R_\Delta$ does not exceed $2k + 1$. Therefore, $C_\Delta = c_\nabla$ and $r_\nabla \circ L = L \circ R_\Delta$.

The second part of the theorem follows from Theorem 2.3.1. ■

2.4 Linear symmetries of self-adjoint operators

Recall that the differential operator

$$L^t = (-1)^{k+1}\partial^{k+1} + \sum_{i=0}^{k}(-1)^i \partial^i \circ A_i$$

is said to be *adjoint* to the operator

$$L = \partial^{k+1} + \sum_{i=0}^{k} A_i \partial^i.$$

The conjugation operation is an involution in the algebra of differential operators, that is

$$(A^t)^t = A, \quad (A \circ B)^t = B^t \circ A^t.$$

A differential operator L is said to be *self-adjoint* if $L^t = L$ and *skew-adjoint* if $L^t = -L$.

Example 2.4.1

1. *One has only skew-adjoint operators in order one. They have the form*

$$2f(x)\partial + f'(x).$$

2. *In order two one has only self-adjoint operators. They have the form*

$$f(x)\partial^2 + f'(x)\partial + g(x).$$

3. *In order three one has only skew-adjoint operators. They have the form*

$$2f(x)\partial^3 + 3f'(x)\partial^2 + (f''(x) + 2g(x))\partial + g'(x).$$

Theorem 2.4.1 *The correspondence $\Delta_\phi \longleftrightarrow \nabla_\phi^t$ establishes an isomorphism between linear symmetries of the differential equation $L(h) = 0$ and linear symmetries of the adjoint equation $L^t(h) = 0$.*

Let us now assume that L is self-adjoint or skew-adjoint. Then if $\Delta \in \mathfrak{S}(L)$ so ∇^t does.

Using this involution we can decompose $\mathfrak{S}(L)$ into the direct sum:

$$\mathfrak{S}(L) = \mathfrak{S}_0(L) \oplus \mathfrak{S}_1(L)$$

where

$$\mathfrak{G}_0(L) = \left\{ \Delta \,|\, L \circ \Delta = -\Delta^t \circ L \right\},$$

$$\mathfrak{G}_1(L) = \left\{ \Delta \,|\, L \circ \Delta = \Delta^t \circ L \right\}.$$

We will define $\mathbb{Z}_2$-parity $\varepsilon(\Delta) = 0 \in \mathbb{Z}_2$ for $\Delta \in \mathfrak{G}_0(L)$ and $\varepsilon(\Delta) = 1 \in \mathbb{Z}_2$ for $\Delta \in \mathfrak{G}_1(L)$, and will consider the above decomposition as $\mathbb{Z}_2$-grading on $\mathfrak{G}(L)$.

Theorem 2.4.2 *Let L be a self- or skew-adjoint differential operator.*

1. *Then the commutator of operators determines a Lie algebra structure on* $\mathfrak{G}(L)$, *such that*

$$[\Delta_a, \Delta_b] \in \mathfrak{G}_{a+b}(L)$$

   if $\Delta_a \in \mathfrak{G}_a(L)$, $\Delta_b \in \mathfrak{G}_b(L)$, $a, b \in \mathbb{Z}_2$.
2. *Let* $\mathrm{Sym}(L)$ *be the Lie algebra of linear symmetries of operator L, and let* $\mathrm{Sym}_{\mathbf{a}}(L) = R(\mathfrak{G}_{\mathbf{a}}(L))$ *for* $\mathbf{a} \in \mathbb{Z}_2$. *Then* $\mathrm{Sym}(L) = \mathrm{Sym}_{\mathbf{0}}(L) \oplus \mathrm{Sym}_{\mathbf{1}}(L)$ *and*

$$[\mathrm{Sym}_{\mathbf{a}}(L), \mathrm{Sym}_{\mathbf{b}}(L)] \subset \mathrm{Sym}_{\mathbf{a+b}}(L).$$

Proof We have

$$L \circ \Delta_a = -(-1)^a \Delta_a^t \circ L.$$

Therefore,

$$L \circ \Delta_a \circ \Delta_b = (-1)^{a+b} \Delta_a^t \circ \Delta_b^t \circ L = (-1)^{a+b} (\Delta_b \circ \Delta_a)^t \circ L$$

and

$$L \circ [\Delta_a, \Delta_b] = -[\Delta_a, \Delta_b]^t \circ L.$$

The second statement of the theorem follows from Theorems 2.3.1 and 2.3.2. ∎

We call elements of $\mathrm{Sym}_{\mathbf{0}}(L)$ by *even symmetries* and elements of $\mathrm{Sym}_{\mathbf{1}}(L)$ by *odd symmetries* of the equation $L(h) = 0$.

2.5 Schrödinger operators

In this section we apply the above results to the *Schrödinger operator* of the form

$$L = \partial^2 + W$$

where $W = W(x)$ is a so-called *potential*.

This operator is self-adjoint. Therefore, the algebra of linear symmetries is $\mathbb{Z}_2$-graded.

Let us begin with $\mathrm{Sym}_{\mathbf{0}}(L)$. If $\Delta = A_0 + A_1\partial \in \mathrm{Sym}_{\mathbf{0}}(L)$ then $L \circ \Delta = -\Delta^t \circ L$.

We have $\Delta^t = (A_0 - A_1') - A_1\partial$ and

$$L \circ \Delta = A_1\partial^3 + (A_0 + 2A_1')\partial^2 + (A_1'' + 2A_0' + WA_1)\partial + WA_0,$$
$$\Delta^t \circ L = -A_1\partial^3 + (A_0 - A_1')\partial^2 - A_1W\partial - (A_1W' + A_1'W - WA_0).$$

Therefore, $\Delta \in \mathrm{Sym}_{\mathbf{0}}(L)$ implies $A_0 = -\frac{1}{2}A_1'$, and the function $z = A_1$ should satisfy the following differential equation:

$$z''' + 4Wz' + 2W'z = 0. \tag{2.8}$$

We denote the differential operator on the right-hand side of this equation by

$$L^{(2)} = \partial^3 + 4W\partial + 2W'.$$

If $\Delta \in \mathrm{Sym}_{\mathbf{1}}(L)$ then $L \circ \Delta = \Delta^t \circ L$ and we obtain $A_1 = 0$, $A_0' = 0$. Thus $\Delta \in \mathrm{Sym}_{\mathbf{1}}(L)$ if and only if Δ proportional to the identity operator.

Finally, we obtain the following theorem.

Theorem 2.5.1 *The Lie algebra of linear symmetries of the Schrödinger operator has the following description:*

1. $\mathrm{Sym}_{\mathbf{0}}(L) = \{-\frac{1}{2}z' + z\partial \mid L^{(2)}(z) = 0\}$;
2. $\mathrm{Sym}_{\mathbf{1}}(L) = \mathbb{R} \cdot \mathrm{id}$;
3. *The Lie algebra* $\mathrm{Sym}_{\mathbf{0}}(L)$ *is isomorphic to the Lie algebra* $\mathbf{sl}_2(\mathbb{R})$.

Proof To prove this theorem we should establish isomorphism between $\mathrm{Sym}_{\mathbf{0}}(L)$ and $\mathbf{sl}_2(\mathbb{R})$. First of all we note that the solution space of the equation $L^{(2)}(z) = 0$ is a Lie algebra with respect to the bracket

$$[z_1, z_2] = z_1z_2' - z_1'z_2$$

induced by the bracket in $\mathrm{Sym}_{\mathbf{0}}(L)$.

This is a three-dimensional Lie algebra. To find an explicit expression for the commutator we take a fundamental system of solutions, say, z_1, z_2, z_3 where

$$z_1(0) = 1, \quad z_1'(0) = 0, \quad z_1''(0) = 0;$$
$$z_2(0) = 0, \quad z_2'(0) = 1, \quad z_2''(0) = 0;$$
$$z_3(0) = 0, \quad z_3'(0) = 0, \quad z_3''(0) = 1.$$

Let $z = [z_1, z_2]$. Then,

$$z(0) = 1, \quad z'(0) = 0, \quad z''(0) = -4W(0).$$

Therefore,

$$[z_1, z_2] = z_1 - 4W(0)z_3.$$

In a similar way we find

$$[z_2, z_3] = z_3, \ [z_1, z_3] = z_2$$

and isomorphism with $\mathbf{sl}_2(\mathbb{R})$. ∎

The straightforward calculations give the following interpretation of the symmetry operator $L^{(2)}$.

Proposition 2.5.1 *The differential operator $L^{(2)}$ is the second* symmetric power *of the Schrödinger operator L. That is, if y_1 and y_2 is a fundamental system of solutions for the Schrödinger equation $L(y) = 0$. Then $L^{(2)}$ is the differential operator of third order with a fundamental system of solutions $z_1 = \frac{1}{2}y_1^2$, $z_2 = y_1y_2$, $z_3 = \frac{1}{2}y_2^2$.*

Corollary 2.5.1 *Let y be a solution of a Schrödinger equation. Then*

$$\Delta = y^2\partial - yy'$$

is a linear symmetry of the equation.

Remark 2.5.1 *Any skew-adjoint operator of third order $\partial^3 + A\partial + B$ is the symmetric power of a Schrödinger operator.*

Remark 2.5.2 *The ratio $z = y_1/y_2$ satisfies the Schwartzian equation*

$$\mathbf{S}[z] = 2W,$$

where

$$\mathbf{S}[z] = \frac{z^{(3)}}{z^{(1)}} - \frac{3}{2}\left(\frac{z^{(2)}}{z^{(1)}}\right)^2$$

is the Schwartzian of z. Therefore linear symmetry of the Schrödinger equation $y'' + Wy = 0$ generates a Möbius transform and a symmetry for the Schwartzian equation

$$2p_3p_1 - 3p_2^2 - 2Wp_1^2 = 0.$$

2.5.1 Integrable potentials

We say that a potential W is *integrable* if for the Schrödinger equation a linear symmetry $z\partial - z'/2$ is known.

In this case the solution space of the Schrödinger equation is also known because the Schrödinger equation has two commuting symmetries p_0 and $\phi_z = zp - \frac{1}{2}z'p_0$.

For example, a potential W is integrable if we know a non-zero solution or a non-zero even symmetry of the equation.

In this section we shall discuss a few methods of finding new integrable potentials from the known ones.

Integration

Assume that the Schrödinger equation

$$y'' + W(x)y = 0 \tag{2.9}$$

has the linear symmetry

$$\phi_z = z(x)p_1 - \frac{z'(x)}{2}p_0,$$

where the function $z = z(x)$ satisfies the following equation:

$$z''' + 4wz' + 2w'z = 0. \tag{2.10}$$

Note that symmetries p_0 and ϕ_z commute and assuming that z given one can find first integrals by quadratures. Namely, taking $f_1 = p_0, f_2 = p_1$ in Theorem 2.1.3 one can obtain two differential 1-forms ω_1 and ω_2, and integrals H_1 and H_2.

The first integral $H = H_1$ can be chosen to be quadratic in p_0, p_1:

$$H = 2zp_1^2 - 2z'p_0p_1 + (z'' + 2wz)p_0^2.$$

We will rewrite this integral in the following way.

Let us note that (2.10) is defined by the skew-adjoint operator

$$L = \frac{d^3}{dx^3} + 4w\frac{d}{dx} + 2w'$$

and the Lagrange formula (see Section 2.5.3) shows that

$$K(z) = 2z(z'' + 2wz) - z'^2$$

is a first integral for (2.10).

We say that the symmetry ϕ_z is *elliptic, hyperbolic* or *parabolic* if $K(z) > 0$, $K(z) < 0$ or $K(z) = 0$ respectively.

Using the generating function of the symmetry we can rewrite the first integral in the following form:

$$H = \frac{2(\phi_z^2 + kp_0^2)}{z},$$

where $4k = K(z)$.

Now taking $f_1 = H, f_2 = p_0$ in Theorem 2.1.3 we find two differential 1-forms with

$$\theta_1 = \frac{dH}{2H},$$

and the restriction θ of the second form θ_2 on levels $H = 2c$ equals

$$\theta = \frac{dp_0 - p_1\,dx}{\phi_z}.$$

Let

$$\alpha = \frac{\phi_z}{\sqrt{|z|}}, \quad \beta = \frac{p_0}{\sqrt{|z|}}.$$

Then,

$$H = 2(\alpha^2 + k\beta^2) = 2c$$

and the restriction θ takes the following form:

$$\theta = \frac{d\beta}{\alpha} - \frac{dx}{z}.$$

Integration of θ gives the following solutions of the Schrödinger equation (see [68]).

Theorem 2.5.2 *Let ϕ_z be a linear symmetry of (2.9). Then solutions of the Schrödinger equation have the following form:*

- *for elliptic symmetry ϕ_z*

$$y(x) = \sqrt{\frac{|cz(x)|}{k}} \sin\left(\sqrt{k}\int \frac{dx}{z(x)}\right),$$

- *for hyperbolic symmetry ϕ_z*

$$y(x) = \sqrt{\frac{|cz(x)|}{-k}} \sinh\left(\sqrt{-k}\int \frac{dx}{z(x)}\right),$$

- *for parabolic symmetry ϕ_z*

$$y(x) = \sqrt{|cz(x)|}\int \frac{dx}{z(x)}.$$

Here $H = 2c$.

Potentials with a common even symmetry

Let us assume that we have a potential W_0 with a non-zero linear symmetry $E_{z_0} = -\frac{1}{2}z_0' + z_0\partial$ and consider (2.8) as an equation for W with given z_0:

$$z_0''' + 4Wz_0' + 2W'z_0 = 0.$$

Then W_0 is a solution of this equation and if W has the same linear symmetry, then

$$W = W_0 + Cz_0^{-2}.$$

Proposition 2.5.2 *Assume that the Schrödinger equation with potential W_0 has an even symmetry E_{z_0}, then the Schrödinger equation with potential $W_0 + cz_0^{-2}$ has the same symmetry for all constants c.*

Corollary 2.5.2 *Assume that the Schrödinger equation with potential W_0 has two solutions y_1 and y_2. Then the Schrödinger equation with potential*

$$W_0 + \frac{c}{(a_{11}y_1^2 + 2a_{12}y_1y_2 + a_{22}y_2^2)^2}$$

has the symmetry E_z with $z = a_{11}y_1^2 + 2a_{12}y_1y_2 + a_{22}y_2^2$ and is therefore integrable.

Example 2.5.1

1. *Take* $W_0 = 0$. *Then the Schrödinger equations with potentials*

$$W = \frac{c}{(a_{11} + 2a_{12}x + a_{22}x^2)^2}$$

are integrable in quadratures.

2. *Take* $W_0 = \omega^2$. *Then the Schrödinger equations with potentials*

$$W = \omega^2 + \frac{c}{(a_{11}\cos^2\omega x + 2a_{12}\cos\omega x\sin\omega x + a_{22}\sin^2\omega x)^2}$$

are integrable in quadratures.

3. *Take* $W_0 = -k^2$. *Then the Schrödinger equations with potentials*

$$W = -k^2 + \frac{c}{(a_{11}e^{kx} + 2a_{12} + a_{22}e^{-kx})^2}$$

are integrable in quadratures.

$\mathbf{sl}_2(\mathbb{R})$-representations

As we have seen $\mathrm{Sym}_0(L)$ gives us a representation of the Lie algebra $\mathbf{sl}_2(\mathbb{R})$ using first-order differential operators E_z, or a representation by vector fields $z\partial$ on $\mathbb{R}$ where z runs over solutions of (2.8). On the other hand, any representation of the Lie algebra by vector fields on $\mathbb{R}$ corresponds to some Schrödinger equation. Indeed, if $z\partial$ is a non-zero vector field from this representation, then solving (2.8) with respect to W we find the Schrödinger equation with even symmetry $z\partial - \frac{1}{2}z'$. This equation can be solved by quadratures and due to (2.5.1) we are able to find all symmetries and therefore the representation of $\mathbf{sl}_2(\mathbb{R})$.

Let us take $W = 0$. Then (2.8) gives the standard representation of $\mathbf{sl}_2(\mathbb{R})$ by vector fields

$$\partial,\ x\partial,\ x^2\partial.$$

Take E_z with $z = a + bx + cx^2$. This is an even symmetry of the equation $y'' = 0$. Therefore, from the previous section, Schrödinger equations with potentials

$$W = \frac{c_1}{(a + bx + cx^2)^2}$$

are integrable in quadratures (see Figure 2.5).

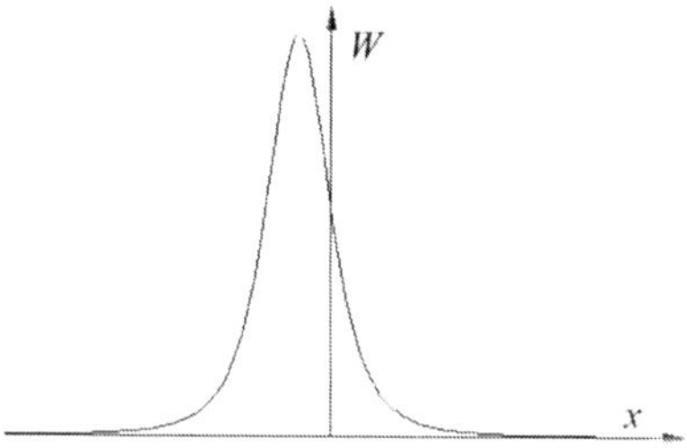

Figure 2.5. The potential $(1+x+x^2)^{-2}$.

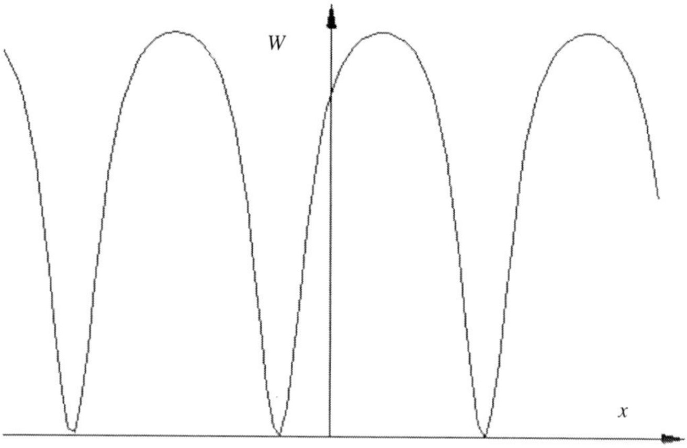

Figure 2.6. The potential $1-(2+\sin 2x)^{-2}$.

Now take $W=\omega^2$. Then (2.8) gives a representation of $\mathbf{sl}_2(\mathbb{R})$ by vector fields

$$\partial, \quad \cos 2\omega x\, \partial, \quad \sin 2\omega x\, \partial$$

and E_z with $z=a+b\cos 2\omega x+c\sin 2\omega x$ is an even symmetry of the harmonic oscillator $y''+\omega^2 y=0$.

Therefore, Schrödinger equations with potentials

$$W=\omega^2+\frac{c_1}{(a+b\cos 2\omega x+c\sin 2\omega x)^2}$$

are integrable in quadratures (see Figure 2.6).

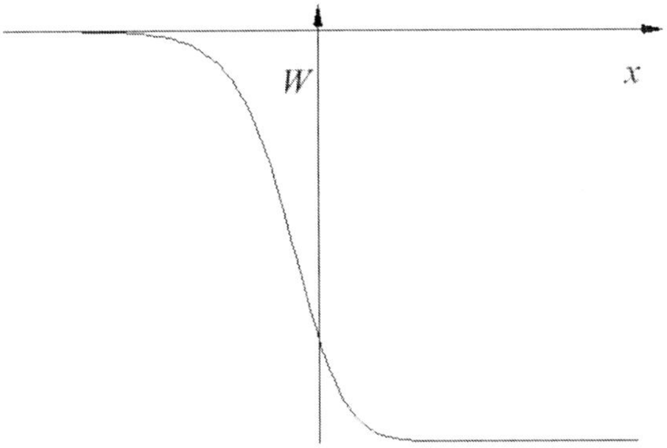

Figure 2.7. The potential $-1 + (1 + e^{2x})^{-2}$.

In the similar way, for the potential $W = -k^2$ we have a representation of $\mathbf{sl}_2(\mathbb{R})$ by vector fields

$$\partial, \quad e^{2kx}\partial, \quad e^{-2kx}\partial$$

and Schrödinger equations with potentials

$$W = -k^2 + \frac{c_1}{(a + be^{2kx} + ce^{-2kx})^2}$$

are also integrable in quadratures (see Figure 2.7).

Relations with the Korteweg–de Vries equation

All previous constructions of integrable potentials have used relations between potentials and even symmetries. Let us make this relation explicit. Namely, let us assume that the potential function W is a function of symmetry:

$$W = w(z).$$

In this case (2.8) takes the following form:

$$z''' + 4wz' + 2w'z'z = 0.$$

We look at a particular case when this equation can be easily integrated.

Namely, let us consider the case of linear dependency

$$W = a\,z + b$$

for some constants $a \neq 0$ and b.

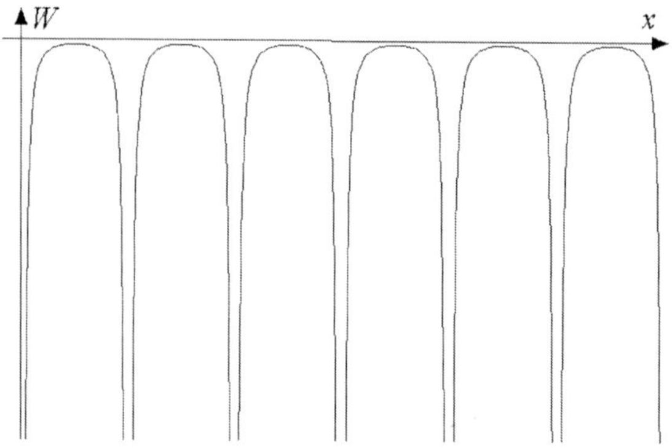

Figure 2.8. The Weierstrass potential.

Then we obtain the Korteweg–de Vries equation

$$z''' + 6azz' + 4bz' = 0$$

for traveling waves.

Integrating, we find

$$z'' + 3az^2 + 4bz = c_1.$$

Multiplying both sides by z' and integrating once more we obtain

$$\tfrac{1}{2}(z')^2 + az^3 + 2bz^2 - c_1 z = c_2$$

and

$$z = -\frac{1}{\sqrt[3]{2a}}\wp(x, g_2, g_3) - \frac{b}{3a}$$

where $\wp(x, g_2, g_3)$ is the Weierstrass function.

Thus Schrödinger equations with potentials

$$W = -\sqrt[3]{\frac{a^2}{2}}\wp(x, g_2, g_3) + \frac{2}{3}b$$

have even linear symmetries E_z with $z = -(1/\sqrt[3]{2a})\wp(x, g_2, g_3) - b/3a$ and therefore can be integrated by quadratures (see Figure 2.8).

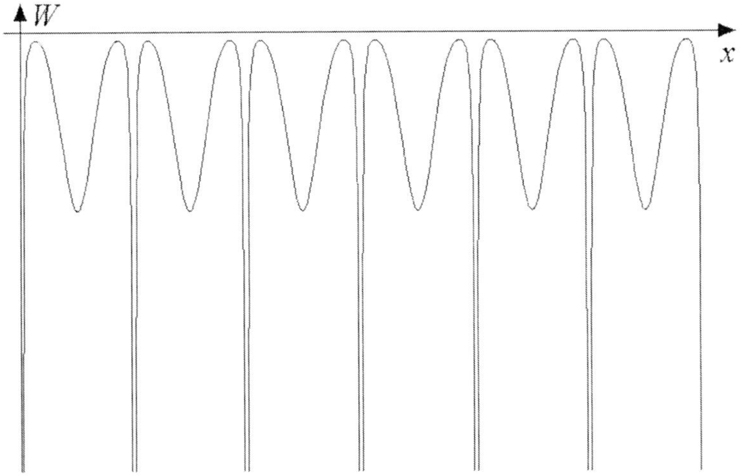

Figure 2.9. The Weierstrass potential with symmetry.

It is also true for potentials (see Figure 2.9)

$$-\sqrt[3]{\frac{a^2}{2}}\wp(x, g_2, g_3) + \frac{2}{3}b + \frac{c}{(-(1/\sqrt[3]{2a})\wp(x, g_2, g_3) - b/3a)^2}.$$

2.5.2 Spectral problems for KdV potentials

In this section we consider potentials $w(x)$ such that the corresponding eigenvalue problem

$$y'' + w(x)y = \lambda y \tag{2.11}$$

possesses linear symmetries $z(x, \lambda)$ which are polynomial in λ.

Let

$$z(x, \lambda) = z_0(x)\lambda^n + z_1(x)\lambda^{n-1} + \cdots + z_{n-1}(x)\lambda + z_n(x)$$

be a linear symmetry for (2.11).

Then the Lie equation gives a polynomial (in λ) of degree $n + 1$

$$z'''(x, \lambda) + 4(w(x) - \lambda)\, z'(x, \lambda) + 2w'(x)\, z(x, \lambda) = 0$$

and a recursive set of equations on z_k:

$$z_0' = 0,$$
$$z_{k+1}' = \tfrac{1}{4}L^{(2)}(z_k), \quad k = 0, \ldots, n-1,$$
$$L^{(2)}(z_n) = 0.$$

Taking $z_0 = 1$, we find inductively functions $z_k = z_k(w)$ by

$$z_{k+1}(w) = \frac{1}{4}\int L^{(2)}(z_k(w))\,dx,$$

($k = 0, 1, \ldots$).

The first functions are as follows:

$$z_1(w) = \frac{w}{2} + c_1,$$

$$z_2(w) = \frac{w'' + 3w^2}{8} + \frac{c_1}{2}w + c_2,$$

$$z_3(w) = \frac{w^{(4)}}{32} + \frac{5}{16}\left(ww'' + \frac{w'^2}{2} + w^3\right) + \frac{c_1}{8}(w'' + 3w^2) + \frac{c_2}{2}w + c_3,$$

$$z_4(w) = \frac{w^{(6)}}{128} + \frac{7w'w^{(3)}}{32} + \frac{7(3w''^2 + 10w^2w'' + 10ww'^2 + 5w^4)}{128}$$
$$+ \frac{c_1}{32}\left(w^{(4)} + 10(ww'' + \frac{w'^2}{2} + w^3\right) + \frac{c_2}{8}(w'' + 3w^2) + \frac{c_3}{2}w + c_4.$$

The conditions $L^{(2)}\,(z_n(w)) = 0$ which can also be reformulated as $z_{n+1}'(w) = 0$ are called nth *KdV stationary equations*.

Below we list the first KdV equations:

zeroth KdV:

$$w' = 0,$$

first KdV:

$$w''' + 6ww' + 4c_1w' = 0,$$

second KdV:

$$w^{(5)} + 10(ww''' + 2w'w'' + 3w^2w')$$
$$+ 4c_1(w'' + 6ww') + 16c_2w' = 0.$$

We conclude that potentials w which satisfy the n-th stationary KdV equation possess linear symmetry ϕ_{S_n} with

$$S_n = \lambda^n + \sum_{k=1}^{n} z_k \lambda^{n-k}.$$

As we have seen, the function $K = 2z(z'' + 2(w - \lambda)z) - z'^2$ is the first integral of the Lie equation and therefore coefficients of the polynomials

$$Q_n = 2S_n(S_n'' + 2(w - \lambda)S_n) - S_n'^2$$

are first integrals for the n-th KdV equation.

For example, for the classical (first) KdV equation, $w''' + 6ww' + 4c_1 w' = 0$, one has

$$S_1 = \lambda + \frac{w}{2} + c_1$$

and

$$Q_1 = -4\lambda^3 - 8c_1\lambda^2 + q_{11}\lambda + q_{10},$$

where

$$q_{11} = w'' + 3w^2 + 4c_1 w - 4c_1^2,$$
$$q_{10} = \frac{2ww'' - w'^2 + 4w^3}{4} + c_1(w'' + 4w^2 + 4c_1 w)$$

are first integrals.

Solving the KdV equation together with equations q_{11} = constant, q_{10} = constant we obtain the first-order ODE for w:

$$w'^2 = -2w^3 - 4c_1 w^2 + 2(q_{11} + 4c_1^2)w + 4(c_1 q_{11} - q_{10} + 4c_1^3)$$

and solutions

$$w = -2\wp(x + c, g_2, g_3) - \frac{2c_1}{3}$$

where the invariants of the Weierstrass elliptic function are equal to

$$g_2 = \frac{4c_1^2 - 6c_1}{3}, \quad g_3 = -\frac{152c_1^3}{27} - \frac{5q_{11}c_1}{3} + 2q_{10}.$$

For the second KdV equation,

$$w^{(5)} + 10(ww''' + 2w'w'' + 3w^2w') + 4c_1(w'' + 6ww') + 16c_2 w' = 0,$$

one has

$$S_2 = \lambda^2 + \left(\frac{w}{2} + c_1\right)\lambda + \frac{w'' + 3w^2}{8} + \frac{c_1}{2}w + c_2$$

and

$$Q_2 = -4\lambda^5 - 8c_1\lambda^4 - 4(c_1^2 + 2c_2)\lambda^3 + q_{22}\lambda^2 + q_{21}\lambda + q_{20},$$

where

$$q_{22} = \frac{10w^3 + 5w'^2 + 10ww'' + w^{(4)}}{10} - 8c_1c_2 + (w'' + 3w^2)c_1 + 4wc_2,$$

$$q_{21} = \frac{2ww^{(4)} - 2w'w''' + w''^2 + 20w^2w'' + 15w^4}{16} + (w'' + 3w^2)c_1^2$$
$$+ 4wc_1c_2 - 4c_2^2 + \frac{4w^{(4)} + 12ww'' + 4w'^2 + 14w^3}{4}c_1 + w^2c_2,$$

$$q_{20} = \frac{2w''w^{(4)} + 6w^2w^{(4)} + 4w(4w''^2 - 3w'w''') - w'''^2 + 12w'^2w'' + 60w^3w'' + 36w^5}{64}$$
$$+ \frac{4w^3 - w'^2 + 2ww''}{4}c_1^2 + (w'' + 4w^2)c_1c_2 + 4wc_2^2$$
$$+ \frac{12w^4 + 13w^2w'' + w''^2 - w'w''' + ww^{(4)}}{8}c_1$$
$$+ \frac{12w^3 + 6w'^2 + 10ww'' + w^{(4)}}{4}c_2$$

are the first integrals for the second KdV.

Using these integrals one can reduce the second KdV equation to the following second-order ODE:

$$25w^8 + 80w^7c_1 + 32w^6(2c_1^2 + 5c_2) - 16w^5(24c_1c_2 + 5Q_{22})$$
$$+ 32w^4(-72c_1^2c_2 + 28c_2^2 + 5Q_{21} - 9c_1Q_{22})$$
$$+ 256(-8c_1^2c_2 + 4c_2^2 + Q_{21} - c_1Q_{22})^2$$
$$+ 256w(8c_1c_2 + Q_{22})(8c_1^2c_2 - 4c_2^2 - Q_{21} + c_1Q_{22})$$
$$- 256w^3(8c_1^3c_2 - c_1(-4c_2^2 + Q_{21}))$$
$$- 256w^3(c_1^2Q_{22} + c_2Q_{22}) + 64w^2(32c_2^3 + Q_{22}^2 + 8c_2(Q_{21} + c_1Q_{22}))$$

$$+\,76w^5(w')^2 + 152w^4c_1(w')^2 + 64w^3(c_1^2 + 3c_2)(w')^2$$

$$-\,64w(4c_2^2 + Q_{21})(w')^2 - 64w^2(6c_1c_2 + Q_{22})(w')^2$$

$$+\,128(8c_1^3c_2 - c_1(20c_2^2 + Q_{21}) + c_1^2Q_{22} + 2(2Q_{20} - c_2Q_{22}))(w')^2$$

$$-\,20w^2(w')^4 - 16wc_1(w')^4 + 16(c_1^2 - 4c_2)(w')^4 + 80w^3(w')^2w''$$

$$+\,96w^2c_1(w')^2w'' + 128wc_2(w')^2w'' - 32(8c_1c_2 + Q_{22})(w')^2w''$$

$$-\,8(w')^4w'' - 10w^4(w'')^2 - 16w^3c_1(w'')^2 - 32w^2c_2(w'')^2$$

$$+\,16w(8c_1c_2 + Q_{22})(w'')^2 - 32(-8c_1^2c_2 + 4c_2^2 + Q_{21} - c_1Q_{22})(w'')^2$$

$$+\,(20w + 8c_1)(w')^2(w'')^2 + (w'')^4 = 0.$$

Two cases when $c = 0$, and $c = 0, q = 0$ give us shorter ODEs:

$c = 0$:

$$\begin{aligned}25w^8 &+ 160w^4Q_{21} + 256Q_{21}^2 - 80w^5Q_{22} - 256wQ_{21}Q_{22} + 64w^2Q_{22}^2\\ &+ 76w^5(w')^2 + 512Q_{20}(w')^2 - 64wQ_{21}(w')^2 - 64w^2Q_{22}(w')^2\\ &- 20w^2(w')^4 + 80w^3(w')^2w'' - 32Q_{22}(w')^2w'' - 8(w')^4w''\\ &- 10w^4(w'')^2 - 32Q_{21}(w'')^2 + 16wQ_{22}(w'')^2\\ &+ 20w(w')^2(w'')^2 + (w'')^4 = 0.\end{aligned}$$

$c = 0, q = 0$:

$$\begin{aligned}25w^8 &+ 76w^5(w')^2 - 20w^2(w')^4 + 80w^3(w')^2w''\\ &- 8(w')^4w'' - 10w^4(w'')^2 + 20w(w')^2(w'')^2 + (w'')^4 = 0.\end{aligned}$$

Note that these equations have a two-dimensional commutative symmetry Lie algebra generated by translations and the first KdV, and therefore can be solved in quadratures.

Now one can apply Theorem 2.5.2 to spectral problems for the Schrödinger ordinary differential equations in which potentials are solutions of n-th KdV equations. This gives complete and explicit solutions of the spectral problems.

We illustrate this method for potentials which satisfy the first KdV equation (this is the case of a special Lamé equation) and for the following boundary problem on an interval $[a, b]$:

$$y(a) = y(b) = 0.$$

Then Theorem 2.5.2 shows that smooth eigenfunctions

$$y(x) = 2\sqrt{\frac{|S_1(x)|}{Q_1(\lambda)}} \sin\left(\frac{\sqrt{Q_1(\lambda)}}{2}\int_a^x \frac{d\tau}{S_1(\tau)}\right)$$

do exist if:

- $S_1 = \lambda + w/2 + c_1 \neq 0$ on the interval,
- $Q_1(\lambda) > 0$,

$$\int_a^b \frac{d\tau}{S_1(\tau)} = \frac{2\pi n}{\sqrt{Q_1(\lambda)}}$$

for $n \in \mathbb{Z}$.

Summarizing we obtain the following result.

Theorem 2.5.3 *Let the potential w satisfies the classical KdV equation,*

$$w''' + 6ww' + 4c_1 w' = 0,$$

then spectral values λ for the boundary problem (2.11) given by the formula

$$\lambda = \wp(\alpha, g_2, g_3) - 2c_1/3$$

where α are solutions of the equations

$$2(b-a)\zeta(\alpha) + \ln\frac{\sigma(b+c-\alpha)\sigma(a+c+\alpha)}{\sigma(a+c-\alpha)\sigma(b+c+\alpha)} = \iota 2\pi n \quad (n \in \mathbb{Z},\ \iota = \sqrt{-1})$$

such that $Q_1(\lambda) > 0$ and

$$\lambda > -c_1 - \frac{1}{2}\min_{a \le x \le b} w(x)$$

or

$$\lambda < -c_1 - \frac{1}{2}\min_{a \le x \le b} w(x).$$

Here

$$Q_1(\lambda) = -4\lambda^3 - 8c_1\lambda^2 + q_{11}(w)\lambda + q_{10}(w),$$

and constants $q_{11}(w)$ and $q_{10}(w)$ are values of first integrals q_{10} and q_{11} on the solution w. The function $\zeta(\alpha)$ is the Weierstrass zeta function and $\sigma(z)$ is the Weierstrass sigma function with invariants

$$g_2 = \frac{4c_1^2 - 6c_1}{3}, \quad g_3 = -\frac{152c_1^3}{27} - \frac{5q_{11}(w)c_1}{3} + 2q_{10}(w).$$

The eigenfunctions corresponding to the eigenvalue λ has the following form:

$$y_\lambda(x) = 2\sqrt{\frac{|\lambda + \frac{1}{2}w(x) + c_1|}{Q_1(\lambda)}}$$
$$\times \sin\left(\frac{\sqrt{Q_1(\lambda)}}{2}\left(2(x-a)\zeta(\alpha) + \ln\frac{\sigma(x-\alpha)\sigma(a+\alpha)}{\sigma(x+\alpha)\sigma(a-\alpha)}\right)\right).$$

Proof Since

$$w = -2\wp(x+c, g_2, g_3) - \frac{2c_1}{3}$$

we have

$$I = \int_a^b \frac{d\tau}{\lambda + \frac{1}{2}w + c_1} = \int_a^b \frac{d\tau}{\lambda - \wp(\tau + c, g_2, g_3) + 2c_1/3}.$$

Let α be such that

$$\wp(\alpha, g_2, g_3) = \lambda + 2c_1/3.$$

Then,

$$I = -\int_{a+c}^{b+c} \frac{d\tau}{\wp(\tau, g_2, g_3) - \wp(\alpha, g_2, g_3)}$$
$$= -\frac{1}{\wp'(\alpha, g_2, g_3)}\left(2z\xi(\alpha) + \ln\frac{\sigma(z-\alpha)}{\sigma(z+\alpha)}\right)\Bigg|_{a+c}^{b+c}.$$

Because

$$\wp'^2(x, g_2, g_3) = 4\wp^3(x, g_2, g_3) - g_2\wp(x, g_2, g_3) - g_3$$

and

$$\wp''(x, g_2, g_3) = 6\wp^2(x, g_2, g_3) - g_2/2,$$

we obtain the following values of the first integrals q_{11} and q_{10}:

$$q_{11}(w) = w'' + 3w^2 + 4c_1 w - 4c_1^2 = g_2 - 16/3c_1^2 = -2c_1 - 4c_1^2,$$
$$q_{10}(w) = \frac{2ww'' - w'^2 + 4w^3}{4} + c_1(w'' + 4w^2 + 4c_1 w)$$
$$= 2/3c_1 g_2 + g_3 - 32/27c_1^3.$$

Note that for

$$\wp(\alpha, g_2, g_3) = \lambda + 2c_1/3$$

we have

$$\wp'^2(\alpha, g_2, g_3) = 4\wp^3(\alpha, g_2, g_3) - g_2\wp(\alpha, g_2, g_3) - g_3 = -Q_1(\lambda).$$

Therefore

$$\wp'(\alpha, g_2, g_3) = \pm\sqrt{-Q_1(\lambda)}$$

and we find that

$$I = \pm\frac{1}{\sqrt{-Q_1(\lambda)}}\left(2z\xi(\alpha) + \ln\frac{\sigma(z-\alpha)}{\sigma(z+\alpha)}\right)\Bigg|_{a+c}^{b+c} = \frac{2\pi n}{\sqrt{Q_1(\lambda)}}.$$

So, we have

$$\pm.\left((2z\zeta(\alpha) + \ln\frac{\sigma(z-\alpha)}{\sigma(z+\alpha)})\right)\Bigg|_{a+c}^{b+c} = \iota 2\pi n,$$

where $\zeta(\alpha)$ is the Weierstrass zeta function and $\sigma(z)$ is the Weierstrass sigma function. Finally, we obtain the following equations for α:

$$2(b-a)\zeta(\alpha) + \ln\frac{\sigma(b+c-\alpha)\sigma(a+c+\alpha)}{\sigma(a+c-\alpha)\sigma(b+c+\alpha)} = \iota 2\pi n. \qquad \blacksquare$$

In the similar way one finds the following result.

Theorem 2.5.4 *Let a potential w satisfies the n-th KdV equation, then spectral values λ for the boundary problem (2.11) given by solutions of the equation*

$$\int_a^b \frac{d\tau}{S_n(\tau)} = \frac{2\pi m}{\sqrt{Q_n(\lambda)}}, \qquad m \in \mathbb{Z},$$

such that $Q_n(\lambda) > 0$ and $S_n(\tau) \neq 0$ on the interval $[a, b]$.

The eigenfunctions corresponding to the eigenvalue λ have the following form:

$$y_\lambda(x) = 2\sqrt{\frac{|S_n(x)|}{Q_n(\lambda)}}\sin\left(\frac{\sqrt{Q_n(\lambda)}}{2}\int_a^x \frac{d\tau}{S_n(\tau)}\right).$$

2.5.3 Lagrange integrals

Recall that a linear differential operator

$$L = \partial^{k+1} + \sum_{i=0}^{k} A_i \partial^i$$

and its adjoint

$$L^t = (-1)^{k+1} + \sum_{i=0}^{k} (-1)^i \partial^i \circ A_i$$

are connected by the *Lagrange formula*

$$vL(u) - uL^t(v) = \partial P_L(u, v)$$

where

$$P_L(u, v) = \sum_{i=1}^{k+1} \sum_{j=0}^{i-1} (-1)^j u^{(i-j-1)} \partial^j (A_i v)$$

is the Lagrange form.

In this section we shall use the Lagrange forms to obtain first integrals from linear symmetries.

Example 2.5.2

1. *For anti-self-adjoint operators* $L = 2A\partial + A'$ *of first order the Lagrange form is*

$$P_L(u, v) = 2Auv.$$

2. *For self-adjoint operators* $L = A\partial^2 + A'\partial + B$ *of second order the Lagrange form is*

$$P_L(u, v) = A(u'v - uv').$$

3. *For anti-self-adjoint operators* $L = 2A\partial^3 + 3A'\partial^2 + (A'' + 2B)\partial + B'$ *of third order the Lagrange form is*

$$P_L(u, v) = 2A(u''v + uv'' - u'v') + A'(u'v + uv') + 2Buv.$$

Let $\mathrm{Sol}(L)$ be the solution space of the homogeneous equation $L(h) = 0$, and let L be self- or anti-self-adjoint operator. It follows from the Lagrange formula that $\partial P_L(h_1, h_2) = 0$ for any $h_1, h_2 \in \mathrm{Sol}(L)$, or $P_L(h_1, h_2) = \text{const}$.

In other words,

$$P_L : \mathrm{Sol}(L) \otimes \mathrm{Sol}(L) \to \mathbb{R},$$

$$P_L : h_1 \otimes h_2 \longmapsto P_L(h_1, h_2)$$

determines a bilinear form on the space of solutions.

This form is symmetric if L is skew-adjoint, and skew-symmetric if L is self-adjoint. We call it the *Lagrange form.*

Lemma 2.5.1 *P_L is a non-degenerate bilinear form.*

Proof Let $h_0 \in \mathrm{Sol}(L)$ belong to the kernel of the form, and $h_0 \neq 0$, say, $h_0(0) \neq 0$. Then $P_L(h_0, h) = 0$ for all $h \in \mathrm{Sol}(L)$. Take the solution h such that $h(0) = 0, \ldots, h^{(k-1)}(0) = 0, h^{(k)}(0) = 1$. Then the Lagrange formula gives $P_L(h_0, h) = h(0) \neq 0$. ■

Summarizing we obtain the following theorem.

Theorem 2.5.5 *The Lagrange form P_L determines*

(*a*) *a symplectic structure on* $\mathrm{Sol}(L)$ *if L is self-adjoint, and*
(*b*) *a pseudo-Euclidean structure if L is anti-self-adjoint.*

Note that the space $\mathrm{Sol}(L)$ can be described, for example, by the initial conditions:

$$x_0 = h(0), \quad x_1 = h'(0), \ldots, x_k = h^{(k)}(0).$$

This gives an isomorphism between $\mathrm{Sol}(L)$ and $\mathbb{R}^{k+1}$, and the Lagrange formula allows one to compute P_L explicitly.

Example 2.5.3 *For the Schrödinger operator $L = \partial^2 + W$ we obtain*

$$P_L(x, y) = x_1 y_0 - x_0 y_1$$

or

$$P_L(f, g) = f'g - g'f$$

for $f, g \in \mathrm{Sol}(L)$.

Example 2.5.4 *For the symmetric power* $L^{(2)} = \partial^3 + 4W\partial + 2W'$ *we obtain*

$$P_{L^{(2)}}(x, y) = x_2 y_0 + y_2 x_0 - x_1 y_1 + 4W(0) x_0 y_0$$

and the first integral

$$K(z) = P_{L^{(2)}}(z, z) = 2zz'' - (z')^2 + 4Wz^2$$

on the solution space, $z \in \mathrm{Sol}(L^{(2)})$.

Let Δ be an even symmetry of the equation, and let L be a self-adjoint operator, then $L \circ \Delta = -\Delta^t \circ L$, or, in other words, $L \circ \Delta$ is a skew-self adjoint operator and $L \circ \Delta(h) = 0$ for all solutions $h \in \mathrm{Sol}(L)$. Therefore, $P_{L\circ\Delta}$ is a symmetric form on the space of solutions $\mathrm{Sol}(L)$. Similar reasoning proves the following.

Theorem 2.5.6

1. *Let* L *be a self-adjoint operator. Then the Lagrange form* P_L *determines a symplectic structure on the solution space* $\mathrm{Sol}(L)$, *and*
   (a) $P_{L\circ\Delta} \in \Lambda^2(\mathrm{Sol}(L))$ *is a skew-symmetric form on the space of solutions* $\mathrm{Sol}(L)$ *for any* $\Delta \in \mathrm{Sym}_0(L)$.
   (b) $P_{L\circ\Delta} \in S^2(\mathrm{Sol}(L))$ *is a symmetric form on the space of solutions* $\mathrm{Sol}(L)$ *for any* $\Delta \in \mathrm{Sym}_1(L)$.
2. *Let* L *be an anti-self-adjoint operator. Then the Lagrange form* P_L *determines a pseudo-Eucledean structure on the solution space* $\mathrm{Sol}(L)$, *and*
   (a) $P_{L\circ\Delta} \in S^2(\mathrm{Sol}(L))$ *is a symmetric form on the space of solutions* $\mathrm{Sol}(L)$ *for any* $\Delta \in \mathrm{Sym}_0(L)$.
   (b) $P_{L\circ\Delta} \in S^2(\mathrm{Sol}(L))$ *is a skew-symmetric form on the space of solutions* $\mathrm{Sol}(L)$ *for any* $\Delta \in \mathrm{Sym}_1(L)$.

Example 2.5.5 *Let* $L = \partial^2 + W$ *be the Schrödinger operator. Then the identity operator* $\Delta = \mathrm{id} \in \mathrm{Sym}_1(L)$ *determines the symplectic structure on* $\mathrm{Sol}(L)$. *Moreover, as we have seen any solution* $y \in \mathrm{Sol}(L)$ *determines the symmetry* $E_y = y^2\partial - yy' \in \mathrm{Sym}_0(L)$ *and* $P_{L^{(2)}}(x^2, y^2) = 2(x'y - y'x)^2$ *for all* $x, y \in \mathrm{Sol}(L)$.

3
Model differential equations and the Lie superposition principle

3.1 Symmetry reduction

3.1.1 Reductions by symmetry ideals

We begin with reduction of distributions by ideals of symmetry algebra.

Let P be a completely integrable distribution on manifold M and let $\mathfrak{g}$ be a Lie algebra of shuffling symmetries. For any ideal $\mathfrak{j}$ of $\mathfrak{g}$ we define a new distribution $P_{\mathfrak{j}}$ as follows: the subspace $P_{\mathfrak{j}}(a) \subset T_aM$ is generated by vectors of the former distribution $P(a)$ and values of representatives of $\mathfrak{j}$ at the point.

Theorem 3.1.1 *The distribution $P_{\mathfrak{j}}$ is completely integrable, and $\mathfrak{g}/\mathfrak{j}$ is a shuffling symmetry algebra of $P_{\mathfrak{j}}$.*

Proof The following two types of vector fields generate $P_{\mathfrak{j}}$:

(a) vector fields $X \in D(P)$ that belong to the distribution P, and
(b) representatives $\overline{Z}$ of shuffling symmetries $Z \in \mathfrak{j}$.

For all of them we have the following commutation relations:

$$X_1 \in D(P),\, X_2 \in D(P) \Longrightarrow [X_1, X_2] \in D(P),$$
$$\overline{Z_1} \in D(P_{\mathfrak{j}}),\, \overline{Z_2} \in D(P_{\mathfrak{j}}) \Longrightarrow [\overline{Z_1}, \overline{Z_2}] = \overline{[Z_1, Z_2]} \in D(P_{\mathfrak{j}}),$$
$$\overline{Z_1} \in D(P_{\mathfrak{j}}),\, X_1 \in D(P) \Longrightarrow [\overline{Z_1}, X_1] \in D(P)$$

where the first relation is true because P is completely integrable, the second one follows from the condition that $\mathfrak{j}$ is closed with respect to the bracket and the third one follows from $\mathfrak{j} \subset \mathrm{Sym}(P)$.

The last statement follows from the construction of $P_{\mathfrak{j}}$ and the fact that $\mathfrak{j}$ is an ideal of the Lie algebra $\mathfrak{g}$. ∎

This theorem shows that integration of a completely integrable distribution can be decomposed into two steps:

1. integration of the completely integrable distribution $P_{\mathfrak{j}}$ with symmetry Lie algebra $\mathfrak{g}/\mathfrak{j}$, and
2. integration on integral manifolds M_c of the restrictions P_c with symmetry algebra $\mathfrak{j}$.

First, applying this procedure to the radical $\mathfrak{r}$ of algebra $\mathfrak{g}$ we decompose the integration problem into two parts: the integration of the distribution with the semi-simple algebra $\mathfrak{g}_0 = \mathfrak{g}/\mathfrak{r}$, and then the integration of the restrictions P_c with the solvable symmetry algebra $\mathfrak{r}$.

We have seen that the last step can be performed by quadratures. Moreover, every semi-simple algebra Lie $\mathfrak{g}_0$ is a direct sum of simple ones which are ideals in $\mathfrak{g}_0$. Thus, the above theorem together with the Lie–Bianchi theorem reduces the integration problem to integrating completely integrable distributions equipped with simple algebras of shuffling symmetries.

Before proceeding with the integration of this type of distribution we consider a reduction procedure that gives us a way to reduce essentially dimensions.

3.1.2 Reductions by symmetry subalgebras

Together with actions of Lie groups we need actions of Lie algebras, where by an *action of a Lie algebra* $\mathfrak{g}$ on a manifold N we mean an injective Lie algebra homomorphism $\lambda : \mathfrak{g} \to D(N)$ of the Lie algebra $\mathfrak{g}$ to the Lie algebra $D(N)$ of vector fields on N. We say that an action λ is transitive if values of vector fields $\lambda(X)$, $X \in \mathfrak{g}$, at any point $a \in N$ generate the whole tangent space T_aN.

By $\mathfrak{g}_a = \{X \in \mathfrak{g} \,|\, \lambda(X)_a = 0\}$ we denote the *stability subalgebra* of the point $a \in N$.

Let P be a completely integrable distribution on M with a Lie algebra $\mathfrak{g}$ of shuffling symmetries of P and let N be a manifold with a transitive action $\lambda : \mathfrak{g} \to D(N)$.

We will assume that $\dim \mathfrak{g} = \operatorname{codim} P$.

Let us construct a new distribution $\hat{P}$ on the manifold $M \times N$ as follows. For any point $x = (a, b) \in M \times N$ denote by $\hat{P}_x$ the subspace of $T_x(M \times N) = T_aM \oplus T_bN$ generated by vectors in $P(a)$ and, in addition, by vectors of the form $\overline{X}_a + \lambda(X)_b$ for all $X \in \mathfrak{g}$.

Lemma 3.1.1 *The distribution* $\hat{P}$ *is completely integrable, and* codim $\hat{P} = \dim \mathfrak{g}$.

Proof Vector fields in $D(\hat{P})$ are linear combinations of the following vector fields: $\overline{X} + \lambda(X)$ for $X \in \mathfrak{g}$, and $Y \in D(P)$.

For them we have

$$[\overline{X_1} + \lambda(X_1), \overline{X_1} + \lambda(X_1)] = \overline{[X_1, X_2]} + \lambda[X_1, X_2] \operatorname{mod} D(P)$$

and

$$[\overline{X_1} + \lambda(X_1), Y] = [\overline{X_1}, Y] \in D(P).$$

■

We say that a surjection

$$h : M \to N$$

is an *N-integral of P* if the graph $L_h = \{(a, h(a)) | \forall a \in M\}$ is an integral manifold for the distribution $\hat{P}$.

It is easy to see that any maximal integral manifold of $\hat{P}$ has the form of N-integral.

Theorem 3.1.2 *Let h be an N-integral of P and let $M_b = h^{-1}(b) \subset M, b \in N$. Then*

1. $P(a) \subset T_a(M_b)$ *for any* $a \in M_b$,
2. *the restriction* P_b *of* P *on* M_b *is a completely integrable distribution with shuffling symmetry algebra* $\mathfrak{g}_b$, *and*
3. $\operatorname{codim} P_b = \dim \mathfrak{g}_b$.

Proof It follows from the construction $\hat{P}$ that $P(a)$ belongs to $T_a M_b$. Further, vector fields $\overline{X}, X \in \mathfrak{g}$, tangent to M_b if $\lambda(X)_b = 0$. Therefore, $\mathfrak{g}_b$ is a symmetry algebra for P_b. The last statement of theorem follows from transitivity λ. ■

3.2 Model differential equations

To find an integral $h : M \to N$ we need solutions of some ordinary differential equations on the homogeneous space N. Let us fix points $a \in M$ and $b \in N$. To find the value of h at $x \in M$ providing $h(a) = b$, we take a path $x(t)$ on M such that $x(0) = a, x(1) = x$. Then tangent vectors to $(x(t), y(t) = h(x(t)))$ should be in $\hat{P}(y(t))$. To find this system of ODEs we consider the local description of the problem. Let differential 1-forms $\omega_1, \ldots, \omega_m$ be a local basis for P, let differential 1-forms $\alpha_1, \ldots, \alpha_k$ be a local basis for the cotangent bundle T^*N, and let $\overline{X}_1, \ldots, \overline{X}_m$ be representatives of a basis $X_1, \ldots, X_m$ in $\mathfrak{g}$. We also assume that $\omega_j(\overline{X}_i) = \delta_{ij}$.

Then differential 1-forms

$$\beta_1 = \alpha_1 - \langle \alpha_1, \lambda(X_1)\rangle \omega_1 - \cdots - \langle \alpha_1, \lambda(X_m)\rangle \omega_m,$$

$$\vdots \qquad \vdots \qquad \vdots \qquad \vdots \qquad \vdots$$

$$\beta_k = \alpha_k - \langle \alpha_k, \lambda(X_1)\rangle \omega_1 - \cdots - \langle \alpha_k, \lambda(X_m)\rangle \omega_m$$

determine $\hat{P}$.

Indeed, if $Y \in D(P)$ then $\beta_i(Y) = 0$ and $\omega_j(Y) = 0$, therefore $\alpha_i(Y) = 0$, and

$$\beta_i(\overline{X_j} + \lambda(X_j)) = \langle \alpha_i, \lambda(X_j)\rangle - \langle \alpha_i, \lambda(X_j)\rangle = 0.$$

Therefore, the ODEs system for $\dot{y}(t)$ takes the form

$$\langle \alpha_1, \dot{y}(t)\rangle = \langle \alpha_1, \lambda(X_1)\rangle \langle \omega_1, \dot{x}(t)\rangle - \cdots - \langle \alpha_1, \lambda(X_m)\rangle \langle \omega_m, \dot{x}(t)\rangle,$$

$$\vdots \qquad \vdots \qquad \vdots \quad \vdots \qquad \vdots \quad \vdots \qquad \vdots$$

$$\langle \alpha_k, \dot{y}(t)\rangle = \langle \alpha_k, \lambda(X_1)\rangle \langle \omega_1, \dot{x}(t)\rangle - \cdots - \langle \alpha_k, \lambda(X_m)\rangle \langle \omega_m, \dot{x}(t)\rangle.$$

Let us denote $\langle \omega_i, \dot{x}(t)\rangle$ by $A_i(t)$. Then the above system gives

$$\langle \alpha_i, \dot{y}(t) - A_1(t)\lambda(X_1) - \cdots - A_m(t)\lambda(X_m)\rangle = 0$$

for all $i = 1, \ldots, r$, and therefore

$$\dot{y}(t) = A_1(t)\lambda(X_1) + \cdots + A_m(t)\lambda(X_m). \qquad (3.1)$$

In other words, to find the N-integral h one should solve (3.1).

Differential equations of this form are extremely important to us. We call them *model differential equations* .

Formally these equations can be described as follows. Let $N = G/H$ be a homogeneous space and let $\lambda : \mathfrak{g} \to D(N)$ be the natural representation of the Lie algebra of group G by vector fields on the homogeneous space. Let X_t be a path on the Lie algebra $\mathfrak{g}, X_t \in \mathfrak{g}$. Then the system of ODEs corresponding to the vector field

$$\frac{\partial}{\partial t} + \lambda(X_t) \qquad (3.2)$$

on $\mathbb{R} \times N$ is called the *model differential equations corresponding to model* $N = G/H$.

Theorem 3.2.1 *Let $N = G/H$ be a homogeneous space. A system of ODEs $Z = \partial/\partial t + Y$ on $\mathbb{R} \times N$ has the form (3.2) for some path $X_t \in \mathfrak{g}$ if and only if there is a path $g(t) \in G, g(0) = e$, on G such that any trajectory $y(t)$ of Z can be written in the following form:*

$$y(t) = g(t)y(0). \tag{3.3}$$

Proof Let $g(t)$ be given. Then identifying $\mathfrak{g}$ with T_eG, where e is the unit of G, we define the path

$$X_t = g_*(t)^{-1}(\dot{g}(t)) \tag{3.4}$$

on $\mathfrak{g}$, and differentiating (3.3) we obtain $\dot{y}(t) = \lambda(X_t)_{y(t)}$.

Since $y(0)$ is arbitrary the relation (3.3) gives all solutions of the model ODEs with path (3.4).

Conversely, assume that the path X_t is given. Then considering (3.4) as ODEs for $g(t)$ with initial conditions $g(0) = e$ we get $g(t)$. ■

3.2.1 One-dimensional model equations

Model ODEs on one-dimensional homogeneous spaces correspond to commonly used scalar valued integrals $h : M \to \mathbb{R}$.

We begin with a local description (at $0 \in \mathbb{R}$) of finite-dimensional Lie algebras which can be realized as transitive Lie algebras of vector fields on $\mathbb{R}$.

Let $\lambda : \mathfrak{g} \to D(\mathbb{R})$ be a transitive action, and let

$$\lambda(X) = f_X(t)\frac{d}{dt}$$

for $X \in \mathfrak{g}$.

Transitivity of the action means that there is $X_1 \in \mathfrak{g}$ such that $f_{X_1}(0) \neq 0$, and assume that the coordinate t is chosen so that

$$\lambda(X_1) = \frac{d}{dt}.$$

Denote by $\operatorname{ord}(Y)$, $Y \in \mathfrak{g}$, the order of zero $\lambda(Y)$ at $0 \in \mathbb{R}$, and let Z be the vector of maximal order k. Then the commutator $[Y, Z]$ has order $\leq l + k - 1$ if $\operatorname{ord}(Y) = l$. Therefore, $l + k - 1 \leq k$, or $l \leq 1$, and $\dim \mathfrak{g} \leq 3$.

Suppose that $\dim \mathfrak{g} = 2$. Then $\mathfrak{g} = \langle X_1, X_2 \rangle$ where

$$\lambda(X_1) = \frac{d}{dt}, \quad \lambda(X_2) = f(t)\frac{d}{dt}.$$

The commutator

$$[\lambda(X_1), \lambda(X_2)] = \lambda([X_1, X_2]) = f'(t)\frac{d}{dt}$$

is a linear combination of $\lambda(X_1)$ and $\lambda(X_2)$.

Therefore,

$$f' = a + bf$$

for some constants $a, b \in \mathbb{R}$, and $f = a + Ce^{bt}, b \neq 0$.

Therefore, we can change the coordinate $t \longmapsto e^{bt}$ and the basis in $\mathfrak{g}$ in such a way that

$$\lambda(X_1) = \frac{d}{dt}, \quad \lambda(X_2) = t\frac{d}{dt}.$$

Let $\dim \mathfrak{g} = 3$ and let X_1, X_2, X_3 be a basis X_1, X_2, X_3 in $\mathfrak{g}$ such that $\operatorname{ord}(X_i) = i - 1$.

Then as above we can assume that $\lambda(X_1) = d/dt, \lambda(X_2) = td/dt$ and

$$\lambda(X_3) = g(t)\frac{d}{dt}$$

where $g(t)$ has a second-order zero at $t = 0$.

Moreover, g' and $tg' - g$ are linear combinations of $1, t, g$. From this we conclude that the third basis vector X_3 can be chosen in such a way that

$$\lambda(X_3) = t^2\frac{d}{dt}.$$

Summarizing we get the following.

Theorem 3.2.2 (Sophus Lie) *Suppose that a Lie algebra* $\lambda : \mathfrak{g} \to D(\mathbb{R})$ *acts transitively on* $\mathbb{R}$. *Then* $\dim \mathfrak{g} \leq 3$ *and there exists a local coordinate t in some neighborhood of* $0 \in \mathbb{R}$ *and a basis* $\langle X_i \rangle$ *in* $\mathfrak{g}$ *such that the representation* λ *has the following form:*

1. $\dim \mathfrak{g} = 1$*:*

$$\lambda(X_1) = \frac{d}{dt};$$

2. $\dim \mathfrak{g} = 2$*:*

$$\lambda(X_1) = \frac{d}{dt}, \quad \lambda(X_2) = t\frac{d}{dt};$$

3. $\dim \mathfrak{g} = 3$*:*

$$\lambda(X_1) = \frac{d}{dt}, \quad \lambda(X_2) = t\frac{d}{dt}, \quad \lambda(X_3) = t^2\frac{d}{dt}.$$

Corollary 3.2.1 *Model ODEs of dimension one have one of the following forms:*

1. $y' = a(t)$;
2. $y' = a(t) + b(t)y$;
3. $y' = a(t) + b(t)y + c(t)y^2$.

3.2.2 Riccati equations

Here we discuss the geometry that produces the third model or Riccati equation.

First of all, let us note that the one-dimensional realization of the three-dimensional Lie algebra is a representation of the Lie algebra $\mathfrak{g} = \mathfrak{sl}_2(\mathbb{R})$ by vector fields.

Let $G = \mathbf{SL}_2(\mathbb{R})$ be the Lie group of 2×2 matrices with determinant equal to 1. This group acts on $\mathbb{R}^2$ in the natural way. Let us fix a vector $v_0 \in \mathbb{R}^2$, say $v = (1, 0)$, and consider the following subgroup:

$$H = \{g \in \mathbf{SL}_2(\mathbb{R}) | gv_0 = \lambda v_0 \text{ for some } \lambda\}$$

or

$$H = \left\{ \begin{Vmatrix} a_{11} & a_{12} \\ 0 & a_{22} \end{Vmatrix} \in \mathbf{SL}_2(\mathbb{R}) \right\}.$$

The homogeneous space $N = G/H$ isomorphic to the projective line $\mathbb{R}P^1$ by

$$g = \begin{Vmatrix} a_{11} & a_{12} \\ a_{21} & a_{22} \end{Vmatrix} \in \mathbf{SL}_2(\mathbb{R}) \longmapsto [a_{11} : a_{21}] \in \mathbb{R}P^1.$$

Under this identification the natural action of $\mathbf{SL}_2(\mathbb{R})$ on N corresponds to the projective transformations

$$[x_1 : x_2] \longmapsto [a_{11}x_1 + a_{12}x_2 : a_{21}x_1 + a_{22}x_2].$$

The differential of this action determines the representation λ of $\mathfrak{sl}_2(\mathbb{R})$. To find this action consider the canonical basis

$$X_+ = \begin{Vmatrix} 0 & 1 \\ 0 & 0 \end{Vmatrix}, \quad X_- = \begin{Vmatrix} 0 & 0 \\ 1 & 0 \end{Vmatrix}, \quad H = \begin{Vmatrix} 1 & 0 \\ 0 & -1 \end{Vmatrix}$$

in $\mathfrak{sl}_2(\mathbb{R})$.

Then,

$$\begin{aligned} \exp(tH) : [x_1 : x_2] &\longmapsto [e^t x_1 : e^{-t} x_2], \\ \exp(tX_+) : [x_1 : x_2] &\longmapsto [(x_1 + t x_2) : x_2], \\ \exp(tX_-) : [x_1 : x_2] &\longmapsto [x_1 : (x_2 + t x_1)]. \end{aligned}$$

Let us identify $\mathbb{R}$ with the affine part of $\mathbb{R}P^1$ by $x \longmapsto [x : 1]$, then

$$\begin{aligned} \exp(tH) : x &\longmapsto e^{2t} x, \\ \exp(tX_+) : x &\longmapsto x + t, \\ \exp(tX_-) : x &\longmapsto \frac{x}{1 + tx}. \end{aligned}$$

Therefore,

$$\begin{aligned} \lambda : H &\longmapsto 2x\partial, \\ \lambda : X_+ &\longmapsto \partial, \\ \lambda : X_- &\longmapsto -x^2\partial, \end{aligned}$$

and the corresponding model equation is the Riccati equation.

3.3 Model equations: the series A_k, D_k, C_k

In this and the next two sections we discuss model equations that correspond to the simple Lie algebras of the series A_k, B_k and D_k.

3.3.1 Series A_k

Let E be a vector space of dimension $n + k$ equipped with a volume form and let $\mathrm{Gr}_{\mathbf{n+k,k}}$ be the Grassmanian of k-dimensional subspaces in E.

The Lie group $\mathbf{SL}_{n+k}(\mathbb{R}) = \mathbf{SL}(E)$ acts in the natural way on $\mathrm{Gr}_{\mathbf{n+k,k}}$:

$$g : L \subset E \longmapsto g(L) \subset E$$

for all $g \in \mathbf{SL}(E),\ L \in \mathrm{Gr}_{\mathbf{n+k,k}}$.

This action is obviously transitive, and the stationary subgroup $H = H_L$ of an element $L \in \mathrm{Gr}_{\mathbf{n+k,k}}$ is

$$H = \{g \in \mathbf{SL}(E),\ g(L) = L\}.$$

Thus,

$$\mathrm{Gr}_{\mathbf{n+k,k}} = \mathbf{SL}_{n+k}(\mathbb{R})/H.$$

To find a coordinate description of the above action and therefore a model equation of the series A we introduce the canonical coordinates into $\mathrm{Gr}_{\mathbf{n+k,k}}$.

Let us fix an element $L_0 \in \mathrm{Gr}_{\mathbf{n+k,k}}$ and let $L_0^{\perp}$ be a complementary n-dimensional subspace:

$$E = L_0 \oplus L_0^{\perp}.$$

Then all elements $L \in \mathrm{Gr}_{\mathbf{n+k,k}}$ that belong to a sufficiently small neighborhood of L_0 are transversal to $L_0^{\perp}$ and can be represented by graphs of linear operators $A : L_0 \to L_0^{\perp}$, in other words, we obtain local coordinates

$$A \in \hom(L_0, L_0^{\perp}) \longmapsto L_A = \{x \oplus Ax | x \in L_0\} \in \mathrm{Gr}_{\mathbf{n+k,k}}\,.$$

Let us write down the $\mathbf{SL}_{n+k}(\mathbb{R})$-action in these coordinates.

To this end we represent the vectors $x \in E$ as pairs

$$x = \begin{Vmatrix} x_1 \\ x_2 \end{Vmatrix}$$

where $x_1 \in L_0$ and $x_2 \in L_0^{\perp}$, and operators $g \in \mathbf{SL}(E)$ as matrices

$$g = \begin{Vmatrix} g_{11} & g_{12} \\ g_{21} & g_{22} \end{Vmatrix}$$

with $g_{11} \in \hom(L_0, L_0)$, $g_{12} \in \hom(L_0^{\perp}, L_0)$, $g_{21} \in \hom(L_0, L_0^{\perp}), g_{22} \in \hom(L_0^{\perp}, L_0^{\perp})$, and elements of the stability subgroup H_{L_0} are operators of the form

$$\begin{Vmatrix} h_{11} & h_{12} \\ 0 & h_{22} \end{Vmatrix}.$$

The action in these coordinates has the form

$$x = \left\| \begin{matrix} x_1 \\ x_2 = Wx_1 \end{matrix} \right\| \stackrel{g}{\longmapsto} \left\| \begin{matrix} g_{11}x_1 + g_{12}Wx_1 \\ g_{21}x_1 + g_{22}Wx_1 \end{matrix} \right\| = \left\| \begin{matrix} x_1' \\ g(W)x_1' \end{matrix} \right\|,$$

where

$$g : W \longmapsto g(W) = (g_{21} + g_{22}W) \circ (g_{11} + g_{12}W)^{-1}.$$

Let us find the corresponding representation $\lambda : \mathfrak{sl}_{n+k}(\mathbb{R}) \to D(\mathrm{Gr}_{\mathbf{n+k,k}})$ of the Lie algebra $\mathfrak{sl}_{n+k}(\mathbb{R})$ into vector fields on $\mathrm{Gr}_{\mathbf{n+k,k}}$. First of all, we note that elements of $\mathfrak{sl}_{n+k}(\mathbb{R})$ correspond to matrices

$$X = \left\| \begin{matrix} x_{11} & x_{12} \\ x_{21} & x_{22} \end{matrix} \right\|,$$

where

$$\operatorname{tr} x_{11} + \operatorname{tr} x_{22} = 0.$$

To find $\lambda(X)$ let us denote by $x_{ij}(t)$ the components of $\exp(tX)$:

$$\exp(tX) = \left\| \begin{matrix} x_{11}(t) & x_{12}(t) \\ x_{21}(t) & x_{22}(t) \end{matrix} \right\|.$$

Then,

$$\left.\frac{dx_{ij}(t)}{dt}\right|_{t=0} = x_{ij}, \quad \text{and} \quad x_{ij}(0) = \delta_{ij}.$$

We have

$$W(t) = \exp(tX)(W) = (x_{21}(t) + x_{22}(t)W) \circ (x_{11}(t) + x_{12}(t)W)^{-1}$$

and therefore

$$\lambda(X) = (x_{21} + x_{22}W - W(x_{11} + x_{12}W))\frac{\partial}{\partial W}$$

where $\operatorname{tr} x_{11} + \operatorname{tr} x_{22} = 0$.

The A-model differential equation therefore takes the form of the *matrix* $\mathfrak{sl}$*-Riccati equation*

$$\dot{W} = A_{21}(t) + A_{11}(t)W - WA_{22}(t) - WA_{12}(t)W$$

with

$$\operatorname{tr} A_{11} + \operatorname{tr} A_{22} = 0.$$

Example 3.3.1 *Take* $n = 1, k = 2$. *Then* $\mathfrak{g} = \mathfrak{sl}_3(\mathbb{R})$ *and* $\mathrm{Gr}_{\mathbf{3,2}} = \mathbb{R}P^2$. *The corresponding Riccati equation is a system of the following form:*

$$\dot{x}_1 = a_1(t) + (2a_{11}(t) + a_{22}(t))x_1 + a_{12}(t)x_2 + Q_1(x_1, x_2),$$
$$\dot{x}_2 = a_2(t) + a_{21}(t)x_1 + (a_{11}(t) + 2a_{22}(t))x_2 + Q_2(x_1, x_2),$$

where Q_1 *and* Q_2 *are quadrics in* x_1 *and* x_2.

3.3.2 Series D_k

Let E be a Euclidean vector space of dimension $n + k$ with a metric g, and let $\mathbf{SO}(g) = \mathbf{SO}(n+k)$ be the group of orthogonal transformations preserving the orientation. Because $\mathbf{SO}(n+k) \subset \mathbf{SL}(n+k)$, the group $G = \mathbf{SO}(g)$ acts on the Grassmanian $N = \mathrm{Gr}_{\mathbf{n+k,k}}$ and the action is transitive. To find the corresponding representation of the Lie algebra $\mathfrak{so}_{n+k}$ we use the same coordinate systems as above with $L_0^{\perp}$ being the orthogonal complement of $L_0 \in \mathrm{Gr}_{\mathbf{n+k,k}}$. In these coordinates the operator

$$X = \begin{Vmatrix} x_{11} & x_{12} \\ x_{21} & x_{22} \end{Vmatrix}$$

belongs to $\mathfrak{so}_{n+k}$ if and only if

$$x_{11}^t + x_{11} = 0, \quad x_{22} + x_{22}^t = 0, \quad x_{12} + x_{21}^t = 0.$$

Therefore

$$\lambda(X) = (x_{21} + x_{22}W - W(x_{11} - x_{21}^t W))\frac{\partial}{\partial W}$$

and the **D**-model differential equations are *matrix* $\mathfrak{so}$*-Riccati equations* of the form

$$\dot{W} = A_{12}(t) + A_{11}(t)W - WA_{22}(t) + WA_{12}^t(t)W$$

where A_{11} and A_{22} are skew-symmetric matrices.

Example 3.3.2 *Take* $n = 1, k = 2, \mathfrak{g} = \mathfrak{so}_3(\mathbb{R})$ *and* $\mathrm{Gr}_{3,2} = \mathbb{R}P^2$. *The corresponding* $\mathfrak{so}$*-Riccati equation is the system of the following form:*

$$\dot{x}_1 = a_1(t) - a(t)x_2 + a_1(t)x_1^2 + a_1(t)a_2(t)x_1x_2,$$
$$\dot{x}_2 = a_2(t) + a(t)x_1 + a_2(t)x_2^2 + a_1(t)a_2(t)x_1x_2.$$

3.3.3 Series C_k

Let E be a symplectic space (see), of dimension $2n$ with a structure of the form ω, and let $G = \mathbf{Sp}(E) = \mathbf{Sp}(2n)$ be the group of symplectic transformations of (E, ω):

$$\mathbf{Sp}(E) = \{A \in \mathbf{GL}(E), \omega(Ax, Ay) = \omega(x, y)\}.$$

Owing to Darbouxus, theorem there exists a canonical basis $e_1, \ldots, e_n, e_{n+1}, \ldots, e_{2n}$ in E such that

$$\omega(x, y) = (x_1 y_{n+1} - x_{n+1} y_1) + \cdots + (x_n y_{2n} - x_{2n} y_n).$$

In this basis the matrix

$$A = \begin{Vmatrix} a_{11} & a_{12} \\ a_{21} & a_{22} \end{Vmatrix}$$

of an operator A, where a_{ij} are $n \times n$ matrices, belongs to $\mathbf{Sp}(E)$ if and only if

$$A\Omega A^t = \Omega$$

where

$$\Omega = \begin{Vmatrix} 0 & 1 \\ -1 & 0 \end{Vmatrix}$$

is the matrix of ω in the canonical basis.

Elements of the Lie algebra $\mathfrak{sp}_{2n}$ can therefore be represented by matrices

$$X = \begin{Vmatrix} x_{11} & x_{12} \\ x_{21} & x_{22} \end{Vmatrix}$$

where

$$x_{11} + x_{22}^t = 0, \quad x_{12} = x_{12}^t, \quad x_{21} = x_{21}^t, \tag{3.5}$$

or by matrices

$$X = \begin{Vmatrix} x_{11} & x_{12} \\ x_{21} & -x_{11}^t \end{Vmatrix}$$

where x_{12} and x_{21} are symmetric matrices.

Consider the action $\mathbf{Sp}(E)$ on Grassmanians $\mathrm{Gr}_{\mathbf{2n,k}}$. This action is not transitive if $1 < k < 2n$ because, due to the Witt theorem, two k-dimensional subspaces L and L' belong to the same $\mathbf{Sp}$-orbit if and only if restrictions of the structure form $\omega|_L$ and $\omega|_{L'}$ have the same rank.

We consider model equations that correspond to the case of the Grassmanian of Lagrangian subspaces.

A subspace $L \subset V$ is called *Lagrangian if*

1. $\dim L = n$;
2. L is an *isotropic subspace*, that is, $\omega(x, y) = 0$ for all $x, y \in L$.

Denote by LGr_{2n} the Grassmanian of Lagrangian subspaces. Then the action of the Lie group $G = \mathbf{Sp}(E)$ on $N = \mathrm{LGr}_{2n}$ is transitive.

To find the coordinate description of the corresponding model equations. We shall use the same coordinate system as in series A with the extra requirement that $L_0^{\perp}$ is a Lagrangian subspace too.

Then $L_0^{\perp}$ is isomorphic to the dual space L_0^* by the correspondence

$$X \in L_0^{\perp} \longmapsto \hat{X} = \omega(X, \cdot) \in L_0^*.$$

In this case a graph linear map $W : L_0 \to L_0^{\perp} = L_0^*$ determines a Lagrangian subspace

$$L_W = \{x \oplus Wx | x \in L_0\}$$

if and only if W is self-adjoint

$$W = W^*.$$

Indeed, if $x, y \in L_0$, then

$$\omega(x + Wx, y + Wy) = \omega(x, Wy) + \omega(Wx, y)$$

$$= -(\widehat{Wy})(x) + \hat{y}(Wx) = (-\widehat{Wy} + \widehat{W^*(y)})(x) = 0.$$

Using (3.5) we obtain

$$\lambda(X) = (x_{21} + x_{11}W - W(-x_{11}^t + x_{12}W))\frac{\partial}{\partial W}$$

where x_{12} and x_{21} are symmetric.

Finally, the corresponding *matrix* $\mathfrak{sp}_{2n}$*-Riccati equation* has the form

$$\dot{W} = A_{21}(t) + A_{11}(t)W + WA_{11}^t(t) - WA_{12}(t)W$$

where W and A_{12}, A_{21} are symmetric $n \times n$ matrices.

3.4 The Lie superposition principle

Sophus Lie found that solutions of model differential equations on homogeneous spaces possess some superposition rules.

Let us consider a homogeneous space $N = G/H$ and a model system

$$\frac{\partial}{\partial t} + \lambda(X_t), \quad X_t \in \mathfrak{g}. \tag{3.6}$$

As we have seen in (3.2.1) any such equation determines a path $g(t), g(0) = e$, on the group G such that any solution $x(t)$ of the model equation can be obtained using the formula

$$x(t) = g(t)x(0). \tag{3.7}$$

Therefore, instead of finding solutions of (3.6) we can try to find the path $g(t)$ and then obtain all solutions by (3.7).

Remarkably $g(t)$ can be found by using some solutions and the geometry of the homogeneous space.

Let us consider the diagonal action of the group Lie G on the direct product $N^k = N \times \cdots \times N$:

$$g : (a_1, \ldots, a_k) \longmapsto g(a_1, \ldots, a_k) = (ga_1, \ldots, ga_k).$$

We say that the homogeneous space N is *k-stiff* if the set of points $(a_1, \ldots, a_k)$ with trivial stationary subgroup is dense in N^k.

Denote by G_a the stationary subgroup of $a \in N$. Then the stationary subgroup of $(a_1, \ldots, a_k) \in N^k$ is equal to the intersection $G_{a_1} \cap G_{a_2} \cap \cdots \cap G_{a_k}$.

Let us call a point $(a_1, \dots, a_k) \in N^k$ *regular* if

$$G_{a_1} \cap G_{a_2} \cap \dots \cap G_{a_k} = e.$$

On orbits of regular points we obtain a function $F(a_1, \dots, a_k; b_1, \dots, b_k) \in G$ such that

$$(b_1, \dots, b_k) = F(a_1, \dots, a_k; b_1, \dots, b_k) \cdot (a_1, \dots, a_k)$$

for all regular points $(b_1, \dots, b_k) \in G \cdot (a_1, \dots, a_k)$.

A family $y_1(t), \dots, y_k(t)$ of solutions of the model equation is called a *fundamental solution* if $(y_1(0), \dots, y_k(0))$ is a regular point in N^k.

The fundamental solution allows us to find the path $g(t)$ as follows:

$$g(t) = F(y_1(0), \dots, y_k(0); y_1(t), \dots, y_k(t))$$

and solutions of the model equation are

$$x(t) = F(y_1(0), \dots, y_k(0); y_1(t), \dots, y_k(t)) \cdot x(0). \tag{3.8}$$

The function F is called the *superposition function*, and the way of obtaining solutions by (3.8) is called the *Lie superposition principle.*

Theorem 3.4.1 (The Lie superposition principle) *Let $\partial/\partial t + \lambda(X_t)$, $X_t \in \mathfrak{g}$, be a model equation on a homogeneous space G/H. Let k be the stiff number of G/H and $F(a_1, \dots, a_k; b_1, \dots, b_k)$ be the superposition function. Then for any fundamental solution $y_1(t), \dots, y_k(t)$ (3.8) gives all solutions of the model equation.*

Example 3.4.1 *Let us consider the case of one-dimensional homogeneous spaces (3.2.2). Then*

1. *If $\mathfrak{g} = \mathbb{R}$, then $k = 1$, and $F(x, y) = y - x$.*
   The fundamental solution is a solution of the equation $y' = a(t)$. Then the superposition principle gives

   $$x(t) = x(0) + (y(t) - y(0))$$

   for any solution $x(t)$.
2. *If $\dim \mathfrak{g} = 2, G$ is the "$ax + b$"-group of affine transformations of the line. Therefore $k = 2$, and $F(x_1, x_2; y_1, y_2)$ corresponds to the transformation*

   $$x \longmapsto \frac{(x - x_2)y_1 - (x - x_1)y_2}{x_1 - x_2}.$$

The fundamental solution of the model equation $y' = a(t)y + b(t)$ is a pair of solutions $(y_1(t), y_2(t))$ such that $y_1(0) \neq y_2(0)$. The superposition principle gives

$$x(t) = \frac{(x(0) - y_2(0))y_1(t) - (x(0) - y_1(0))y_2(t)}{y_1(0) - y_2(0)}.$$

3. *If* $\dim \mathfrak{g} = 3, G = \mathbf{SL}_2(\mathbb{R})$ *is the group of projective transformations of the form*

$$x \longmapsto \frac{a_{11}x + a_{12}}{a_{21}x + a_{22}}.$$

It is known in projective geometry that every projective transformation is determined by the images of three different points. Therefore the stiffness of $\mathbb{R}P^1$ *is equal to* 3. *It is also known that the double ratio*

$$\frac{x - x_1}{x - x_2}\frac{x_3 - x_2}{x_3 - x_1}$$

is invariant in the projective transformations. Therefore,

$$y = F(x_1, x_2, x_3; y_1, y_2, y_3) \cdot x$$

can be found from the equation

$$\frac{x - x_1}{x - x_2}\frac{x_3 - x_2}{x_3 - x_1} = \frac{y - y_1}{y - y_2}\frac{y_3 - y_2}{y_3 - y_1}.$$

The fundamental solution is a triple $(y_1(t), y_2(t), y_3(t))$ of solutions of the Riccati equation $y' = a(t) + b(t)y + c(t)y^2$ such that $y_1(0), y_2(0), y_3(0)$ are distinct. In this case any solution $x(t)$ can be found from the double ratio:

$$\frac{x(0) - y_1(0)}{x(0) - y_2(0)}\frac{y_3(0) - y_2(0)}{y_3(0) - y_1(0)} = \frac{x(t) - y_1(t)}{x(t) - y_2(t)}\frac{y_3(t) - y_2(t)}{y_3(t) - y_1(t)}.$$

4. *Stiffness of* $\mathbb{R}P^n$ *is equal to* $n + 2$.
5. *Stiffness of* $\mathrm{Gr}_{2n,n}$ *for* $\mathfrak{sl}$*-Riccati equations is equal to* 5.
6. *Stiffness of* $\mathrm{Gr}_{2n,n}$ *for* $\mathfrak{so}$*-Riccati equations is equal to* 4.
7. *Stiffness of* LGr_{2n} *for* $\mathfrak{sp}$*-Riccati equations is equal to* 4 *if* $n \geq 2$.

3.4.1 Bianchi equations

Bianchi classified all primitive and symplectic actions of Lie algebras on the plane. They are:

1. $\dim \mathfrak{g} = 5$, and

$$\mathfrak{g} = \left\langle \frac{\partial}{\partial x_1}, \frac{\partial}{\partial x_2}, x_1 \frac{\partial}{\partial x_2}, x_2 \frac{\partial}{\partial x_1}, x_1 \frac{\partial}{\partial x_1} - x_2 \frac{\partial}{\partial x_2} \right\rangle$$

is the Lie algebra of all affine symplectic transformations on $\mathbb{R}^2$. The corresponding model equations has the form

$$\dot{x}_1 = a_1(t) + a_{11}(t)x_1 + a_{12}(t)x_2,$$
$$\dot{x}_2 = a_2(t) + a_{21}(t)x_1 - a_{11}(t)x_2$$

and the stiffness equals to 3.

2. $\dim \mathfrak{g} = 3$, and

$$\mathfrak{g} = \left\langle \frac{\partial}{\partial x_1}, \frac{\partial}{\partial x_2}, x_1 \frac{\partial}{\partial x_2} - x_2 \frac{\partial}{\partial x_1} \right\rangle$$

is the Lie algebra of the Lie group of motions on $\mathbb{R}^2$. The corresponding model equations are

$$\dot{x}_1 = a_1(t) + a(t)x_2,$$
$$\dot{x}_2 = a_2(t) - a(t)x_1$$

and the stiffness equals 2.

3. $\dim \mathfrak{g} = 3$, and

$$\mathfrak{g} = \left\langle x_1 \frac{\partial}{\partial x_2} - x_2 \frac{\partial}{\partial x_1}, (1 + x_1^2 - x_2^2) \frac{\partial}{\partial x_1} + 2x_1x_2 \frac{\partial}{\partial x_2}, \right.$$
$$\left. 2x_1x_2 \frac{\partial}{\partial x_1} + (1 + x_2^2 - x_1^2) \frac{\partial}{\partial x_2} \right\rangle$$

is the Lie algebra of motions on the sphere equipped with the metric

$$g = \frac{dx_1^2 + dx_2^2}{(1 + x_1^2 + x_2^2)^2}.$$

The corresponding model equations are

$$\dot{x}_1 = -a_1(t)x_2 + a_2(t)(1 + x_1^2 - x_2^2) + 2a_3(t)x_1x_2,$$
$$\dot{x}_2 = a_1(t)x_1 + 2a_2(t)x_1x_2 + a_3(t)(1 + x_2^2 - x_1^2)$$

and the stiffness equals 2.

4. $\dim \mathfrak{g} = 3$, and

$$\mathfrak{g} = \left\langle x_1\frac{\partial}{\partial x_2} - x_2\frac{\partial}{\partial x_1}, (1 + x_2^2 - x_1^2)\frac{\partial}{\partial x_1} - 2x_1x_2\frac{\partial}{\partial x_2}, \right.$$
$$\left. - 2x_1x_2\frac{\partial}{\partial x_1} + (1 + x_1^2 - x_2^2)\,\frac{\partial}{\partial x_2}\right\rangle$$

is the Lie algebra of motions on the pseudosphere equipped with the metric

$$g = \frac{dx_1^2 + dx_2^2}{(1 - x_1^2 - x_2^2)^2}.$$

The corresponding model equations are

$$\dot{x}_1 = -a_1(t)x_2 + a_2(t)(1 + x_2^2 - x_1^2) - 2a_3(t)x_1x_2,$$
$$\dot{x}_2 = a_1(t)x_1 - 2a_2(t)x_1x_2 + a_3(t)(1 + x_1^2 - x_2^2)$$

and the stiffness equals 2.

5. $\dim \mathfrak{g} = 3$, and

$$\mathfrak{g} = \left\langle \frac{\partial}{\partial x_2}, x_1\frac{\partial}{\partial x_2} - x_2\frac{\partial}{\partial x_1}, 2x_1x_2\frac{\partial}{\partial x_1} + (x_1^{-2} - x_2^2)\frac{\partial}{\partial x_2}\right\rangle$$

is the Lie algebra of motions on the sphere equipped with the metric

$$g = x_1^{-2}\,dx_1^2 + x_1^2\,dx_2^2.$$

The corresponding model equations are

$$\dot{x}_1 = a_2(t)x_2 + 2a_3(t)x_1x_2,$$
$$\dot{x}_2 = a_1(t) - a_2(t)x_2 + a_3(t)(x_1^{-2} - x_2^2)$$

and the stiffness equals 2.

3.5 $\mathcal{AP}$-structures and their invariants

3.5.1 Decomposition of the de Rham complex

A *real $\mathcal{AP}$-structure* on a smooth real manifold M is an ordered set $\mathcal{P}$ of r distributions $P_1, \dots, P_r, \mathcal{P} = (P_1, \dots, P_r)$, such that at each point $a \in M$ the tangent space T_aM splits into the direct sum

$$T_aM = \bigoplus_{i=1}^{r} P_i(a).$$

A *complex $\mathcal{AP}$-structure* on a smooth real manifold M is an ordered set $\mathcal{P}$ of r complex distributions $P_1, \dots, P_r, \mathcal{P} = (P_1, \dots, P_r)$, such that at each point $a \in M$ the complexification of the tangent space T_aM splits into the direct sum

$$T_aM^{\mathbb{C}} = \bigoplus_{i=1}^{r} P_i(a).$$

Let us consider a real $\mathcal{AP}$-structure on M. The vector space $\Lambda^s(T_a^*M)$ of exterior s-forms on T_aM falls into the direct sum

$$\Lambda^s(T_a^*M) = \bigoplus_{|\mathbf{k}|=s} \Lambda^{\mathbf{k}}(T_a^*M),$$

where $\mathbf{k}$ is a multi-index, $\mathbf{k} = (k_1, \dots, k_r), k_i \in \{0, 1, \dots, \dim P_i\}, |\mathbf{k}| = k_1 + \dots + k_r$, and

$$\Lambda^{\mathbf{k}}(T_a^*M) \stackrel{\text{def}}{=} \bigotimes_{i=1}^{r} \Lambda^{k_i}(P_{i,a}).$$

Here

$$\Lambda^{k_i}(P_{i,a}) \stackrel{\text{def}}{=} \{\alpha \in \Lambda^{k_i}(T_a^*M) | X \rfloor \alpha = 0, \forall X \in P_{1,a} \oplus \dots \oplus \widehat{P_{i,a}} \oplus \dots \oplus P_{r,a}\}.$$

We obtain the following decomposition of the de Rham complex the $C^\infty(M)$-modules of differential s-forms $\Omega^s(M)$ split into the direct sums

$$\Omega^s(M) = \bigoplus_{|\mathbf{k}|=s} \Omega^{\mathbf{k}}(M), \tag{3.9}$$

where

$$\Omega^{\mathbf{k}}(M) \stackrel{\text{def}}{=} \bigotimes_{i=1}^{r} \Omega^{k_i}(P_i)$$

and

$$\Omega^{k_i}(P_i)=\left\{\alpha \in \Omega^{k_i}(M)|X\rfloor\alpha = 0, \quad \forall X \in D(P_1) \oplus \cdots \oplus \widehat{D(P_i)} \oplus \cdots \oplus D(P_r)\right\}.$$

The de Rham differential d splits into the following direct sum:

$$d = \bigoplus_{|\sigma|=1} d_\sigma,$$

where $\sigma_j \in I_j \stackrel{\text{def}}{=} \{z \in \mathbb{Z} | |z| \leq \dim P_j\}$ and

$$d_\sigma : \Omega^{\mathbf{k}}(M) \to \Omega^{\mathbf{k}+\sigma}(M).$$

The following theorem allows one to construct tensor invariants of a $\mathcal{AP}$-structure.

Theorem 3.5.1 *If one of the components t_i of a multi-index $\mathbf{t}$ is negative, then the operator $d_{\mathbf{t}}$ is a $C^\infty(M)$-homomorphism.*

Proof It is sufficient to prove that $d_{\mathbf{t}}(h\alpha) = hd_{\mathbf{t}}(\alpha)$ for any function $h \in C^\infty(M)$. For $\alpha \in \Omega^{\mathbf{k}}(M)$ we obtain

$$d(h\alpha) = \sum_{|\sigma|=1} d_\alpha(h\alpha) = \sum_{i=1}^{r} d_{\mathbf{1}_i}(h\alpha) + \sum_{|\mathbf{t}|=1} d_{\mathbf{t}}(h\alpha),$$

where $\mathbf{1}_i = (0, \ldots, 1_i, \ldots, 0)$ and the multi-indexes $\mathbf{t}$ have negative components. On the other hand,

$$d(h\alpha) = dh \wedge \alpha + hd\alpha = \left(\sum_{i=1}^{r} d_{\mathbf{1}_i} h\right) \wedge \alpha + h \sum_{|\mathbf{t}|=1} d_{\mathbf{t}}\alpha.$$

Therefore $d_{\mathbf{1}_i} h \wedge \alpha \notin \Omega^{\mathbf{k}+\mathbf{t}}(M)$ for each $i = 1, \ldots, r$ and one finds that $d_{\mathbf{t}}(h\alpha) = hd_{\mathbf{t}}\alpha$ for any $h \in C^\infty(M)$. ∎

In other words, if one of the components t_i of the multi-index $\mathbf{t}$ is negative, then the operator $d_{\mathbf{t}}$ is a tensor which acts from $\Omega^{\mathbf{k}}(M)$ to $\Omega^{\mathbf{k}+\mathbf{t}}(M)$.

Suppose that $d_{\mathbf{t}}$ is such a tensor. Then the difference between the number of non-negative components s and the number of negative components in $\mathbf{t}$ is 1. The tensor $d_{\mathbf{t}}$ is a sum of the tensors of the type $\theta \otimes X$, where θ is a differential s-form and X is the $(s-1)$-vector field on M.

Recall that the tensor $\theta \otimes X$ acts on a differential form α as

$$(\theta \otimes X)(\alpha) = \theta \wedge (X \rfloor \alpha).$$

Remark 3.5.1 *The constructed tensors $d_{\mathfrak{t}}$ are analogs of the Nijenhuis tensor [26]. Recall that the Nijenhuis tensor arises in a bi-graded complex (see [75]).*

A tensor field $T_{\mathcal{P}}$ on M that is associated with $\mathcal{P}$ is called a *tensor invariant* of $\mathcal{P}$ if $F^*(T_{\mathcal{P}}) = T_{F_*(\mathcal{P})}$ for any diffeomorphism $F : M \to M$. Here $F_*(\mathcal{P}) = (F_*(P_1), \ldots, F_*(P_r))$ and a diffeomorphism F acts on a tensor $\beta \otimes X$ by the following rule: $F^*(\beta \otimes X) = F^*(\alpha) \otimes F_*^{-1}(X)$.

Constructed tensors $d_{\mathfrak{t}}$ are invariants with respect to diffeomorphisms of the manifolds M.

In the case of a complex $\mathcal{AP}$-structure one can construct similar tensor invariants. In this case instead of the decomposition (3.9) one must consider the decomposition

$$\Omega^s(M)^{\mathbb{C}} = \bigoplus_{|\mathbf{k}|=s} \Omega^{\mathbf{k}}(M)^{\mathbb{C}},$$

where $\mathbf{k}$ is a multi-index, $\mathbf{k} = (k_1, \ldots, k_r)$, $k_i \in \{0, 1, \ldots, \dim_{\mathbb{C}} P_i\}$,

$$\Omega^{\mathbf{k}}(M)^{\mathbb{C}} \stackrel{\text{def}}{=} \bigotimes_{i=1}^{r} \Omega^{k_i}(P_i),$$

and

$$\Omega^{k_i}(P_i)$$
$$= \{\alpha \in \Omega^{k_i}(M)^{\mathbb{C}} \mid X \rfloor \alpha = 0, \quad \forall X \in D(P_1) \oplus \cdots \oplus \widehat{D(P_i)} \oplus \cdots \oplus D(P_r)\},$$

where

$$D(P_i) = \{X \in D(M)^{\mathbb{C}} \mid X_a \in P_a, \quad \forall a \in M\}$$

is a $C^\infty(M)^{\mathbb{C}}$-module of complex vector fields from the distribution P_i $(i = 1, \ldots, r)$.

3.5.2 Classical almost product structures

Recall that a field of endomorphisms (a tensor field of $(1, 1)$-type) A on a $2n$-dimensional smooth manifold M is called a classical *almost product structure* if $A^2 = 1$.

If on a smooth manifold M an almost product structure is defined, at each point $a \in M$ the tangent space T_aM splits into the direct sum of two eigensubspaces of A: $T_aM = V_+(a) \oplus V_-(a)$:

$$V_\pm(a) \stackrel{\text{def}}{=} \{X \in T_aM \mid A_aX = \pm X\}.$$

In the other words, an almost product structure on a $2n$-dimensional smooth manifold M generates the pair of n-dimensional distributions V_+ and V_- on M.

Let us consider an almost product structure A on a four-dimensional smooth manifold M. Then we obtain $\mathcal{P} =(P_1, P_2)$, where $\dim P_1 = \dim P_1 = 2$. The module $\Omega^s(M)$ falls into the direct sum

$$\Omega^s(M) = \bigoplus_{p+q=s} \Omega^{p,q}(M),$$

where

$$\Omega^{p,q}(M) \stackrel{\text{def}}{=} \Omega^p(P_1) \otimes \Omega^q(P_2).$$

Here

$$\Omega^p(P_1) = \{\alpha \in \Omega^p(M) | X \rfloor \alpha = 0 \text{ for any vector field } X \in D(P_2)\},$$

$$\Omega^q(P_2) = \{\alpha \in \Omega^q(M) | X \rfloor \alpha = 0 \text{ for any vector field } X \in D(P_1)\}.$$

Let us write these decompositions for each s:

$$\Omega^0(M) = C^\infty(M),$$

$$\Omega^1(M) = \Omega^{1,0}(M) \oplus \Omega^{0,1}(M),$$

$$\Omega^2(M) = \Omega^{2,0}(M) \oplus \Omega^{1,1}(M) \oplus \Omega^{0,2}(M),$$

$$\Omega^3(M) = \Omega^{2,1}(M) \oplus \Omega^{1,2}(M),$$

$$\Omega^4(M) = \Omega^{2,2}(M).$$

Moreover, we obtain the following decomposition of the exterior differential $d : \Omega^s(M) \to \Omega^{s+1}(M)$:

$$d = d_{1,0} \oplus d_{0,1} \oplus d_{2,-1} \oplus d_{-1,2}.$$

The components $d_{2,-1}$ and $d_{-1,2}$ in this sum are tensors.

Proposition 3.5.1

(1) The distribution P_1 is completely integrable if and only if $d_{2,-1} = 0$.
(2) The distribution P_2 is completely integrable if and only if $d_{-1,2} = 0$.

Proof We prove the first statement only. Suppose that the distribution P_1 is generated by a pair of 1-forms $\alpha, \beta \in \Omega^{0,1}(M), \alpha \wedge \beta \neq 0$. This distribution is completely integrable if and only if $d\alpha = \alpha \wedge \gamma_1 + \beta \wedge \delta_1$, and $d\beta = \alpha \wedge \gamma_2 + \beta \wedge \delta_2$ for some 1-forms $\gamma_i, \delta_i \in \Omega^1(M)$, $(i = 1, 2)$. This means that $d\alpha, d\beta \notin \Omega^{2,0}(M)$. ∎

3.5.3 Almost complex structures

A field of real endomorphisms A on a $2n$-dimensional smooth manifold M is called an *almost complex structure* if $A^2 = -1$.

An almost product structure defines at each point $a \in M$ a splitting of the complexification of the tangent space T_aM in the direct sum of two eigensubspaces of A: $T_aM^{\mathbb{C}} = V_+(a) \oplus V_-(a)$:

$$V_\pm(a) \overset{\text{def}}{=} \{X \in T_aM^{\mathbb{C}} \mid A_aX = \pm\iota X\},$$

where $\iota = \sqrt{-1}$. This means that an almost complex structure generates the pair of complex distributions V_+ and V_- on a real smooth manifold M.

Suppose that $n = 2$. We obtain the following decomposition of the de Rham complex:

$$\Omega^s(M)^{\mathbb{C}} = \bigoplus_{p+q=s} \Omega^{p,q}(M),$$

$$d = d_{1,0} \oplus d_{0,1} \oplus d_{2,-1} \oplus d_{-1,2}$$

where $\Omega^{p,q}(M) = \Omega^p(V_+) \otimes \Omega^q(V_-)$ and $d_{i,j} : \Omega^{p,q}(M) \to \Omega^{p+i,q+k}(M)$.

Following the above results, $d_{-1,2}$ and $d_{2,-1}$ are tensor invariants of the Monge – Ampère equation E_ω. Since A is real, the tensors $d_{-1,2}$ and $d_{2,-1}$ are complex conjugated: $\overline{d_{-1,2}} = d_{2,-1}$.

3.5.4 $\mathcal{AP}$-structures on five-dimensional manifolds

Let M be a five-dimensional smooth manifold and $\mathcal{P} = (P_1, P_2, P_3)$ a real $\mathcal{AP}$-structure. Here $\dim P_1 = \dim P_3 = 2$ and $\dim P_2 = 1$.

The module of differential s-forms on M falls into the direct sum

$$\Omega^s(M) = \bigoplus_{r+p+q=s} \Omega^{p,r,q}(M),$$

where

$$\Omega^{p,r,q}(M) \stackrel{\text{def}}{=} \Omega^p(P_1) \otimes \Omega^r(P_2) \otimes \Omega^q(P_3)$$

and

$$\Omega^p(P_1) = \{\alpha \in \Omega^p(M) | X \rfloor \alpha = 0 \text{ for any vector field } X \in D(P_2) \oplus D(P_3)\},$$
$$\Omega^r(P_2) = \{\alpha \in \Omega^r(M) | X \rfloor \alpha = 0 \text{ for any vector field } X \in D(P_1) \oplus D(P_3)\},$$
$$\Omega^q(P_3) = \{\alpha \in \Omega^q(M) | X \rfloor \alpha = 0 \text{ for any vector field } X \in D(P_1) \oplus D(P_2)\}.$$

Let us write these decompositions for each s:

$$\Omega^0(M) = C^\infty(M),$$
$$\Omega^1(M) = \Omega^{1,0,0}(M) \oplus \Omega^{0,1,0}(M) \oplus \Omega^{0,0,1}(M),$$
$$\Omega^2(M) = \Omega^{2,0,0}(M) \oplus \Omega^{1,1,0}(M) \oplus \Omega^{1,0,1}(M) \oplus \Omega^{0,1,1}(M) \oplus \Omega^{0,0,2}(M),$$
$$\Omega^3(M) = \Omega^{2,1,0}(M) \oplus \Omega^{2,0,1}(M) \oplus \Omega^{1,1,1}(M) \oplus \Omega^{1,0,2}(M) \oplus \Omega^{0,1,2}(M),$$
$$\Omega^4(M) = \Omega^{2,1,1}(M) \oplus \Omega^{2,0,2}(M) \oplus \Omega^{1,1,2}(M),$$
$$\Omega^5(M) = \Omega^{2,1,2}(M).$$

Moreover, we have the following decomposition of the exterior differential:

$$d = \bigoplus_{\substack{i+j+k=1 \\ i,k \in I_1; j \in I_2}} d_{i,j,k},$$

where $I_1 = \{-2, -1, 0, 1, 2\}$, $I_2 = \{-1, 0, 1\}$ and

$$d_{i,j,k} : \Omega^{p,r,q}(M) \to \Omega^{p+i,r+j,q+k}(M).$$

The following operators are tensors: $d_{-1,1,1}$, $d_{1,1,-1}$, $d_{1,-1,1}$, $d_{0,-1,2}$, $d_{2,-1,0}$, $d_{2,1,-2}$ and $d_{-2,1,2}$.

These tensors will be used later on for classification of hyperbolic Monge – Ampère equations.

For the complex $\mathcal{AP}$-structure $\mathcal{P} = (P_1, P_2, P_3)$, $\dim_{\mathbb{C}} P_1 = \dim_{\mathbb{C}} P_3 = 2$, $\dim_{\mathbb{C}} P_2 = 1$ one has similar complex tensor invariants.

PART II

Symplectic Algebra

4

Linear algebra of symplectic vector spaces

4.1 Symplectic vector spaces

4.1.1 Bilinear skew-symmetric forms on vector spaces

Let V be a real finite-dimensional vector space and let $\Omega : V \times V \to \mathbb{R}$ be a bilinear skew-symmetric form on V.

Denote by

$$\Gamma : V \to V^*$$

a linear map from V to its dual space V^* defined by the formula

$$\Gamma(x) = \iota_x \Omega,$$

in other words

$$\langle \Gamma(x), y \rangle = \Omega(x, y)$$

for all $x, y \in V$.

The dimension of the image of the map Γ is called a *rang* of Ω, and the kernel of Γ, which we denote by $\ker \Omega$, is called a *kernel* of Ω.

Note that

$$\ker \Omega = \{x \in V \mid \Omega(x, y) = 0, \text{ for all } y \in V\}.$$

The adjoint map Γ^* is equal to $-\Gamma$ because of the skew-symmetry of Ω:

$$\langle \Gamma^* x, y \rangle = \langle x, \Gamma y \rangle = -\Omega(x, y) = -\langle \Gamma x, y \rangle$$

and one could identify skew-symmetric 2-forms Ω with skew-adjoint maps $\Gamma : V \to V^*$.

Let $V_0 = V/\ker\Omega$ be the factor space. Then Ω determines a skew-symmetric form Ω_0 on V_0 in the natural way:

$$\Omega_0(x \bmod \ker\Omega, y \bmod \ker\Omega) = \Omega(x, y)$$

and this 2-form is non-degenerate: $\ker\Omega_0 = 0$.

Moreover,

$$V_0^* = \mathrm{Ann}\ \ker\Omega \subset V^*.$$

Therefore, we decomposed any skew-symmetric form in the composition

$$\begin{array}{ccc} V & \xrightarrow{\ \Gamma\ } & V^* \\ \downarrow & & \uparrow \\ V_0 & \xrightarrow{\ \Gamma_0\ } & V_0^* \end{array}$$

where the vertical arrows are the natural projection and the embedding.

4.1.2 Symplectic structures on vector spaces

A non-degenerate skew-symmetric bilinear form Ω is called a *symplectic form* or a *symplectic structure* on the vector space V.

A vector space V equipped with a symplectic form Ω is called a *symplectic vector space* and is denoted by (V, Ω).

Let (V_1, Ω_1) and (V_2, Ω_2) be symplectic vector spaces. Then $V_1 \oplus V_2$ has an obvious symplectic structure defined by

$$\Omega(x_1 \oplus y_1, x_2 \oplus y_2) = \Omega_1(x_1, x_2) + \Omega_2(y_1, y_2),$$

where $x_1, x_2 \in V_1$ and $y_1, y_2 \in V_2$.

We call this structure the *direct sum of* (V_1, Ω_1) and (V_2, Ω_2) and write

$$\Omega = \Omega_1 \oplus \Omega_2.$$

Let (V, Ω) be a symplectic space and let $\Gamma : V \to V^*$ be the corresponding map. Then the operator $\Gamma^{-1} : V^* \to V = (V^*)^*$ defines a symplectic structure on the dual space V^*. We denote this structure by (V^*, Ω^*) and call it the *dual structure*.

Note that

$$\Omega^*(x^*, y^*) = \Omega(\Gamma^{-1}x^*, \Gamma^{-1}y^*)$$

for all $x^*, y^* \in V^*$.

Example 4.1.1 *Let V be a two-dimensional space. Then the space $\Lambda^2(V^*)$ of exterior 2-forms on V is one-dimensional, and therefore any non-zero exterior 2-form Ω determines a symplectic structure on V. Using the direct sum operation we can construct a series of symplectic spaces:*

$$\underbrace{V \oplus \cdots \oplus V}_{k \text{ times}}$$

equipped with the symplectic form

$$\Omega = \Omega_1 \oplus \cdots \oplus \Omega_k$$

where $\Omega_1, \ldots, \Omega_k$ are non-zero 2-forms on V.

Example 4.1.2 *Let E be a vector space, and let*

$$V = E \oplus E^*.$$

Define an exterior 2-form on V as follows:

$$\Omega(x \oplus x^*, y \oplus y^*) = \langle x^*, y \rangle - \langle y^*, x \rangle$$

for all $x, y \in E$, and $x^*, y^* \in E^*$.

Then, $(E \oplus E^*, \Omega)$ is a symplectic space.

Let (V, Ω) be a symplectic vector space. We say that two vectors $x, y \in V$ are *skew-orthogonal* if

$$\Omega(x, y) = 0$$

and write $x \perp y$.

Let $W \subset V$ be a vector subspace and let

$$W^\perp = \{x \in V \mid \Omega(x, y) = 0 \text{ for all } y \in W\}$$

be the space of all vectors skew-orthogonal to W.

We call $W^\perp$ the *skew-orthogonal of W*.

We collect some properties of the skew-orthogonality in the following.

Proposition 4.1.1 *Let W, W_1, W_2 be subspaces of a symplectic vector space. Then*

$$\begin{aligned}
\dim W^{\perp} &= \operatorname{codim} W, \\
\ker \Omega_W &= W \cap W^{\perp}, \\
(W^{\perp})^{\perp} &= W, \\
(W_1 + W_2)^{\perp} &= W_1^{\perp} \cap W_2^{\perp}, \\
(W_1 \cap W_2)^{\perp} &= W_1^{\perp} + W_2^{\perp}, \\
W_1 \subset W_2 &\Longrightarrow W_1^{\perp} \supset W_2^{\perp},
\end{aligned}$$

where $\Omega_W \in \Lambda^2(W^*)$ is the restriction of the structure form Ω on W.

The proof of this proposition is straightforward.

Note that Ω_W is non-degenerate if and only if

$$W \cap W^{\perp} = 0.$$

We call the subspace W *regular, or symplectic* if Ω_W is non-degenerate. In this case (W, Ω_W) is also symplectic, and we get a splitting

$$V = W \oplus W^{\perp}.$$

Moreover, $W^{\perp}$ is also regular and therefore

$$(V, \Omega) = (W, \Omega_W) \oplus (W^{\perp}, \Omega_{W^{\perp}})$$

is a decomposition of the symplectic structure into a direct sum of two structures.

Remark also that any symplectic space contains a regular plane W, $\dim W = 2$, and thus, applying this remark and the above splitting to $W^{\perp}$ we decompose V into a direct sum of symplectic planes.

Summing up we get the following.

Theorem 4.1.1 *Any symplectic space (V, Ω) is a direct sum of symplectic planes:*

$$(V, \Omega) = (W_1, \Omega_1) \oplus \cdots \oplus (W_n, \Omega_n)$$

where $W_1, \ldots, W_n$ are symplectic planes in W, $\Omega_i = \Omega_{W_i}$, $W_i \subset W_j^{\perp}$ for $i = 1, \ldots, n$, and $j \neq i$.

4.1.3 Canonical bases and coordinates

We begin with a two-dimensional symplectic space (W, Ω). If e_1, e_2 is a basis in W and e_1^*, e_2^* is the dual basis in W^* then

$$\Omega = \lambda e_1^* \wedge e_2^*$$

for some $\lambda \in \mathbb{R}, \lambda \neq 0$.

The scale transformation $e = \lambda^{-1} e_1$, $f = e_2$ gives a basis in W such that

$$\Omega = e^* \wedge f^*.$$

We call this basis *canonical.*

Note that a basis $e' = a_{11}e + a_{12}f$, $f' = a_{21}e + a_{22}f$ is canonical if and only if $\det \|a_{ij}\| = 1$.

Now let us take an arbitrary symplectic vector space (V, Ω). From decomposition Theorem 4.1.1 we get canonical bases: (e_1, f_1) in the plane $W_1, \ldots,$ and (e_n, f_n) in the plane W_n.

We call such a basis $(e_1, \ldots, e_n, f_1, \ldots, f_n)$ *canonical.*

The basic property of the canonical basis is

$$\Omega(e_i, e_j) = \Omega(f_i, f_j) = 0 \quad \text{and} \quad \Omega(e_i, f_j) = \delta_{ij}.$$

Therefore, in the basis

$$(e_1^*, \ldots, e_n^*, f_1^*, \ldots, f_n^*)$$

dual to the canonical basis

$$(e_1, \ldots, e_n, f_1, \ldots, f_n)$$

the structure 2-form Ω can be written as

$$\Omega = e_1^* \wedge f_1^* + \cdots + e_n^* \wedge f_n^*.$$

Let $(x_1, \ldots, x_{2n})$ and $(y_1, \ldots, y_{2n})$ be coordinates of vectors X and Y in the canonical basis, then

$$\Omega(X, Y) = \sum_{i=1}^{n} (x_i y_{i+n} - x_{i+n} y_i).$$

One can define the symplectic structure by the above formula. This structure is usually called the *standard symplectic structure* on $\mathbb{R}^{2n}$.

Note that the corresponding linear operator $\Gamma : V \to V^*$ in the canonical basis acts as follows:

$$\Gamma : e_i \longmapsto f_i^*,$$
$$\Gamma : f_i \longmapsto -e_i^*.$$

4.2 Symplectic transformations

Let (V_1, Ω_1) and (V_2, Ω_2) be symplectic spaces. A linear operator $A : V_1 \to V_2$ is called *symplectic* if

$$A^*(\Omega_2) = \Omega_1$$

or, in other words, if

$$\Omega_2(Ax, Ay) = \Omega_1(x, y)$$

for all $x, y \in V_1$.

Note that $\ker A = \ker \Omega_1 = 0$, and A is an isomorphism if $\dim V_1 = \dim V_2$.

Example 4.2.1 *An embedding $W \hookrightarrow V$ is symplectic if and only if W is a symplectic subspace of V.*

Example 4.2.2 *The decomposition theorem shows that any two symplectic spaces (V_1, Ω_1) and (V_2, Ω_2) of the same dimension are symplectically isomorphic.*

Symplectic isomorphisms $A : V \to V$ are called *symplectic transformations*. They form a group with respect to the composition. This group is called the *symplectic group* of the symplectic vector space (V, Ω) and we denote it by $\mathrm{Sp}\,(V, \Omega)$.

The group of symplectic transformations of the standard symplectic structure is denoted by $\mathrm{Sp}\,(n)$ or $\mathrm{Sp}\,(n, \mathbb{R})$.

The group $\mathrm{Sp}\,(V, \Omega)$ is a Lie group, and the corresponding Lie algebra $\mathfrak{sp}(V, \Omega)$ consists of the linear operators $B : V \to V$ that satisfy the following relation:

$$\Omega(Bx, y) + \Omega(x, By) = 0$$

for all $x, y \in V$.

We call them *Hamiltonian operators*.

Note that $\exp(tB) \in \mathrm{Sp}\,(V, \Omega)$, $t \in \mathbb{R}$, if $B \in \mathfrak{sp}(V, \Omega)$.

Proposition 4.2.1 *1. Let $B \in \mathfrak{sp}\,(V, \Omega)$, then*

$$q_B(x, y) \stackrel{\text{def}}{=} \Omega\,(Bx, y)$$

is a symmetric bilinear form on V.

2. Let $q : V \times V \to \mathbb{R}$ be a symmetric bilinear form on V, then the linear operator $B_q : V \to V$ given by

$$q\,(x, y) \stackrel{\text{def}}{=} \Omega\,(B_q x, y)$$

belongs to $\mathfrak{sp}(V, \Omega)$.

Proof One has

$$q_B(x, y) = \Omega\,(Bx, y) = -\Omega\,(x, By) = \Omega\,(By, x) = q_B(x, y)$$

and

$$\Omega\,(B_q x, y) = q(x, y) = q(y, x) = \Omega\,(B_q y, x) = -\Omega\,(x, B_q y). \qquad \blacksquare$$

Corollary 4.2.1 $\dim \mathrm{Sp}(V, \Omega) = (\dim V(\dim V + 1))/2$.

This proposition establishes an isomorphism between the Lie algebra of Hamiltonian operators and the space $\mathbf{S}^2(V^*)$ of bilinear symmetric forms on V, and therefore induces a Lie bracket on $\mathbf{S}^2(V^*)$.

Namely, let $q_1, q_2 \in \mathbf{S}^2(V^*)$ then the bracket

$$[q_1, q_2] \in \mathbf{S}^2(V^*)$$

corresponds to the commutator $[B_{q_1}, B_{q_2}]$.

In other words,

$$[q_1, q_2](x, y) = \Omega\,([B_{q_1}, B_{q_2}]x, y) = q_1(x, B_{q_2} y) - q_2(x, B_{q_1} y).$$

Let $\alpha \in V^*$ and let $q = \alpha^2 \in \mathbf{S}^2(V^*)$ be the symmetric square of α:

$$\alpha^2(x, y) \stackrel{\text{def}}{=} \alpha(x)\alpha(y).$$

The corresponding Hamiltonian operator $B_q \in \mathfrak{sp}(V, \Omega)$ is called a *symplectic transvection* and is denoted by τ_α.

We have

$$\alpha^2(x,y) = \alpha(x)\alpha(y) = \Omega(\Gamma^{-1}\alpha, x)\Omega(\Gamma^{-1}\alpha, y)$$

or

$$\tau_\alpha(x) = \alpha(x)\,\widehat{\alpha},$$

where $\widehat{\alpha} = \Gamma^{-1}\alpha \in V$.

Note that

$$\tau_\alpha^2 = 0$$

and

$$\ker \tau_a = \ker \alpha, \quad \operatorname{Im} \tau_a = \mathbb{R}\widehat{\alpha}.$$

The corresponding symplectic transformations act as follows:

$$\exp(t\tau_\alpha) = 1 + t\tau_\alpha,$$

where $t \in \mathbb{R}$.

Proposition 4.2.2 (E. Artin [3])

1. *The symplectic transvections $\tau_\alpha, \alpha \in V^*$ generate the Lie algebra $\mathfrak{sp}(V,\Omega)$, and*

$$[\tau_\alpha, \tau_\beta] = \frac{\Omega(\widehat{\alpha}, \widehat{\beta})}{2}(\tau_{\alpha+\beta} - \tau_{\alpha-\beta}).$$

2. *The symplectic transformations $1 + t\tau_\alpha$ $(\alpha \in V^*, t \in \mathbb{R})$ generate the Lie group* $\mathrm{Sp}(V,\Omega)$.

4.2.1 Matrix representation of symplectic transformations

The symplectic transformations of the standard symplectic structure in $\mathbb{R}^{2n}$ are represented by matrices $A \in \mathrm{GL}(2n,\mathbb{R})$.

Let us write vectors $X \in \mathbb{R}^{2n}$ as

$$X = \begin{Vmatrix} X_1 \\ X_2 \end{Vmatrix},$$

where

$$X_1 = \sum_{i=1}^{n} x_i e_i \quad \text{and} \quad X_2 = \sum_{i=1}^{n} x_{i+n} f_i$$

and let

$$\langle X, Y \rangle = \sum_{i=1}^{2n} x_i y_i.$$

Then,

$$\Omega(x, y) = \sum_{i=1}^{n} (x_i y_{i+n} - x_{i+n} y_i) = \langle Jx, y \rangle,$$

where

$$J = \begin{Vmatrix} 0 & 1 \\ -1 & 0 \end{Vmatrix}$$

is the matrix of the standard symplectic structure.

Let

$$A = \begin{Vmatrix} A_{11} & A_{12} \\ A_{21} & A_{22} \end{Vmatrix}$$

be a matrix of a symplectic transformation $A : \mathbb{R}^{2n} \to \mathbb{R}^{2n}$.

Then the condition $\Omega(AX, AY) = \Omega(X, Y)$ means

$$A^t J A = J,$$

or, in coordinates,

$$A_{21}^t A_{11} - A_{11}^t A_{21} = 0,$$
$$A_{22}^t A_{12} - A_{12}^t A_{22} = 0,$$
$$A_{11}^t A_{22} - A_{21}^t A_{12} = 1.$$

Example 4.2.3 *The matrices of the form*

$$A = \begin{Vmatrix} a & 0 \\ 0 & b \end{Vmatrix}$$

are symplectic if and only if

$$a^t b = 1.$$

Example 4.2.4 *The matrices of the form*

$$A = \begin{Vmatrix} 1 & a \\ 0 & 1 \end{Vmatrix}$$

are symplectic if and only if

$$a^t = a.$$

Hamiltonian operators $B \in \mathfrak{sp}_n$ have matrices

$$B = \begin{Vmatrix} B_{11} & B_{12} \\ B_{21} & B_{22} \end{Vmatrix}$$

that satisfy the following condition:

$$B^t J + JB = 0,$$

or

$$B_{12} - B_{12}^t = 0,$$
$$B_{21} - B_{21}^t = 0,$$
$$B_{11} + B_{22}^t = 0.$$

We call the matrices of such a type *Hamiltonian.*

Example 4.2.5 *Matrices of the form*

$$A = \begin{Vmatrix} a & 0 \\ 0 & b \end{Vmatrix}$$

are Hamiltonian if and only if

$$a^t + b = 0.$$

Example 4.2.6 *Matrices of the form*

$$A = \begin{Vmatrix} 0 & a \\ b & 0 \end{Vmatrix}$$

are Hamiltonian if and only if

$$a^t = a, \quad b^t = b.$$

4.3 Lagrangian subspaces

We have called a subspace $W \subset V$ of a symplectic space (V, Ω) *regular* if the restriction Ω_W of the structure 2-form Ω on W is non-degenerate or symplectic.

It follows that dimensions of regular subspaces are even, and $V = W \oplus W^\perp$ for any regular W.

On the other hand, any subspace W of $\operatorname{codim} W = 1$ can be defined by a covector α, that is, $W = \ker\alpha$, and the corresponding vector $\widehat{\alpha} = \Gamma^{-1}(\alpha) \in V$ belongs to $\ker\alpha$, because of

$$\alpha(\widehat{\alpha}) = \Omega(\widehat{\alpha}, \widehat{\alpha}) = 0.$$

Therefore, in this case $W^\perp \subset W$.

In general, we say that a subspace $W \subset V$ is *involutive* if the above inclusion holds:

$$W^\perp \subset W.$$

Note that a subspace W is involutive if and only if

$$\Omega_{W^\perp} = 0.$$

A subspace E of V is said to be *isotropic* if $\Omega_E = 0$.

In other words, the notions of involutive and isotropic subspaces are dual to each other with respect to the operation $W \Longrightarrow W^\perp : W$ is isotropic, or involutive if and only if the orthogonal $W^\perp$ is involutive or isotropic respectively.

Example 4.3.1 *Any one-dimensional subspace is isotropic and any one-codimensional subspace is involutive.*

A *Lagrangian subspace* is a maximal isotropic subspace; that is, if $L \subset V$ is Lagrangian and $W \supset L$ is isotropic then $W = L$.

One could also define Lagrangian subspaces as minimal involutive subspaces.

Theorem 4.3.1

1. *Lagrangian subspaces do exist.*
2. *A subspace L is Lagrangian if and only if $L^\perp = L$.*
3. *An isotropic subspace $L \subset V$ is Lagrangian if and only if* $\dim L = \frac{1}{2} \dim V$.

Proof As we have seen, any one-dimensional subspace is isotropic and V is not. Therefore any chain $W_0 \subset W_1 \subset \cdots \subset W_k \subset \cdots$ of isotropic subspaces has a proper maximal element which corresponds to a Lagrangian subspace. On the other hand, if W is isotropic, then $W^\perp \supset W$ and therefore $\dim V - \dim W \geq \dim W$ or $\dim W \leq n$. If $W^\perp \neq W$ then there is a non-zero vector $x \in W^\perp$ such that $x \notin W$.

Therefore, $W \oplus \mathbb{R}x \supset W$ is isotropic too, and W is Lagrangian if and only if $W^\perp = W$. ■

Example 4.3.2 *Let* (e,f) *be a canonical basis of* V. *Then* $\mathrm{Lin}(e_1, \ldots, e_n)$ *and* $\mathrm{Lin}(f_1, \ldots, f_n)$ *are Lagrangian subspaces, but* $\mathrm{Lin}(e_1, f_1, e_3, \ldots, e_n)$ *is not.*

Example 4.3.3 *Let* $V = E \oplus E^*$ *be the symplectic space, and let* $L_A \subset V$ *be a graph of a linear operator* $A : E \to E^*$. *Then* L_A *is Lagrangian if and only if* A *is a self-adjoint operator.*

Example 4.3.4 *The subspaces* $(e_1, \ldots, e_k, f_{k+1}, \ldots, f_n)$, $1 \leq k \leq n$ *are Lagrangian.*

Theorem 4.3.2 *For any two Lagrangian subspaces* $L, L' \subset V$, *and any linear isomorphism* $A_0 : L \to L'$ *there exists a symplectic linear transformation* $A : V \to V$ *such that the restriction* $A|_L$ *coincides with* A_0.

Proof Let us consider a filtration

$$L = L_n \supset L_{n-1} \supset \cdots \supset L_1 \supset 0$$

of the Lagrangian subspace L by subspaces L_i, where $\dim L_i = i$, and let $L'_i = A_0(L_i)$ be the corresponding filtration of L':

$$L' = L'_n \supset L'_{n-1} \supset \cdots \supset L'_1 \supset 0.$$

Then the corresponding orthogonal $L_i^\perp$ together with L_i gives a filtration of V:

$$V = 0^\perp \supset L_1^\perp \supset \cdots \supset L_{n-1}^\perp \supset L = L_n \supset L_{n-1} \supset \cdots \supset L_1 \supset 0.$$

We obtain a similar filtration with L':

$$V = 0^\perp \supset L_1'^\perp \supset \cdots \supset L_{n-1}'^\perp \supset L' = L'_n \supset L'_{n-1} \supset \cdots \supset L'_1 \supset 0.$$

To construct A we shall use an induction in $n - i$. To find the symplectic extension A_0 on $L_{n-1}^\perp$ we pick a non-zero vector $e_n \in L$ such that $e_n \notin L_{n-1}$ and a vector $f_n \in L_{n-1}^\perp$ but $f_n \notin L$, and such that $\Omega(e_n, f_n) = 1$. Then the image of vector f_n under operator A shall be a vector $f'_n \in L'^{\perp}_{n-1}$ such that $\Omega(Ae_n, f'_n) = \Omega(e_n, f_n) = 1$. This vector is defined up to vectors from L'. Let us pick one of them. This choice gives us an extension of A_0 on $L^{\perp}_{n-1}$ such that $A : L^{\perp}_{n-1} \to L'^{\perp}_{n-1}$ sends $\Omega_{L'^{\perp}_{n-1}}$ to $\Omega_{L^{\perp}_{n-1}}$. A similar procedure shows that one can extend A from $L^{\perp}_i$ to $L^{\perp}_{i-1}$ in such a way that A transforms $\Omega_{L'^{\perp}_{i-1}}$ to $\Omega_{L^{\perp}_{i-1}}$ if it does for $\Omega_{L'^{\perp}_i}$ to $\Omega_{L^{\perp}_i}$. For $i = 0$ we find the required A. ■

Corollary 4.3.1 *For any two isotropic subspaces W and W',* $\dim W = \dim W'$, *of a symplectic space V and any linear isomorphism $A_0 : W \to W'$ there exists a symplectic linear transformation $A : V \to V$ such that the restriction $A|_W$ coincides with A_0.*

Proof It is enough to embed W and W' into Lagrangian subspaces L and L' and consider any extension of A_0 to a linear isomorphism $L \to L'$. ■

Theorem 4.3.3 (Witt) *Let W and W' be subspaces of a symplectic vector space (V, Ω), and let $A_0 : W \to W'$ be a linear isomorphism such that*

$$A_0^*(\Omega_{W'}) = \Omega_W.$$

Then A_0 can be extended to a symplectic isomorphism A, such that the restriction $A|_W$ coincides with A_0.

Proof Let $W_0 = W \cap W^{\perp}$ and let W_1 be a complementary subspace of W_0 in W; that is,

$$W = W_0 \oplus W_1.$$

Let

$$W' = W'_0 \oplus W'_1$$

be the corresponding decomposition of W' with $W'_1 = A_0(W_1)$.

The 2-form $\Omega|_{W_1}$ is non-degenerate. Therefore, W_1 and $W_1^{\perp}$ are regular,

$$V = W_1 \oplus W_1^{\perp}$$

and $A_0 : W_1 \to W'_1$ is a symplectic isomorphism.

Moreover, $W_0 \subset W_1^{\perp}$ and $W'_0 \subset W_1'^{\perp}$ are isotropic subspaces, and due to the above theorem there is an extension of $A_0 : W_0 \to W'_0$ to symplectic isomorphism $A_1 : W_1^{\perp} \to W_1'^{\perp}$. Then the symplectic isomorphism

$$A_0 \oplus A_1 : W_1 \oplus W_1^{\perp} \to W_1' \oplus W_1'^{\perp}$$

is the required extension. ∎

Theorem 4.3.4 *Let $W \subset V$ be a subspace of a symplectic space (V, Ω). Then,*

1. *W contains a Lagrangian subspace if and only if W is involutive;*
2. *if W is involutive and $L \subset V$ is a Lagrangian subspace such that $L \subset W$, then $L \supset W^{\perp}$.*

Proof Indeed, let L be Lagrangian and let $L \subset W$. Then $L^{\perp} \supset W^{\perp}$ and

$$W \supset L = L^{\perp} \supset W^{\perp}.$$

On the other hand, let W be involutive. Then $W^{\perp}$ is isotropic, and therefore there exists a Lagrangian subspace L such that $L \supset W^{\perp}$ and $L = L^{\perp} \subset (W^{\perp})^{\perp} = W$. ∎

Let us now consider the decomposition

$$(V, \Omega) = (\Pi_1, \Omega_1) \oplus \cdots \oplus (\Pi_n, \Omega_n)$$

into a direct sum of symplectic planes, $\Omega_i = \Omega|_{\Pi_i}$, and let

$$\pi_i : V \to \Pi_i$$

be the corresponding projectors.

Note that all embeddings

$$\Pi_{i_1} \oplus \cdots \oplus \Pi_{i_k} \hookrightarrow V$$

are symplectic.

Let $L \subset V$ be a Lagrangian subspace. Then images $L_i = \pi_i(L) \neq 0$ for any $i = 1, \ldots, n$.

Indeed, if, say, $\pi_1(L) = 0$, then $L \subset \Pi_2 \oplus \cdots \oplus \Pi_n$ is isotropic and $\dim L = n > n - 1$.

This observation gives us a method to choose coordinates on Lagrangian subspaces.

Theorem 4.3.5 *Let* $(e_1,\ldots,e_n,f_1,\ldots,f_n)$ *be a canonical basis of* V*,* L *be a Lagrangian subspace. Then there is a subset* $\mathbf{I} \subset \{1,\ldots,n\}$*,* $\#\mathbf{I} = k$*, such that linear functions*

$$e^*_{i_1},\ldots,e^*_{i_k}, \quad i_s \in \mathbf{I},$$
$$f^*_{j_1},\ldots,f^*_{j_{n-k}}, \quad j_l \notin \mathbf{I}$$

are coordinates on L*.*

4.3.1 Symplectic and Kähler spaces

Recall that a *Hermitian space* is a complex vector space (V,I) equipped with a positive Hermitian form h.

Let

$$g(x,y) = \operatorname{Re} h(x,y)$$

be the real part and

$$\Omega(x,y) = \operatorname{Im} h(x,y)$$

be the imaginary part of h.

Then

$$h(x,y) = g(x,y) + \imath\Omega(x,y),$$

$\imath = \sqrt{-1}$, and the Hermitian properties of h:

$$h(x,y) = \overline{h(y,x)},$$
$$h(Ix,y) = \imath h(x,y)$$

implies that:

1. $g(x,y)$ is a Euclidian structure on V;
2. $\Omega(x,y)$ is a symplectic structure on V; and
3. $g(x,y) = \Omega(Ix,y)$ or $I \in \mathfrak{sp}(V,\Omega)$.

We can reformulate the definition of Hermitian space in terms of the symplectic structure (V,Ω) equipped with a complex structure $I: V \to V$.

To do this we say that a complex structure $I : V \to V$ and a symplectic structure are *compatible* if:

1. $I \in \mathfrak{sp}(V, \Omega)$; and
2. $\Omega(Ix, x) > 0$ if $x \neq 0$.

A triple (V, Ω, I) with compatible symplectic and complex structures is called a *Kähler space.*

If (V, Ω, I) is a Kähler space then the form

$$h(x, y) = g(x, y) + \imath\Omega(x, y)$$

determines a Hermitian space structure on V.

Example 4.3.5 *If the Hermitian form h has the form*

$$h = \sum_{k=1}^{n} z_k \overline{z_k}$$

in coordinates $z_k = p_k + \imath q_k$*, then the symplectic form that corresponds to the imaginary part of h has the form*

$$\Omega = \sum_{k=1}^{n} p_k \wedge q_k.$$

Let us discuss briefly relations between symplectic structures and compatible complex structures.

First of all, let us note that for any unitary transformation $A : V \to V$ the condition $h(Ax, Ay) = h(x, y)$ implies $g(Ax, Ay) = g(x, y)$ and $\Omega(Ax, Ay) = \Omega(x, y)$, and therefore the unitary group $\mathrm{U}(V, h)$ belongs to the intersection of the symplectic group $\mathrm{Sp}(V, \Omega)$ and the orthogonal group $\mathrm{O}(V, g)$, and the relation $g(x, y) = \Omega(Ix, y)$ shows that

$$\mathrm{U}(V, h) = \mathrm{Sp}(V, \Omega) \cap \mathrm{O}(V, g).$$

Note also that Hermitian orthogonality $h(x, y) = 0$ implies both the Euclidian $g(x, y) = 0$ and skew-orthogonality $\Omega(x, y) = 0$.

On the other hand, if W is an isotropic subspace and $e_1, \ldots, e_k$ is an orthonormal basis in W with respect to g, then these vectors are orthonormal with respect to h, and therefore give us an orthonormal $\mathbb{C}$-basis in the $\mathbb{C}$-vector space $W \oplus IW$. If W is a Lagrangian subspace then any g-orthonormal basis $e_1, \ldots, e_n$ in W gives an h-orthonormal basis in V. In other words, Lagrangian subspaces are real enveloping spaces of h-orthogonal $\mathbb{C}$-bases in V.

5

Exterior algebra on symplectic vector spaces

Let (V, Ω) be a symplectic vector space, $\dim V = 2n$, and let $\Lambda^*(V^*)$ be the exterior algebra of forms on V. The structure form Ω generates an ideal $\mathcal{I}_C = \Omega \wedge \Lambda^*(V^*)$.

We call it the *Cartan ideal* of the symplectic space. The corresponding factor algebra

$$\Lambda_\varepsilon(V^*) = \Lambda^*(V^*)/\mathcal{I}_C$$

is called the *algebra of effective forms*.

This algebra has great importance for the partial differential equation (PDE) theory, and in this chapter we give a detailed description of effective forms as some special exterior forms on the symplectic space V.

5.1 Operators $\perp$ and $\top$

We extend the isomorphism $\Gamma : V \to V^*$,

$$\Gamma(X) \stackrel{\text{def}}{=} \iota_X(\Omega),$$

to isomorphisms

$$\Gamma_k : \Lambda^k(V) \to \Lambda^k(V^*)$$

between k-vectors and k-forms on V as follows:

$$\Gamma_k(X_1 \wedge \cdots \wedge X_k) \stackrel{\text{def}}{=} \Gamma(X_1) \wedge \cdots \wedge \Gamma(X_k),$$

and denote by $X_\alpha \in \Lambda^k(V)$ the k-vector that corresponds to a k-form $\alpha \in \Lambda^k(V^*)$:

$$X_\alpha \stackrel{\text{def}}{=} \Gamma_k^{-1}(\alpha).$$

Example 5.1.1 *Let $k = 1$ and $\alpha = \sum_{i=1}^n (a_i e_i^* + b_i f_i^*)$ in a canonical basis $(e_1, \ldots, e_n, f_1, \ldots, f_n)$. Then,*

$$X_\alpha = \sum_{i=1}^n (b_i e_i - a_i f_i).$$

Example 5.1.2 *For the structure 2-form $\Omega = \sum_{i=1}^n e_i^* \wedge f_i^*$ we obtain*

$$X_\Omega = \sum_{i=1}^n e_i \wedge f_i.$$

Define two operators on the algebra of exterior forms $\Lambda^*(V^*)$.
The first operator

$$\top : \Lambda^k(V^*) \to \Lambda^{k+2}(V^*) \tag{5.1}$$

is the operator of exterior multiplication by the structure 2-form Ω:

$$\top(\omega) \stackrel{\text{def}}{=} \omega \wedge \Omega.$$

The second operator

$$\perp : \Lambda^i(V^*) \to \Lambda^{i-2}(V^*) \tag{5.2}$$

is an operator of inner multiplication by the bivector X_Ω:

$$\perp(\omega) \stackrel{\text{def}}{=} \iota_{X_\Omega}(\omega).$$

Lemma 5.1.1 *Let*

$$\Omega_k = \frac{1}{k!} \underbrace{\Omega \wedge \cdots \wedge \Omega}_{k \text{ times}}.$$

Then

$$\perp(\Omega_k) = (n - k + 1)\Omega_{k-1}.$$

Proof Let $e_1, \ldots, e_n, f_1, \ldots, f_n$ be a canonical basis of the vector space V. Then,

$$\iota_{e_i}(\Omega) = f_i^*, \quad \iota_{f_i}(\Omega) = -e_i^*$$

and

$$\iota_{e_i}(\Omega_k) = f_i^* \wedge \Omega_{k-1}, \quad \iota_{f_i}(\Omega_k) = -e_i^* \wedge \Omega_{k-1}.$$

Therefore

$$\begin{aligned}\iota_{e_i \wedge f_i}(\Omega_k) &= \iota_{f_i}(\iota_{e_i}(\Omega_k)) = \iota_{f_i}(f_i^* \wedge \Omega_{k-1}) \\ &= \Omega_{k-1} - f_i^* \wedge \iota_{f_i}(\Omega_{k-1}) \\ &= \Omega_{k-1} + f_i^* \wedge e_i^* \wedge \Omega_{k-2}\end{aligned}$$

and

$$\begin{aligned}\perp (\Omega_k) &= \sum_{i=1}^{n} \iota_{e_i \wedge f_i}(\Omega_k) = \sum_{i=1}^{n} (\Omega_{k-1} + f^i \wedge e^i \wedge \Omega_{k-2}) \\ &= n\Omega_{k-1} - \Omega \wedge \Omega_{k-2} \\ &= n\Omega_{k-1} - (k-1)\Omega_{k-1} = (n-k+1)\Omega_{k-1}.\end{aligned}$$

■

Lemma 5.1.2 *If $\theta \in V^*$ and $\omega \in \Lambda^k(V^*)$, then*

$$\perp (\theta \wedge \omega) = \theta \wedge (\perp \omega) - \iota_{X_\theta}(\omega).$$

Proof If $e_1, \ldots, e_n, f_1, \ldots, f_n$ is a canonical basis of V, then

$$\begin{aligned}\perp(\theta \wedge \omega) = \iota_{X_\Omega}(\theta \wedge \omega) &= \sum_{i=1}^{n} \iota_{f_i}(\iota_{e_i}(\theta \wedge \omega)) \\ &= \sum_{i=1}^{n} \iota_{f_i}(\iota_{e_i}(\theta)\omega - \theta \wedge \iota_{e_i}(\omega)) \\ &= \sum_{i=1}^{n} (\theta(e_i)\iota_{f_i}(\omega) - \theta(f_i)\iota_{e_i}(\omega) + \theta \wedge \iota_{e_i \wedge f_i}(\omega)) \\ &= \sum_{i=1}^{n} (\theta(e_i)\iota_{f_i}(\omega) - \theta(f_i)\iota_{e_i}(\omega)) + \theta \wedge \perp(\omega).\end{aligned}$$

Note that

$$\theta = \sum_{i=1}^{n} (\theta(e_i)e_i^* + \theta(f_i)f_i^*)$$

and

$$X_\theta = \sum_{i=1}^{n} (\theta(f_i)e_i - \theta(e_i)f_i).$$

Therefore, the last formula reads as

$$\perp(\theta \wedge \omega) = -\iota_{X_\theta}(\omega) + \theta \wedge (\perp(\omega)).$$ ■

Lemma 5.1.3 *For any k-form $\omega \in \Lambda^k(V)$ we have*

$$[\perp, \top](\omega) = (n - k)\omega.$$

Proof The proof follows by induction on k.

If $k = 1$, then $\perp(\omega) = 0$, and due to Lemma 5.1.1, we have

$$\begin{aligned}[\perp, \top](\omega) &= \perp(\top(\omega)) = \iota_{X_\Omega}(\Omega \wedge \omega) \\ &= \iota_{X_\Omega}(\Omega)\omega = \perp(\Omega_1)\omega \\ &= (n-1)\omega.\end{aligned}$$

Suppose that for any $\omega \in \Lambda^{k-1}(V)$ we have

$$[\perp, \top](\omega) = (n - (k-1))\omega.$$

Note that any k-form can be represented as a sum of forms $\theta \wedge \omega$ where θ is an exterior 1-form, and ω is an exterior $(k-1)$-form. For these forms we have

$$\begin{aligned}[\perp, \top](\theta \wedge \omega) &= \perp(\Omega \wedge \theta \wedge \omega) - \top(\perp(\theta \wedge \omega)) \\ &= \perp(\theta \wedge \Omega \wedge \omega) - \top(\theta \wedge \perp(\omega)) + \top(\iota_{X_\theta}(\omega)) \\ &= \perp(\theta \wedge \top(\omega)) - \Omega \wedge \theta \wedge \perp(\omega) + \Omega \wedge \iota_{X_\theta}(\omega) \\ &= \theta \wedge \perp\top(\omega) - \iota_{X_\theta}(\Omega \wedge \omega) - \theta \wedge \top\perp(\omega) + \Omega \wedge \iota_{X_\theta}(\omega) \\ &= \theta \wedge [\perp, \top](\omega) - \iota_{X_\theta}(\Omega) \wedge \omega - \Omega \wedge \iota_{X_\theta}\omega + \Omega \wedge \iota_{X_\theta}(\omega) \\ &= \theta \wedge [\perp, \top](\omega) - \iota_{X_\theta}(\Omega) \wedge \omega \\ &= (n - (k-1))\theta \wedge \omega - \theta \wedge \omega = (n-k)\theta \wedge \omega.\end{aligned}$$ ■

Let

$$\top_k \stackrel{\text{def}}{=} \frac{1}{k!}\top^k$$

and

$$\perp_k \stackrel{\text{def}}{=} \frac{1}{k!}\perp^k.$$

Lemma 5.1.4 *Let $\omega \in \Lambda^k(V)$, then*

$$[\perp, \top_s](\omega) = (n-k-s+1)\top_{s-1}(\omega),$$

$$[\perp_s, \top](\omega) = (n-k+s-1)\perp_{s-1}(\omega)$$

and

$$\begin{aligned}\top^s\perp_s(\omega) &= (-1)^s(n-k+s-1)(n-k+s-2)\cdots(n-k)\omega,\\ &\quad \textit{if } \top\omega = 0,\\ \perp^s\top_s(\omega) &= (n-k-s+1)(n-k-s+2)\cdots(n-k)\omega,\\ &\quad \textit{if } \perp\omega = 0.\end{aligned}$$

Proof The proof is induction on s. We prove only the first formula; the second one can be proved in the similar way.

Suppose that

$$[\perp, \top_{s-1}](\omega) = (n-k-s+2)\top_{s-2}(\omega)$$

and

$$[\perp_{s-1}, \top](\omega) = (n-k+s-2)\perp_{s-2}(\omega).$$

Then,

$$\begin{aligned}[\perp, \top_s](\omega) &= \perp\top_s(\omega) - \top_s\perp(\omega)\\ &= \frac{1}{s}(\perp\top_{s-1}(\top(\omega)) - \top_{s-1}(\top\perp(\omega)))\\ &= \frac{1}{s}(\top_{s-1}\perp(\top(\omega)) + [\perp, \top_{s-1}](\top(\omega)) - \top_{s-1}(\top\perp(\omega)))\end{aligned}$$

$$
\begin{aligned}
&= \frac{1}{s}(\top_{s-1}([\bot, \top_{s-1}](\omega)) + (n - (k+2) - s + 2)\top_{s-2}(\top(\omega)))\\
&= \frac{1}{s}((n-k)\top_{s-1}(\omega) + (s-1)(n-k-s)\top_{s-1}(\omega))\\
&= \frac{1}{s}(n - k + (s-1)(n-k) - (s-1)s)\top_{s-1}(\omega)\\
&= (n-k-s+1)\top_{s-1}(\omega).
\end{aligned}
$$

For the last two formulas we have: if $s = 1$, then it follows from the previous formula, and for $s+1$ we have

$$
\begin{aligned}
\top^{s+1}\bot_{s+1}(\omega) &= \top^s \top \bot_{s+1}(\omega) = \top^s([\top, \bot_{s+1}](\omega) - \bot_{s+1}\top(\omega))\\
&= -(n-k+s)\top^s\bot_s(\omega)\\
&= (-1)^{s+1}(n-k+s)(n-k+s-1)\cdots(n-k)\omega.
\end{aligned}
$$

The last formula can be proved in a similar way. ■

Lemma 5.1.5

1. *The operators*

$$\top : \Lambda^k(V) \to \Lambda^{k+2}(V)$$

   are injective if $k \leq n-1$.
2. *The operators*

$$\bot : \Lambda^k(V) \to \Lambda^{k-2}(V)$$

   are injective if $k \geq n+1$.

Proof Suppose $k \leq n-1$ and $\top(\omega) = 0$ for some form $\omega \in \Lambda^k(V)$. Then,

$$\top^{n+1}\bot_{n+1}(\omega) = (-1)^{n+1}(2n-k)(2n-k-1)\cdots(n-k)\omega.$$

Since $k \leq n-1$ then $\bot_{n+1}(\omega) = 0$. Therefore,

$$(2n-k)(2n-k-1)\cdots(n-k)\omega = 0$$

and $\omega = 0$.

To prove the second part of the statement suppose that $k \geq n+1$ and $\perp(\omega) = 0$ for some form $\omega \in \Lambda^k(V)$.

Then,

$$\perp^n \top_n(\omega) = (1-k)(2-k)\cdots(n-k)\omega.$$

Since $k \geq n+1$, we see that $\top_n(\omega) = 0$,

$$(1-k)(2-k)\cdots(n-k)\omega = 0.$$

and $\omega = 0$. ■

5.2 Effective forms and the Hodge–Lepage theorem

A form $\omega \in \Lambda^k(V^*)$ is called *effective* if

$$\perp(\omega) = 0.$$

We denote by $\Lambda^k_\varepsilon(V^*)$ the space of all effective k-forms on V. Note that non-trivial effective k-forms exist only if $k \leq n$.

Theorem 5.2.1 (Hodge–Lepage) *Any k-form $\omega \in \Lambda^k(V)$ can be decomposed as*

$$\omega = \omega_0 + \top(\omega_1) + \top_2(\omega_2) + \cdots$$

where $\omega_i \in \Lambda^{k-2i}(V^*)$ are uniquely determined effective forms.

Proof The proof is by induction on k. The case of $k = 1$ is trivial.

Assume that the proposition holds for forms of degree less then k.

Let $\omega \in \Lambda^k(V^*)$. Then $\perp(\omega) \in \Lambda^{k-2}(V^*)$ and by the inductive hypothesis,

$$\perp(\omega) = \alpha_0 + \top(\alpha_1) + \top_2(\alpha_2) + \cdots,$$

where $\alpha_0, \alpha_1, \alpha_2, \ldots$ are uniquely determined effective forms.

Hence, if we set

$$\omega = x_0 + \top(x_1) + \top_2(x_2) + \cdots,$$

then under the assumption that $x_0, x_1, \ldots$ are effective, we obtain

$$\perp(\omega) = (n-k+2)x_1 + (n-k+3)\top(x_2) + \cdots.$$

Consequently, if we take

$$x_1 = \frac{1}{n-k+2}\alpha_0, \quad x_2 = \frac{1}{n-k+3}\alpha_1, \quad \ldots,$$

then

$$\bot\,(\omega - \top(x_1) - \top_2(x_2) - \cdots) = 0,$$

i.e.,

$$\omega_0 = \omega - \top(x_1) - \top_2(x_2) - \cdots$$

is an effective form. ■

Suppose $\omega \in \Lambda^k(V^*)$ and $\omega = \sum_{s\geq 0} \top_s(\omega_s)$ is the Hodge–Lepage decomposition. Straightforward calculations lead us to the following expressions:

$$\bot_r\,(\omega) = \sum_{s\geq r} C^r_{n-k+s+r} \top_{s-r}(\omega_s),$$

$$(\top_r \circ \bot_r)(\omega) = \sum_{s\geq r} C^r_{n-k+s+r} C^r_s \top_s(\omega_s)$$

and to the following expression

$$\omega_0 = \left(\sum_{s\geq 0}(-1)^s \frac{1}{s+1} \top_s \bot_s\right)(\omega)$$

for the effective part of ω.

Theorem 5.2.2 *The operators*

$$\bot_k\,: \Lambda^{n+k}(V^*) \to \Lambda^{n-k}(V^*)$$

and

$$\top_k : \Lambda^{n-k}(V^*) \to \Lambda^{n+k}(V^*),$$

$k = 1, \ldots, n-1$, *are isomorphisms and for effective forms* ω *we have*

$$\bot_k \circ \top_k(\omega) = \omega.$$

Proof Let $\omega \in \Lambda^{n-k}(V^*)$ and $\top_k(\omega) = 0$.

Then from the Hodge–Lepage decomposition we obtain

$$\perp_k \circ \top_k(\omega) = \omega_0 + (C^1_{k+1})^2 \top(\omega_1) + \cdots .$$

From the uniqueness of the Hodge–Lepage decomposition we have $\omega_0 = \omega, \omega_1 = 0, \ldots$. ■

Example 5.2.1 *Let* (V, Ω) *be a symplectic space,* $\dim V = 4$, *and let* e_1, e_2, f_1, f_2 *be a canonical basis.*

Then,

$$\Omega = e_1^* \wedge f_1^* + e_2^* \wedge f_2^*$$

and

$$X_\Omega = e_1 \wedge f_1 + e_2 \wedge f_2.$$

Let

$$\omega = a_1 e^1 + a_2 e^2 + b_1 f^1 + b_2 f^2 \in \Lambda^1(V^*),$$

be a 1-form. Note that $\Lambda^1(V^*) = \Lambda^1_\varepsilon(V^*)$,

$$\top(\omega) = a_1 e_1^* \wedge e_2^* \wedge f_2^* - a_2 e_1^* \wedge e_2^* \wedge f_2^* - b_1 e_2^* \wedge f_1^* \wedge f_2^* + b_2 e_1^* \wedge f_1^* \wedge f_2^*$$

and $\top : \Lambda^1(V^*) \to \Lambda^3(V^*)$ *is an isomorphism.*

If

$$\omega = A e_1^* \wedge e_2^* + B e_1^* \wedge f_1^* + C e_1^* \wedge f_2^* + D e_2^* \wedge f_1^* + E e_2^* \wedge f_2^* + F f_1^* \wedge f_2^* \in \Lambda^2(V^*),$$

is a 2-form, then

$$\perp(\omega) = B + E$$

and

$$\top(\omega) = (B + E) e^1 \wedge f^1 \wedge e^2 \wedge f^2 = \frac{B+E}{2} \Omega^2.$$

Hence, $\Lambda^2_\varepsilon(V^*)$ is the space of 2-forms such that

$$B + E = 0$$

and

$$\ker \top = \ker \perp .$$

Theorem 5.2.3 *The form $\omega \in \Lambda^{n-k}(V)$ is effective if and only if $\top_{k+1}(\omega) = 0$.*

Proof First of all, note that for the form $\omega \in \Lambda^{n-k}(V)$ we have

$$\perp \top_{k+1}(\omega) = \top_{k+1} \perp (\omega).$$

Therefore, if $\perp (\omega) = 0$, then $\perp \top_{k+1}(\omega) = 0$ and $\top_{k+1}(\omega) = 0$. Conversely, if $\top_{k+1}(\omega) = 0$, then $\top_{k+1} \perp (\omega) = \perp \top_{k+1}(\omega) = 0$. So, $\perp (\omega) = 0$. ∎

Theorem 5.2.4 *Let $\omega_1, \omega_2 \in \Lambda^{n-k}_\varepsilon(V^*)$, $0 \le k \le n$, be effective $(n-k)$ forms, and let the form ω_2 vanish on every isotropic subspace $L \subset V$, $\dim L = n - k$, such that $\omega_1|_L = 0$.*

Then $\omega_2 = \lambda\omega_1$ for some $\lambda \in \mathbb{R}$.

Proof The proof follows by induction on $\frac{1}{2}\dim V$. The case $n = 1$ is trivial.

Let V be a symplectic space of dimension $2n$. Choose a pair of covectors $\alpha, \alpha^+ \in V^*$ such that $\iota_{X_\alpha}(\alpha^+) = 1$. Then the restriction Ω' of the form Ω to the subspace

$$V' = \ker\alpha \cap \ker\alpha^+$$

defines a symplectic structure. Denote by $\top'$ and $\perp'$ the corresponding operators. Since

$$V = V' \oplus \mathbb{R}X_\alpha \oplus \mathbb{R}X_{\alpha^+}$$

we can identify forms on V' with forms on V that are degenerated on $\mathbb{R}X_\alpha \oplus \mathbb{R}X_{\alpha^+}$.

An arbitrary form $\beta \in \Lambda^s(V^*)$ can be decomposed as

$$\beta = \bar{\beta}_0 + \alpha \wedge \beta_1 + \alpha^+ \wedge \beta_2 + \alpha^+ \wedge \alpha \wedge \beta_3$$

where $\bar{\beta}_0, \beta_1, \beta_2$ and β_3 are uniquely determined forms on V' and therefore

$$\iota_\alpha(\beta_i) = \iota_{\alpha^+}(\beta_i) = 0.$$

Let us now assume that β is an effective form. From

$$X_\Omega = X_{\alpha^+} \wedge X_\alpha + X_{\Omega'}$$

we obtain

$$0 = \perp\beta = \perp'(\bar{\beta}_0) + \alpha \wedge \perp'(\beta_1) + \alpha^+ \wedge \perp'(\beta_2) + \alpha^+ \wedge \alpha \wedge \perp'(\beta_3) + \beta_3.$$

Thus, β_1, β_2 and β_3 are effective forms, and

$$\perp'(\bar{\beta}_0) = -\beta_3.$$

Let

$$\bar{\beta}_0 = \beta_0 + \top'(x_1) + (\top')^2(x_2) + \cdots$$

be the Hodge–Lepage decomposition. Then

$$\perp'(\bar{\beta}_0) = (n - s + 1)x_1 + 2(n - s + 2)\top'(x_2) + \cdots$$

and hence

$$x_1 = -\frac{1}{n - s + 1}\beta_3, \quad x_2 = 0, \quad \ldots .$$

Finally, we obtain the following decomposition of effective forms:

$$\beta = \beta_0 + \alpha \wedge \beta_1 + \alpha^+ \wedge \beta_2 + \alpha^+ \wedge \alpha \wedge \beta_3 - \frac{1}{n - s + 1}\top'(\beta_3)$$

in which all the β_i are effective.

In particular, for effective forms $\omega_i \in \Lambda^{n-k}(V^*)$, $i = 1, 2$, we find

$$\omega_i = \omega_{0i} + \alpha \wedge \omega_{1i} + \alpha^+ \wedge \omega_{2i} + \alpha^+ \wedge \alpha \wedge \omega_{3i} - \frac{1}{k + 1}\top'(\omega_{3i}).$$

Let us now go back to the conditions of the theorem. We construct an isotropic subspace **annihilating** the form ω_i in the following way. We choose arbitrary vectors $Z_1, \ldots, Z_{n-k-2} \in V'$ which are linear independent and are in involution:

$$\Omega'(Z_i, Z_j) = 0,$$

and add to them vectors $u + X_\alpha,\ v + X_{\alpha^+},\ u, v \in V'$, in such a way that vectors

$$u + X_\alpha, \quad v + X_{\alpha^+}, \quad Z_1, \dots, Z_{n-k-2}$$

are in involution also. To do this it is enough to require that

$$\Omega'(Z_i, u) = \Omega'(Z_i, v) = 0 \quad \text{and} \quad \Omega'(u, v) = 1.$$

Elementary computations show that

$$\begin{aligned}&(u + X_\alpha) \wedge (v + X_{\alpha^+}) \wedge Z \rfloor \omega_i \\ &\quad = u \wedge v \wedge Z \rfloor \omega_{0i} + u \wedge Z \rfloor \omega_{1i} + v \wedge Z \rfloor \omega_{2i} - \frac{k+2}{k+1} Z \rfloor \omega_{3i},\end{aligned}$$

where $Z = Z_1 \wedge \cdots \wedge Z_{n-k-2}$. Therefore, for a fixed vector v the linear equation

$$u \rfloor (v \wedge Z \rfloor \omega_{02} + Z \rfloor \omega_{12}) = \frac{k+2}{k+1} Z \rfloor \omega_{32} - v \wedge Z \rfloor \omega_{22}$$

on u is satisfied whenever

$$\begin{cases} u \rfloor (V \rfloor Z \rfloor \omega_{01} + Z \rfloor \omega_{11}) = \dfrac{k+2}{k+1} Z \rfloor \omega_{31} - v \rfloor Z \rfloor \omega_{21}, \\ u \rfloor (v \rfloor \Omega') = -1, \quad u \rfloor (Z_i \rfloor \Omega') = 0. \end{cases}$$

Hence, the forms

$$\begin{cases} \gamma_0 = v \wedge Z \rfloor \omega_{02} + Z \rfloor \omega_{12}, \quad \gamma_1 = v \wedge Z \rfloor \omega_{01} + Z \rfloor \omega_{11} \\ \gamma_2 = \Gamma'(v), \quad \gamma_{i+2} = \Gamma'(Z_i) \quad (i = 1, \dots, n-k-2) \end{cases}$$

must be linear dependent for any choice of vectors $v, Z_1, \dots, Z_{n-k-2}$ in involution. Therefore,

$$\gamma_0 \wedge \gamma_1 \wedge \gamma_2 \wedge \Pi = 0,$$

where $\Pi = \gamma_3 \wedge \cdots \wedge \gamma_{n-k}$. We consider the last equation as an equation for v:

$$\begin{aligned}&(v \wedge Z \rfloor \omega_{02}) \wedge (v \wedge Z \rfloor \omega_{01}) \wedge \Gamma'(v)\Pi + (v \wedge Z \rfloor \omega_{02}) \wedge (Z \rfloor \omega_{11}) \\ &\quad + (Z \rfloor \omega_{12}) \wedge (v \wedge Z \rfloor \omega_{01}) \wedge \Gamma'(v) \wedge \Pi + (Z \rfloor \omega_{12}) \wedge (Z \wedge \omega_{11}) \\ &\quad \times \wedge \Gamma'(v) \wedge \Pi = 0. \end{aligned} \tag{5.3}$$

Let us fix a set $Z_1, \dots, Z_{n-k-2}$ of vectors in involution. The equation should hold for all vectors v for which

$$\Omega'(Z_i, v) = 0, \quad i = 1, \dots, n-k-2.$$

Let us now assume that $\omega_{11} \neq 0$ and $Z \rfloor \omega_{11} \neq 0$. Replace v by tv in (5.3) and differentiate with respect to t at $t = 0$. We obtain the following equation:

$$(Z \rfloor \omega_{12}) \wedge (Z \rfloor \omega_{11}) \wedge \Gamma'(v) \wedge \Pi = 0,$$

for all $v, \Omega'(v, Z_i) = 0$. In other words, the restrictions of $Z \rfloor \omega_{12}$ and $Z \rfloor \omega_{11}$ on $\cap \ker \Gamma'(Z_i)$ for any involutive set $Z_1, \dots, Z_{n-k-2}$ lead to linear dependent forms. In particular, this gives us the fact that ω_{12} vanishes on all isotropic subspaces that annihilate ω_{11}. Hence,

$$\omega_{12} = \lambda \omega_{11}$$

by induction.

We now consider terms of second order in v in (5.3). To do this we replace v by tv, and differentiate twice with respect to t at $t = 0$.

We obtain

$$(v \wedge Z \rfloor \omega_{02} - \lambda v \wedge Z \rfloor \omega_{01}) \wedge (Z \rfloor \omega_{11}) \wedge \Gamma'(v) \wedge \Pi = 0$$

and thus on $\cap \ker \Gamma'(Z_i)$ the forms $v \wedge Z \rfloor (\omega_{02} - \lambda \omega_{01})$ and $\Gamma'(v)$ are linear dependent. Therefore, $\omega_{02} - \lambda \omega_{01}$ vanishes on all isotropic subspaces, and since $\omega_{02} - \lambda \omega_{01}$ is effective, we find that

$$\omega_{02} - \lambda \omega_{01} = 0.$$

Similarly, replacing v by u, we prove that $\omega_{12} = \lambda' \omega_{21}$.

If $\omega_{01} \neq 0$, $\lambda \neq \lambda'$, then for any involutive set $Z_1, \dots, Z_{n-k-2}$, we obtain

$$Z \rfloor (\omega_{32} - \lambda \omega_{31}) = 0.$$

Since $\omega_{32} - \lambda \omega_{31}$ is effective we obtain $\omega_{32} = \lambda \omega_{31}$ and $\omega_2 = \lambda \omega_1$. ■

Corollary 5.2.1 (Lepage's theorem) *A k-form $\omega \in \Lambda^k(V^*)$ belongs to the image of $\top$ if and only if $\omega|_L = 0$ for any Lagrangian subspace $L \subset V$.*

Corollary 5.2.2 *Let $\omega \in \Lambda^k(V^*)$ be an effective k-form, and let $\theta \in \Lambda^s(V^*)$ be an s-form, $s < k$, such that $\theta|_L = 0$ for any isotropic k-dimensional subspace $W \subset V$ on with $\omega|_L = 0$. Then $\theta \in \operatorname{Im} \top$.*

5.2.1 $\mathfrak{sl}_2$-method

In this section we will discuss a more conceptual approach to the Hodge–Lepage decomposition. This approach is based on the following observation.

Let us consider the following three operators acting in the exterior algebra $\Lambda^*(V^*)$ of a symplectic vector space (V, Ω): operators $\perp, \top$ and

$$\mathbb{I} : \Lambda^*(V^*) \to \Lambda^*(V^*),$$
$$\mathbb{I} : \omega \in \Lambda^k(V^*) \longmapsto (n-k)\omega \in \Lambda^k(V^*).$$

As we have seen

$$[\perp, \top] = \mathbb{I}.$$

Moreover,

$$[\mathbb{I}, \perp] = 2\perp,$$
$$[\mathbb{I}, \top] = -2\top.$$

In other words, these three operators give a representation of the Lie algebra $\mathfrak{sl}_2$ in the exterior algebra of the symplectic vector space. We will use this representation in order to obtain the Hodge–Lepage decomposition.

$\mathfrak{sl}_2$-representations

We remind the reader of some standard facts from the representation theory for the Lie algebra $\mathfrak{sl}_2$.

Let $\mathfrak{sl}_2$ be the Lie algebra of 2×2 matrices with zero trace. This is a three-dimensional simple Lie algebra of rank 1.

We choose the Chevalley–Cartan basis E_+, E_-, H in $\mathfrak{sl}_2$, where

$$E_+ = \begin{Vmatrix} 0 & 1 \\ 0 & 0 \end{Vmatrix}, \quad E_- = \begin{Vmatrix} 0 & 0 \\ 1 & 0 \end{Vmatrix}, \quad H = \begin{Vmatrix} 1 & 0 \\ 0 & -1 \end{Vmatrix}.$$

The commutation relations takes the form

$$[E_+, E_-] = H, \quad [H, E_+] = 2E_+, \quad [H, E_-] = -2E_-.$$

Let $\rho : \mathfrak{sl}_2 \to \operatorname{End} W$ be a representation of the Lie algebra. A vector v is said to be of *height* λ if

$$Hv = \lambda v.$$

It is easy to see that the vector $E_+ v$ is of height $(\lambda + 2)$ and the vector $E_- v$ is of height $(\lambda - 2)$ if the vector v has height λ.

Assume that the vector space W is finite dimensional, then beginning with an eigenvector H and repeatedly applying operator E_+ we should arrive at an eigenvector of H that belongs to $\ker E_+$. Such vectors are called *primitive* .

Let $v \in W$ be a primitive vector of height λ, and let

$$v_n = E_-^n v$$

for $n \geq 0$ and $v_{-1} = 0$.

Then straightforward computations show that

$$\begin{aligned} Hv_n &= (\lambda - 2n)v_n, \\ E_+ v_n &= (\lambda - n + 1)v_{n-1}, \\ E_- v_n &= (n+1)v_{n+1}. \end{aligned}$$

The first relation shows that vectors v_n are linear independent if they are non-zero vectors. Therefore, $v_{n+1} = 0$ for some n, and assume that $v_n \neq 0$. Then the second relation gives $\lambda = n$. In other words, the heights of primitive vectors are natural numbers.

These relations also show how to construct an irreducible representation in any dimension n.

Namely, let us denote by W_n an $(n+1)$-dimensional vector space with a basis $(v_0, v_1, \ldots, v_n)$ and $\mathfrak{sl}_2$ action given by the above formulas. It is clear that W_n is irreducible, and any $\mathfrak{sl}_2$-representation is a sum of modules of W_n-type.

To obtain such a decomposition one should remark that the subspace $\ker E_+$ is invariant with respect to the operator H. As we have seen, the spectrum of the restriction of H on $\ker E_+$ consist, of natural numbers. Therefore,

$$\ker E_+ = \oplus_i Z_{n_i},$$

where Z_{n_i} is a subspace of eigenvectors with eigenvalue $n_i \in \mathbb{N}$.

Each non-zero vector $v \in Z_{n_i}$ gives an irreducible

$$W_{n_i} = \{v = v_0, v_1 = E_- v, \ldots, v_{n_i} = E_-^{n_i} v\}$$

$\mathfrak{sl}_2$-representation of dimension $n_i + 1$, and W is a direct sum of W_{n_i} with multiplicities $\dim Z_{n_i}$.

Effective forms and primitive vectors

Let us apply the description of $\mathfrak{sl}_2$-representations to the representation given by operators $\perp$, $\top$ and $\mathbb{I}$.

As we have seen, the operator $\perp$ corresponds to E_+, the operator $\top$ correspond to E_- and $\mathbb{I}$ corresponds to H.

Therefore, primitive vectors are exactly effective forms. They corresponds to non-negative eigenvalues of $\mathbb{I}$ and therefore belong to $\Lambda^k(V^*)$ with $k \leq n$.

Any non-zero effective form $\omega \in \Lambda^k_\varepsilon(V^*)$ has height $n-k$ and thus generates an irreducible $\mathfrak{sl}_2$-module $(\omega, \top(\omega), \ldots, \top^{n-k}(\omega))$ of dimension $n-k+1$.

Note that we obtain the relation

$$\top^{n-k+1}(\omega) = 0$$

once more.

Effective forms in the middle dimension $\omega \in \Lambda^n_\varepsilon(V^*)$ have height 0. Therefore they generate the trivial $\mathfrak{sl}_2$-modules.

Finally note, that the Hodge–Lepage decomposition corresponds to the decomposition of the $\mathfrak{sl}_2$-module $\Lambda^*(V^*)$ into a sum of irreducibles.

6
A symplectic classification of exterior 2-forms in dimension 4

6.1 Pfaffian

Let (V, Ω) be a four-dimensional symplectic vector space with a structure 2-form Ω, and let $\omega \in \Lambda^2(V^*)$ be a 2-form.

Define a linear operator $A_\omega : V \to V$ by the following condition:

$$X \rfloor \omega = A_\omega X \rfloor \Omega \tag{6.1}$$

for any vector $X \in V$.

Proposition 6.1.1 *Operator A_ω is uniquely determined by ω, and*

1. *A_ω is symmetric with respect to Ω, i.e.,*

$$\Omega(A_\omega X, Y) = \Omega(X, A_\omega Y)$$

   for any $X, Y \in V$;
2. *vectors X and $A_\omega X$ are skew-orthogonal, i.e.,*

$$\Omega(A_\omega X, X) = 0$$

   for any vector $X \in V$;
3. *the map $A : \Lambda^2(V) \to \mathrm{End}(V), \omega \longmapsto A_\omega$, is $\mathbb{R}$-linear, i.e.,*

$$A_{h_1\omega_1 + h_2\omega_2} = h_1 A_{\omega_1} + h_2 A_{\omega_2}$$

   for any $\omega_1, \omega_2 \in \Lambda^2(V)$ and any $h_1, h_2 \in \mathbb{R}$;
4. *let $L \subset V$ be a Lagrangian subspace such that $\omega|_L = 0$. Then L is invariant under A_ω:*

$$A_\omega L = L.$$

Proof We prove the last property only.

Let L be the Lagrangian subspace. Then, $L = L^{\perp}$. We have $\Omega(A_\omega X, X) = 0$ for each vector $X \in V$. If $X \in L$, then $AX \in L^{\perp} = L$. Therefore L is an invariant space of A_ω. ■

For any 2-form $\omega \in \Lambda^2(V^*)$ the 4-form $\omega^2 = \omega \wedge \omega$ is proportional to the volume form $\Omega^2 = \Omega \wedge \Omega$, and we define the *Pfaffian* $\mathrm{Pf}(\omega)$ to be the coefficient of proportionality:

$$\omega^2 = \mathrm{Pf}(\omega)\Omega^2,$$

Lemma 6.1.1 *Let ω be an effective 2-form on V. Then,*

$$A_\omega^2 + \mathrm{Pf}(\omega) = 0. \tag{6.2}$$

Proof The effectivity condition for 2-form in four-dimensional symplectic space takes the form

$$\omega \wedge \Omega = 0$$

and we obtain

$$\begin{aligned}
0 &= (A_\omega X \wedge X)\rfloor(\omega\wedge\Omega) \\
&= X\rfloor((A_\omega X\rfloor\omega) \wedge \Omega + \omega\wedge(A_\omega X\rfloor\Omega)) \\
&= \omega(A_\omega X, X)\Omega - (A_\omega X\rfloor\omega)\wedge(X\rfloor\Omega) + (X\rfloor\omega)\wedge(A_\omega X\rfloor\Omega) + \Omega(A_\omega X, X)\omega \\
&= (A_\omega^2 X\rfloor\Omega)\wedge(X\rfloor\Omega) + (A_\omega X\rfloor\Omega)\wedge(A_\omega X\rfloor\Omega) = (A_\omega^2 X\rfloor\Omega)\wedge(X\rfloor\Omega)
\end{aligned}$$

for any vector $X \in V$.

This means that 1-forms $A_\omega^2 X\rfloor\ \Omega$ and $X\rfloor\ \Omega$ are linearly dependent and since Ω is non-degenerate, we find that the vectors $A_\omega^2 X$ and X are linearly dependent for all vectors X. Therefore $A_\omega^2 + \alpha = 0$ for some $\alpha \in \mathbb{R}$. Now we have

$$(X \wedge A_\omega X)\rfloor\ (\Omega \wedge \Omega) = -2X\rfloor\omega \wedge X\rfloor\Omega$$

and

$$(X \wedge A_\omega X)\rfloor\ (\omega \wedge \omega) = 2\alpha X\rfloor\omega \wedge X\rfloor\Omega.$$

Therefore

$$(\alpha - \mathrm{Pf}(\omega))(X\rfloor\omega \wedge X\rfloor\Omega) = 0$$

for all vectors X. Hence $\alpha = \mathrm{Pf}(\omega)$ or $X\rfloor\omega \wedge X\rfloor\Omega = 0$ for all vectors X.

In the last case all vectors are linear dependent and therefore $A_\omega = \lambda$, and $\omega = \lambda\Omega$, that impossible. ■

Corollary 6.1.1 $\det A_\omega = (\mathrm{Pf}(\omega))^2$.

Let us give a coordinate representation for $\mathrm{Pf}(\omega)$ and A_ω.

Suppose that $e_1,\ e_2,\ f_1,\ f_2$ is a canonical basis in V. Then 2-forms

$$e_1^* \wedge e_2^*,\ e_1^* \wedge f_1^* - e_2^* \wedge f_2^*,$$

$$e_1^* \wedge f_2^*,\ e_2^* \wedge f_1^*,\ f_1^* \wedge f_2^*$$

constitute a basis in the space of effective 2-forms on V.

Let

$$\omega = Ee_1^* \wedge e_2^* + B(e_1^* \wedge f_1^* - e_2^* \wedge f_2^*) + Ce_1^* \wedge f_2^* - Ae_2^* \wedge f_1^* + Df_1^* \wedge f_2^*.$$

Then

$$\omega^2 = 2(AC - B^2 - DE)\, e_1^* \wedge f_1^* \wedge e_2^* \wedge f_2^* = (AC - B^2 - DE)\Omega^2$$

and

$$\mathrm{Pf}(\omega) = AC - B^2 - DE.$$

Simple and straightforward calculations show that

$$A_\omega = \left\| \begin{matrix} B & 0 & -A & -D \\ 0 & B & E & C \\ C & D & -B & 0 \\ -E & -A & 0 & -B \end{matrix} \right\|$$

in the canonical basis.

6.2 Normal forms

In this section we consider the problem of symplectic equivalence for effective forms.

We say that effective 2-forms ω_1 and ω_2 on V are *equivalent* if there exists a symplectic isomorphism $\Phi \in \mathrm{Sp}(V, \Omega)$ such that

$$\Phi^*(\omega_1) = \omega_2.$$

Proposition 6.2.1 *The Pfaffian is an invariant of the symplectic group action*

$$\mathrm{Pf}(\Phi^*(\omega)) = \mathrm{Pf}(\omega)$$

for all $\Phi \in \mathrm{Sp}(V, \Omega)$.

Proof Indeed,

$$\begin{aligned}\mathrm{Pf}(\Phi^*(\omega))\Omega \wedge \Omega = \Phi^*(\omega) \wedge \Phi^*(\omega) = \Phi^*(\omega \wedge \omega) = \Phi^*(\mathrm{Pf}(\omega)\Omega \wedge \Omega)\\ = \mathrm{Pf}(\omega)\Phi^*(\Omega) \wedge \Phi^*(\Omega) = \mathrm{Pf}(\omega)\Omega^2.\end{aligned}$$ ■

Moreover

$$\mathrm{Pf}(t\omega) = t^2\,\mathrm{Pf}(\omega)$$

for $t \in \mathbb{R}$.

Therefore the Pfaffian divides the space of effective forms into three parts corresponding to $\mathrm{sign}(\mathrm{Pf}(\omega))$.

We say that an effective 2-form $\omega \neq 0$ on a four-dimensional symplectic space is

1. *hyperbolic,* if $\mathrm{Pf}(\omega) < 0$,
2. *elliptic,* if $\mathrm{Pf}(\omega) > 0$, and
3. *parabolic,* if $\mathrm{Pf}(\omega) = 0$.

Theorem 6.2.1 *Let* $\omega \in \Lambda^2_\varepsilon(V)$ *be an effective form on four-dimensional symplectic space.*

Then there is a canonical basis (e_1, e_2, f_1, f_2) *in* V *such that*

1.

$$\omega = \sqrt{\mathrm{Pf}(\omega)}(e_1^* \wedge f_2^* - e_2^* \wedge f_1^*)$$

if ω *is elliptic,*

2.

$$\omega = \sqrt{-\,\mathrm{Pf}(\omega)}(e_1^* \wedge f_1^* - e_2^* \wedge f_2^*)$$

if ω *is hyperbolic, and*

3.

$$\omega = e_1^* \wedge f_2^*$$

if ω *is parabolic.*

Proof Let us consider first the hyperbolic case and let

$$\lambda = \sqrt{-\operatorname{Pf}(\omega)}.$$

Then the forms $\omega \pm \lambda\Omega$ are decomposable because

$$(\omega \pm \lambda\Omega) \wedge (\omega \pm \lambda\Omega) = \omega \wedge \omega + \lambda^2 \Omega \wedge \Omega = 0.$$

Therefore,

$$\omega + \lambda\Omega = \gamma_1 \wedge \gamma_2,$$

$$\omega - \lambda\Omega = \gamma_3 \wedge \gamma_4$$

for some 1-forms $\gamma_1, \gamma_2, \gamma_3, \gamma_4$, and

$$\Omega = \frac{1}{2\lambda}(\gamma_1 \wedge \gamma_2 - \gamma_3 \wedge \gamma_4),$$

$$\omega = \frac{1}{2}(\gamma_1 \wedge \gamma_2 + \gamma_3 \wedge \gamma_4).$$

Therefore the basis dual to

$$e_1^* = \frac{1}{2\lambda}\gamma_1, \quad e_2^* = -\frac{1}{2\lambda}\gamma_3, \quad f_1^* = \gamma_2, \quad f_2^* = \gamma_4$$

is a canonical basis. In this basis we have

$$\Omega = e_1^* \wedge f_1^* + e_2^* \wedge f_2^*,$$

$$\omega = \lambda(e_1^* \wedge f_1^* - e_2^* \wedge f_2^*).$$

In the elliptic case we put

$$\omega_0 = \frac{\omega}{\sqrt{\operatorname{Pf}(\omega)}}.$$

Then the operator

$$I = A_{\omega_0}$$

defines a complex structure on V and a complex-valued 2-form

$$\Theta(x, y) = \Omega(x, y) - \imath\omega_0(x, y)$$

on the complex space V:

$$\Theta(Ix, y) = \Theta(x, Iy) = \imath\Theta(x, y).$$

In other words, (V, Θ) is a complex symplectic plane. Let e_1, f_1 be a canonical basis for Θ, and let $e_2 = Ie_1, f_2 = -If_1$. Then,

$$\Theta(e_1,f_1) = 1 \quad \Longrightarrow \quad \Omega(e_1,f_1) = 1,\ \omega_0(e_1,f_1) = 0,$$
$$\Theta(e_2,f_1) = \imath\Theta(e_1,f_1) = \imath \quad \Longrightarrow \quad \Omega(e_2,f_1) = 0,\ \omega_0(e_2,f_1) = -1,$$
$$\Theta(e_2,f_2) = \Theta(e_1,f_1) = 1 \quad \Longrightarrow \quad \Omega(e_2,f_2) = 1,\ \omega_0(e_2,f_2) = 0,$$
$$\Theta(e_1,f_2) = -\imath\Theta(e_1,f_1) = -\imath \quad \Longrightarrow \quad \Omega(e_1,f_2) = 0,\ \omega_0(e_1,f_2) = 1,$$
$$\Theta(e_1,e_2) = \imath\Theta(e_1,e_1) = 0 \quad \Longrightarrow \quad \Omega(e_1,e_2) = 0,\ \omega_0(e_1,e_2) = 0,$$
$$\Theta(f_1,f_2) = -\imath\Theta(f_1,f_1) = 0 \quad \Longrightarrow \quad \Omega(f_1,f_2) = 0,\ \omega_0(f_1,f_2) = 0.$$

Therefore e_1, e_2, f_1, f_2 is a canonical basis in (V, Ω) and ω has the required form in the basis.

If $\omega \neq 0$ is parabolic, then

$$\omega \wedge \omega = 0 \quad \Longrightarrow \quad \omega = \alpha \wedge \beta$$

for some linear independent 1-forms $\alpha, \beta \in V^*$.

The condition $\perp \omega = 0$ gives

$$\alpha(X_\beta) = \beta(X_\alpha) = \Omega(X_\alpha, X_\beta) = 0.$$

In other words, the plane

$$\Pi_\omega = \ker\alpha \cap \ker\beta$$

corresponding to ω has a basis X_α, X_β and is Lagrangian. ■

Remark 6.2.1 *Note that in hyperbolic and elliptic cases for normalized 2-forms*

$$\widetilde{\omega} = \frac{1}{\sqrt{|\mathrm{Pf}\ (\omega)|}}\omega$$

we have

$$A^2_{\widetilde{\omega}} + \mathrm{sign}(\mathrm{Pf}(\widetilde{\omega})) = 0.$$

This means that elliptic forms define complex structures on V and hyperbolic forms give a splitting of V into direct sums of two symplectic planes. Parabolic forms correspond to Lagrangian planes in V.

We say that a Lagrangian plane $L \subset V$ is an *ω-plane* if

$$\omega|_L = 0.$$

The following theorem gives a description of ω-planes.

Theorem 6.2.2

1. ***Elliptic forms.*** *Let ω be an elliptic 2-form. Then L is an ω-plane if and only if*

$$A_\omega(L) = L.$$

2. ***Hyperbolic forms.*** *Let ω be an hyperbolic 2-form, and let*

$$V = V_+ \oplus V_-$$

be the decomposition of V into sum of symplectic planes

$$V_\pm = \ker(A_\omega \mp \lambda)$$

where $\lambda = \sqrt{-\mathrm{Pf}(\omega)}$.
Then L is an ω-plane if and only if

$$\dim(L \cap V_\pm) = 1$$

and

$$L = L \cap V_+ \oplus L \cap V_-.$$

3. ***Parabolic forms.*** *Let ω be a parabolic 2-form and let*

$$L_\omega = \mathrm{Im}\, A_\omega$$

be the corresponding Lagrangian plane. Then L is an ω-plane if and only if

$$\dim(L \cap L_\omega) = 1.$$

Proof **Elliptic forms.** For any non-zero vector $X \in V$ we have

$$\Omega(X, A_\omega X) = -\omega(X, X) = 0$$

and

$$\omega(X, A_\omega X) = \Omega(A_\omega X, A_\omega X) = 0.$$

Therefore, the plane $L = \mathrm{span}(X, A_\omega X)$ is the ω-plane.

Let L be an ω-plane, and let $X, Y \in L$ be a basis. Then

$$\Omega(A_\omega X, X) = \omega(X, X) = 0$$

and

$$\Omega(A_\omega X, Y) = \omega(X, Y) = 0.$$

Therefore,

$$A_\omega X \in L^\perp = L.$$

Hyperbolic forms. As we have seen in the proof of the classification theorem

$$V = V_+ \oplus V_-$$

splits into a sum of two regular planes and

$$A_\omega|_{V_\pm} = \pm\lambda.$$

Let L be an ω-plane. Then the above reasons show that

$$A_\omega(L) = L.$$

$A_\omega|_L$ is not scalar operator because $V_\pm$ are regular, but L is Lagrangian. Therefore $A_\omega|_L$ has two distinct eigenvalues and $L = L \cap V_+ \oplus L \cap V_-$.

Parabolic forms. It follows from the representation

$$L_\omega = \ker\alpha \cap \ker\beta$$

if $\omega = \alpha \wedge \beta$. ■

6.3 Jacobi planes

By *Jacobi planes* we mean two-dimensional subspaces $\Pi \subset \Lambda^2(V^*)$ in the space of 2-forms on a four-dimensional vector space V over $\mathbb{R}$.

Such planes arise in the geometrical approach to systems of non-linear partial differential equations on two-dimensional manifolds.

6.3.1 Classification of Jacobi planes

We classify Jacobi planes with respect to the group $\mathrm{GL}(V)$ of linear transformations of V.

Let $\Pi \subset \Lambda^2(V^*)$ be a two-dimensional plane and let us fix a volume 4-form $\mu \in \Lambda^4(V^*)$.

One can define a symmetric bilinear form q on Π:

$$q : \Pi \times \Pi \ni (\alpha, \beta) \longmapsto q(\alpha, \beta) \in \mathbb{R}$$

where

$$\alpha \wedge \beta = q(\alpha, \beta)\mu.$$

By analogy with effective 2-forms on four-dimensional symplectic spaces we distinguish Jacobi planes by restrictions of q on Π.

We say that a Jacobi plane Π is

1. *elliptic*, if $q|_\Pi$ is a non-degenerate definite quadratic form,
2. *hyperbolic*, if $q|_\Pi$ is a non-degenerate sign indefinite quadratic form,
3. *parabolic*, if $q|_\Pi$ is a degenerate non-zero quadratic form, and
4. *Euler*, if $q|_\Pi = 0$.

Let us take a basis ω_1, ω_2 in Π and let

$$Q = \begin{Vmatrix} q(\omega_1, \omega_1) & q(\omega_1, \omega_2) \\ q(\omega_2, \omega_1) & q(\omega_2, \omega_2) \end{Vmatrix}$$

be the matrix of q, and

$$\varepsilon(\Pi) = \operatorname{sign} \det Q.$$

The following result follows directly from the structure of quadratic forms on the plane.

Proposition 6.3.1 *Let $\Pi \subset \Lambda^2(V^*)$ be a Jacobi plane. Then*

1. Π is elliptic $\iff \varepsilon(\Pi) = 1$, and there exists a basis ω_1, ω_2 in Π, such that

$$\omega_1 \wedge \omega_2 = 0, \qquad \omega_1 \wedge \omega_1 = \omega_2 \wedge \omega_2 \neq 0;$$

2. Π is hyperbolic $\iff \varepsilon(\Pi) = -1$, and there exists a basis ω_1, ω_2 on Π, such that

$$\omega_1 \wedge \omega_2 = 0, \qquad \omega_1 \wedge \omega_1 = -\omega_2 \wedge \omega_2 \neq 0;$$

3. *Π is parabolic* $\Longleftrightarrow \varepsilon(\Pi) = 0$, *and there exists a basis* ω_1, ω_2 *on* Π, *such that*

$$\omega_1 \wedge \omega_2 = 0, \quad \omega_1 \wedge \omega_1 \neq 0, \quad \omega_2 \wedge \omega_2 = 0;$$

4. *Π is Euler* $\Longleftrightarrow q|_\Pi = 0$, *and there exists a basis* ω_1, ω_2 *on* Π, *such that*

$$\omega_1 \wedge \omega_2 = \omega_1 \wedge \omega_1 = \omega_2 \wedge \omega_2 = 0.$$

Note that ω_1, ω_2 for elliptic, hyperbolic and parabolic cases are orthogonal with respect to q and the 2-form ω_1 determines a symplectic structure on V, and the 2-form ω_2 is effective. Therefore, we can use the above classification of effective forms to obtain the following theorem.

Theorem 6.3.1 *Let* $\Pi \subset \Lambda^2(V^*)$ *be a Jacobi plane, and let* ω_1, ω_2 *be the orthogonal basis in* Π. *Then there is a basis* e_1, e_2, e_3, e_4 *in* V *such that* ω_1, ω_2 *take one of the following normal forms.*

1. *Elliptic Jacobi planes:*

$$\omega_1 = e_1^* \wedge f_1^* + e_2^* \wedge f_2^*,$$
$$\omega_2 = e_1^* \wedge f_2^* - e_2^* \wedge f_1^*.$$

2. *Hyperbolic Jacobi planes:*

$$\omega_1 = e_1^* \wedge f_1^* + e_2^* \wedge f_2^*,$$
$$\omega_2 = e_1^* \wedge f_1^* - e_2^* \wedge f_2^*.$$

3. *Parabolic Jacobi planes:*

$$\omega_1 = e_1^* \wedge f_1^* + e_2^* \wedge f_2^*,$$
$$\omega_2 = e_1^* \wedge f_2^*.$$

4. *Euler planes:*

$$\omega_1 = e_1^* \wedge f_1^*,$$
$$\omega_2 = e_1^* \wedge f_2^*.$$

Corollary 6.3.1 *There are four orbits of the group* GL(V) *on the set of Jacobi planes: elliptic, hyperbolic, parabolic and Euler planes.*

6.3.2 Operators associated with Jacobi planes

Let $b \stackrel{\text{def}}{=} (\omega_1, \omega_2)$ be an oriented orthogonal basis of Π, where ω_1 is a non-degenerate 2-form. We can consider $\omega_2 = \Omega$ as a symplectic structure on V and $\omega_1 = \omega$ as an effective 2-form.

Consider the linear operator $A_b = A_\omega : V \to V$, where

$$X \rfloor \omega_2 = A_b X \rfloor \omega_1. \tag{6.3}$$

As we have seen A_b satisfies the equation

$$A_b^2 + \mathrm{Pf}(\omega) = 0$$

and $\mathrm{Pf}(\omega) = \varepsilon(\Pi)$ in our case.

Let Π be an elliptic or a hyperbolic plane, then transformations between oriented orthogonal bases, say (ω_1, ω_2) and (ω_1', ω_2'), are given by *elliptic* or *hyperbolic similitudes*.

That is,

$$\omega_1 = (t \cos s)\omega_1' - (t \sin s)\omega_2',$$
$$\omega_2 = (t \sin s)\omega_1' + (t \cos s)\omega_2'$$

for elliptic planes and

$$\omega_1 = (t \cosh s)\omega_1' + (t \sinh s)\omega_2',$$
$$\omega_2 = (t \sinh s)\omega_1' + (t \cosh s)\omega_2'$$

for hyperbolic planes.

Here $s \in \mathbb{R}, t \in \mathbb{R} \setminus \{0\}$.

Proposition 6.3.2 *The operator A_b does not depend on the choice of oriented orthogonal basis in elliptic Jacobi planes and on the orthogonal basis for the hyperbolic ones.*

Proof We prove the proposition for the elliptic case only.

Let $A_{b'}$ corresponds to the orthogonal basis $b' = (\omega_1', \omega_2')$. We show that $A_b = A_{b'}$.

Indeed, the definition of A_b, $X \rfloor \omega_2 = A_b X \rfloor \omega_1$ reads as

$$X \rfloor (\sin s\, \omega_1' + \cos s\, \omega_2') = A_b X \rfloor (\cos s\, \omega_1' - \sin s\, \omega_2')$$

or

$$(-\sin s\, X + \cos s\, A_b X)\rfloor \omega_1' = (\cos s\, X + \sin s\, A_b X)\rfloor \omega_2'.$$

Therefore

$$A_{b'} = (-\sin s + \cos s\, A_b)(\cos s + \sin s\, A_b)^{-1} = A_b.$$

The proof for the hyperbolic case goes in the same way, with an additional remark: bases (ω_1, ω_2) and (ω_2, ω_1) define the same operator. ∎

Since the operator A_b does not depend on the choice of (oriented) orthogonal basis in Π, we will denote the operator A_b by A_Π. Thus, any (oriented) elliptic or hyperbolic Jacobi plane Π generates a complex structure in the elliptic case or a product structure in the hyperbolic case on V.

Moreover, operators A_Π inherit all the properties of the operators A_ω. For example, if Π is hyperbolic, the space V splits into a direct sum of two-dimensional planes $V = V_- \oplus V_+$ and $A_\Pi|_{V_\pm} = \pm 1$.

7

Symplectic classification of exterior 2-forms

In this chapter we classify exterior 2-forms on symplectic spaces with respect to the symplectic group.

7.1 Pfaffians and linear operators associated with 2-forms

We have introduced the Pfaffian of 2-forms on symplectic four-dimensional spaces. The same idea can be used to find this notion for symplectic spaces of arbitrary dimension.

Namely, we note that the space $\Lambda^{2n}(V^*)\dim V = 2n$ has dimension one and the volume n-form is fixed by Ω^n. Then for all $\omega \in \Lambda^2(V^*)$ there is a unique scalar λ_ω such that

$$\omega^n = \lambda_\omega \Omega^n.$$

This scalar is called a *Pfaffian* of the form ω and is denoted by $\mathrm{Pf}(\omega)$:

$$\omega^n = \mathrm{Pf}(\omega)\Omega^n. \tag{7.1}$$

The Pfaffian is obviously an invariant of the $\mathrm{Sp}(V)$-action, i.e., if $S \in \mathrm{Sp}(V, \Omega)$, then

$$\mathrm{Pf}(S^*(\omega)) = \mathrm{Pf}(\omega).$$

In fact,

$$S^*(\omega^n) = S^*(\mathrm{Pf}(\omega)\Omega^n) = \mathrm{Pf}(\omega)S^*(\Omega^n) = \mathrm{Pf}(\omega)\ \Omega^n,$$

and

$$S^*(\omega^n) = (S^*(\omega))^n = \mathrm{Pf}(S^*(\omega))\Omega^n.$$

We also define a *characteristic polynomial* of an exterior 2-form $\omega \in \Lambda^2(V^*)$, as

$$P_\omega(\lambda) = \mathrm{Pf}(\omega - \lambda\Omega),$$

$\lambda \in \mathbb{R}$.

The polynomial $P_\omega(\lambda)$ is an invariant of the $\mathrm{Sp}(V, \Omega)$-action:

$$P_\omega(\lambda) = P_{S^*(\omega)}(\lambda). \tag{7.2}$$

Therefore the coefficients $I_0(\omega), \dots, I_{n-1}(\omega)$ of the characteristic polynomial

$$P_\omega(\lambda) = I_0(\omega) - \lambda I_1(\omega) + \cdots + (-1)^{n-1} I_{n-1}(\omega)\lambda^{n-1} + (-1)^n \lambda^n$$

are invariants of the symplectic group $\mathrm{Sp}(V, \Omega)$.

The roots of the polynomial P_ω are called *characteristic numbers* of the 2-form ω. We denote by $\mathrm{char}\,\omega \subset \mathbb{C}$ the set of all characteristic numbers of ω.

Let $\lambda \in \mathbb{R}$ be a characteristic number of ω, then $(\omega - \lambda\Omega)^n = 0$ and therefore $\omega - \lambda\Omega$ is the degenerate 2-form.

Denote by $K(\lambda) \subset V$ the kernel of this form. This is an even dimensional subspace and $\omega - \lambda\Omega$ is non-degenerate on the factor-space $V / K(\lambda)$.

If $\lambda \in \mathbb{C}$ is a complex characteristic number of ω then the above holds for the complexifications of ω and V.

Let us denote by $A_\omega : V \to V$ the linear operator defined by the 2-form ω:

$$\iota_X \omega = \iota_{A_\omega X} \Omega$$

for all $X \in V$.

Lemma 7.1.1

1. *The operator A_ω is symmetric with respect to the structure form Ω, i.e.*

$$\Omega(A_\omega X, Y) = \Omega(X, A_\omega Y)$$

for all $X, Y \in V$.

2. *For any vector $X \in V$ the vectors X and $A_\omega X$ are skew-orthogonal:*

$$\Omega(A_\omega X, X) = 0.$$

3.

$$A_{S^*(\omega)} = S^{-1} \circ A_\omega \circ S$$

for all $S \in \mathrm{Sp}(V, \Omega)$.

4. *λ is a characteristic number for ω if and only if λ is an eigenvalue of A_ω.*

Proof Indeed, for X and $Y \in V$ we have

$$\Omega(X, A_\omega Y) = -\Omega(A_\omega Y, X) = -\omega(Y, X) = \omega(X, Y) = \Omega(A_\omega X, Y)$$

and

$$\Omega(A_\omega X, X) = \omega(X, X) = 0.$$

In addition,

$$\begin{aligned} \iota_X S^*(\omega)(Y) &= \omega(SX, SY) \\ &= \Omega(A_\omega SX, SY) = \Omega(S^{-1} A_\omega SX, Y) \\ &= \iota_{S^{-1} A_\omega S(X)}(\Omega)(Y). \end{aligned}$$

Assume that $\lambda \in \mathbb{R}$ is a real characteristic number for ω and let $X \in K(\lambda)$ be a non-zero vector. Then

$$(\omega - \lambda\Omega)(X, Y) = 0$$

for all $Y \in V$.

But

$$(\omega - \lambda\Omega)(X, Y) = \Omega((A_\omega - \lambda)X, Y).$$

Therefore, $(A_\omega - \lambda)X = 0$. ∎

Remark 7.1.1 *It follows that all eigenvalues of A_ω have even multiplicities, and, as we will see below, (see also, [3]) :*

$$(\mathrm{Pf}(\omega))^2 = \det(A_\omega).$$

7.2 Symplectic classification of 2-forms with distinct real characteristic numbers

Proposition 7.2.1 *Let λ_1 and λ_2 be distinct real characteristic numbers of a 2-form ω. Then $K(\lambda_1)$ and $K(\lambda_2)$ are skew-orthogonal with respect to Ω:*

$$\Omega(X, Y) = 0,$$

as well as with respect to ω,

$$\omega(X, Y) = 0$$

for all $X \in K(\lambda_1), Y \in K(\lambda_2)$.

Moreover,

$$K(\lambda_1) \cap K(\lambda_2) = 0.$$

Proof We have

$$\omega(X, Y) = \Omega(A_\omega X, Y) = \lambda_1 \Omega(X, Y)$$

and

$$\omega(X, Y) = \Omega(X, A_\omega Y) = \lambda_2 \Omega(X, Y).$$

Therefore, $(\lambda_1 - \lambda_2)\Omega(X, Y) = 0$ and $\Omega(X, Y) = 0$. This also implies that $\omega(X, Y) = 0$.

Subspaces $K(\lambda_1)$ and $K(\lambda_2)$ have zero intersection because they are eigen subspaces of the operator A_ω corresponding to distinct eigenvalues. ■

Let us assume now that characteristic numbers of the 2-form ω are real and distinct, say, $\lambda_1, \ldots, \lambda_n$, where $\dim V = 2n$.

Then

$$V = K(\lambda_1) \oplus \cdots \oplus K(\lambda_n)$$

is a skew-orthogonal decomposition symplectic space V into the direct sum of planes. These planes are symplectic too because the structure form Ω has no kernel.

The restrictions $\omega_i = \omega|_{K(\lambda_i)}$ are proportional to the restrictions $\Omega_i = \Omega|_{K(\lambda_i)}$, and

$$\omega_i(X, Y) = \Omega_i(A_\omega X, Y) = \lambda_i \Omega_i(X, Y)$$

for all $X, Y \in K(\lambda_i)$.

Therefore,

$$\omega_i = \lambda_i \Omega_i$$

for all $i = 1, \ldots, n$ and

$$\omega = \lambda_1 \Omega_1 \oplus \cdots \oplus \lambda_n \Omega_n.$$

Summarizing we obtain the following result.

Theorem 7.2.1 *Let $\omega \in \Lambda^2(V^*)$ be a 2-form on the symplectic vector space (V, Ω). Assume that all characteristic numbers $\lambda_1, \dots, \lambda_n$ of ω are distinct and real. Then*

1. *the symplectic vector space V is the direct sum of the symplectic planes $K(\lambda_i), \lambda_i \in \operatorname{char} \omega$:*

$$V = K(\lambda_1) \oplus \cdots \oplus K(\lambda_n);$$

2. *all planes $K(\lambda_i)$ and $K(\lambda_j), i \neq j$, are skew-orthogonal with respect to both forms ω and Ω;*
3. *the form ω is the direct sum of plane 2-forms $\lambda_i \Omega_i$, where $\Omega_i = \Omega|_{K(\lambda_i)}$ are the restrictions of the structure form Ω on the planes*

$$\omega = \lambda_1 \Omega_1 \oplus \cdots \oplus \lambda_n \Omega_n$$

and

$$\Omega = \Omega_1 \oplus \cdots \oplus \Omega_n.$$

Corollary 7.2.1 *Let $\omega \in \Lambda^2(V^*)$ be a 2-form on the symplectic vector space (V, Ω) having all characteristic numbers $\lambda_1, \dots, \lambda_n$ of ω distinct and real. Then there is a canonical basis $e_1, \dots, e_n, f_1, \dots, f_n$ in the symplectic vector space V such that*

$$\omega = \lambda_1 e_1^* \wedge f_1^* + \cdots + \lambda_n e_n^* \wedge f_n^*.$$

Corollary 7.2.2

$$\operatorname{Pf}(\omega) = \lambda_1 \cdots \lambda_n.$$

Corollary 7.2.3 *Two 2-forms $\omega \in \Lambda^2(V^*)$ and $\omega' \in \Lambda^2(V^*)$ on the symplectic vector space (V, Ω) with distinct and real characteristic numbers are equivalent with respect to $\mathrm{Sp}(V, \Omega)$ if and only if their characteristic numbers, or equally the values of invariants $I_0, \dots, I_{n-1}$, are coincident:*

$$I_0(\omega) = I_0(\omega'), \dots, I_{n-1}(\omega) = I_{n-1}(\omega').$$

7.3 Symplectic classification of 2-forms with distinct complex characteristic numbers

In this section we consider the case of complex characteristic numbers.

Let us denote by $V^{\mathbb{C}}$ the complexification of the symplectic vector space V, and let $\Omega^{\mathbb{C}}$ and $\omega^{\mathbb{C}}$ be the complexifications of forms Ω and ω respectively. Obviously, all of the above results are valid for 2-forms $\omega^{\mathbb{C}}$ in the complex symplectic space $(V^{\mathbb{C}}, \Omega^{\mathbb{C}})$. Assume now that the complex characteristic numbers $\lambda_1, \ldots, \lambda_n$ of $\omega^{\mathbb{C}}$ are distinct. Then one has the direct decomposition of the symplectic space

$$V^{\mathbb{C}} = K(\lambda_1) \oplus \cdots \oplus K(\lambda_n)$$

and forms

$$\omega^{\mathbb{C}} = \lambda_1 \Omega_1 \oplus \cdots \oplus \lambda_n \Omega_n,$$
$$\Omega^{\mathbb{C}} = \Omega_1 \oplus \cdots \oplus \Omega_n$$

where $\Omega_i = \Omega^{\mathbb{C}}|_{K(\lambda_i)}, \omega_i = \omega^{\mathbb{C}}|_{K(\lambda_i)}$.

For $v = x + \iota y \in V^{\mathbb{C}}$ $(x, y \in V)$ denote its complex conjugate by $\overline{v} = x - \iota y \in V^{\mathbb{C}}$. Then elements of V (real vectors) are fixed vectors for the conjugation:

$$v \in V \Longleftrightarrow \overline{v} = v.$$

Note also that, for example, $\omega^{\mathbb{C}}(\overline{v}, \overline{w}) = \overline{\omega^{\mathbb{C}}(v, w)}$, and therefore,

$$\overline{K(\lambda)} = K(\overline{\lambda}).$$

Hence, for the characteristic number λ with non-trivial imaginary part, $\lambda \neq \overline{\lambda}$, the subspace

$$K(\lambda) \oplus K(\overline{\lambda})$$

is a four-dimensional (over $\mathbb{C}$) symplectic vector space with respect to the restriction $\Omega_\lambda \oplus \Omega_{\overline{\lambda}}$.

The real part

$$K(\lambda, \overline{\lambda}) = (K(\lambda) \oplus K(\overline{\lambda})) \cap V$$

is a four-dimensional (over $\mathbb{R}$) symplectic vector space with structure form $\Omega_{\lambda,\overline{\lambda}} = \Omega|_{K(\lambda,\overline{\lambda})}$.

This observation produces the decomposition

$$V = \bigoplus_{\operatorname{Im}\lambda=0} K(\lambda) \oplus \bigoplus_{\operatorname{Im}\lambda\neq 0} K(\lambda,\bar{\lambda})$$

of the symplectic vector space into the direct sum of symplectic planes $K(\lambda)$ and symplectic four-dimensional spaces $K(\lambda,\bar{\lambda})$.

They are pairwise skew-orthogonal with respect to Ω and ω, and

$$\Omega = \bigoplus_{\lambda\in\operatorname{char}\omega,\operatorname{Im}\lambda=0} \Omega_\lambda \oplus \bigoplus_{\lambda\in\operatorname{char}\omega,\operatorname{Im}\lambda\neq 0} \Omega_{\lambda,\bar{\lambda}},$$

$$\omega = \bigoplus_{\lambda\in\operatorname{char}\omega,\operatorname{Im}\lambda=0} \omega_\lambda \oplus \bigoplus_{\lambda\in\operatorname{char}\omega,\operatorname{Im}\lambda\neq 0} \omega_{\lambda,\bar{\lambda}},$$

where $\omega_{\lambda,\bar{\lambda}} = \omega|_{K(\lambda,\bar{\lambda})}$.

Note that on the symplectic four-dimensional space $K(\lambda,\bar{\lambda})$ we have

$$I_0(\omega_{\lambda,\bar{\lambda}}) = \lambda\bar{\lambda} = |\lambda|^2, \quad I_1(\omega_{\lambda,\bar{\lambda}}) = 2\operatorname{Re}\lambda.$$

Therefore the 2-form

$$\theta_{\lambda,\bar{\lambda}} = \omega_{\lambda,\bar{\lambda}} - \operatorname{Re}\lambda\ \Omega_{\lambda,\bar{\lambda}}$$

is an effective form of elliptic type with

$$\operatorname{Pf}(\theta_{\lambda,\bar{\lambda}}) = (\operatorname{Im}\lambda)^2.$$

It follows from the above classification of 2-forms on four-dimensional vector spaces that in a canonical basis

$$\omega_{\lambda,\bar{\lambda}} = \operatorname{Re}\lambda\ (e_1^* \wedge f_1^* + e_2^* \wedge f_2^*) + \operatorname{Im}\lambda\ (e_1^* \wedge f_2^* - e_2^* \wedge f_1^*).$$

Finally, we obtain the following theorem.

Theorem 7.3.1 *Let $\omega \in \Lambda^2(V^*)$ be a 2-form on the symplectic vector space (V,Ω) with district characteristic numbers $\lambda_1,\ldots,\lambda_n$. Then,*

1. *the symplectic vector space V is the direct sum of the symplectic planes $K(\lambda)$ corresponding to real characteristic numbers λ and four-dimensional symplectic spaces $K(\lambda,\bar{\lambda})$ corresponding to the complex ones:*

$$V = \bigoplus_{\lambda\in\operatorname{char}\omega,\operatorname{Im}\lambda=0} K(\lambda) \oplus \bigoplus_{\lambda\in\operatorname{char}\omega,\operatorname{Im}\lambda\neq 0} K(\lambda,\bar{\lambda});$$

2. *all of these subspaces are pairwise skew-orthogonal with respect to forms* ω *and* Ω, *and*

$$\omega = \bigoplus_{\lambda \in \operatorname{char}\omega, \operatorname{Im}\lambda = 0} \lambda\Omega_\lambda \oplus \bigoplus_{\lambda \in \operatorname{char}\omega, \operatorname{Im}\lambda \neq 0} \omega_{\lambda,\bar{\lambda}},$$

$$\Omega = \bigoplus_{\lambda \in \operatorname{char}\omega, \operatorname{Im}\lambda = 0} \Omega_\lambda \oplus \bigoplus_{\lambda \in \operatorname{char}\omega, \operatorname{Im}\lambda \neq 0} \Omega_{\lambda,\bar{\lambda}},$$

where

$$\omega_{\lambda,\bar{\lambda}} = \operatorname{Re}\lambda\ \Omega_{\lambda,\bar{\lambda}} + \operatorname{Im}\lambda\ \theta_{\lambda,\bar{\lambda}}$$

and $\theta_{\lambda,\bar{\lambda}}$ *is an elliptic effective form on the four-dimensional symplectic space* $(K(\lambda,\bar{\lambda}), \Omega_{\lambda,\bar{\lambda}})$ *with* $\operatorname{Pf}(\theta_{\lambda,\bar{\lambda}}) = (\operatorname{Im}\lambda)^2$.

Corollary 7.3.1 *Let* $\omega \in \Lambda^2(V^*)$ *be a 2-form on the symplectic vector space* (V,Ω) *with distinct characteristic numbers* $\lambda_1,\ldots,\lambda_n$. *Then there is a canonical basis* $e_1,\ldots,e_k,f_1,\ldots,f_k,e_{k+1},\tilde{e}_{k+1},f_k,\tilde{f}_{k+1},\ldots$ *in the symplectic vector space* V *such that*

$$\omega = \sum_{\operatorname{Im}\lambda_i = 0} \lambda_i e_1^* \wedge f_1^*$$
$$+ \sum_{\operatorname{Im}\lambda_i \neq 0} (\operatorname{Re}\lambda_i\ (e_i^* \wedge f_i^* + \tilde{e}_i^* \wedge \tilde{f}_i^*) + \operatorname{Im}\lambda_i\ (e_i^* \wedge \tilde{f}_i^* - \tilde{e}_i^* \wedge f_i^*)).$$

Corollary 7.3.2 $\operatorname{Pf}(\omega) = \lambda_1 \cdots \lambda_n$.

Corollary 7.3.3 *Two 2-forms* $\omega \in \Lambda^2(V^*)$ *and* $\omega' \in \Lambda^2(V^*)$ *on the symplectic vector space* (V,Ω) *with distinct characteristic numbers are equivalent with respect to* $\operatorname{Sp}(V,\Omega)$ *if and only if their characteristic numbers, or equally the values of invariants* $I_0,\ldots,I_{n-1}$, *coincide:*

$$I_0(\omega) = I_0(\omega'),\ldots,I_{n-1}(\omega) = I_{n-1}(\omega').$$

7.4 Symplectic classification of 2-forms with multiple characteristic numbers

Let

$$K_i(\lambda) = \ker(A_\omega - \lambda)^i$$

and

$$K_*(\lambda) = \bigcup_{i \geq 1} K_i(\lambda)$$

be subspaces of vectors associated with a characteristic number λ.

Proposition 7.4.1 *$K_*(\lambda_1)$ and $K_*(\lambda_2)$ are skew-orthogonal with respect to ω and Ω if $\lambda_1 \neq \lambda_2$.*

Proof The proof is based on the following formula:

$$\Omega((A_\omega - \lambda)X, Y) - \Omega(X, (A_\omega - \nu)Y) = (\nu - \lambda)\Omega(X, Y). \qquad (7.3)$$

It follows immediately from this formula that $K_1(\lambda_1)$ and $K_1(\lambda_2)$ are skew-orthogonal if $\lambda_1 \neq \lambda_2$.

Assume that we have proved that $K_i(\lambda_1)$ and $K_j(\lambda_2)$ are skew-orthogonal under the condition $\lambda_1 \neq \lambda_2$ for all i and j such that $i + j \leq k - 1$. Let us take $X \in K_i(\lambda_1)$ and $Y \in K_j(\lambda_2)$, with $i + j = k$.

Then,

$$\Omega((A_\omega - \lambda_1)X, Y) = 0 \quad \text{and} \quad \Omega(X, (A_\omega - \lambda_2)Y) = 0$$

by the induction hypothesis. Therefore, it follows from (7.3) that $\Omega(X, Y) = 0$. In the same way one obtains ω-skew orthogonality from the similar formula

$$\omega((A_\omega - \lambda)X, Y) - \omega(X, (A_\omega - \nu)Y) = (\nu - \lambda)\omega(X, Y)$$

which follows from (7.3). ■

Theorem 7.4.1 *Let $\omega \in \Lambda^2(V^*)$ be a 2-form on the symplectic vector space (V, Ω). Then,*

$$V = \bigoplus_{\lambda \in \operatorname{char} \omega} K_*(\lambda)$$

is the decomposition of V into the direct sum of ω- and Ω-skew orthogonal symplectic subspaces.

We shall investigate the nilpotent case only.

Proposition 7.4.2 *Let $\omega \in \Lambda^2(V^*)$ be a 2-form on the symplectic vector space (V, Ω) such that the operator A_ω is a nilpotent, $A_\omega^k = 0$, but $A_\omega^{k-1} \neq 0$. Let*

$X \in V$ be a vector of height k, $A_\omega^{k-1}X \neq 0$. Then,

1. *the subspace generated by the vectors $X, A_\omega X, \ldots, A_\omega^{k-1}X$ is isotropic;*
2. *there exists a symplectic $2k$-dimensional subspace $W \subset V$ such that $X \in W$ and $A_\omega W \subset W$;*
3. *the vector X can be included in a basis*

$$X = X_k, \quad X_{k-1} = A_\omega X_k, \quad \ldots, \quad X_1 = A_\omega^{k-1} X_k,$$
$$Y_k, \quad Y_{k-1} = A_\omega Y_k, \quad \ldots, \quad Y_1 = A_\omega^{k-1} Y_k$$

   of W in such a way that

$$\Omega(X_i, X_j) = \Omega(Y_i, Y_j) = 0$$

   and

$$\Omega(X_i, Y_j) = \delta_{i+j,\, k+1};$$

4. *the skew-orthogonal subspace $W^\perp$ is an A_ω-invariant;*
5. *there is a canonical basic $e_1, \ldots, e_k, f_1, \ldots, f_k$ such that the 2-form $\omega|_W$ can be written in the basis as*

$$\omega|_W = \sum_{i=1}^{k-1} e_i^* \wedge f_{i+1}^*. \tag{7.4}$$

Proof First of all let us prove that

$$\Omega(A_\omega^i X, X) = 0$$

for all $i = 1, 2, \ldots$.

Indeed, if i is even, say $i = 2l$, then

$$\Omega(A_\omega^i X, X) = \Omega(A_\omega^{2l} X, X) = \Omega(A_\omega^l X, A_\omega^l X) = 0,$$

and if i is odd, say $i = 2l + 1$, then

$$\Omega(A_\omega^i X, X) = \Omega(A_\omega^{2l+1} X, X) = \Omega(A_\omega A_\omega^l X, A_\omega^l X) = \omega(A_\omega^l X, A_\omega^l X) = 0.$$

Let Z be a such that

$$\Omega(X_1, Z) = 1.$$

We obtain

$$1 = \Omega(X_1, Z) = \Omega(A_\omega^{k-1}X, Z) = \Omega(X, A_\omega^{k-1}Z).$$

Therefore, $A_\omega^{k-1}Z \neq 0$ and Z is a vector of height k. We define vectors

$$Z_i = A_\omega^{k-i}Z$$

for $i = 0, 1, \ldots, k-1$. Let us take $Y_1 = Z_1$ and use induction to construct the vectors Y_j such that

$$A_\omega(Y_j) = Y_{j-1} \tag{7.5}$$

for all $j = 2, \ldots, k$, and

$$\Omega(X_k, Y_j) = 0.$$

To obtain Y_2 we change Z_2 for

$$Y_2 = Z_2 - aZ_1$$

for some $a \in \mathbb{R}$. One has

$$A_\omega(Z_2 - aZ_1) = A_\omega(Z_2) = Y_1$$

and

$$\begin{aligned}\Omega(X_k, Z_2 - aZ_1) &= \Omega(X_k, Z_2 - aZ_1) = \Omega(X_k, Z_2) - a\Omega(X_k, Z_1)\\ &= \Omega(X_k, Z_2) - a\Omega(X_k, A_\omega^{k-1}Z_k) = \Omega(X_k, Z_2) - a\Omega(X_1, Z_k)\\ &= \Omega(X_k, Z_2) - a.\end{aligned}$$

Therefore, the vector

$$Y_2 = Z_2 - \Omega(X_k, Z_2)\, Z_1$$

satisfies the above requirement.

Assume that we have constructed the vectors $Y_2, \ldots, Y_{j-1}$ such that (7.5) is satisfied and $\Omega(X_k, Y_i) = 0$ for $i = 2, \ldots, j-1$.

To find Y_j let us consider the following vector:

$$v = Z_j - a_1 Y_1 - \cdots - a_{j-1} Y_{j-1}.$$

We have

$$A_\omega v = Z_{j-1} - a_2 Y_1 - \cdots - a_{j-1} Y_{j-2}.$$

Let

$$Y_{j-1} = Z_{j-1} + b_1 Y_1 + b_2 Y_2 + \cdots + b_{j-2} Y_{j-2}.$$

If we take

$$a_2 + b_1 = 0, \ldots, a_{j-1} + b_{j-2} = 0$$

and arbitrary a_1, then

$$A_\omega v = Y_{j-1}.$$

Finally, the condition $\Omega(X_k, v) = 0$ defines a_1 because $\Omega(X_k, Y_1) = 1$. ■

Theorem 7.4.2 *Let $\omega \in \Lambda^2(V^*)$ be a 2-form on a symplectic vector space (V, Ω) such that operator A_ω is nilpotent. Then there is a decomposition of the symplectic vector space V into the direct sum of symplectic subspaces*

$$V = \bigoplus_{j=1}^{1/2 \dim \ker A_\omega} W_j$$

that are ω-skew-orthogonal (or A_ω-invariant), and the restrictions $\omega|_{W_j}$ can be written in normal forms (7.4) for all $j = 1, \ldots, 1/2 \dim \ker A_\omega$.

Summarizing all of these results we obtain the following.

Theorem 7.4.3 *Let $\omega \in \Lambda^2(V^*)$ be a 2-form on the symplectic vector space (V, Ω). Then there exists a decomposition of V into a direct sum of symplectic subspaces $V = \oplus_{\lambda \in \mathrm{char}\, \omega} V(\lambda)$ in such a way that:*

1. *$V(\lambda)$ are Ω- and ω-skew orthogonal;*
2. *$V(\lambda)$ are symplectic planes $K(\lambda)$ for distinct real characteristic numbers and four-dimensional symplectic subspaces $K(\lambda, \overline{\lambda})$ for complex distinct real characteristic numbers;*

3. *if λ is a multiple characteristic number then there is an extra decomposition of $V(\lambda)$ into a direct sum of symplectic subspaces*

$$V(\lambda) = \bigoplus_{j=1}^{1/2 \dim \ker(A_\omega - \lambda)} W_j(\lambda)$$

which are skew-orthogonal with respect to forms ω and Ω;

4. *in all three cases there are canonical basises (e,f) such that the restrictions ω have one of the following normal forms:*

   - $$\lambda e_1^* \wedge f_1^*$$

     if λ is a real and simple characteristic number,

   - $$\operatorname{Re} \lambda\, (e_1^* \wedge f_1^* + e_2^* \wedge f_2^*) + \operatorname{Im} \lambda\, (e_1^* \wedge f_2^* - e_2^* \wedge f_1^*)$$

     if λ is a complex and simple characteristic number,

   - $$\lambda \sum_{i=1}^{k} e_i^* \wedge f_i^* + \sum_{i=1}^{k-1} e_i^* \wedge f_{i+1}^*$$

     if λ is a real characteristic number of multiplicity k.

   - $$\operatorname{Re} \lambda \sum_{i=1}^{2k} e_i^* \wedge f_i^* + \operatorname{Im} \lambda \sum_{i=1}^{k} (e_i^* \wedge f_{i+k}^* - e_{i+k}^* \wedge f_i^*)$$
     $$+ \sum_{i=1}^{k-1} (e_i^* \wedge f_{i+1}^* + e_{k+i}^* \wedge f_{k+i+1}^*)$$

     if λ is a complex characteristic number of multiplicity k.

Proof For any real and distinct complex characteristic numbers the theorem follows from the above results and the formula $\Omega(A_\omega X, Y) = \omega(X, Y)$. The proof for complex multiple characteristic numbers is as follows. Let $\lambda \in \mathbb{C}$ be a complex multiple characteristic number, then after complexification we obtain $\overline{V(\lambda)} = V(\bar{\lambda}) \subset V^{\mathbb{C}}$ and $W_j(\lambda)$ can be chosen in such a way that

$\overline{W_j(\lambda)} = W_j(\overline{\lambda})$. Moreover, if $(X_k, \ldots, X_1, Y_k, \ldots, Y_1)$ is the canonical basis in $W_j(\lambda)$, then $(\overline{X_k}, \ldots, \overline{X_1}, \overline{Y_k}, \ldots, \overline{Y_1})$ is a canonical basis in $W_j(\overline{\lambda})$. Then the basis

$$\begin{aligned}
e_1 &= \frac{X_1 + \overline{X_1}}{2}, \quad \ldots, \quad e_k = \frac{X_k + \overline{X_k}}{2}, \\
e_{k+1} &= \frac{X_1 - \overline{X_1}}{2\imath}, \quad \ldots, \quad e_{2k} = \frac{X_k - \overline{X_k}}{2\imath}, \\
f_1 &= \frac{Y_1 + \overline{Y_1}}{2}, \quad \ldots, \quad f_k = \frac{Y_k + \overline{Y_k}}{2}, \\
f_{k+1} &= \frac{Y_1 - \overline{Y_1}}{2\imath}, \quad \ldots, \quad f_{2k} = \frac{Y_k - \overline{Y_k}}{2\imath}
\end{aligned}$$

is the required basis for ω on $(W_j(\lambda) \oplus W_j(\overline{\lambda})) \cap V$. ■

Corollary 7.4.1 *Two 2-forms ω and ω' on a symplectic vector space V are* Sp-*equivalent if and only if the corresponding linear operators A_ω and $A_{\omega'}$ are* GL-*equivalent.*

Corollary 7.4.2 $\det A_\omega = (\mathrm{Pf}\,\omega)^2$.

7.5 Symplectic classification of effective 2-forms in dimension 6

Let V be a symplectic vector space of dimension 6, and let ω be an effective 2-form on V. Then $\omega \wedge \Omega^2 = 0$ and $I_2(\omega) = 0$. Therefore, the characteristic equation for this case takes the following form:

$$\mathrm{Pf}\,(\omega - \lambda\Omega) = -\lambda^3 - I_1(\omega)\lambda + I_0(\omega) = 0.$$

There are four possible cases for the characteristic numbers $\lambda_1, \lambda_2, \lambda_3$.

1. Three real and distinct characteristic numbers with

$$\lambda_1 + \lambda_2 + \lambda_3 = 0.$$

2. Three real characteristic numbers with one real number, say λ_1, of multiplicity 2:

$$\lambda_1 = \lambda_2 = \lambda, \quad \lambda_3 = -2\lambda.$$

3. One real number, say λ_1, and two complex characteristic numbers:

$$\lambda_1 = 2\lambda, \quad \lambda_2 = -\lambda + \imath\mu, \quad \lambda_3 = -\lambda - \imath\mu.$$

4. One real characteristic number of multiplicity 3:

$$\lambda_1 = \lambda_2 = \lambda_3 = 0.$$

Applying the classification theorem for these four cases we obtain the following theorem.

Theorem 7.5.1 *Let $\omega \in \Lambda^2_\varepsilon(V_6^*)$ be an effective 2-form on a symplectic real space V of dimension 6. Then there is a canonical basis $e_1, e_2, e_3, f_1, f_2, f_3$ such that ω can be written in one of the following normal forms:*

1.

$$\lambda_1 e_1^* \wedge f_1^* + \lambda_2 e_2^* \wedge f_2^* + \lambda_3 e_3^* \wedge f_3^*$$

with $\lambda_1 + \lambda_2 + \lambda_3 = 0$;

2.

$$2\lambda e_1^* \wedge f_1^* - \lambda(e_2^* \wedge f_2^* + e_3^* \wedge f_3^*) + \mu(e_3^* \wedge f_2^* - e_2^* \wedge f_3^*);$$

3.

$$2\lambda e_1^* \wedge f_1^* - \lambda(e_2^* \wedge f_2^* + e_3^* \wedge f_3^*) + e_2^* \wedge f_3^*;$$

4.

$$e_1^* \wedge f_2^* + e_2^* \wedge f_3^*.$$

8

Classification of exterior 3-forms on a six-dimensional symplectic space

This chapter is devoted to a classification of effective 3-forms in six-dimensional symplectic space under the natural Sp-action. The question has a surprisingly long history. Being in the spirit of the classical problems of geometric invariants theory, the solution of this problem over $\mathbb{C}$ was found in [40, 90]. The case of real symplectic spaces which has great importance in classification of Monge–Ampère equations was investigated in [77] and completed in [5].

8.1 A symplectic invariant of effective 3-forms

Let V be a six-dimensional symplectic vector space and let $\omega \in \Lambda^3_\varepsilon(V)$ be an effective 3-form.

We associate with ω the following quadratic form $q_\omega \in \mathbf{S}^2(V^*)$:

$$q_\omega(X) = -\frac{1}{2}\perp_2 \omega_X^2,$$

where

$$\omega_X \stackrel{\text{def}}{=} \iota_X \omega \in \Lambda^2(V^*).$$

This form is clearly Sp-invariant:

$$q_{S^*(\omega)} = S^*(q_\omega),$$

for all $S \in \mathrm{Sp}(V)$.

Indeed, we have $(S^*\omega)_X = S^*(\omega_{S(X)})$ for any operators S, and

$$S^* \circ \perp = \perp \circ S^*$$

if $S \in \mathrm{Sp}(V)$. Therefore,

$$q_{S^*(\omega)}(X) = -\frac{1}{2}\bot_2(S^*(\omega))^2_X = -\frac{1}{2}\bot_2 S^*(\omega_{S(X)})^2 = -\frac{1}{2}\bot_2(\omega_{S(X)})^2$$
$$= q_\omega(SX) = S^*(q_\omega)(X).$$

Consider now the characteristic polynomial of ω_X:

$$(\omega_X - \lambda\Omega)^3 = \omega_X^3 - 3\lambda\omega_X^2 \wedge \Omega + 3\lambda^2\omega_X \wedge \Omega^2 - \lambda^3\Omega^3.$$

We have $\omega_X^3 = 0$ because ω_X is a degenerated 2-form, $X \in \ker\omega_X$, and $\omega_X \wedge \Omega^2 = 0$ because ω_X is an effective form. Moreover,

$$\omega_X^2 \wedge \Omega = -\frac{1}{3}q_\omega(X)\,\Omega^3,$$

because

$$\bot_3\,(\omega_X^2 \wedge \Omega) = \bot_3 \circ \top(\omega_X^2) = [\bot_3, \top](\omega_X^2) = \bot_2\,(\omega_X^2) = -2q_\omega(X)$$

and

$$\bot_3\,(\Omega^3) = 6.$$

Therefore,

$$\mathrm{Pf}(\omega_X - \lambda\Omega) = -\lambda^3 + q_\omega(X)\,\lambda$$

and the characteristic numbers of ω_X are

$$\lambda_1 = 0, \quad \lambda_2 = \sqrt{q_\omega(X)}, \quad \lambda_3 = -\sqrt{q_\omega(X)}.$$

Below we give another interpretation of the invariant q_ω.

To this end we choose two vectors X and Y in V in such a way that $\Omega(X, Y) = 1$ and denote by W a symplectic subspace which is skew-orthogonal to the symplectic plane $\langle X, Y\rangle$. Let Ω' be the restriction of Ω on W, and $\top'$ and $\bot'$ be the corresponding operators.

Let $\widehat{X} \in V^*$ be the covector dual to the vector $X \in V$, $\widehat{X} = \iota_X\Omega$. The 2-form ω can be uniquely decomposed as

$$\omega = \omega_0 \wedge \widehat{X} \wedge \widehat{Y} + \omega_1 \wedge \widehat{X} + \omega_2 \wedge \widehat{Y} + \omega_3,$$

with $\omega_0 \in W^*, \omega_1, \omega_2 \in \Lambda^2(W^*), \omega_3 \in \Lambda^3(W^*)$. Using the decomposition $V = W \oplus \langle X, Y \rangle$ and $\perp = \perp' + X \wedge Y$ we obtain

$$0 = \perp(\omega) = \omega_0 + \perp'(\omega_1) \wedge \widehat{X} + \perp'(\omega_2) \wedge \widehat{Y} + \perp'(\omega_3).$$

Therefore,

$$\perp'(\omega_1) = \perp'(\omega_2) = 0$$

and

$$\omega_0 + \perp'(\omega_3) = 0.$$

In other words, 2-forms ω_1 and ω_2 are effective, but ω_0 and ω_3 satisfy the above relation.

Note also that

$$\perp'(\top'\omega_0 + \omega_3) = \omega_0 + \perp'(\omega_3) = 0$$

and $\perp' : \Lambda^3(W^*) \to W^*$ is an isomorphism. Therefore,

$$\omega_3 = -\top'\omega_0,$$

and finally we obtain the decomposition

$$\omega = \omega_0 \wedge (\widehat{X} \wedge \widehat{Y} - \Omega') + \omega_1 \wedge \widehat{X} + \omega_2 \wedge \widehat{Y}$$

where all forms ω_0, ω_1 and ω_2 are effective on W.

The invariant q_ω provides us with some information about this decomposition. Namely, we find

$$\omega_X = -\omega_0 \wedge \widehat{X} - \omega_2$$

and

$$q_\omega(X) = -\frac{1}{2}\perp_2'(\omega_2 \wedge \omega_2) = -\operatorname{Pf}(\omega_2).$$

Thus, if we denote by $q_\omega(X, Y)$ the symmetric bilinear form associated with q_ω, then

$$q_\omega(X, Y) = -\frac{1}{2}\perp_2'(\omega_1 \wedge \omega_2).$$

Therefore, we arrive at the following result.

Proposition 8.1.1 *The skew-orthogonal decomposition $V = W \oplus \langle X, Y\rangle$ gives rise to the decomposition of an effective 3-form,*

$$\omega = \omega_0 \wedge (\widehat{X} \wedge \widehat{Y} - \Omega') + \omega_1 \wedge \widehat{X} + \omega_2 \wedge \widehat{Y}$$

with effective forms ω_1, ω_2 and ω_0 on W.

Moreover,

$$q_\omega(X) = -\operatorname{Pf}(\omega_2)$$

and

$$q_\omega(X, Y) = -\frac{1}{2}\perp_2'(\omega_1 \wedge \omega_2).$$

8.1.1 The case of trivial invariants

In this section we investigate a structure of effective 3-forms with $q_\omega = 0$.

Lemma 8.1.1 *Let $\omega \in \Lambda^3_\varepsilon(V)$ satisfy $\omega \wedge \omega_X = 0$ for all vectors $X \in V$. Then there exists a non-zero vector v such that $\omega_v = 0$.*

Proof Let $\omega \wedge \omega_X = 0$ for all vectors X. Then also

$$\omega_X \wedge \omega_X = 0,$$

for all vectors X.

Note that this relation implies that $\omega_{X+Y} \wedge \omega_{X+Y} = 0$ and therefore

$$\omega_X \wedge \omega_Y = 0$$

for all vectors X and Y.

We shall consider two cases:

1. there is a pair of vectors X and Y such that $\Omega(X, Y) = 1$ and $Y \in \ker \omega_X$;
2. for all non-zero vectors $X \in V$, $\Omega(X, Y) = 0$ for all $Y \in \ker \omega_X$.

In the latter case one has

$$\omega_X \wedge \widehat{X} = 0$$

for all vectors X.

Let us now take vectors X and Z such that $\Omega(X, Z) = 1$. It follows from the decomposition $\omega = \omega_0 \wedge (\widehat{X} \wedge \widehat{Z} - \Omega') + \omega_1 \wedge \widehat{X} + \omega_2 \wedge \widehat{Z}$ that

$$\omega_X = -\omega_0 \wedge \widehat{X} - \omega_2$$

and

$$\omega_X \wedge \widehat{X} = -\omega_2 \wedge \widehat{X} = 0$$

if and only if $\omega_2 = 0$. For the same reasons we have $\omega_1 = 0$ and finally

$$\omega = \omega_0 \wedge (\widehat{X} \wedge \widehat{Z} - \Omega').$$

Taking $v = X_{\omega_0} \in W$ we obtain

$$\omega_v = \omega_0 \wedge \omega_0 = 0.$$

Now let us consider the first case, and let X, Y be the pair of vectors such that $\Omega(X, Y) = 1$ and $Y \in \ker \omega_X$. Then for the symplectic decomposition $V = W \oplus \langle X, Y \rangle$ we obtain

$$\omega = \omega_1 \wedge \widehat{X} + \omega_2 \wedge \widehat{Y}$$

with effective 2-forms ω_1 and ω_2. It follows from $\omega_Y \wedge \omega_Y = 0$ and $\omega_Y \wedge \omega_Y = 0$ that $\omega_1 \wedge \omega_1 = 0$, and $\omega_2 \wedge \omega_2 = 0$. Moreover, the relation $\omega_X \wedge \omega_Y = 0$ gives $\omega_1 \wedge \omega_2 = 0$.

We shall assume that both $\omega_1 = \omega_Y$ and $\omega_2 = -\omega_X$ are non-zero forms. In the other case the lemma is proved. Since both 2-forms ω_1 and ω_2 are effective then the corresponding planes $\ker \omega_1$ and $\ker \omega_2$ are Lagrangian, and the relation $\omega_1 \wedge \omega_2 = 0$ implies that these Lagrangian planes have a non-trivial intersection. Let $v \in \ker \omega_1 \cap \ker \omega_2$ be a non-zero vector. Then

$$\omega_v = \omega_{1v} \wedge \widehat{X} + \omega_{2v} \wedge \widehat{Y} = 0. \qquad \blacksquare$$

Theorem 8.1.1 *Let $\omega \in \Lambda^3_\varepsilon(V)$ be a non-zero effective 3-form on a six-dimensional symplectic vector space having the trivial invariant q_ω, ν_i. Then* *there exists a canonical basis* $e_1, e_2, e_3, f_1, f_2, f_3$ *such that*

$$\omega = e_1^* \wedge e_2^* \wedge e_3^*.$$

Proof As we have seen, $q_\omega = 0$ implies

$$\Omega \wedge \omega_X \wedge \omega_X = 0,$$

and therefore

$$\Omega \wedge \omega_X \wedge \omega_Y = 0$$

for all vectors X and Y.

Note also that the $\omega \in \Lambda^3_\varepsilon(V)$ means $\omega \wedge \Omega = 0$, and

$$\Omega \wedge \omega_X = -\Omega_X \wedge \omega$$

for all vectors X.

Therefore,

$$\Omega_X \wedge \omega \wedge \omega_Y = 0,$$

for all $X, Y \in V$ and the relation $\omega \wedge \omega_Y = 0$ holds for all $Y \in V$.

Applying the above lemma, we find a non-zero vector X such that $\omega_X = 0$.

Then let $Y \in V$ be such that $\Omega(X, Y) = 1$. In the symplectic decomposition $W \oplus \langle X, Y \rangle$ we obtain

$$\omega = \omega_1 \wedge \widehat{X},$$

where ω_1 is an effective 2-form on W.

Moreover, ω_1 is a parabolic effective 2-form on W, since $q_\omega(Y) = 0$.

From the classification of effective 2-forms on four-dimensional symplectic spaces we conclude that there is a canonical basis (e_1, e_2, f_1, f_2) in W such that

$$\omega_1 = e_1^* \wedge e_2^*.$$

■

8.1.2 The case of non-trivial invariants

In this section we shall assume that $q_\omega \neq 0$.

Lemma 8.1.2 *For any non-zero vector* X *such that* $q_\omega(X) \neq 0$ *the subspace* $\ker \omega_X$ *is a symplectic plane.*

Proof One has $\dim \ker \omega_X = 2$ because ω_X is a degenerate 2-form with $\omega_X^2 \neq 0$. Assume now that $\ker \omega_x$ is an isotropic plane and pick a vector $Z \in \ker \omega_X$ such that $X \wedge Z \neq 0$. Let Y be a vector such that $\Omega(X, Y) = 1$ and $\Omega(X, Z) = 0$.

Let us take the decomposition $\omega = \omega_0 \wedge (\widehat{X} \wedge \widehat{Y} - \Omega') + \omega_1 \wedge \widehat{X} + \omega_2 \wedge \widehat{Y}$ and compute $\omega_Z = 0$.

We obtain

$$\omega_Z = \omega_0(Z)(\widehat{X} \wedge \widehat{Y} - \Omega') + \omega_0 \wedge \widehat{Z} + (\omega_1)_Z \wedge \widehat{X} + (\omega_2)_Z \wedge \widehat{Y} = 0$$

and therefore

$$(\omega_1)_Z = (\omega_2)_Z = 0,$$

$$\omega_0(Z) = 0, \qquad \omega_0 \wedge \widehat{Z} = 0.$$

However, since $q_\omega(X) = -\operatorname{Pf}(\omega_2) \neq 0$, the 2-form ω_2 is non-degenerate on W and then $Z = 0$. ■

Proposition 8.1.2 *Let $\omega \in \Lambda^3_\varepsilon(V)$ be an effective 3-form and let $X, Y \in \ker \omega_X$ be a canonical basis, such that $q_\omega(X) \neq 0$ and $q_\omega(X, Y) = 0$. Then in the decomposition $V = W \oplus \langle X, Y \rangle$, where $\langle X, Y \rangle \stackrel{\text{def}}{=} \operatorname{span}(X, Y)$, the form ω can be written as*

$$\omega = \omega_1 \wedge \widehat{X} + \omega_2 \wedge \widehat{Y},$$

where ω_1 and ω_2 are effective 2-forms on the symplectic space (W, Ω') and $\omega_1 \wedge \omega_2 = 0$.

Proof The decomposition shall take the above form because

$$\omega_0 = (\omega)_{XY} = 0.$$

As we have seen both forms ω_1 and ω_2 are effective and $q_\omega(X) = -\operatorname{Pf}(\omega_2) \neq 0$.

The property $\omega_1 \wedge \omega_2 = 0$ follows from $q_\omega(X, Y) = 0$ in the following way. As we have seen $q_\omega(X, Y) = -\frac{1}{2}\perp_2'(\omega_1 \wedge \omega_2)$ and hence $\perp_2(\omega_1 \wedge \omega_2) = -2q_\omega(X, Y) = 0$. This is equivalent to the fact that

$$\top \circ \perp_2 (\omega_1 \wedge \omega_2) = \perp_2 \circ \top(\omega_1 \wedge \omega_2) = 0$$

because $\top$ and $\perp_2$ commute on $\Lambda^4(V)$ for six-dimensional symplectic vector spaces.

Furthermore, $\perp_2 \colon \Lambda^6(V) \to \Lambda^4(V)$ is a monomorphism and therefore

$$\top(\omega_1 \wedge \omega_2) = \Omega \wedge \omega_1 \wedge \omega_2 = 0.$$

Substituting the vector X into $\Omega \wedge \omega_1 \wedge \omega_2$ we obtain

$$\widehat{X} \wedge \omega_1 \wedge \omega_2 = 0$$

and therefore

$$\omega_1 \wedge \omega_2 = 0. \qquad \blacksquare$$

Theorem 8.1.2 *Any non-zero effective 3-form on a six-dimensional symplectic vector space V can be written in a canonical basis $e_1, e_2, e_3, f_1, f_2, f_3$ in one of the following forms.*

1.

$$\omega = e_1^* \wedge e_2^* \wedge e_3^* + \gamma f_1^* \wedge f_2^* \wedge f_3^*$$

with the invariant

$$q_\omega = \frac{\gamma}{2}(e_1^* f_1^* + e_2^* f_2^* + e_3^* f_3^*)$$

and $\gamma \neq 0$.

2.

$$\omega = f_1^* \wedge e_2^* \wedge e_3^* + f_2^* \wedge e_1^* \wedge e_3^* + + f_3^* \wedge e_1^* \wedge e_2^* + \nu^2 f_1^* \wedge f_2^* \wedge f_3^*$$

with the invariant

$$q_\omega = (e_1^*)^2 - (e_2^*)^2 + (e_3^*)^2 + \nu^2((f_1^*)^2 - (f_2^*)^2 + (f_3^*)^2)$$

and $\nu \neq 0$.

3.

$$\omega = f_1^* \wedge e_2^* \wedge e_3^* - f_2^* \wedge e_1^* \wedge e_3^* + f_3^* \wedge e_1^* \wedge e_2^* - \nu^2 f_1^* \wedge f_2^* \wedge f_3^*$$

with the invariant

$$q_\omega = -(e_1^*)^2 - (e_2^*)^2 - (e_3^*)^2 - \nu^2((f_1^*)^2 + (f_2^*)^2 + (f_3^*)^2)$$

and $\nu \neq 0$

4.

$$\omega = f_1^* \wedge e_2^* \wedge e_3^* + f_2^* \wedge e_1^* \wedge e_3^* + f_3^* \wedge e_1^* \wedge e_2^*$$

with the invariant

$$q_\omega = (e_1^*)^2 - (e_2^*)^2 + (e_3^*)^2.$$

5.

$$\omega = f_1^* \wedge e_2^* \wedge e_3^* - f_2^* \wedge e_1^* \wedge e_3^* + f_3^* \wedge e_1^* \wedge e_2^*$$

with the invariant

$$q_\omega = -(e_1^*)^2 - (e_2^*)^2 - (e_3^*)^2.$$

6.

$$\omega = f_3^* \wedge e_1^* \wedge e_2^* + f_2^* \wedge e_1^* \wedge e_3^*$$

with the invariant

$$q_\omega = (e_1^*)^2.$$

7.

$$\omega = f_3^* \wedge e_1^* \wedge e_2^* - f_2^* \wedge e_1^* \wedge e_3^*$$

with the invariant

$$q_\omega = -(e_1^*)^2.$$

8.

$$\omega = e_1^* \wedge e_2^* \wedge e_3^*$$

with the invariant

$$q_\omega = 0.$$

Proof We shall consider the case of non-trivial invariants only. Let us take a pair of vectors X, Y as in the above proposition and use the decomposition

$$\omega = \omega_1 \wedge \widehat{X} + \omega_2 \wedge \widehat{Y}.$$

Then the problem of finding normal forms for ω shall be reduced to a study of 2-forms $\Omega', \omega_1, \omega_2$ on the symplectic four-dimensional space W.

The forms ω_1, ω_2 are effective and can be hyperbolic, elliptic, parabolic and ω_2 can be trivial. A careful study of all cases allows us to obtain all possible orbits. This study is a little bit long and tiresome, so we will only discuss the details of one case.

Namely, the case when ω_2 and Ω are elliptic in the symplectic vector space (W, ω_1). Then there exists a canonical basis e_1, e_2, f_1, f_2 in (W, ω_1) such that

$$\omega_2 = \lambda(e_1^* \wedge f_2^* + f_1^* \wedge e_2^*)$$

for some $\lambda \neq 0$.

From the relations $\Omega' \wedge \omega_1 = \Omega' \wedge \omega_2 = 0$, we conclude that

$$\Omega' = p f_1^* \wedge f_2^* + q e_1^* \wedge e_2^* + r(e_1^* \wedge f_2^* + e_2^* \wedge f_1^*) + s(e_1^* \wedge f_1^* - e_2^* \wedge f_2^*)$$

where $pq + r^2 + s^2 < 0$, since Ω' is elliptic.

In particular, q cannot be equal to zero.

Let A_t and B_t be transformations of W that depend on the real parameter t and in the basis e_1, e_2, f_1, f_2 has the following forms:

$$A_t = \begin{Vmatrix} 1 & 0 & 0 & t \\ 0 & 1 & t & 0 \\ 0 & 0 & 1 & 0 \\ 0 & 0 & 0 & 1 \end{Vmatrix}, \qquad B_t = \begin{Vmatrix} 1 & 0 & t & 0 \\ 0 & 1 & 0 & -t \\ 0 & 0 & 1 & 0 \\ 0 & 0 & 0 & 1 \end{Vmatrix}.$$

Then A_t and B_t preserve ω_1, ω_2 and act on Ω in the following way:

$$A_t : (p, q, r, s) \xrightarrow{A_t} (p - qt^2 - 2st, q, r, s + qt),$$

$$B_t : (p, q, r, s) \xrightarrow{B_t} (p - qt^2 + 2rt, q, r, r - qt).$$

If we apply the transformation B_u and then the transformation A_v with $u = -r/q$ and $v = -s/q$, we obtain the following 2-forms:

$$\omega_1 = x_1 \wedge y_1 + x_2 \wedge y_2,$$

$$\omega_2 = \lambda(x_1 \wedge y_2 - x_2 \wedge y_1),$$

$$\Omega' = px_1 \wedge x_2 + qy_1 \wedge y_2,$$

where $\lambda \neq 0$ and $pq < 0$.

After applying the following transformation

$$F = \begin{Vmatrix} e^t & 0 & 0 & 0 \\ 0 & e^t & 0 & 0 \\ 0 & 0 & e^{-t} & 0 \\ 0 & 0 & 0 & e^{-t} \end{Vmatrix},$$

with $e^{4t} = -q/p > 0,$ we obtain the following expression for Ω':

$$\Omega' = \mu(x_1 \wedge x_2 - y_1 \wedge y_2)$$

and F does not change ω_1 and ω_2.

In the canonical basis

$$e_1' = X, \quad e_2' = \mu e_1, \quad e_3' = \mu f_1, \quad f_1' = Y, \quad f_2' = e_2, \quad f_3' = -f_2$$

of (V, Ω) one obtains

$$\omega = \frac{1}{\mu^2} f_1'^* \wedge f_2'^* \wedge f_3'^* - f_1'^* \wedge e_2'^* \wedge e_3'^* + \frac{\lambda}{\mu} e_1'^* \wedge f_2'^* \wedge e_3'^* + \frac{\lambda}{\mu} e_1'^* \wedge e_2'^* \wedge f_3'^*$$

and in the canonical basis $e_1, e_2, e_3, f_1, f_2, f_3$, where

$$e_1 = \mu e_1', \qquad e_2 = \mu\nu f_2', \qquad e_3 = \nu f_3',$$

$$f_1 = \frac{1}{\mu} f_1', \quad f_2 = \frac{1}{\mu\nu} f_2', \quad f_3 = -\frac{1}{\nu} e_3',$$

with $\mu/\lambda = \varepsilon\nu^2$, $\varepsilon = \pm 1$, we have

$$\omega = e_2^* \wedge f_1^* \wedge f_3^* - e_3^* \wedge f_1^* \wedge f_2^* - \varepsilon e_1^* \wedge f_2^* \wedge f_3^* + \varepsilon\nu^2 e_1^* \wedge e_2^* \wedge e_3^*. \quad \blacksquare$$

8.1.3 Hitchin's results on the geometry of 3-forms

Here we describe some results of N. Hitchin on the geometry of 3-forms in six-dimensional vector spaces and their relations to the above symplectic geometry of 3-forms.

Let V be a six-dimensional real vector space and let θ be a (fixed) volume form on V. With a given 3-form $\omega \in \Lambda^3(V^*)$ we associate a linear operator $H_\omega : V \to V$ by

$$H_\omega(X) \rfloor \theta \stackrel{\text{def}}{=} \iota_X(\omega) \wedge \omega.$$

The *Hitchin Pfaffian* of a 3-form $\omega \in \Lambda^3(V^*)$ is

$$h(\omega) \stackrel{\text{def}}{=} \frac{1}{6} \operatorname{Tr} H_\omega^2.$$

If $h(\omega)$ is non-zero, then we say that ω is *non-degenerate*.

Proposition 8.1.3 (N. Hitchin [38]) *Let $\omega \in \Lambda^3(V^*)$ be non-degenerate. Then,*

1. $H_\omega^2 = h(\omega)$;
2. $h(\omega) > 0$ *if and only if* $\omega = \alpha + \beta$, *where* α *and* β *are decomposable 3-forms on* V. *Furthermore, if we suppose that*

$$\frac{\alpha \wedge \beta}{\theta} > 0,$$

then α *and* β *are unique and*

$$2\alpha = \omega + |h(\omega)|^{-3/2}(H_\omega)^*(\omega)$$
$$2\beta = \omega - |h(\omega)|^{-3/2}(H_\omega)^*(\omega);$$

3. $h(\omega) < 0$ *if and only if* $\omega = \alpha + \bar{\alpha}$, *where* α *is a complex decomposable 3-form on* V. *Furthermore, if we suppose that*

$$\frac{\alpha \wedge \bar{\alpha}}{\iota\theta} > 0,$$

then α is a unique:

$$2\alpha = \omega + \iota |h(\omega)|^{-3/2} (H_\omega)^* (\omega).$$

Remark 8.1.1 *Let $e_1, \ldots, e_6$ be a base of V and form the volume*

$$\theta = e_1^* \wedge \cdots \wedge e_6^*,$$

then,

1. $h(\omega) > 0$ if and only if ω is in the $\mathrm{GL}(V)$-orbit of

$$e_1^* \wedge e_2^* \wedge e_3^* + e_4^* \wedge e_5^* \wedge e_6^*;$$

2. $h(\omega) < 0$ if and only if ω is in the $\mathrm{GL}(V)$-orbit of

$$(e_1^* + \iota e_4^*) \wedge (e_2^* + \iota e_5^*) \wedge (e_3^* + \iota e_6^*) + (e_1^* - \iota e_4^*) \wedge (e_2^* - \iota e_5^*) \wedge (e_1^* - \iota e_6^*).$$

That is why the $\mathrm{GL}(V)$-action on $\Lambda^3(V^*)$ has two open orbits separated by the quartic hypersurface $h = 0$.

Let ω be a non-degenerate 3-form on V.
The *dual* (by N. Hitchin) form $\hat{\omega}$ is defined as follows:

$$\text{if } \omega = \alpha + \beta, \text{ then } \hat{\omega} \stackrel{\text{def}}{=} \alpha - \beta,$$

$$\text{if } \omega = \alpha + \bar{\alpha}, \text{ then } \hat{\omega} \stackrel{\text{def}}{=} \iota(\bar{\alpha} - \alpha).$$

Hitchin also noted that the space $\Lambda^3(V^*)$ is a symplectic vector space with respect to the skew-form $\Theta \in \Lambda^2(\Lambda^3(V^*))$:

$$\Theta(\omega, \omega') \stackrel{\text{def}}{=} \frac{\omega \wedge \omega'}{\theta}.$$

Proposition 8.1.4 (N. Hitchin [38]) *The action of $\mathrm{SL}(V, \theta)$ on $\Lambda^3(V^*)$ enabled with the symplectic structure Θ is a Hamiltonian with the moment map $H : \Lambda^3(V^*) \to \mathfrak{sl}_6(V)$.*

Suppose now that V is a symplectic vector space with the symplectic form Ω and we fix the volume form in such a way that $\theta = -16\Omega^3$.

The space of all effective forms is a symplectic subspace in $(\Lambda^3(V^*), \Theta)$. It follows immediately from Hodge–Lepage decomposition.

Lemma 8.1.3 *The action of* Sp(V) *on the symplectic linear variety* $(\Lambda^3_\epsilon(V^*), \Theta)$ *is Hamiltonian with momentum map*

$$H : \Lambda^3_\varepsilon(V^*) \to \mathfrak{sp}(V).$$

It is a natural question to compare this invariant of the Sp(V)-action with our quadratic invariant $q_\omega \in S^2(V^*) = \mathfrak{sp}(V)$.

We have the following proposition.

Proposition 8.1.5 (B. Banos [4]) *Let ω be an effective 3-form on V. Then for all $X \in V$*

$$q_\omega(X) = \Omega(H_\omega X, X).$$

Hence, the invariant quadratic form q_ω is a moment map of the Hamiltonian action of Sp(V) on $\Lambda^3_\epsilon(V^*)$):

$$q_\omega : \Lambda^3_\varepsilon(V^*) \to S^2(V^*),$$

and the operator H_ω is the dual to q_ω with respect to the symplectic form Ω.

8.2 The stabilizers of orbits and their prolongations

8.2.1 Stabilizers

In this section we describe the Lie algebras for stabilizers of effective 3-forms. Let ω be an effective 3-form on the six-dimensional symplectic vector space (V, Ω).

We denote by

$$\mathfrak{g}_\omega \overset{\text{def}}{=} \{A \in \mathfrak{sp}(V, \Omega) \mid \omega(AX, Y) + \omega(X, AY) = 0, \quad \forall X, Y \in V\}$$

the stabilizer of ω.

In order to find stabilizers we shall use the normal forms and the canonical bases for effective 3-forms.

In these bases elements of $\mathfrak{g}_\omega$ will have the matrix form

$$\begin{Vmatrix} A & B \\ C & -A^t \end{Vmatrix}$$

with $B^t = B, C^t = C$.

A straightforward computation (using Maple) shows the following.

Theorem 8.2.1

1. *Let* $\omega = e_1^* \wedge e_2^* \wedge e_3^* + \gamma f_1^* \wedge f_2^* \wedge f_3^*$, *then* $\mathfrak{g}_\omega = \mathfrak{sl}(3,\mathbb{R})$, *and elements of* $\mathfrak{g}_\omega$ *have the following matrix form:*

$$\mathfrak{g}_\omega = \left\{ \begin{Vmatrix} B & 0 \\ 0 & -B^t \end{Vmatrix} \right\}$$

with $B \in \mathfrak{sl}(3,\mathbb{R})$.

2. *Let* $\omega = f_1^* \wedge e_2^* \wedge e_3^* + f_2^* \wedge e_1^* \wedge e_3^* + f_3^* \wedge e_1^* \wedge e_2^* + \nu^2 f_1^* \wedge f_2^* \wedge f_3^*$, *then* $\mathfrak{g}_\omega = \mathfrak{su}(2,1)$, *and elements of* $\mathfrak{g}_\omega$ *have the following matrix form:*

$$\mathfrak{g}_\omega = \left\{ \begin{Vmatrix} 0 & \alpha & \beta & \lambda_1 & \xi_1 & \xi_2 \\ \alpha & 0 & \gamma & \xi_1 & \lambda_2 & \xi_3 \\ -\beta & \gamma & 0 & \xi_2 & \xi_3 & \lambda_3 \\ -\nu^2\lambda_1 & \nu^2\xi_1 & -\nu^2\xi_2 & 0 & -\alpha & \beta \\ \nu^2\xi_1 & -\nu^2\lambda_2 & \nu^2\xi_3 & -\alpha & 0 & -\gamma \\ -\nu^2\xi_2 & \nu^2\xi_3 & -\nu^2\lambda_3 & -\beta & -\gamma & 0 \end{Vmatrix} \right\}.$$

3. *Let* $\omega = f_1^* \wedge e_2^* \wedge e_3^* - f_2^* \wedge e_1^* \wedge e_3^* + f_3^* \wedge e_1^* \wedge e_2^* - \nu^2 f_1^* \wedge f_2^* \wedge f_3^*$, *then* $\mathfrak{g}_\omega = \mathfrak{su}(3)$, *and elements of* $\mathfrak{g}_\omega$ *have the following matrix form:*

$$\mathfrak{g}_\omega = \left\{ \begin{Vmatrix} 0 & \alpha & \beta & \lambda_1 & \xi_1 & \xi_2 \\ -\alpha & 0 & \gamma & \xi_1 & \lambda_2 & \xi_3 \\ -\beta & -\gamma & 0 & \xi_2 & \xi_3 & \lambda_3 \\ -\nu^2\lambda_1 & -\nu^2\xi_1 & -\nu^2\xi_2 & 0 & \alpha & \beta \\ -\nu^2\xi_1 & -\nu^2\lambda_2 & -\nu^2\xi_3 & -\alpha & 0 & \gamma \\ -\nu^2\xi_2 & -\nu^2\xi_3 & -\nu^2\lambda_3 & -\beta & -\gamma & 0 \end{Vmatrix} \right\}$$

where $\lambda_1 + \lambda_2 + \lambda_3 = 0$.

4. *If* $\omega = f_1^* \wedge e_2^* \wedge e_3^* + f_2^* \wedge e_1^* \wedge e_3^* + f_3^* \wedge e_1^* \wedge e_2^*$, *then the stabilizer algebra is the semidirect product* $\mathfrak{g}_\omega = \mathfrak{so}(2,1) \times \mathbf{H}_2(2,1)$, *where* $\mathbf{H}_2(2,1)$ *is the space of pseudo-harmonic symmetric 2-tensors and elements of* $\mathfrak{g}_\omega$, *have the following matrix form:*

$$\mathfrak{g}_\omega = \left\{ \begin{Vmatrix} 0 & \alpha & \beta & \lambda_1 & \xi_1 & \xi_2 \\ \alpha & 0 & \gamma & \xi_1 & \lambda_2 & \xi_3 \\ -\beta & \gamma & 0 & \xi_2 & \xi_3 & \lambda_3 \\ 0 & 0 & 0 & 0 & -\alpha & \beta \\ 0 & 0 & 0 & -\alpha & 0 & -\gamma \\ 0 & 0 & 0 & -\beta & -\gamma & 0 \end{Vmatrix} \right\}$$

with $\lambda_1 - \lambda_2 + \lambda_3 = 0$.

5. *Let* $\omega = f_1^* \wedge e_2^* \wedge e_3^* - f_2^* \wedge e_1^* \wedge e_3^* + f_3^* \wedge e_1^* \wedge e_2^*$. *Then the stabilizer algebra is the semidirect product* $\mathfrak{g}_\omega = \mathfrak{so}(3) \rtimes \mathbf{H}_2(3)$, *where* $\mathbf{H}_2(3)$ *is the space of harmonic symmetric 2-tensors, and elements of* $\mathfrak{g}_\omega$ *have the following matrix form:*

$$\mathfrak{g}_\omega = \left\{ \begin{Vmatrix} 0 & \alpha & \beta & \lambda_1 & \xi_1 & \xi_2 \\ -\alpha & 0 & \gamma & \xi_1 & \lambda_2 & \xi_3 \\ -\beta & -\gamma & 0 & \xi_2 & \xi_3 & \lambda_3 \\ 0 & 0 & 0 & 0 & \alpha & \beta \\ 0 & 0 & 0 & -\alpha & 0 & \gamma \\ 0 & 0 & 0 & -\beta & -\gamma & 0 \end{Vmatrix} \right\}$$

with $\lambda_1 + \lambda_2 + \lambda_3 = 0$.

6. *Let* $\omega = f_3^* \wedge e_1^* \wedge e_2^* + f_2^* \wedge e_1^* \wedge e_3^*$. *Then elements of the stabilizer algebra have the following matrix form:*

$$\mathfrak{g}_\omega = \left\{ \begin{Vmatrix} 0 & 0 & 0 & 0 & 0 & 0 \\ \lambda_1 & \mu & \nu & 0 & \alpha_1 & \alpha_2 \\ \lambda_2 & -\nu & \mu & 0 & \alpha_2 & -\alpha_1 \\ \gamma_1 & \gamma_2 & \gamma_3 & 0 & -\lambda_1 & -\lambda_2 \\ \gamma_2 & \gamma_4 & \gamma_5 & 0 & -\mu & \nu \\ \gamma_3 & \gamma_5 & -\gamma_4 & 0 & -\nu & -\nu \end{Vmatrix} \right\}.$$

7. *Let* $\omega = f_3^* \wedge e_1^* \wedge e_2^* - f_2^* \wedge e_1^* \wedge e_3^*$. *Then elements of the stabilizer algebra have the following matrix form:*

$$\mathfrak{g}_\omega = \left\{ \begin{Vmatrix} 0 & 0 & 0 & 0 & 0 & 0 \\ \lambda_1 & \mu & \nu & 0 & \alpha_1 & \alpha_2 \\ \lambda_2 & \nu & \mu & 0 & \alpha_2 & \alpha_1 \\ \gamma_1 & \gamma_2 & \gamma_3 & 0 & -\lambda_1 & -\lambda_2 \\ \gamma_2 & \gamma_4 & \gamma_5 & 0 & -\mu & -\nu \\ \gamma_3 & \gamma_5 & -\gamma_4 & 0 & -\nu & -\nu \end{Vmatrix} \right\}.$$

8. *Let* $\omega = e_1^* \wedge e_2^* \wedge e_3^*$. *Then the stabilizer algebra is the semidirect product* $\mathfrak{g}_\omega = \mathfrak{sl}(3, \mathbb{R}) \rtimes \mathbf{S}^2(3)$, *where* $\mathbf{S}^2(3)$ *is the space of symmetric 2-tensors on three-dimensional vector space, and elements of* $\mathfrak{g}_\omega$ *have the following matrix form:*

$$\mathfrak{g}_\omega = \left\{ \begin{Vmatrix} A & 0 \\ B & -A \end{Vmatrix}, \ A \in \mathfrak{sl}(3, \mathbb{R}),\ B^t = B \right\}.$$

8.2.2 Prolongations

Recall that the prolongation of a linear subspace g of $S^2(V^*)$ is a subspace $g^{(1)}$ of $S^3(V^*)$ such that

$$\theta \in g^{(1)} \Longleftrightarrow v(\theta) \in g$$

for all $v \in V$.

Here we denoted by $v(\theta) \in S^2(V^*)$ the derivative of $\theta \in S^3(V^*)$ along a vector $v \in V$.

We identify the Lie algebra $\mathfrak{g}_\omega$ with a subspace in $S^2(V^*)$ by taking Hamiltonians $A \in \mathfrak{g}_\omega \longmapsto h_A \in S^2(V^*)$, where $h_A(x, y) = \Omega(Ax, y)$.

Theorem 8.2.2 *Let ω be an effective 3-form on a six-dimensional symplectic vector space. Then*

$$\mathfrak{g}_\omega^{(1)} = 0$$

for all ω that are equivalent to one of the normal forms (1)–(5) in Theorem 8.2.1.

Proof To illustrate the proof we consider the first normal form only. If the matrix

$$A = \begin{Vmatrix} B & 0 \\ 0 & -B^t \end{Vmatrix} \in \mathfrak{g}_\omega,$$

then the corresponding quadric form is

$$h_A = Be^* f^*.$$

Assume that $\theta \in S^3(V^*)$ belongs to $\mathfrak{g}_\omega^{(1)}$ and let

$$\theta = \alpha\, f^{*3} + \beta\, f^{*2} e^* + \gamma\, f^* e^{*2} + \delta\, e^{*3}.$$

Then $e(\theta) = \beta\, f^{*2} + 2\gamma\, f^* e^* + 3\delta\, e^{*2} \in \mathfrak{g}_\omega$ implies $\beta = 0$, $\delta = 0$, and $f(\theta) = 3\alpha\, f^{*2} + 2\beta\, f^* e^* + \gamma\, e^{*2} \in \mathfrak{g}_\omega$ implies $\alpha = 0$, $\gamma = 0$. Therefore, $\theta = 0$. ■

PART III

Monge–Ampère Equations

9

Symplectic manifolds

9.1 Symplectic structures

A smooth manifold M is called *symplectic*, if there is a non-degenerate closed differential 2-form defined on M.

That is, there is a differential 2-form $\Omega \in \Omega^2(M)$ such that $d\Omega = 0$, and $\Omega_a \in \Lambda^2(T_a^*M)$ is a symplectic structure on T_aM for any point $a \in M$.

A smooth map $\varphi : (M_1, \Omega_1) \to (M_2, \Omega_2)$ of symplectic manifolds is called *symplectic* if

$$\varphi^*(\Omega_2) = \Omega_1.$$

Symplectic diffeomorphisms are also called *symplectomorphisms*, and, in the case where $M_1 = M_2$, *symplectic or canonical transformations.*

Theorem 9.1.1 (Darboux) *Let (M_1, Ω_1) and (M_2, Ω_2) be symplectic manifolds of the same dimension. Then for any points $a \in M_1$ and $b \in M_2$ there are neighborhoods $\mathcal{O}_1 \ni a, \mathcal{O}_2 \ni b$ and a diffeomorphism $\varphi : \mathcal{O}_1 \to \mathcal{O}_2$ such that $\varphi(a) = b$ and $\varphi^*(\Omega_2) = \Omega_1$ in $\mathcal{O}_1$.*

Proof Without loss of generality one can assume that two symplectic forms, say Ω_0 and Ω_1, are given in a neighborhood of a point $a \in M$, where in addition $\Omega_{0,a} = \Omega_{1,a}$. Let us take a path $\Omega_t = (1-t)\Omega_0 + t\Omega_1$ connecting Ω_0 and Ω_1, and find a path ϕ_t in local diffeomorphisms of M, such that

$$\phi_t^*(\Omega_t) = \Omega_0 \tag{9.1}$$

and $\phi_0 = 1,\ \phi_t(a) = a$.

Note that forms Ω_t are symplectic in a neighborhood of a, and

$$\frac{d\Omega_t}{dt} = \Omega_1 - \Omega_0 = d\theta$$

in a smaller neighborhood.

Denote by X_t the vector field corresponding to ϕ_t, that is $\phi^*_{t+\varepsilon} = \phi^*_t(1 + \varepsilon X_t + o(\varepsilon))$. Then (9.1) is equivalent to

$$\frac{d\phi^*_t(\Omega_t)}{dt} = 0$$

or

$$L_{X_t}(\Omega_t) + \frac{d\Omega_t}{dt} = 0.$$

The last equation can be rewritten as

$$d\iota_{X_t}\Omega_t + d\theta = 0.$$

Therefore the vector field should be choosen such, that

$$\iota_{X_t}\Omega_t + \theta = dS$$

for some function S.

Let S be a function such that $\theta_a = d_a S$, then we obtain a uniquie vector field X_t satisfying the equation and $X_{t,a} = 0$. ■

Corollary 9.1.1 *Let (M, Ω) be a $2n$-dimensional symplectic manifold. Then for any point $a \in M$ there are local canonical coordinates $(q_1, \ldots, q_n, p_1, \ldots, p_n)$ such that $q_i(a) = p_i(a) = 0$ $(i = 1, \ldots, n)$ and Ω has the following* canonical form*:*

$$\omega = dq_1 \wedge dp_1 + \cdots + dq_n \wedge dp_n.$$

9.1.1 The cotangent bundle and the standard symplectic structure

Let $\tau^* : T^*B \to B$ be the cotangent bundle of a manifold B. A differential 1-form ρ on T^*B defined by

$$\rho_a(X_a) \stackrel{\text{def}}{=} a(\tau_* X_a),$$

for any covector $a \in T^*B$ and any vector $X_a \in T_a(T^*B)$, is called *universal* or a *Liouville form*.

If $q_1, \ldots, q_n$ are local coordinates in B and $(q, p) = (q_1, \ldots, q_n, p_1, \ldots, p_n)$ are the induced local coordinates on T^*B, then

$$\rho = \sum_{i=1}^{n} p_i \, dq_i.$$

The 2-form

$$\Omega = -d\rho,$$

or in local coordinates (q, p)

$$\Omega = \sum_{i=1}^{n} dq_i \wedge dp_i,$$

is closed and non-degenerate, and therefore defines a symplectic structure on T^*M.

For $B = \mathbb{R}^n$ this 2-form Ω defines the so-called *standard symplectic structure* on the arithmetic space $\mathbb{R}^{2n} = T^*\mathbb{R}^n$.

One can use the standard structure as a model. Thus, due to the Darboux theorem, for any symplectic manifold M, and any $a \in M$, there are local coordinates $(x_1, \ldots, x_n, y_1, \ldots, y_n)$ called *canonical* such that

$$\Omega = \sum_{i=1}^{n} dx_i \wedge dy_i$$

in some neighborhood of a.

Example 9.1.1 *Any diffeomorphism $f : B \to B$ can be lifted to automorphism $f^{(1)} : T^*B \to T^*B$ of the cotangent bundle as*

$$f^{(1)}|_{T_a^*B} = (f^*)^{(-1)} : T_a^*B \to T_{f(a)}^*B$$

and because of the property

$$(f^{(1)})^*\rho = \rho$$

this automorphism is a symplectic transformation.

Example 9.1.2 *Any smooth function $h \in C^\infty(B)$ determines a shift*

$$S_h : T^*B \to T^*B$$

by

$$S_h : (x, \lambda) \longmapsto (x, \lambda + d_x h)$$

for all $x \in B, \lambda \in T_x^*B$.

Then

$$S_h^*(\rho) = \rho + d\tau^*(h)$$

and therefore S_h is symplectic.

9.1.2 Kähler manifolds

A *Kähler manifold* is a complex manifold equipped with a compatible symplectic structure.

That is, at each point $a \in M$ of the Kähler manifold M the tangent space T_aM has the complex structure $I_a : T_aM \to T_aM$, induced by the complex structure on M, and the symplectic structure Ω_a, induced by the symplectic form $\Omega \in \Omega^2(M)$, and they are compatible in the sense that

$$g_a(x, y) = \Omega_a(I_a x, y)$$

defines a Riemannian structure on M.

Locally Kähler structure has a potential, called the *Kähler potential*, i.e. a smooth function K, such that

$$\Omega = \frac{i}{2}\partial\bar{\partial}K = \frac{i}{2}\sum_{s,t}\frac{\partial^2 K}{\partial z_s \partial \bar{z}_t} dz_s \wedge d\bar{z}_t.$$

The potential functions are strictly plurisubharmonic in the sense that the matrix

$$\left\| \frac{\partial^2 K}{\partial z_l \partial \bar{z}_s} \right\|$$

is positive definite in any local complex coordinates.

Moreover, any such function defines (locally) Kähler structure.

The advantage of using Kähler manifolds is based on the fact that complex submanifolds of Kähler manifolds are also Kähler.

Example 9.1.3 *A complex vector space $\mathbb{C}^n$ is Kähler with*

$$\Omega = \frac{i}{2}\sum_{k=1}^{n} dz_k \wedge d\bar{z}_k.$$

Therefore every complex submanifold of $\mathbb{C}^n$ is Kähler.

Example 9.1.4 *The potential*

$$K = \log(1 + |z|^2)$$

defines Kähler structure by the Fubini–Study form

$$\Omega = \frac{\iota}{2} \partial \bar{\partial} K$$

on $\mathbb{C}^n$.

Let $\mathbb{C}\mathbf{P}^n$ *be the complex projective space and one use the standard affine charts*

$$[z_0 : \cdots : z_n] \rightarrow \left(\frac{z_0}{z_k}, \ldots, \frac{z_{k-1}}{z_k}, \frac{z_{k+1}}{z_k}, \ldots, \frac{z_n}{z_k} \right)$$

then the Fubini–Study forms defines a Kähler structure on $\mathbb{C}\mathbf{P}^n$.

Example 9.1.5 *Smooth projective manifolds are Kähler.*

9.1.3 Orbits and homogeneous symplectic spaces

Let (M, Ω) be a symplectic manifold and let G be a Lie group that acts on the manifold

$$G \times M \rightarrow M,$$

$$g : x \longmapsto gx$$

by symplectomorphisms

$$g^*(\Omega) = \Omega,$$

for all $g \in G$.

The symplectic space is called *homogeneous* if the action is transitive.

We recall the construction of homogeneous symplectic spaces by A. Kirillov [47].

Any Lie group G acts on itself by conjugations:

$$\mathrm{Ad}_g : G \rightarrow G,$$

$$\mathrm{Ad}_g : x \longmapsto gxg^{-1}.$$

Let us identify the Lie algebra $\mathfrak{g}$ of G with the tangent space T_eG at unit $e \in G$, and let

$$\mathrm{Ad}_* : G \to \mathrm{GL}(\mathfrak{g})$$

be the *adjoint representation* given by the differentials of conjugations:

$$\mathrm{Ad}_*(g) = (\mathrm{Ad}_g)_{*,e} : T_eG \to T_eG.$$

The corresponding Lie algebra representation is the adjoint representation

$$\mathrm{ad} : \mathfrak{g} \to \mathrm{End}\, \mathfrak{g},$$

$$\mathrm{ad}_X : Y \longmapsto [X, Y].$$

Denote by Ad^* and ad^* the corresponding dual and *coadjoint representations:*

$$\mathrm{Ad}^* : G \to \mathrm{GL}(\mathfrak{g}^*),$$

$$\mathrm{Ad}^*(g) = (\mathrm{Ad}_{g^{-1}})_{*,e} : T_e^*G \to T_e^*G,$$

and

$$\mathrm{ad}^* : \mathfrak{g} \to \mathrm{End}\, \mathfrak{g}^*,$$

$$\langle \mathrm{ad}_X^*\, \alpha, Y \rangle = \langle \alpha, [Y, X] \rangle$$

where $X, Y \in \mathfrak{g}$, $\alpha \in \mathfrak{g}^*$. Then any vector $X \in \mathfrak{g}$ generates a vector field $\hat{X}$ on $\mathfrak{g}^*$ corresponding to the one-parameter group $\exp(t\, \mathrm{ad}_X^*) : \mathfrak{g}^* \to \mathfrak{g}^*$. The value $\hat{X}_\alpha$ of $\hat{X}$ at the point $\alpha \in \mathfrak{g}^*$ can be identified with a vector in $\mathfrak{g}^*$ and

$$\langle \hat{X}_\alpha, Y \rangle = \langle \alpha, [Y, X] \rangle.$$

Let us define

$$\Omega_\alpha(\hat{X}_\alpha, \hat{Y}_\alpha) = \langle \alpha, [Y, X] \rangle.$$

One can easily check that Ω is determined a closed and non-degenerated 2-form on orbits of coadjoint representation.

This structure is called the *Konstant–Kirillov symplectic structure.*

Example 9.1.6 *Let* $G = \mathrm{SU}(n+1)$ *be the special unitary group. Then the Lie algebra* $\mathfrak{g} = \mathfrak{su}(n+1)$ *is the Lie algebra of skew-Hermitian traceless* $(n+1) \times (n+1)$ *matrices. The Killing form*

$$\kappa(X, Y) = \mathrm{Re}(\mathrm{tr}(X^t \bar{Y}))$$

identifies $\mathfrak{g}$ *with* $\mathfrak{g}^*$ *by the isomorphism* $X \to \kappa(X, \cdot)$ *and the (co)adjoint action is the conjugation* $g(X) = gXg^{-1}$*. Therefore, eigenvalues describe the orbits. Let* X_0 *be a traceless Hermitian matrix with eigenvalues* $\lambda_1 = \cdots = \lambda_n = 1/n$*, and* $\lambda_{n+1} = -1$*. Then the Konstant–Kirillov form defines the symplectic structure on the orbit* $M_{X_0} = \mathrm{SU}(n+1)/\mathrm{U}(n) \simeq \mathbb{C}\mathbf{P}^n$.

The Konstant–Kirillov structures describe practically all homogeneous symplectic spaces.

Theorem 9.1.2 (Kostant–Souriau) *Let G be a Lie group and* $\mathfrak{g}$ *be the Lie algebra of G. Then for Lie algebras with trivial first and second cohomology groups,*

$$H^1(\mathfrak{g}) = H^2(\mathfrak{g}) = 0,$$

there is, up to covering, a one-to-one correspondence between homogeneous symplectic spaces with structure group G and coadjoint orbits.

9.2 Vector fields on symplectic manifolds

9.2.1 Poisson bracket and Hamiltonian vector fields

Let (M, Ω) be a symplectic manifold and $X \in D(M)$ be a vector field on M. We say that X is *canonical* if the flow A_t corresponding to X consists of symplectic diffeomorphisms.

Proposition 9.2.1 *The following conditions are equivalent:*

(*a*) *X is a canonical vector field;*
(*b*) $L_X(\Omega) = 0$;
(*c*) $\iota_X(\Omega)$ *is closed.*

Proof

(a) $\Longleftrightarrow$ (b) If X is canonical and A_t is the corresponding flow, then

$$A_t^*(\Omega) = \Omega$$

and therefore

$$L_X(\Omega) = \frac{d}{dt}\bigg|_{t=0} A_t^*(\Omega) = 0.$$

On the other hand, if $L_X(\Omega) = 0$, then

$$\frac{d}{dt}\bigg|_{t=s} A_t^*(\Omega) = A_s^*(L_X\Omega) = 0$$

and therefore

$$A_t^*(\Omega) = A_0^*(\Omega) = \Omega.$$

(b) $\Longleftrightarrow$ (c) It follows from the formula for the Lie derivative:

$$L_X(\Omega) = (\iota_X \circ d + d \circ \iota_X)(\Omega) = d(\iota_X\Omega). \qquad \blacksquare$$

Let $\alpha \in \Omega^1(M)$ be a differential 1-form and let $X_\alpha \in D(M)$ be the vector field corresponding to α under isomorphism

$$\Gamma : D(M) \to \Omega^1(M),$$

$$\Gamma : X \longmapsto \iota_X(\Omega).$$

That is,

$$X_\alpha \rfloor \Omega = \alpha.$$

Moreover,

$$L_{X_\alpha}(\Omega) = d(X_\alpha \rfloor \Omega) = d\alpha.$$

Define the *Poisson bracket* $[\alpha, \beta]$ of differential 1-forms α and β as follows:

$$X_{[\alpha,\beta]} = [X_\alpha, X_\beta].$$

Theorem 9.2.1 *The Poisson bracket determines a Lie algebra structure on* $\Omega^1(M)$ *and*

$$[\alpha, \beta] = X_\alpha \rfloor d\beta - X_\beta \rfloor d\alpha + d\Omega(X_\beta, X_\alpha).$$

Proof Indeed,

$$\begin{aligned} \iota_{[X_\alpha, X_\beta]}(\Omega) &= [L_{X_\alpha}, \iota_{X_\beta}](\Omega) = L_{X_\alpha}(\beta) - X_\beta \rfloor d\alpha \\ &= X_\alpha \rfloor d\beta + d(\beta(X_\alpha)) - X_\beta \rfloor d\alpha. \end{aligned}$$

■

We have seen that canonical vector fields correspond to closed 1-forms, and the above formula shows that the Poisson bracket of closed 1-forms is an exact 1-form.

Canonical vector fields that correspond to exact forms are called *Hamiltonian*.

If $\alpha = df$ then we will denote the Hamiltonian vector field X_α by X_f and call the function f the *Hamiltonian* of X_f.

The above formula for the Poisson bracket gives us

$$[df, dg] = d\Omega(X_g, X_f).$$

The function $\Omega(X_g, X_f)$ is called a *Poisson bracket* of functions f and g and is denoted by

$$\{f, g\} \stackrel{\text{def}}{=} \Omega(X_g, X_f).$$

In this notation the formula for Poisson brackets reads as

$$[X_f, X_g] = X_{\{f,g\}}.$$

Note also that

$$\{f, g\} = \Omega(X_g, X_f) = dg(X_f) = X_f(g).$$

The Poisson bracket defines a Lie algebra structure on the space of functions $C^\infty(M)$.

9.2.2 Canonical coordinates

Let $(x_1, \ldots, x_n, y_1, \ldots, y_n)$ be canonical coordinates for Ω, and let

$$\alpha = a_1(x, y)\, dx_1 + \cdots + a_n(x, y)\, dx_n + \widetilde{a}_1(x, y)\, dy_1 + \cdots + \widetilde{a}_n(x, y)\, dy_n$$

in these coordinates.

Then,

$$X_\alpha = \widetilde{a}_1(x,y)\frac{\partial}{\partial x_1} + \cdots + \widetilde{a}_n(x,y)\frac{\partial}{\partial x_n} - a_1(x,y)\frac{\partial}{\partial y_1} - \cdots - a_n(x,y)\frac{\partial}{\partial y_n}$$

and

$$X_f = \frac{\partial f}{\partial y_1}\frac{\partial}{\partial x_1} + \cdots + \frac{\partial f}{\partial y_n}\frac{\partial}{\partial x_n} - \frac{\partial f}{\partial x_1}\frac{\partial}{\partial y_1} - \cdots - \frac{\partial f}{\partial x_n}\frac{\partial}{\partial y_n}.$$

The Poisson bracket in the canonical coordinates has the form

$$\{f,g\} = \left(\frac{\partial f}{\partial y_1}\frac{\partial g}{\partial x_1} - \frac{\partial f}{\partial x_1}\frac{\partial g}{\partial y_1}\right) + \cdots + \left(\frac{\partial f}{\partial y_n}\frac{\partial g}{\partial x_n} - \frac{\partial f}{\partial x_n}\frac{\partial g}{\partial y_n}\right).$$

Example 9.2.1 *Let $M = T^2 = \mathbb{R}^2/\mathbb{Z}^2$ be the two-dimensional torus with the coordinates x and y, and let $\Omega = dx \wedge dy$ be the symplectic structure on T^2. Then*

$$X_\alpha = \widetilde{a}(x,y)\frac{\partial}{\partial x} - a(x,y)\frac{\partial}{\partial y}$$

corresponds to $\alpha = a(x,y)\,dx + \widetilde{a}(x,y)\,dy$.

Differential forms $c_1\,dx + c_2\,dy$ with constant coefficients are closed but not exact. Therefore, the vector fields

$$c_2\frac{\partial}{\partial x} - c_1\frac{\partial}{\partial y}$$

are canonical, but not Hamiltonian.

9.3 Submanifolds of symplectic manifolds

9.3.1 Presymplectic manifolds

Let M be a manifold and let ω be a differential 2-form on M.

At each point $a \in M$ one has a subspace

$$\ker \omega_a = \{X \in T_aM \mid X \rfloor \omega_a = 0\}.$$

Assume that $a \longmapsto \dim \ker \omega_a$ is a locally constant function, that is, the 2-form has a constant rank.

Then the distribution

$$\ker \omega : a \in M \longmapsto \ker \omega_a$$

is called the *characteristic distribution of* ω.

Lemma 9.3.1 *The characteristic distribution* $\ker \omega$ *is completely integrable if*

$$d\omega = 0.$$

Proof Let vector fields X and Y belong to $\ker \omega$. Then

$$L_X \omega = \iota_X \, d\omega + d\iota_X \, \omega = 0$$

and

$$\iota_{[X,Y]}\omega = [L_X, \iota_Y](\omega) = 0. \qquad \blacksquare$$

A *presymplectic structure* on the manifold M is a closed differential 2-form $\omega \in \Omega^2(M)$ of constant rank.

For presymplectic structures $\ker \omega$ is completely integrable. Let us denote by $S = M/\ker \omega$ the set of all maximal integral manifolds of the distribution. Assume that there is a smooth manifold structure on S such that the projection $\pi : M \to S$ is a smooth bundle. In this case vertical vector fields v, that is vector fields tangent to fibres of π, are characteristics and therefore satisfy the property

$$\iota_v \omega = 0, \quad L_v \omega = 0,$$

and thus the 2-form ω defines a 2-form $\widetilde{\omega}$ on the base S as

$$\widetilde{\omega}_s(X_s, Y_s) = \omega_a(\bar{X}_a, \bar{Y}_a)$$

where $s \in S, a \in M, \pi(a) = s$, and $\bar{X}_a, \bar{Y}_a \in T_a M$ are arbitrary lifts of $X_s, Y_s \in T_s S$:

$$\pi_*(\bar{X}_a) = X_s, \quad \pi_*(\bar{Y}_a) = Y_s.$$

Moreover, the form $\widetilde{\omega}$ is closed and non-degenerate also.

Indeed, the non-degeneracy of $\widetilde{\omega}$ follows from the construction and if X, Y and Z are vector fields on S and $\bar{X}, \bar{Y}$ and $\bar{Z}$ are their lifts on M, then

$$\begin{aligned} d\widetilde{\omega}(X,Y,Z) &= X(\widetilde{\omega}(Y,Z)) - Y(\widetilde{\omega}(X,Z)) + Z(\widetilde{\omega}(X,Y)) \\ &\quad - \widetilde{\omega}([X,Y],Z) + \widetilde{\omega}([X,Z],Y) - \widetilde{\omega}([Y,Z],X) \\ &= \bar{X}(\omega(\bar{Y},\bar{Z})) - \bar{Y}(\omega(\bar{X},\bar{Z})) + \bar{Z}(\omega(\bar{X},\bar{Y})) \\ &\quad - \omega([\bar{X},\bar{Y}],\bar{Z}) + \omega([\bar{X},\bar{Z}],\bar{Y}) - \omega([\bar{Y},\bar{Z}],\bar{X}) \\ &= d\omega(\bar{X},\bar{Y},\bar{Z}) = 0. \end{aligned}$$

The symplectic manifold $(S, \widetilde{\omega})$ is called *reduction* of the presymplectic manifold (M, ω).

One particular case of the reduction is very important for applications. Namely, let us assume that $\Gamma \subset M$ is a section of π; that is, a submanifold such that $\pi : \Gamma \to S$ is a diffeomorphism, then the symplectic manifold $(S, \widetilde{\omega})$ is equivalent to the symplectic manifold $(\Gamma, \omega|_\Gamma)$.

9.3.2 Lagrangian submanifolds

We say that a submanifold L of a symplectic manifold (M, Ω) is *Lagrangian* if the tangent space T_aL is a Lagrangian subspace of the symplectic vector space (T_aM, Ω_a) at every point $a \in L$.

Example 9.3.1 *Let $(T^*B, \Omega = d\rho)$ be the phase space with the standard symplectic structure. Any differential 1-form $\alpha \in \Omega^1(B)$ can be viewed as the section*

$$S_\alpha : B \to T^*B,$$

$$S_\alpha : x \in B \longmapsto \alpha_x \in T_x^*B$$

of the cotangent bundle $\tau^* : T^*B \to B$.

Moreover, by the definition of ρ, one gets

$$S_\alpha^*(\rho) = \alpha.$$

Namely this property defines ρ uniquely and for this reason ρ is called the universal 1-form.

Denote by $B_\alpha = S_\alpha(B) \subset T^*B$ the graph of α. Then the universality property reads as

$$\rho|_{B_\alpha} = \alpha,$$

and B_α is a Lagrangian submanifold if and only if α is closed.

If this form is exact, $\alpha = df$, for some function $f \in C^\infty(B)$, then we denote B_{df} by L_f and call f the generating function of L_f.

Note that a Lagrangian submanifold $L \subset T^*B$ has the form B_α or L_f if and only if: (1) $\tau^* : L \to B$ is a diffeomorphism and (2) $\rho|_L$ is closed or exact respectively.

In the canonical coordinates $(q_1, \ldots, q_n, p_1, \ldots, p_n)$ in T^*B the Lagrangian submanifolds L_f have the form

$$p = \frac{\partial f}{\partial q}.$$

Example 9.3.2 *Let $B_0 \subset B$ be a submanifold. Define $\hat{B}_0 \subset T^*B$ as follows:*

$$\hat{B}_0 = \bigcup_{x \in B_0} \operatorname{Ann} T_x B_0.$$

Then $\rho|_{\hat{B}_0} = 0$ and $\dim \hat{B}_0 = \dim B$. Therefore $\hat{B}_0$ is Lagrangian.

Let $L \subset M$ be a Lagrangian submanifold, and let $(x_1, \ldots, x_n, y_1, \ldots, y_n)$ be canonical coordinates in a neighborhood $\mathcal{O}$ of $a \in L$.

If $\mathcal{O}$ is sufficiently small, there is a subset $I = \{i_1, \ldots, i_k\} \subset \{1, \ldots, n\}$ such that $x_I = (x_{i_1}, \ldots, x_{i_k})$ and $y_{I^c} = (y_{j_1}, \ldots, y_{j_{n-k}})$, where $j_s \notin I$ are local coordinates in $L \cap \mathcal{O}$.

Let us write the structure form Ω as follows:

$$\Omega = dx \wedge dy = dx_I \wedge dy_I + d(-y_{I^c}) \wedge dx_{I^c}.$$

Assume that $L \cap \mathcal{O}$ is simply connected. Then due to the above example $L \cap \mathcal{O} = L_f$ for some function

$$f = f(x_I, y_{I^c})$$

and

$$y_I = \frac{\partial f}{\partial x_I},$$

$$x_{I^c} = -\frac{\partial f}{\partial y_{I^c}}$$

on L.

The function $f(x_I, y_{I^c})$ is called the *generating function* for the Lagrangian manifold L.

The above description of Lagrangian submanifolds can be used for a local description of canonical transformations. This is based on the following observation.

Let $\phi : (M_1, \Omega_1) \to (M_2, \Omega_2)$ be a diffeomorphism, and let $G_\phi \subset M_1 \times M_2$ be the graph of ϕ. Define a differential 2-form on $M = M_1 \times M_2$ as

$$\Omega = \pi_1^* \Omega_1 - \pi_2^* \Omega_2,$$

where $\pi_i : M \to M_i,\ i = 1, 2$ are the projections.

Then Ω defines the symplectic structure on M and

$$\Omega|_{G_\phi} = \Omega_1 - \phi^*(\Omega_2).$$

Proposition 9.3.1 *A diffeomorphism $\phi : M_1 \to M_2$ is symplectic if and only if the graph G_ϕ is a Lagrangian submanifold.*

The local description of Lagrangian submanifolds gives us the following description of symplectic diffeomorphisms in terms of generating functions.

Let (x, y) be canonical coordinates on M_1, (u, v) be canonical coordinates on M_2 and let

$$\phi : (x, y) \to (u = U(x, y), v = V(x, y))$$

be a local symplectic diffeomorphism.

Then

$$\Omega = dx \wedge dy - du \wedge dv$$

and therefore canonical coordinates on G_ϕ can be taken as follows:

$$x_I, u_J, y_{I^c}, v_{J^c}$$

where $\#I + \#J = n$.

Let

$$f = f(x_I, u_J, y_{I^c}, v_{J^c})$$

be the generating function for G_ϕ. Then the equations

$$y_I = \frac{\partial f}{\partial x_I}, \qquad x_{I^c} = -\frac{\partial f}{\partial y_{I^c}},$$
$$v_J = -\frac{\partial f}{\partial u_J}, \quad u_{J^c} = \frac{\partial f}{\partial v_{J^c}}$$

determine ϕ.

To find ϕ one should solve this system with respect to $u = U(x, y)$ and $v = V(x, y)$.

Example 9.3.3 *Let $M = (T^*B, d\rho)$, and let (p, q) be the canonical coordinates, $\rho = p\,dq$. Taking $f = f(q, q')$ we obtain the following symplectic transformations:*

$$p = \frac{\partial f}{\partial q}, \quad p' = -\frac{\partial f}{\partial q'}.$$

General symplectic diffeomorphisms can be obtained from generating functions of the form

$$f = f(q_I, p_{I^c}, q'_J, p'_{J^c})$$

as follows:

$$p_I = \frac{\partial f}{\partial q_I}, \quad q_{I^c} = -\frac{\partial f}{\partial p_{I^c}},$$

$$p'_J = \frac{\partial f}{\partial q'_J}, \quad q'_{J^c} = -\frac{\partial f}{\partial p'_{J^c}}.$$

9.3.3 Involutive submanifolds

A submanifold E of a symplectic manifold (M, Ω) is said to be *involutive* if the tangent space T_aE is an involutive subspace of the symplectic vector space (T_aM, Ω_a) at every point $a \in L$.

Note that the restriction $\Omega_E = \Omega|_E$ of the structure form has constant rank because

$$\dim \ker \Omega_E = \operatorname{codim} E,$$

and therefore an involutive submanifold E determines a presymplectic structure (E, Ω_E).

The characteristic distribution of Ω_E coincides with the distribution of skew-orthogonal complements:

$$\operatorname{char}_E : a \in E \longmapsto (\ker \Omega_E)(a) = (T_aE)^\perp$$

and we obtain the following property of involutive submanifolds.

Theorem 9.3.1 *Let E be an involutive submanifold of a symplectic manifold (M, Ω). Then,*

1. *the characteristic distribution* char_E *is completely integrable, and*
2. *any Lagrangian submanifold $L \subset M$ such that $L \subset E$ consists of leaves of* char_E*; that is,*

$$T_a L \supset \operatorname{char}_E(a)$$

for any $a \in L$.
Moreover, if $\pi : E \to E/\operatorname{char}_E$ is a smooth bundle then $\pi(L)$ is a Lagrangian submanifold of the symplectic manifold E/char_E, and for any Lagrangian submanifold $L' \subset E/\operatorname{char}_E$ the preimage $L = \pi^{-1}(L') \subset E$ is a Lagrangian submanifold of M.

Example 9.3.4 *Any submanifold E of codimension 1 is involutive. Let $E = f^{-1}(0)$ for some function $f \in C^\infty(M)$, and $d_a f$ does not vanish at points $a \in E$. Then* char_E *generates by the Hamiltonian vector field X_f, and therefore any Lagrangian submanifold $L \subset E$ is invariant with respect to X_f.*

9.3.4 Lagrangian polarizations

A *Lagrangian polarization* on a symplectic manifold (M, Ω) is a completely integrable distribution P such that $P_a \subset T_a M$ is a Lagrangian subspace for each $a \in M$.

Example 9.3.5 (Standard polarizations) *Let $(M = T^*B, d\rho)$ be the symplectic manifold. Then fibres of the projection $\tau^* : T^*B \to B$ are Lagrangian because $\rho = 0$ on the fibres, and therefore define a Lagrangian polarization on T^*B.*

Theorem 9.3.2 *Let P be a Lagrangian polarization. Then*

1. $X_f \in D(P)$,
2. $\{f, g\} = 0$,

   for any first integrals f and g of the distribution.

Proof Let f be a first integral of P. Then $\ker d_a f$ is an involutive subspace and $P_a \subset \ker d_a f$. Therefore, $(\ker d_a f)^\perp \subset P_a^\perp = P_a$. But $(\ker d_a f)^\perp = \mathbb{R} \cdot X_{f,a}$ and therefore

$$X_{f,a} \in P_a$$

for all $a \in M$.

If g is a first integral too, then $\{f, g\} = X_f(g) = dg(X_f) = 0$ because $X_f \in D(P)$ and $dg = 0$ on P. ∎

The following theorem shows that Lagrangian polarizations are locally equivalent.

Theorem 9.3.3 *Let P and P' be Lagrangian polarizations on a connected symplectic manifold (M, Ω). Then for any points a, a' in M there are neighborhoods $\mathcal{O} \ni a$ and $\mathcal{O}' \ni a'$ and a local symplectomorphism $\phi : \mathcal{O} \to \mathcal{O}'$ such that $\phi(a) = a'$ and $\phi_*(P) = P'$.*

Proof To prove this theorem we show a method of finding canonical coordinates in some neighborhood of a such that P will coincide with a standard polarization. Let $\dim M = 2n$, and let us take n independent first integrals, say $f_1, \ldots, f_n$, for P in a neighborhood of a. The corresponding Hamiltonian vector fields $X_{f_1}, \ldots, X_{f_n}$ commute due to the above theorem. Therefore, we can find a function g_1 such that

$$X_{f_1}(g_1) = 1, \quad X_{f_2}(g_1) = 0, \quad \ldots, \quad X_{f_n}(g_1) = 0$$

in a neighborhood of a.

Then Hamiltonian vector fields $X_{f_1}, \ldots, X_{f_n}, X_{g_1}$ commute because

$$[X_{f_i}, X_{g_1}] = X_{\{f_i, g_1\}}$$

and $\{f_i, g_1\} = \text{constant}$.

Note also that $df_1, \ldots, df_n, dg_1$ are linear independent because $dg_1(X_{f_1}) = 1$.

In the same way we can find a function g_2 such that

$$X_{f_1}(g_2) = 0, \; X_{f_2}(g_2) = 1, \; X_{f_3}(g_2) = 0, \; \ldots, \; X_{f_n}(g_2) = 0, \quad X_{g_1}(g_2) = 0$$

in a neighborhood of a.

For the same reasons Hamiltonian vector fields $X_{f_1}, \ldots, X_{f_n}, X_{g_1}, X_{g_2}$ commute and $df_1, \ldots, df_n, dg_1, dg_2$ are linear independent.

Finally, we construct n functions $g_1, \ldots, g_n$ such that $df_1, \ldots, df_n$, $dg_1, \ldots, dg_n$ are linear independent and

$$\{f_i, g_j\} = \delta_{ij} \text{ but } \{g_i, g_j\} = 0.$$

These $2n$ functions provide local canonical coordinates for Ω and integral manifolds of P are

$$f_1 = c_1, \quad \ldots, \quad f_n = c_n.$$

∎

Remark 9.3.1 *The above method is not applicable in practice. It is much better to use the method of Section 2.1.3. Namely, let us choose functions $h_1, \dots, h_n$ in a domain D such that functions $f_1, \dots, f_n, h_1, \dots, h_n$ are independent in D, and form the matrix*

$$W = \begin{Vmatrix} \{f_1, h_1\} & \{f_2, h_1\} & \cdots & \{f_n, h_1\} \\ \{f_1, h_2\} & \{f_2, h_2\} & \cdots & \{f_n, h_2\} \\ \vdots & \vdots & \vdots & \vdots \\ \{f_1, h_n\} & \{f_2, h_n\} & \cdots & \{f_n, h_n\} \end{Vmatrix}.$$

Then differential 1-forms $\theta_1, \dots, \theta_n$, where

$$\theta = W^{-1}\, dh,$$

and

$$\theta = \begin{Vmatrix} \theta_1 \\ \theta_2 \\ \vdots \\ \theta_n \end{Vmatrix}, \quad dh = \begin{Vmatrix} dh_1 \\ dh_2 \\ \vdots \\ dh_n \end{Vmatrix},$$

are closed on submanifolds $M_c = \{f_1 = c_1, \dots, f_n = c_n\} \cap D$. If the forms are exact, say $\theta_k = dg_k$ on M_c, $k = 1, 2, \dots, n$, then $2n$ functions $f_1, \dots, f_n, g_1, \dots, g_n$ provide us with local canonical coordinates for Ω in domain D.

10
Contact manifolds

10.1 Contact structures

A *contact structure* on a manifold M is a distribution C of codimension 1 such that the curvature form Ω_C is non-degenerate.

In other words, if C is locally defined by a differential 1-form ω, $C = \ker \omega$, then the restriction $d\omega|_C$ is non-degenerate.

A *contact manifold* is a manifold equipped with a contact structure C.

A *contact differential* 1*-form* is a form ω such that the distribution $\ker \omega$ is contact. A *strict contact manifold* is a pair (M, ω) where ω is a contact form on M.

By definition any contact structure locally admits a contact form, and therefore is strict. A global form ω exists if and only if the one-dimensional bundle TM/C has a non-zero section, i.e. is orientable.

At each point $a \in M$ of a strict contact manifold one has the splitting

$$T_aM = \ker \omega_a \oplus \ker d_a\omega$$

and $d_a\omega$ is a symplectic structure on $C_a = \ker \omega_a$.

Therefore, the dimension of C is even, say $2n$, and $\dim M = 2n + 1$ is odd. Note that the restriction of $(d_a\omega)^n$ on C_a is a volume form, and therefore $\omega \wedge (d\omega)^n$ is a volume form on M.

One could rephrase the definition of contact forms as follows: a differential 1-form ω is contact if and only if

$$\omega \wedge (d\omega)^n \neq 0$$

i.e., $\omega \wedge (d\omega)^n$ is a volume form.

Note that the Frobenius theorem gives the following integrability condition

$$\omega \wedge d\omega = 0$$

for distribution $\ker \omega$.

This shows that completely integrable and contact distributions provide two extreme cases for distributions in codimension 1.

10.1.1 Examples

Standard contact structure

Let $M = \mathbb{R}^{2n+1}$ be the arithmetic space with coordinates $(x_1, \ldots, x_n, y_1, \ldots, y_n, z)$ and let

$$\omega = dz - \sum_{i=1}^{n} y_i \, dx_i.$$

Then

$$\omega \wedge (d\omega)^n = dz \wedge (dx \wedge dy)^n \neq 0.$$

The distribution $\ker \omega$ is called the *standard contact structure*.

The Cartan distribution

Recall that 1-jet of function $f \in C^\infty(M)$ at a point $a \in M$ is a pair $[f]_a^1 = (f(a), d_a f)$.

Denote by $J^1 M$ the space of all 1-jets. Then $J^1 M = \mathbb{R} \times T^* M$, and we define the *Cartan form* on $J^1 M$ as

$$\omega = du - \rho,$$

where $u : J^1 M = \mathbb{R} \times T^* M \to \mathbb{R}$ is the first projection and ρ is the universal 1-form on $T^* M$.

If $q_1, \ldots, q_n$ are local coordinates on M and $(q_1, \ldots, q_n, p_1, \ldots, p_n)$ are the corresponding coordinates on $T^* M$ then $(q_1, \ldots, q_n, p_1, \ldots, p_n, u)$ are local coordinates on $J^1 M$ and

$$\omega = du - p \, dq$$

in these coordinates.

Therefore ω is a contact form and

$$C = \ker \omega$$

is a contact distribution on J^1M called the *Cartan distribution*.

The space of contact elements

A *contact element at a point* $a \in M$ is a hyperplane $H_a \subset T_aM$.

A hyperplane H_a can be defined by $\operatorname{Ann} H_a$ which is a one-dimensional subspace of T_a^*M. Thus the set of all contact elements at point $a \in M$ is the projectivization $\mathbb{P}T_a^*M$ of the cotangent space and the space of all contact elements is the projectivization of the cotangent bundle

$$\pi : \mathbb{P}T^*M = \bigcup_{a \in M} \mathbb{P}T_a^*M \to M.$$

Let $\alpha \in T_a^*M$, $\alpha \neq 0$, be a covector and let $[\alpha] \in \mathbb{P}T_a^*M$ be the one-dimensional subspace generated by α.

Define a hyperplane

$$C_{[\alpha]} \subset T_{[\alpha]}(\mathbb{P}T^*M)$$

to be a space of vectors $X \in T_{[\alpha]}(\mathbb{P}T^*M)$ such that their projections $\pi_*(X)$ belong to the contact element $\ker \alpha$, that is

$$C_{[\alpha]} = \pi_*^{-1}(\ker \alpha).$$

This distribution can be obtained in a different way.

Let us consider the manifold of non-zero covectors $T^*M \setminus \{0\}$ and distribution $\ker \rho$ on it. This distribution has characteristics; namely, if we define the vector field X_ρ dual to ρ as above

$$X_\rho \rfloor d\rho = \rho$$

then X_ρ generates the kernel of restriction $d\rho$ on $\ker \rho$, and

$$X_\rho \rfloor \rho = 0 \quad \text{and} \quad L_{X_\rho}(\rho) = \rho.$$

The one-parameter group of shifts along X_ρ corresponds to the standard action of the multiplicative group $\mathbb{R}^+ = \{\lambda > 0\}$

$$\lambda \in \mathbb{R}^+ : (x, p) \longmapsto (x, \lambda p),$$

on $T^*M \setminus \{0\}$.

The orbit space $\mathbb{S}T^*M$ is called the *spherization* of the cotangent bundle.

The distribution $\ker\rho$ defines a distribution C on $\mathbb{S}T^*M$. Note, that the vector field X_ρ preserves this distribution and scales $d\rho$, $L_{X_\rho}(d\rho) = d\rho$. Moreover X_ρ generates the kernel of $d\rho$, and therefore $d\rho$ defines a non-degenerate 2-form on the factor $\ker\rho/X_\rho$. This means that the distribution C on the spherization is contact.

The projectivization $\mathbb{P}T^*M$ can be viewed as an orbit space of the multiplicative group $\mathbb{R}^* = \mathbb{R}\setminus\{0\}$ action on

$$\lambda \in \mathbb{R}^* : (x,p) \longmapsto (x,\lambda p),$$

where $a \in M, p \in T_a^*M \setminus \{0\}$.

As above the distribution $\ker\rho$ is invariant with respect to actions and orbits correspond to the kernel of the restriction $d\rho$ on $\ker\rho$. This means that the distribution $\ker\rho$ can be pushed down and we obtain a contact distribution on $\mathbb{P}T^*M$. The definition of the universal 1-form ρ shows that this distribution coincides with the distribution C on the space of contact elements.

To find a local representation of this distribution we introduce the local homogeneous coordinates $(q_1,\ldots,q_n; p_1:\cdots:p_n)$ on $\mathbb{P}T^*M$, then C given by

$$\sum_{i=1}^{n} p_i\, dq_i = 0.$$

If, say $p_1 \neq 0$, then in the affine coordinates $(q_1,\ldots,q_n; p_2/p_1,\ldots,p_n/p_1)$ our distribution can be given by the 1-form

$$dq_1 + \sum_{i=2}^{n} \frac{p_i}{p_1}\, dq_i = 0,$$

and it shows that C is contact.

Exact symplectic manifolds

The above constructions can be used to construct more examples of contact manifolds.

Assume that (M,Ω) is an *exact symplectic manifold*; that is, $\Omega = d\beta$ for some 1-form β, and $\beta_a \neq 0$ for all $a \in M$.

Let X_β be the vector field corresponding to $\beta : X_\beta \rfloor \Omega = \beta$, and let us assume that the space B of trajectories of X_β is a smooth manifold and $\pi : M \to B$ is a smooth bundle. Then the above discussion shows that the distribution $\ker\beta$ on M defines a contact distribution on B.

Example 10.1.1 *Let* $M = \mathbb{R}^{2n}\backslash\{0\}$ *with coordinates* $(x_1, \ldots, x_n, y_1, \ldots, y_n)$ *and*

$$\beta = \sum_{i=1}^{n}(x_i\, dy_i - y_i\, dx_i).$$

Then $d\beta = 2dx \wedge dy$ *and*

$$X_\beta = \frac{1}{2}\sum_{i=1}^{n}\left(x_i\frac{\partial}{\partial x_i} + y_i\frac{\partial}{\partial y_i}\right).$$

The space of orbits can be identified with the sphere S^{2n-1}

$$\sum_{i=1}^{n}(x_i^2 + y_i^2) = 1$$

and restriction of β *on the sphere gives a contact structure.*

One-dimensional bundles

At first, we briefly recall basic constructions on connections in a form suitable for us.

Let $\pi : E \to B$ be a bundle. A *connection* in the bundle is a distribution P, $\dim P = \dim B$, transversal to fibres of π. In other words, if $T_a(\pi)$ is the tangent space to the fibre passing through $a \in E$, then a connection gives a splitting

$$T_aE = T_a(\pi) \oplus P_a$$

at each point $a \in E$.

Vectors of $T_a(\pi)$ are said to be *vertical* and vectors of P_a are *horizontal.*

As we have seen the curvature of distribution P defines a 2-form Ω_P. Owing to the splitting the normal space $\nu_{P,a}$ can be identified with the vertical space $T_a(\pi)$ and the curvature form can be identified with the 2-form on P with values in $T(\pi)$. Under this identification our construction of the curvature coincides with the standard one.

Namely, let us denote by $\overline{X}$ the horizontal lifting of the vector field $X \in D(B)$ into E, i.e. $\overline{X}_a \in P_a$ and the differential $\pi_* : P_a \to T_bB$ sends $\overline{X}_a$ to X_b, where $b = \pi(a)$.

Then the commutator $[\overline{X}, \overline{Y}]$ of two liftings projects on $[X, Y]$ and therefore $[\overline{X}, \overline{Y}] - \overline{[X, Y]}$ is a vertical vector field on E coinciding with $\Omega_P(X, Y)$.

Thus,

$$\Omega_P(X, Y) = [\overline{X}, \overline{Y}] - \overline{[X, Y]}$$

for all vector fields $X, Y \in D(B)$.

Let A_t be a flow of X and let $\overline{A}_t$ be a flow of $\overline{X}$. Then $\overline{A}_t$ give automorphisms of the bundle such that the following diagram

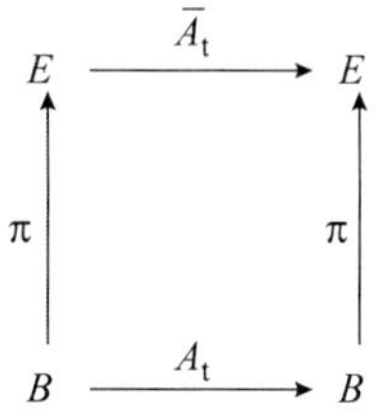

commutes.

If the bundle under consideration is equipped with an extra structure then the reasonable class of connections should provide us with automorphisms $\overline{A}_t$ that preserves the structure.

For example, let $\pi : E \to B$ be a vector bundle. Then a connection P is said to be *linear* if $\overline{A}_t$ are linear automorphisms for any vector field X on B.

In the last case the connection form Ω_P of the linear connection reduces to an operator-valued 2-form. Namely, the value $\Omega_{P,a}(X_b, Y_b)$, $a \in \pi^{-1}(b)$ depends on a in a linear way; that is,

$$\Omega_{P,a}(X_b, Y_b) = \hat{\Omega}_{P,b}(X_b, Y_b)(a)$$

for some linear operator

$$\hat{\Omega}_{P,b}(X_b, Y_b) : \pi^{-1}(b) \to \pi^{-1}(b).$$

This gives us a 2-form $\hat{\Omega}_P \in \Omega^2(B) \otimes \mathrm{End}\, \pi$ on the base manifold B with values in endomorphisms of π.

Let $\pi : E \to B$ be a one-dimensional vector bundle, then $\hat{\Omega}_P \in \Omega^2(B)$ for any linear connection P, and $\hat{\Omega}_P$ is closed due to the Bianchi identity.

The case $\hat{\Omega}_P = 0$ corresponds to flat connections and the case when $\hat{\Omega}_P$ is non-degenerate is the second extreme class of connections. This corresponds to linear connections P in line bundles $\pi : E \to B$ over symplectic manifolds (B, Ω) such that $\hat{\Omega}_P = \Omega$. The corresponding horizontal distributions in E are contact.

Example 10.1.2 *Let $\pi : B \times \mathbb{R} \to B$ be the trivial line bundle. A linear connection in π can be given by the differential 1-form $\theta \in \Omega^1(B)$.*

Namely,

$$P = P_\theta = \ker(du - \theta)$$

where $u : B \times \mathbb{R} \to \mathbb{R}$ is the projection.

In this case

$$\hat{\Omega}_P = -d\theta.$$

Therefore, if $(B, -d\theta)$ is an exact symplectic structure, then P_θ defines a contact structure on $B \times \mathbb{R}$.

Example 10.1.3 *The projection $\pi : J^1M \to T^*M$ is the trivial line bundle over the exact symplectic manifold $(T^*M, d\rho)$. The above example gives us the Cartan distribution on J^1M.*

Hermitian line bundles

Let $\pi : E \to B$ be a Hermitian complex line bundle, and let P be a Hermitian connection. In this case values of the curvature $\hat{\Omega}_P \in \Omega^2(B) \otimes \operatorname{End}\pi$ belong to the space of skew-Hermitian operators, and because of $\dim \pi = 1$, $\iota\hat{\Omega}_P$ is a real-valued closed 2-form on M. This form can be obtained in a different and more geometrical way.

Let h be an Hermitian form on π and let

$$S_x(\pi) = \{v \in \pi^{-1}(x) | h_x(v, v) = 1\}$$

be the circle of unit vectors at $x \in M$.

Consider the subbundle

$$\pi_S : S(E) = \bigcup_{x \in M} S_x(\pi) \to B$$

of the bundle π.

This is a one-dimensional bundle over M. Fibres of π_S are circles, the abelian group $\mathbf{S}^1 = \{z \in \mathbb{C} \mid |z| = 1\}$ acts on $S(E)$ and fibres $\pi_S^{-1}(x)$ are orbits of the action. Thus π_S is a principal $\mathbf{S}^1$-bundle.

Now let P be a linear connection on π, then P is Hermitian if and only if P is tangent to $S(E)$:

$$P_a \subset T_a(S(E))$$

for all $a \in S(E)$.

Denote by P_S the restriction of P on $S(E)$. This is a connection in π_S, and moreover P_S is invariant with respect to $\mathbf{S}^1$-actions. Therefore the curvature form Ω_{P_S} is $\mathbf{S}^1$-invariant. Using the standard identification of the vertical vectors with elements of the Lie algebra of $\mathbf{S}^1$, that is $\mathbb{R}$, we obtain the closed 2-form $\iota\hat{\Omega}_P$ on the base manifold B.

This form is non-degenerate if and only if the distribution P_S is contact.

Note also that the symplectic structure $\iota\hat{\Omega}_P$ just obtained is not arbitrary. The cohomology class $[\iota\hat{\Omega}_P]$ of the form should be integral:

$$[\iota\hat{\Omega}_P] \in H^2(B, \mathbb{Z}).$$

It is known that for any integral 2-cocycle θ there is a Hermitian bundle π and a Hermitian connection P such that $\theta = \iota\hat{\Omega}_P$. Thus for any symplectic manifold (M, Ω) with integral structure form there is a Hermitian line bundle π and corresponding contact manifold that produce (M, Ω). Boothby and Wang [10] have proved that all strict compact and regular contact structures can be obtained in this way.

10.2 Contact transformations and contact vector fields

Let (M, C) be a contact manifold. A diffeomorphism $\varphi : M \to M$ is called a *contact transformation* if

$$\varphi_*(C) = C.$$

If (M, ω) is a strict contact manifold then contact transformations are diffeomorphisms φ such that

$$\varphi^*(\omega) = \lambda_\varphi \omega$$

for some smooth function $\lambda_\varphi \in C^\infty(M)$.

If in addition

$$\varphi^*(\omega) = \omega$$

the contact diffeomorphism is called *strict*.

10.2.1 Examples

The standard contact manifolds

Let $M = \mathbb{R}^{2n+1}$ and let

$$\omega = dz - \sum_{i=1}^{n} y_i \, dx_i$$

be the standard contact structure C on M.

Then the following transformations are contact.

Proposition 10.2.1

1. Translations*:*

$$T_c : (x_1, \ldots, x_n, z, y_1, \ldots, y_n) \longmapsto (x_1 + c_1, \ldots, x_n + c_n, z + c_0, y_1, \ldots, y_n),$$

where $c = (c_0, c_1, \ldots, c_n) \in \mathbb{R}^{n+1}$.

2. Scale transformations*:*

$$S_\lambda : (x_1, \ldots, x_n, z, y_1, \ldots, y_n) \\ \longmapsto (e^{\lambda_1} x_1, \ldots, e^{\lambda_n} x_n, e^{\lambda_0} z, e^{(\lambda_0 - \lambda_1)} y_1, \ldots, e^{(\lambda_0 - \lambda_n)} y_n).$$

3. The Legendre transformation*:*

$$\mathcal{L} : (x_1, \ldots, x_n, z, y_1, \ldots, y_n) \mapsto \left(y_1, \ldots, y_n, z - \sum_{i=1}^{n} y_i x_i, -x_1, \ldots, -x_n, \right).$$

4. The Euler *(or* partial Legendre*)* transformation*:*

$$\mathcal{L}_k : (x_1, \ldots x_k, x_{k+1}, \ldots, x_n, z, y_1, \ldots, y_k, y_{k+1}, \ldots, y_n) \\ \longmapsto \left(y_1, \ldots, y_k, x_{k+1}, \ldots, x_n, z - \sum_{i=1}^{k} y_i x_i, -x_1, \ldots, -x_k, y_{k+1}, \ldots, y_n \right).$$

5. Shifts*:*

$$\mathcal{S}_h : (x_1, \ldots, x_n, z, y_1, \ldots, y_n) \\ \mapsto \left(x_1, \ldots, x_n, z + h(x_1, \ldots, x_n), y_1 + \frac{\partial h}{\partial x_1}, \ldots, y_n + \frac{\partial h}{\partial x_n} \right),$$

where $h \in C^\infty(\mathbb{R}^n)$.

The Cartan distribution

1. Let $\phi : M \to M$ be a diffeomorphism, and let $p_1(\phi) : J^1M \to J^1M$ be the *prolongation of* ϕ on 1-jets:

$$p_1(\phi) : [h]_a^1 \longmapsto [h \circ \phi^{-1}]_{\phi^{-1}(a)}^1.$$

Then

$$p_1(\phi)^*(\omega) = \omega,$$

i.e. $p_1(\phi)$ is a strict contact transformation.

2. Let $\phi : T^*M \to T^*M$ be a symplectomorphism. Then $\phi^*(d\rho) = d\rho$ implies that $\phi^*(\rho) - \rho$ is a closed 1-form on T^*M. Assume that this form is exact, then

$$\phi^*(\rho) - \rho = dS$$

for some function S. Define a lifting $\tilde{\phi} : J^1M \to J^1M$ as follows

$$\tilde{\phi} : (u, x) \longmapsto (u - S(x), \phi(x))$$

where $x \in T^*M$. Then,

$$\tilde{\phi}^*(du - \rho) = d(u - S) - \phi^*(\rho) = du - \rho.$$

3. Any function $h \in C^\infty(M)$ determines a strict contact transformation:

$$\mathcal{S}_h : [f]_a^1 \longmapsto [f + h]_a^1.$$

The space of contact elements

Let $\phi : M \to M$ be a diffeomorphism. Define a prolongation $\overline{\phi} : \mathbb{P}T^*M \to \mathbb{P}T^*M$ on contact elements as follows:

$$\overline{\phi} : [\alpha] \longmapsto [(\phi^{-1})^*(\alpha)]$$

where $[\alpha]$ is a class of non-zero covector $\alpha \in T^*M$.

Then $\overline{\phi}$ is a projectivization of the symplectic transformation $\phi^{(1)} : T^*M \to T^*M$, and because $(\phi^{(1)})^*(\rho) = \rho$, $\overline{\phi}$ is a contact diffeomorphism. These contact diffeomorphisms are called *point transformations*.

Exact symplectic manifolds

Let $(M, d\beta)$ be an exact symplectic manifold and let us assume that the corresponding contact manifold B exists. Then any *strict symplectic transformation* $\phi : M \to M$, i.e. $\phi^*(\beta) = \beta$ induces a contact transformation B.

Contact transformations of hypersurfaces in classic differential geometry

In this section we discuss the classical coordinate approach to contact transformations.

We are going to investigate different types of transformation of hypersurfaces in $\mathbb{R}^{n+1}$. The fist type is given by diffeomorphisms, or so-called *point transformations*.

Let $\varphi : \mathbb{R}^{n+1} \to \mathbb{R}^{n+1}$ be a diffeomorphism

$$\varphi : (x, u) = (x_1, \ldots, x_n, u) \longmapsto (X, U) = (\phi(x, u), \psi(x, u)),$$

and let

$$S_h = \{u = h(x)\}$$

be a graph of a smooth function.

Assume that the hypersurface $\varphi(S_h)$ is a graph of a new function, say h^φ:

$$\varphi(S_h) = S_{h^\varphi}.$$

This function is a solution $U = h^\varphi(X)$ of equations

$$X = \phi(x, h(x)), \quad U = \psi(x, h(x)). \tag{10.1}$$

Obviously images $\varphi(S)$ and $\varphi(S')$ are tangent at $\varphi(a)$ if S and S' were tangent at $a \in S \cap S'$. Therefore, φ induces a transformation on contact elements (i.e., tangent planes to hypersurfaces).

It is easy to find the corresponding transformation of contact elements; namely, a tangent plane to S_h given by the equation

$$du - \frac{\partial h}{\partial x} dx = 0.$$

To find tangent planes to $\varphi(S_h)$ we differentiate h^φ and obtain

$$\frac{d\phi}{dx} \frac{\partial h^\varphi}{\partial X} = \frac{d\psi}{dx}$$

where, for example,

$$\frac{d\psi}{dx} = \frac{\partial \psi}{\partial x} + \frac{\partial \psi}{\partial u} \frac{\partial h}{\partial x}.$$

Therefore,

$$\frac{\partial h^{\varphi}}{\partial X} = \left(\frac{d\phi}{dx}\right)^{-1} \frac{d\psi}{dx}. \tag{10.2}$$

Denoting coordinates of contact elements by $p = (p_1, \ldots, p_n)$ we obtain the following transformation:

$$(x, u, p) \longmapsto \left(\phi(x,u), \psi(x,u), \left(\frac{\partial \phi}{\partial x} + \frac{\partial \phi}{\partial u} p\right)^{-1} \left(\frac{\partial \psi}{\partial x} + \frac{\partial \psi}{\partial u} p\right)\right).$$

This transformation sends contact elements to contact elements. Moreover it is easy to check that this transformation preserves zeros of the contact form $\omega = du - p\,dx$.

Suppose now that transformation (10.1) in addition depends on contact elements; that is,

$$X = \phi\left(x, h(x), \frac{\partial h}{\partial x}\right),$$
$$U = \psi\left(x, h(x), \frac{\partial h}{\partial x}\right).$$

Then the corresponding transformation of contact elements will be

$$\left(\frac{\partial \phi}{\partial x} + \frac{\partial \phi}{\partial u}\frac{\partial h}{\partial x} + \frac{\partial^2 h}{\partial x^2}\frac{\partial \phi}{\partial p}\right)\frac{\partial h^{\varphi}}{\partial X} = \left(\frac{\partial \psi}{\partial x} + \frac{\partial \psi}{\partial u}\frac{\partial h}{\partial x} + \frac{\partial^2 h}{\partial x^2}\frac{\partial \psi}{\partial p}\right)$$

and therefore it includes the second derivatives of h.

To get transformation on contact elements only we need some extra requirements on ϕ and ψ, such that the above equation on $\partial h^{\varphi}/\partial X$ gives us a solution independent on $\partial^2 h/\partial x^2$. This gives us immediately the conditions:

$$\frac{\partial \phi}{\partial p}\frac{\partial h^{\varphi}}{\partial X} = \frac{\partial \psi}{\partial p},$$
$$\left(\frac{\partial \phi}{\partial x} + \frac{\partial \phi}{\partial u}\frac{\partial h}{\partial x}\right)\frac{\partial h^{\varphi}}{\partial X} = \left(\frac{\partial \psi}{\partial x} + \frac{\partial \psi}{\partial u}\frac{\partial h}{\partial x}\right)$$

and finally

$$\left(\frac{\partial \phi}{\partial x} + \frac{\partial \phi}{\partial u}\frac{\partial h}{\partial x}\right)\left(\frac{\partial \phi}{\partial p}\right)^{-1}\frac{\partial \psi}{\partial p} = \left(\frac{\partial \psi}{\partial x} + \frac{\partial \psi}{\partial u}\frac{\partial h}{\partial x}\right). \tag{10.3}$$

The corresponding transformation has the form

$$(x,u,p) \longmapsto \left(\phi(x,u,p), \psi(x,u,p), \left(\frac{\partial\phi}{\partial p}\right)^{-1}\frac{\partial\psi}{\partial p}\right).$$

An alternative method of obtaining the conditions (10.3) is provided by differential forms.

As we have seen zeros of the differential 1-form $\omega = du - p\,dx$ describe contact elements. Let us try to extend the transformation

$$(x,u) \longmapsto (\phi(x,u,p), \psi(x,u,p))$$

to a transformation of the form

$$(x,u,p) \longmapsto (\phi(x,u,p), \psi(x,u,p), P(x,u,p))$$

in such a way that the form

$$d\psi - P\,d\phi$$

equals zero every time $\omega = 0$.

This gives us

$$\frac{d\psi}{dx} - P\frac{d\phi}{dx} = 0,$$
$$\frac{\partial\psi}{\partial p} - P\frac{\partial\phi}{\partial p} = 0,$$

and therefore (10.3).

Example 10.2.1 (Pedal transformations, see Figures 10.1 and 10.2) *Consider a smooth hypersurface $L \subset \mathbb{R}^{n+1}$ and a fixed point P, say $P = 0$. For an arbitrary point a of L we denote by $S(a)$ the intersection of the tangent hyperplane T_aL and the perpendicular from P to the tangent T_a. Obviously this transformation depends on contact elements and one obtains:*

$$S : (x,u) \longmapsto \left(-\frac{(u-px)}{1+p^2}p, \frac{u-px}{1+p^2}\right).$$

The corresponding contact transformation is called a pedal. *It has the form*

$$(x,u,p) \longmapsto \left(-\frac{(u-px)}{1+p^2}p, \frac{u-px}{1+p^2}, \frac{x+(2u-px)p}{u(1-p^2)-2px}\right).$$

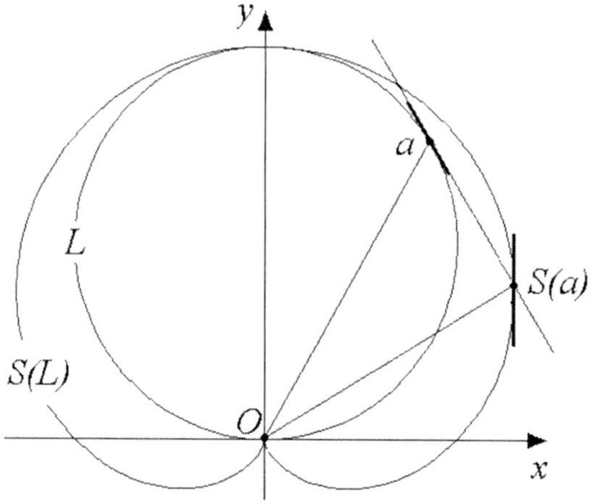

Figure 10.1. The pedal transformation

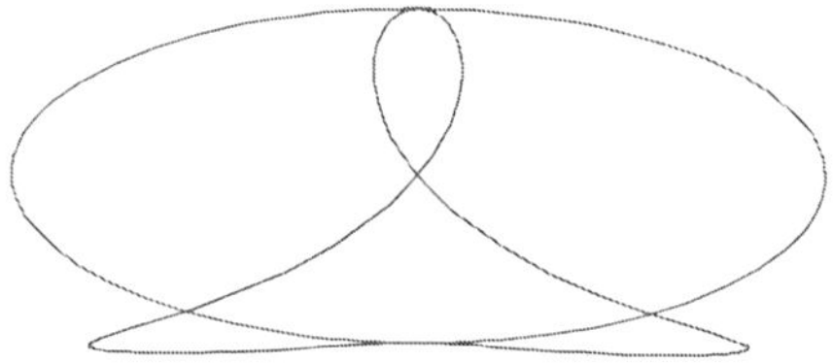

Figure 10.2. The pedal transformation of an ellipse.

Example 10.2.2 (Polar transformations) *Let E be an ellipse in $\mathbb{R}^2$, and let a be a point in the exterior part of the ellipse. Let t_1 and t_2 be tangent lines to E passing through a, say $t_1 = T_{a_1}E, t_2 = T_{a_2}E$. The line $l(a)$ passing through the points a_1 and a_2 is called a* polar of the point *a. A* polar transformation *sends a contact element (a, p) to a point A in such a way that (1) $A \in l(a)$, (2) $a \in l(A)$ and (3) $l(A)$ has slope p. In coordinates the polar transformation with respect to the ellipse $x^2/a^2 + y^2/b^2 = 1$ has the form*

$$(x, u, p) \longmapsto \left(\frac{a^2 p}{px - u}, \frac{b^2}{u - px}, -\frac{b^2 x}{a^2 u} \right).$$

Example 10.2.3 (Dilatations, see Figure 10.3) Dilatations *or* optical transformations *came from geometrical optics. Under this transformation the contact element (x, u, p) moves in the orthogonal direction to p with velocity 1 during*

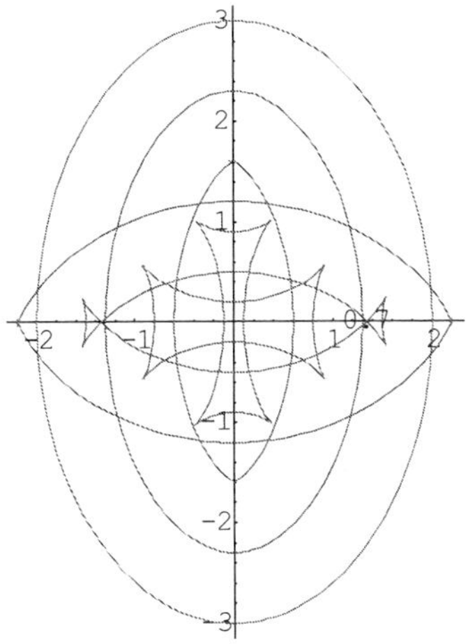

Figure 10.3. Dilitations of an ellipse.

time t, that is

$$(x, u) \longmapsto \left(x - \frac{tp}{\sqrt{1+p^2}}, u + \frac{t}{\sqrt{1+p^2}} \right)$$

and

$$(x, u, p) \longmapsto \left(x - \frac{tp}{\sqrt{1+p^2}}, u + \frac{t}{\sqrt{1+p^2}}, p \right).$$

10.2.2 Contact vector fields

Contact vector fields and their generating functions

An infinitesimal symmetry of a contact distribution is called a *contact vector field*.

For the strict contact structure this means that a vector field X is contact if and only if

$$L_X(\omega) = \lambda_X \omega,$$

for some smooth function $\lambda_X \in C^\infty(M)$.

We can also re-write this condition without using the function λ_X:

$$L_X(\omega) \wedge \omega = 0.$$

The tangent bundle of a strict contact manifold (M, ω) can be spit into a direct sum:

$$T_aM = C_a \oplus \ker d_a\omega$$

because $d_a\omega$ is a symplectic structure on $C_a = \ker \omega_a$.

Therefore, any vector field X on M can be decomposed into the sum

$$X = X^h + X^v,$$

where $X^h \in D(C)$ is called *horizontal* and $X^v \in D(\ker d\omega)$ is the *vertical component* of the vector field.

The distribution $\ker d\omega$ is one dimensional and the 1-form ω can be viewed as a coordinate on this distribution.

Let us denote by X_1 the basis vector field in $\ker d\omega$ with coordinate 1; that is,

$$\iota_{X_1} d\omega = 0 \qquad \text{and} \qquad \iota_{X_1}\omega = 1.$$

Then

$$L_{X_1}(\omega) = \iota_{X_1} d\omega + d(\iota_{X_1}\omega) = 0$$

and, therefore, X_1 is a contact vector field.

Below we show that any contact vector field X on a strict contact manifold is uniquely determined by its vertical component X^v or by a function $f_X = \omega(X)$.

This function is called a *generating function* or a *contact Hamiltonian* of the contact vector field X.

Theorem 10.2.1 *Let (M, ω) be a strict contact manifold. Then for any function $f \in C^\infty(M)$ there exists a unique contact vector field X_f such that:*

$$\begin{aligned} &\omega(X_f) = f, \\ &X_f \rfloor d\omega = X_1(f)\omega - df, \\ &L_{X_f}(\omega) = X_1(f)\omega. \end{aligned}$$

Moreover,

$$\begin{aligned} &X_{f+g} = X_f + X_g, \\ &X_{fg} = f X_g + gX_f - fgX_1, \\ &X_f(g) + X_g(f) = X_1(f)g + X_1(g)f \end{aligned}$$

for all $f, g \in C^\infty(M)$.

The vector field X_1 *is called the* Reeb vector field.

Proof Let X be a contact field on M and $L_X\omega = \lambda_X\omega$. Then,

$$X = X^h + f X_1$$

if

$$f = \omega(X)$$

and

$$L_X(\omega) = X\rfloor d\omega + d(X\rfloor\omega) = X^h\rfloor d\omega + df = \lambda_X\omega.$$

In other words,

$$X^h\rfloor d\omega = \lambda_X\omega - df. \tag{10.4}$$

Since $X_1 \in \ker d\omega$, we have

$$\iota_{X_1}(\lambda_X\omega - df) = \lambda_X - X_1(f) = 0$$

or

$$\lambda_X = X_1(f).$$

Now the differential 1-form

$$X_1(f)\omega - df$$

vanishes on $\ker d\omega$, and therefore can be viewed as a differential 1-form on the contact distribution C. The restriction of $d\omega$ on C is symplectic and therefore (10.4) uniquely determines the horizontal component X^h.

Let us denote the contact vector field X by X_f. It follows that

$$L_{X_f}(\omega) = X_1(f)\omega$$

and

$$\iota_{X_f}d\omega = X_1(f)\omega - df.$$

Consider now values of both 1-forms in the last equality on a contact vector field X_g. We obtain

$$d\omega(X_f, X_g) = X_1(f)g - X_g(f).$$

By the same reasoning

$$d\omega(X_g, X_f) = X_1(g)f - X_f(g)$$

and therefore

$$X_1(f)g - X_g(f) = X_f(g) - X_1(g)f.$$

To finish the proof we denote by Y the following vector field:

$$Y = f\,X_g + gX_f - fgX_1.$$

Then,

$$\begin{aligned}
\omega(Y) &= fg,\\
Y \rfloor d\omega &= f\iota_{X_g}\,d\omega - g\iota_{X_f}\,d\omega = (fX_1(g) + gX_1(f))\omega - f\,dg - g\,df\\
&= X_1(fg)\omega - d(fg)
\end{aligned}$$

and therefore

$$L_Y(\omega) = X_1(fg)\omega.$$

In other words, Y is a contact vector field with generating function fg, and $Y = X_{fg}$. ■

In canonical coordinates contact vector fields have the following form:

$$X_f = -\sum_{i=1}^{n} \frac{\partial f}{\partial p_i}\frac{\partial}{\partial q_i} + \left(f - \sum_{i=1}^{n} p_i \frac{\partial f}{\partial p_i}\right)\frac{\partial}{\partial u} + \sum_{i=1}^{n} \frac{\partial f}{\partial q_i}\frac{\partial}{\partial p_i}$$

or

$$X_f = -\sum_{i=1}^{n} \frac{\partial f}{\partial p_i}\frac{d}{dq_i} + f\frac{\partial}{\partial u} + \sum_{i=1}^{n} \frac{\partial f}{\partial q_i}\frac{\partial}{\partial p_i}.$$

Lagrange brackets

The commutator of two contact vector fields X_f and X_g with generating functions f and g on a contact manifold M is again a contact vector field $[X_f, X_g] = X_h$ with a generating function h, which is defined by

$$h = \omega([X_f, X_g])$$

and is called a *Lagrange bracket* of the functions f and g. We denote the Lagrange bracket $\omega([X_f, X_g])$ by $[f, g]$.

Theorem 10.2.2 *The Lagrange bracket defines a Lie algebra structure on $C^\infty(M)$ and*

$$[f, g] = X_f(g) - X_1(f)g.$$

Proof The fact that $C^\infty(M)$ is a Lie algebra with respect to the Lagrange bracket is an immediate consequence of the definition. To verify the second statement one needs the standard formula:

$$\iota_{[X_f, X_g]}\omega = [L_{X_f}, \iota_{X_g}](\omega) = X_f(g) - X_1(f)g,$$

which completes the proof. ■

Finally, we give expressions for the Lagrange bracket in canonical coordinates:

$$[f, g] = f\frac{\partial g}{\partial u} - g\frac{\partial f}{\partial u} + \sum_{i=1}^{n}\left(\frac{df}{dq_i}\frac{\partial g}{\partial p_i} - \frac{dg}{dq_i}\frac{\partial f}{\partial p_i}\right).$$

10.3 Darboux theorem

The following theorem claims that all contact manifolds are locally equivalent.

Theorem 10.3.1 (Darboux) *Let ω and ϖ be contact forms on a smooth manifold M. Then for any point $a \in M$ there exists a local diffeomorphism φ such that $\varphi(a) = a$ and $\varphi^*(\varpi) = \omega$.*

Proof Without loss of generality it can be assumed that $\omega_a = \varpi_a$ and $(d\omega)_a = (d\varpi)_a$. Consider the one-parameter family of 1-forms

$$\omega_t = t\varpi + (1 - t)\omega,$$

where $t \in [0, 1]$. Then $\omega_0 = \omega$ and $\omega_1 = \varpi$. We will find a one-parameter family of local diffeomorphisms φ_t such that $\varphi_t(a) = a$, $\varphi_0 = \mathrm{id}$ and

$$\varphi_t^*(\omega_t) = \omega_0. \tag{10.5}$$

Differentiating both sides of the last equality we obtain

$$\frac{d}{dt}\varphi_t^*(\omega_t) = \varphi_t^*(L_{X_t}(\omega_t) + \dot{\omega}_t) = 0, \tag{10.6}$$

where $\dot{\omega}_t \stackrel{\text{def}}{=} d\omega_t/dt = \omega - \varpi$ and X_t is a one-parameter family of vector fields defined in some neighborhood of a such that the translations along X_t from $t = 0$ to t coincide with φ_t. From (10.6) it follows that

$$\iota_{X_t}\, d\omega_t + d(\omega_t(X_t)) = -\dot{\omega}_t. \tag{10.7}$$

We will find the vector field X_t in the form $X_t = f_t X_1^t + Y_t$, where $\omega_t(Y_t) = 0$ and X_1^t is the contact vector field such that $\omega_t(X_1^t) = 1$ for all $t \in [0, 1]$. Then from (10.7) it follows that

$$\iota_{Y_t}\, d\omega_t = -\dot{\omega}_t - df_t. \tag{10.8}$$

Let f_t be a one-parameter family of smooth functions on M such that in some neighborhood of the point a we have $(-\dot{\omega}_t - df_t)(X_1^t) = 0$, or

$$X_1^t(f_t) = \omega_t(X_t). \tag{10.9}$$

Thus, choosing functions f_t and the vector field Y_t so that $f_t(a) = 0$, f_t satisfy (10.9) and Y_t satisfies (10.8), we see that the vector fields X_t vanish at the point a and the family φ_t satisfies condition (10.5). Therefore, $\varphi_1^*(\omega_1) = \omega_0$. ■

Corollary 10.3.1 *Suppose that M is a $(2n + 1)$ – dimensional contact manifold with a contact form ω. Then one can find local canonical coordinates $(q_1, \ldots, q_n, u, p_1, \ldots, p_n)$ in a neighborhood $\mathcal{O}_a$ of a point $a \in M$ such that $q_i(a) = p_i(a) = u(a) = 0$ $(i = 1, \ldots, n)$ and ω has the following* canonical form*:*

$$\omega = du - p_1\, dq_1 - \cdots - p_n\, dq_n. \tag{10.10}$$

10.4 A local description of contact transformations

10.4.1 Generating functions of Lagrangian submanifolds

We begin with a local description of integral manifolds of a contact structure. Let $(\mathbb{R}^{2n+1}, \omega = du - p\,dq)$ be the standard contact structure, and let $L \subset \mathbb{R}^{2n+1}$ be an n-dimensional integral manifold, $\omega|_L = 0$. Note that any smooth function $S(q)$ defines the integral manifold $L_S = \{u = S(q), p = \partial S/\partial q\}$, and moreover any integral manifold where functions $q_1, \dots, q_n$ are independent has this form with function $S = u|_L$.

For general integral manifolds a tangent space $T_aL \subset \ker \omega_a$ is a Lagrangian subspace of the symplectic space $(\ker \omega_a, d_a\omega)$. Therefore one can find a permutation $(i_1, \dots, i_k, i_{k+1}, \dots, i_n)$ of the indexes $(1, \dots, n)$ in such a way that

$$d_a q_{i_1}, \dots, d_a q_{i_k}, d_a p_{i_{k+1}}, \dots, d_a p_{i_n}$$

are linear independent on T_aL, and functions $q_I = (q_{i_1}, \dots, q_{i_s})$, $p_{I^c} = (p_{i_{s+1}}, \dots, p_{i_n})$, where $I = (i_1, \dots, i_k)$, $I^c = (i_{k+1}, \dots, i_n)$, can be chosen as local coordinates in some neighborhood of $a \in L$.

Let us rewrite the structure form

$$\omega = du - p_I\,dq_I - p_{I^c}\,dq_{I^c}$$

in the following way:

$$\omega = d(u - q_{I^c}p_{I^c}) - p_I\,dq_I + q_{I^c}\,dp_{I^c}.$$

It follows from the above remark that L can be represented in some neighborhood of the point a as follows:

$$p_I = \frac{\partial S}{\partial q_I}, \quad q_{I^c} = -\frac{\partial S}{\partial p_{I^c}}, \quad u = S - p_{I^c}\frac{\partial S}{\partial p_{I^c}}$$

for some smooth function $S = S(q_I,\ p_{I^c})$.

This function S is called a *generating function* of the integral manifold L.

Let us use the just-obtained local description of integral manifolds for a description of contact transformations.

For this purpose, we will represent graphs of contact transformations as integral manifolds.

Namely, any contact transformation

$$\varphi : (q, u, p) \longmapsto (A(q, u, p), C(q, u, p), B(q, u, p)),$$

defines a function λ such that $\varphi^*(\omega) = \lambda\omega$.

The last equation means that

$$-\lambda\omega + \varphi^*(\omega) = dC - B\,dA - \lambda\,du + \lambda p\,dq = 0. \tag{10.11}$$

Consider now the $(2(2n+1)+1)$-dimensional manifold $\mathbb{R}^{4n+3}$ with coordinates

$$(Q_1, \ldots, Q_n, U, P_1, \ldots, P_n, \lambda, q_1, \ldots, q_n, u, p_1, \ldots, p_n)$$

and the following 1-form:

$$\theta = dU - P\,dQ + \lambda p\,dq - \lambda\,du.$$

It is easy to check that θ determines a contact structure on $\mathbb{R}^{4n+3}$.

A graph of a contact transformation φ is the following $(2n+1)$-dimensional submanifold L_φ:

$$Q = C(q, u, p), \quad U = C(q, u, p), \quad P = B(q, u, p), \quad \lambda = \lambda(q, u, p).$$

This manifold is integral for θ, and therefore one could use generating functions to describe contact transformations.

Example 10.4.1 *Assume that functions Q, q and u are independent on L_φ. Then the contact transformation is defined by the generating function $S = S(Q, q, u)$ in the following way:*

$$P = \frac{\partial S}{\partial Q}, \quad p = -\frac{\partial S}{\partial q} \Big/ \frac{\partial S}{\partial u}, \quad U = S.$$

10.4.2 A description of contact transformations in $\mathbb{R}^3$

We give the complete description of contact transformations in the space of 1-jets $J^1\mathbb{R} = \mathbb{R}^3$ with the standard coordinates (q, u, p) and the Cartan form $\omega = du - p\,dq$.

The description can be divided into the following eight types in accordance with the types of function which can be chosen as coordinates on the graph.

1. The generating function is $S = S(Q, u, q)$. Then the contact transformation is

$$P = \frac{\partial S}{\partial Q}, \quad p = -\frac{\partial S}{\partial q} \Big/ \frac{\partial S}{\partial u}, \quad U = S.$$

2. The generating function is $S = S(P, u, q)$. Then the contact transformation is

$$Q = -\frac{\partial S}{\partial P}, \quad p = -\frac{\partial S}{\partial q}\Big/\frac{\partial S}{\partial u}, \quad U = S - P\frac{\partial S}{\partial P}.$$

3. The generating function is $S = S(Q, \lambda, q)$. Then the contact transformation is

$$P = \frac{\partial S}{\partial Q}, \quad u = -\frac{\partial S}{\partial \lambda}, \quad p = -\lambda^{-1}\frac{\partial S}{\partial q}, \quad V = S - \lambda\frac{\partial S}{\partial \lambda}.$$

4. The generating function is $S = S(P, \lambda, q)$. Then the contact transformation is

$$Q = -\frac{\partial S}{\partial P}, \quad u = -\frac{\partial S}{\partial \lambda}, \quad p = -\lambda^{-1}\frac{\partial S}{\partial q}, \quad V = S - \lambda\frac{\partial S}{\partial \lambda} - P\frac{\partial S}{\partial P}.$$

5. The generating function is $S = S(Q, \lambda, R)$, where $R = -\lambda p$. Then the contact transformation is

$$P = \frac{\partial S}{\partial Q}, \quad u = -\frac{\partial S}{\partial \lambda}, \quad q = -\frac{\partial S}{\partial R}, \quad p = -R/\lambda, \quad V = S - \lambda\frac{\partial S}{\partial \lambda} - R\frac{\partial S}{\partial R}.$$

6. The generating function is $S = S(P, u, R)$, where $R = -\lambda p$. Then the contact transformation is

$$Q = -\frac{\partial S}{\partial P}, \quad q = -\frac{\partial S}{\partial R}, \quad p = -R\Big/\frac{\partial S}{\partial u}, \quad V = S - P\frac{\partial S}{\partial P} - R\frac{\partial S}{\partial R}.$$

7. The generating function is $S = S(P, \lambda, Q)$. Then the contact transformation is

$$Q = -\frac{\partial S}{\partial P}, \quad u = -\frac{\partial S}{\partial \lambda}, \quad p = -\lambda^{-1}\frac{\partial S}{\partial u}, \quad V = S - P\frac{\partial S}{\partial P} - \lambda\frac{\partial S}{\partial \lambda}.$$

8. The generating function is $S = S(P, \lambda, R)$, where $R = -\lambda p$. Then the contact transformation is

$$Q = -\frac{\partial S}{\partial P}, \quad u = -\frac{\partial S}{\partial \lambda}, \quad p = -\lambda^{-1}\frac{\partial S}{\partial R}, \quad V = S - P\frac{\partial S}{\partial P} - \lambda\frac{\partial S}{\partial \lambda}.$$

Example 10.4.2 *Take the following generating function $S(Q, u, q) = Qu + 1/2q^2$. The corresponding contact transformation has the following form:*

$$\varphi : (q, p, u) \to \left(-\frac{q}{p}, u, \frac{q^2}{2} - \frac{qu}{p}\right).$$

11

Monge–Ampère equations

11.1 Monge–Ampère operators

Let M be an n-dimensional smooth manifold and let J^1M be the manifold of 1-jets equipped with the standard contact structure $\omega_0 \in \Omega^1(J^1M)$.

With any differential n-form $\omega \in \Omega^n(J^1M)$ we associate a differential operator (see [69])

$$\Delta_\omega : C^\infty(M) \to \Omega^n(M),$$

which acts as

$$\Delta_\omega(v) = \omega|_{j^1(v)(M)} \tag{11.1}$$

where $v \in C^\infty(M)$ is a smooth function on the manifold M, and $j_1(v)(M) \subset J^1M$ is the graph of 1-jet $j_1(v) : M \to J^1M$. We call such operators *Monge–Ampère.*

To explain this terminology let us consider a few examples of such operators and the corresponding differential forms.

Example 11.1.1 *Let*

$$\omega = 3p^2\,dp - dx \tag{11.2}$$

be the differential 1-form on $J^1\mathbb{R}$. The corresponding operator Δ_ω acts as

$$\Delta_\omega(v) = (3(v')^2v'' - 1)\,dx. \tag{11.3}$$

Indeed,

$$\Delta_\omega(v) = (3p^2\,dp - dx)|_{j^1(v)(M)} = 3(v')^2\,d(v') - dx = (3(v')^2v'' - 1)\,dx.$$

Example 11.1.2 *The differential 2-form*

$$\omega = dp_1 \wedge dp_2$$

on $J^1\mathbb{R}^2$ corresponds to the operator

$$\begin{aligned}\Delta_\omega(v) &= d(v_{q_1}) \wedge d(v_{q_2})\\ &= (v_{q_1q_1}\, dq_1 + v_{q_1q_2}\, dq_2) \wedge (v_{q_2q_1}\, dq_1 + v_{q_2q_2}\, dq_2)\\ &= (v_{q_1q_1}v_{q_2q_2} - v^2_{q_1q_2})\, dq_1 \wedge dq_2\\ &= (\det \operatorname{Hess} v)\, dq_1 \wedge dq_2,\end{aligned}$$

where

$$\operatorname{Hess} v = \begin{Vmatrix} v_{q_1q_1} & v_{q_1q_2} \\ v_{q_1q_2} & v_{q_2q_2} \end{Vmatrix}$$

is the Hessian of the function v.

Thus,

$$\Delta_\omega(v) = (\det \operatorname{Hess} v)\, dq_1 \wedge dq_2. \tag{11.4}$$

Example 11.1.3 *The differential 3-form*

$$\omega = dq_1 \wedge dp_1 \wedge dp_3 \tag{11.5}$$

on $J^1\mathbb{R}^3$ produces the following differential operator:

$$\Delta_\omega(v) = dq_1 \wedge d(v_{q_1}) \wedge d(v_{q_3}) = (v_{q_1q_2}v_{q_3q_3} - v_{q_1q_3}v_{q_2q_3})\, dq_1 \wedge dq_2 \wedge dq_3.$$

Example 11.1.4 *The differential 2-form*

$$\omega = dp_1 \wedge dq_2 - dp_2 \wedge dq_1$$

on $J^1\mathbb{R}^2$ represents the two-dimensional Laplace operator

$$\Delta_\omega(v) = (v_{q_1q_1} + v_{q_2q_2})\, dq_1 \wedge dq_2.$$

In the general case, the form

$$\omega = \sum_{i=1}^{n} dq_1 \wedge \cdots \wedge dq_{i-1} \wedge dp_i \wedge dq_{i+1} \wedge \cdots \wedge dq_n$$

on $J^1\mathbb{R}^n$ generates the Laplace operator

$$\Delta_\omega(v) = \left(\sum_{i=1}^{n} v_{q_i q_i}\right) dq_1 \wedge \cdots \wedge dq_n. \tag{11.6}$$

Example 11.1.5 *Any differential n-form*

$$\omega = \omega_0 \wedge \alpha + d\omega_0 \wedge \beta$$

on J^1M, where $\alpha \in \Omega^{n-1}(J^1M)$, $\beta \in \Omega^{n-2}(J^1M)$ and ω_0 is the Cartan form, gives the zero operator.

Example 11.1.6 *Two differential 2-forms*

$$\omega = dq_1 \wedge du \qquad \text{and} \qquad \varpi = p_2\, dq_1 \wedge dq_2$$

on $J^1\mathbb{R}^2$ generate the same operator:

$$\Delta_\omega(v) = dq_1 \wedge (v_{q_1}\, dq_1 + v_{q_2}\, dq_2) = v_{q_2}\, dq_1 \wedge dq_2,$$
$$\Delta_\varpi(v) = v_{q_2}\, dq_1 \wedge dq_2.$$

11.2 Effective differential forms

Differential forms on J^1M vanishing on any integral manifold, and therefore producing zero differential operators, form a graded ideal of the exterior algebra $\Omega^*(J^1M)$.

We denote this ideal by $I^* = \oplus_{s\geq 0} I^s$, $I^s \subset \Omega^s(J^1M)$.

It follows from the Lepage theorem that the ideal I^* is generated by forms

$$\omega_0 \wedge \alpha + d\omega_0 \wedge \beta. \tag{11.7}$$

Note also that $I^0 = 0$ and $I^s = \Omega^s(J^1M)$ for $s \geq n+1$.

We call elements of the quotient module $\Omega^s(J^1M)/I^s$ *effective* s-forms ($s \leq n$):

$$\Omega^s_\varepsilon(J^1M) \cong \Omega^s(J^1M)/I^s. \tag{11.8}$$

One should use these forms to define Monge–Ampère operators.

The choice of the Cartan form ω_0 on J^1M gives us a more explicit description of effective forms.

First of all, let us note that the choice of ω_0 gives us the decomposition of the tangent space:

$$T_a(J^1M) = C(a) \oplus \mathbb{R}X_{1,a} \tag{11.9}$$

and the cotangent space:

$$T_a^*(J^1M) = C^*(a) \oplus \mathbb{R}\omega_{0a} \tag{11.10}$$

for any point $a \in J^1M$.

So, if we identify the dual space $C^*(a)$ with a subspace of $T_a^*(J^1M)$ consisting of 1-forms degenerating along $X_{1,a}$:

$$C^*(a) \cong \{\alpha \in T_a^*(J^1M) \mid \alpha(X_{1,a}) = 0\} \tag{11.11}$$

then decomposition (11.10) also generates the decomposition of s-forms:

$$\Lambda^s(T_a^*(J^1M)) = \Lambda^s(C^*(a)) \oplus \omega_{0a} \wedge \Lambda^{s-1}(C^*(a)), \tag{11.12}$$

where

$$\Lambda^s(C^*(a)) \stackrel{\text{def}}{=} \{\alpha \in \Lambda^s(T_a^*(J^1M)) \mid X_{1,a} \rfloor \alpha = 0\}.$$

Let us denote by $\Omega^s(C^*)$ the module of differential s-forms on J^1M degenerating along the vector field X_1.

Note that in the canonical local coordinates $(q, u, p) = (q_1, \ldots, q_n, u, p_1, \ldots, p_n)$ elements $\omega \in \Omega^s(C^*)$ have the following representation:

$$\omega = \sum_{I,J} \omega_{I,J}(q, u, p)\, dq^I \wedge dp^J, \tag{11.13}$$

where $\omega_{I,J}$ are smooth functions, I, J are multi-indices: $I = (i_1, \ldots, i_k)$, $J = (j_1, \ldots, j_{s-k})$ and

$$dq^I \stackrel{\text{def}}{=} dq^{i_1} \wedge \cdots \wedge dq^{i_k}, \quad dp^J \stackrel{\text{def}}{=} dp^{j_1} \wedge \cdots \wedge dp^{j_{s-k}}.$$

The corresponding decomposition of modules of differential forms

$$\Omega^s(J^1M) = \Omega^s(C^*) \oplus \omega_0 \wedge \Omega^{s-1}(C^*) \tag{11.14}$$

follows from (11.12).

On the formula level this decomposition can be expressed as

$$\alpha = \widetilde{\alpha} + \omega_0 \wedge \alpha_1 \tag{11.15}$$

where $\alpha \in \Omega^s(J^1M)$, $\widetilde{\alpha} \in \Omega^s(C^*)$ and

$$\alpha_1 = X_1 \rfloor \alpha \in \Omega^{s-1}(C^*).$$

Let us define a projection

$$p : \Omega^s(J^1M) \to \Omega^s(C^*),$$

by

$$p(\alpha) \stackrel{\text{def}}{=} \widetilde{\alpha}$$

or

$$p(\alpha) = \alpha - \omega_0 \wedge (X_1 \rfloor \alpha). \tag{11.16}$$

Example 11.2.1 *If* $\alpha = du$, *then* $p(\alpha) = p\, dq$.

Example 11.2.2 $p(\omega_0) = 0$ *and* $p(d\omega_0) = d\omega_0$.

The module

$$\Omega^*(C^*) \stackrel{\text{def}}{=} \bigoplus_{s=0}^{n} \Omega^s(C^*)$$

is obviously an algebra with respect to the operation of exterior multiplication.

In algebra $\Omega^*(C^*)$ we consider two operators $\top$ and $\bot$. Namely, at each point $a \in J^1M$ the restriction Ω of $d_a\omega_0$ on $C(a)$ determines the symplectic structure, and, therefore, the corresponding operators: $\top(\omega) = \Omega \wedge \omega$ and $\bot(\omega) = X_\Omega \rfloor \omega$ for all $\omega \in \Lambda^*(C^*(a))$. We shall use these operators on differential forms

$$\top : \Omega^s(C^*) \to \Omega^{s+2}(C^*) \qquad \bot : \Omega^s(C^*) \to \Omega^{s-2}(C^*)$$

where

$$\top(\omega) \stackrel{\text{def}}{=} d\omega_0 \wedge \omega,$$

$$\bot(\omega) \stackrel{\text{def}}{=} X_\Omega \rfloor \omega$$

and $X_\Omega \in \Lambda^2(C)$ is a bivector field on the Cartan distribution dual to $d\omega_0$ with respect to the symplectic structure on it.

We call a differential form $\omega \in \Omega^s(C^*)$, $s \leq n$, *effective* if

$$\perp(\omega) = 0.$$

In other words, effective forms in the old sense have representatives in $\Omega^*(C^*)$ which are effective in the above sense and which are effective in correspondence with linear algebra notions.

Note also that a realization of effective forms as forms on the Cartan distribution depends heavily on decomposition (11.10) and therefore depends on the choice of a generator of the Cartan distribution C.

It follows from the Lepage theorem that for any differential s-form $\omega \in \Omega^s(C^*)$, $s \leq n$, there exists a unique effective part $\omega_\varepsilon \in \Omega^s_\varepsilon(J^1M)$, and this part can be found by the formula

$$\omega_\varepsilon = \left(\sum_{s \geq 0} (-1)^s \frac{1}{s+1} \top_s \perp_s \right)(\omega).$$

Example 11.2.3 *Any 1-form on the Cartan distribution is effective.*

Example 11.2.4 *Let $M = \mathbb{R}^2$ and $\omega = dq_1 \wedge dp_1 \in \Omega^2(C^*)$. Then,*

$$\omega_\varepsilon = \omega - \frac{1}{2}\Omega = \frac{1}{2}(dq_1 \wedge dp_1 - dq_2 \wedge dp_2)$$

is the effective part of ω.

Lemma 11.2.1 *Any differential form $\omega \in I^s$ admits the following decomposition:*

$$\alpha = \omega_0 \wedge \alpha_1 + d\omega_0 \wedge \alpha_2, \qquad (11.17)$$

where the form $\alpha_1 \in \Omega^{s-1}(C^*)$ is uniquely determined by α, and the form $\alpha_2 \in \Omega^{s-2}(C^*)$ is also uniquely determined if $s < n+2$.

Proof Indeed, from (11.15) it follows that

$$\alpha = p(\alpha) + \omega_0 \wedge (X_1 \rfloor \alpha).$$

Since $\alpha \in I^s$, we see that $p(\alpha)$ vanishes on all integral manifolds. Therefore, $p(\alpha) = \top \alpha_2$, where α_2 is uniquely determined if $s < n+2$. ■

It follows from this lemma that for any differential form $\alpha \in \Omega^s(J^1M)$ we have

$$\alpha = \alpha_\varepsilon + \omega_0 \wedge \alpha_1 + d\omega_0 \wedge \alpha_2 \tag{11.18}$$

where α_ε is the effective part of $p(\alpha)$ and $\alpha_1 \in \Omega^{s-1}(C^*)$, $\alpha_2 \in \Omega^{s-2}(C^*)$.

Example 11.2.5 *A basis of the module $\Omega^2_\varepsilon(J^1\mathbb{R}^2)$ consists of the following differential 2-forms:*

$$dq_1 \wedge dq_2, \quad dq_1 \wedge dp_2, \quad dq_2 \wedge dp_1, \quad dp_1 \wedge dp_2, \quad dq_1 \wedge dp_1 - dq_2 \wedge dp_2.$$

11.3 Calculus on $\Omega^*(C^*)$

The ideal I^* is differential in the sense that the de Rham operator preserves this ideal, $dI^s \subset I^{s+1}$, and therefore induces an operator $d_\varepsilon : \Omega^{s+1}_\varepsilon(J^1M) \to \Omega^{s+1}_\varepsilon(J^1M)$ on the module of effective forms.

We represent this operator in terms of decomposition given by choice of the Cartan form ω_0. To simplify formulas we shall use L_f and ι_f instead of L_{X_f} and ι_{X_f}.

At first we introduce the operator $d_p : \Omega^s(C^*) \to \Omega^{s+1}(C^*)$ by

$$d_p \overset{\text{def}}{=} p \circ d$$

or $d_p\alpha = d\alpha - \omega_0 \wedge (\iota_1\, d\alpha)$.

It is easy to check that this operator satisfies the following relations.

1. $d_p(\omega_1 \wedge \omega_2) = d_p\omega_1 \wedge \omega_2 + (-1)^{\deg \omega_1}\omega_1 \wedge d_p\omega_2$.

2. $d_p^2 + \top \circ L_1 = 0$.

3. $d_p \circ d = 0$.

4. $\top \circ d_p = d_p \circ \top$.

5. $\iota_f(d\omega_0) = -d_pf$.

6. $L_1 \circ d_p = d_p \circ L_1$.

Using these properties it is easy to find an action d_p in the canonical local coordinates q, u, p on J^1M.

Indeed, for any function $f \in C^\infty(J^1M)$ we have

$$d_pf = df - X_1(f)\omega_0 = \sum_{i=1}^{n} \frac{df}{dq_i} dq_i + \sum_{i=1}^{n} \frac{\partial f}{\partial p_i} dp_i,$$

where

$$\frac{df}{dq_i} \stackrel{\text{def}}{=} \frac{\partial f}{\partial q_i} + p_i \frac{\partial f}{\partial u},$$

and for any s-form

$$\alpha = \sum_{I,J} \alpha_{I,J}(q,u,p)\, dq^I \wedge d^J p.$$

we have

$$d_p\alpha = \sum_{I,J} d_p(\alpha_{I,J}(q,u,p)) \wedge dq^I \wedge d^J p.$$

Lemma 11.3.1 *Let the differential form $\alpha \in I^s$ have the following form:*

$$\alpha = \omega_0 \wedge \alpha_1 + d\omega_0 \wedge \alpha_2.$$

Then,

$$d\alpha = -\omega_0 \wedge d_p(\alpha_1 + d_p\alpha_2) + d\omega_0 \wedge (\alpha_1 + d_p\alpha_2). \tag{11.19}$$

Proof We have

$$d\alpha = -\omega_0 \wedge d\alpha_1 + d\omega_0 \wedge (\alpha_1 + d\alpha_2) \tag{11.20}$$

and

$$d_p\alpha_i = p(d\alpha_i) = d\alpha_i - \omega_0 \wedge \iota_1(d\alpha_i).$$

Replacing $d\alpha_i$ by $d_p\alpha_i + \omega_0 \wedge \iota_1(d\alpha_i)$, we obtain

$$d\alpha = -\omega_0 \wedge d_p\alpha_1 + d\omega_0 \wedge (\alpha_1 + d_p\alpha_2) + d\omega_0 \wedge \omega_0 \wedge \iota_1(d\alpha_2). \tag{11.21}$$

From the above properties of d_p it follows that

$$d_p^2\alpha_2 = -d\omega_0 \wedge L_1(\alpha_2).$$

Since $\alpha_2 \in \Omega^{s-2}(C^*)$, we see that $X_1 \rfloor \alpha_2 = 0,\ L_1(\alpha_2) = \iota_1\, d\alpha_2$ and

$$-d\omega_0 \wedge \iota_1(d\alpha_2) = d_p^2 \alpha_2.$$

Therefore (11.19) now follows from (11.21). ■

Theorem 11.3.1 *The complex*

$$0 \to I^1 \overset{d}{\to} I^2 \to \cdots \to I^s \overset{d}{\to} I^{s+1} \to \cdots \to I^{2n+1} \to 0$$

is exact for all s except the case where $s = \dim M + 1$.

Proof Suppose that the s-form $\alpha \in I^s$ is closed. Then $\alpha = \omega_0 \wedge \alpha_1 + d\omega_0 \wedge \alpha_2$ and using Lemma 11.3.1, we obtain

$$d_p(\alpha_1 + d_p\alpha_2) = 0$$

and

$$\top(\alpha_1 + d_p\alpha_2) = d\omega_0 \wedge (\alpha_1 + d_p\alpha_2) = 0.$$

If $s < n + 1$, then the second relation implies that $\alpha_1 + d_p\alpha_2 = 0$.

Then

$$\alpha_1 = -d_p\alpha_2 = d\alpha_2 - \omega_0 \wedge \iota_1(d\alpha_2)$$

and, therefore, by Lemma 11.2.1,

$$\alpha = -\omega_0 \wedge d\alpha_2 + \alpha_2 \wedge d\omega_0 = d(\omega_0 \wedge \alpha_2).$$

This means that the complex is exact in the term.

Note that if $s = \dim M + 1$, then the equality $\alpha_1 + d_p\alpha_2 = 0$ does not follow from the formula

$$d\omega_0 \wedge (\alpha_1 + d_p\alpha_2) = 0.$$

In dimensions $s \geq \dim M + 1$ we have $I^s = \Omega^s(J^1M)$ and therefore the complex coincides with the de Rham complex, but $H^s(J^1M, \mathbb{R}) = 0$ if $s \geq \dim M + 1$. ■

Corollary 11.3.1 *Cohomology $H_\varepsilon^{\cdot}(J^1M)$ of the factor-complex*

$$C^\infty(J^1M) \overset{d_p}{\to} \Omega_\varepsilon^1(J^1M) \overset{d_\varepsilon}{\to} \Omega_\varepsilon^2(J^1M) \to \cdots$$
$$\overset{d_\varepsilon}{\to} \Omega_\varepsilon^{n-1}(J^1M) \overset{d_\varepsilon}{\to} \Omega_\varepsilon^n(J^1M) \to 0$$

is isomorphic to the de Rham cohomology, $H^s_\varepsilon(J^1M) \cong H^s(M,\mathbb{R})$ *if* $s \neq \dim M$.

11.4 The Euler operator

In this section we shall investigate the highest effective cohomology group

$$H^n_\varepsilon(J^1M) = \Omega^n_\varepsilon(J^1M)/d_\varepsilon(\Omega^{n-1}_\varepsilon(J^1M))$$

where $n = \dim M$.

Let $\alpha \in \Omega^n_\varepsilon(C^*)$ be an effective n-form such that the class $\alpha + I^n$ contains a closed differential n-form, say, $\alpha + \omega_0 \wedge x + d\omega_0 \wedge y$, where $x \in \Omega^{n-1}(C^*)$ and $y \in \Omega^{n-2}(C^*)$.

Then

$$d(\alpha + \omega_0 \wedge x + d\omega_0 \wedge y) = 0,$$

and using Lemma 11.3.1, we obtain

$$d\alpha - \omega_0 \wedge d_p(x + d_p y) + d\omega_0 \wedge (x + d_p y) = 0.$$

Since

$$d\alpha = d_p\alpha + \omega_0 \wedge L_1(\alpha),$$

we obtain

$$d_p\alpha + \omega_0 \wedge (L_1(\alpha) - d_p(x + d_p y)) + d\omega_0 \wedge (x + d_p y) = 0.$$

Note that $\top : \Omega^{n-1}(C^*) \to \Omega^{n+1}(C^*)$ is the isomorphism and therefore $d_p\alpha = d\omega_0 \wedge \alpha'$, where $\alpha' = \perp d_p\alpha$.

Therefore

$$\omega_0 \wedge (L_1(\alpha) - d_p(x + d_p y)) + d\omega_0 \wedge (\alpha' + x + d_p y) = 0,$$

and the class $\alpha + I^n$ contains a closed form if and only if the following system is solvable with respect to forms x and y:

$$d_p(x + d_p y) = L_1(\alpha),$$
$$x + d_p y = -\alpha'.$$

The necessary condition for solvability of this system is

$$L_1(\alpha) + d_p\alpha' = 0$$

and the condition takes the form

$$L_1(\alpha) + d_p \circ \perp \circ d_p(\alpha) = 0.$$

Let us denote by

$$\mathcal{E} \stackrel{\text{def}}{=} L_1 + d_p \circ \perp \circ d_p$$

the last operator. We call $\mathcal{E}$ *Euler's operator.*

We shall see later on that the Euler operator is the main operator in calculation of conservation laws and in the calculus of variations. This will explain our terminology.

To summarize, we obtain the following.

Theorem 11.4.1 *A class $\alpha + I^n$ of an effective form $\alpha \in \Omega^n_\varepsilon(C^*)$ contains a closed form if and only if*

$$\mathcal{E}(\alpha) = 0.$$

Corollary 11.4.1 *Let α be an effective n-form such that $\mathcal{E}(\alpha) = 0$. Then,*

$$d(\alpha - \omega_0 \wedge \perp d_p\alpha) = 0.$$

Proof Indeed, one can take the following solution of the system: $x = -\alpha'$, $y = 0$. ■

We now list the main properties of the Euler operator.

Lemma 11.4.1 *The operator $\mathcal{E}$ maps $\Omega^s_\varepsilon(J^1M)$ to itself.*

Proof First, let us prove that

$$\mathcal{E} : \Omega^n(C^*) \to \Omega^n(C^*).$$

Suppose that $\alpha \in \Omega^s(C^*)$. Then, since $\iota_1 \alpha = 0$, we see that

$$L_1(\alpha) = \iota_1 \, d\alpha,$$

and

$$\iota_1 L_1(\alpha) = 0,$$

i.e., $L_1(\omega) \in \Omega^s(C^*)$.

Furthermore, operators $\perp$ and d_p act on $\Omega^*(C^*)$. Therefore, $\mathcal{E}(\alpha) \in \Omega^s(C^*)$

Secondly, we note that the operator $\top$ commutes with differential d_p, the Lie derivative L_1 and $[\perp, \top] = -1$ on $\Omega^{n+1}(C^*)$. Then $\top(\alpha) = 0$ because α is effective, and

$$\begin{aligned}\top \circ \mathcal{E}(\alpha) &= \top \circ L_1(\alpha) + \top \circ d_p \circ \perp \circ d_p(\alpha) = d_p \circ \top \circ \perp \circ d_p(\alpha)\\ &= d_p \circ \perp \circ \top \circ d_p(\alpha) + d_p^2(\alpha) = d_p \circ \perp \circ d_p \circ \top(\alpha) - \top \circ L_1(\alpha)\\ &= 0.\end{aligned}$$

Therefore, $\mathcal{E}(\alpha)$ is effective. ■

Corollary 11.4.2 *The following sequence*

$$0 \to \ker \mathcal{E}/d(\Omega_\varepsilon^{n-1}(J^1M)) \to H_\varepsilon^n(J^1M) \to \Omega_\varepsilon^n(J^1M)/\ker \mathcal{E} \to 0$$

is exact.

Theorem 11.4.2 *The Euler operator $\mathcal{E} : \Omega_\varepsilon^n(C^*) \to \Omega_\varepsilon^n(C^*)$ satisfies the following relations:*

1. $\mathcal{E} \circ L_1 = L_1 \circ \mathcal{E}$;
2. $d_p \circ \mathcal{E} = 0$;
3. $\mathcal{E}^2 = L_1 \circ \mathcal{E}$;
4. $\mathcal{E}(f\alpha) = f\mathcal{E}(\alpha) + X_1(f)\alpha + d_pf \wedge \perp d_p\alpha + d_p(\iota_f\alpha)$, *where* $\alpha \in \Omega_\varepsilon^n(J^1M)$, *and* $f \in C^\infty(J^1M)$;
5. $p \circ L_f = f\mathcal{E} + d_p \circ (\iota_f - f \perp \circ d_p) + \top \circ \iota_f \circ \perp \circ d_p$ *on* $\Omega_\varepsilon^n(J^1M)$.

Proof The proof of the first property is obvious.

One has

$$\begin{aligned}d_p \circ \mathcal{E} &= d_p \circ L_1 + d_p^2 \circ \perp \circ d_p = d_p \circ L_1 - \top \circ L_1 \circ \perp \circ d_p\\ &= d_p \circ L_1 + [\perp, \top] \circ d_p \circ L_1 - \perp \circ \top \circ d_p \circ L_1\\ &= d_p \circ L_1 - d_p \circ L_1 - \perp \circ d_p \circ L_1 \circ \top = 0.\end{aligned}$$

We have

$$
\begin{aligned}
d_p \circ \perp \circ d_p \circ d_p \circ \perp d_p &= -L_1 \circ d_p \circ \perp \circ \top \circ \perp \circ d_p \\
&= -L_1 \circ d_p \circ [\perp, \top] \circ \perp \circ d_p \\
&= -L_1 \circ d_p \circ \perp \circ d_p,
\end{aligned}
$$

and therefore

$$
\mathcal{E}^2 = L_1^2 + 2L_1 \circ d_p \circ \perp \circ d_p - L_1 \circ d_p \circ \perp \circ d_p = L_1 \circ \mathcal{E}.
$$

One has

$$
\begin{aligned}
d_p \circ \perp \circ d_p(f\omega) &= d_p \circ \perp (d_p f \wedge \omega + f d_p \omega) \\
&= d_p(X_f \rfloor \omega) + d_p(f \perp (d_p\omega)) \\
&= d_p(X_f \rfloor \omega) + d_p f \wedge \perp (d_p\omega) + f d_p \perp (d_p\omega).
\end{aligned}
$$

Using the formula $L_X = d \circ \iota_X + \iota_X \circ d$, we obtain

$$
\begin{aligned}
p(L_f\alpha) &= d_p(\iota_f \alpha) + p(\iota_f d\alpha) = d_p(\iota_f \alpha) + \iota_f d_p\alpha + f L_1(\alpha) \\
&= f\mathcal{E}(\alpha) - f d_p \circ \perp \circ d_p(\alpha) + d_p(\iota_f \alpha) + \iota_f d_p \alpha \\
&= f\mathcal{E}(\alpha) + d_p(\iota_f \alpha - f \perp (d_p\alpha)) + \iota_f d_p\alpha + d_p f \wedge \perp (d_p\alpha)
\end{aligned}
$$

for any effective n-form α.

As we have seen

$$
d_p\alpha = \top \perp d_p\alpha.
$$

Therefore,

$$
\iota_f d_p\alpha = -d_p f \wedge \perp (d_p\alpha) + \top(\iota_f \perp d_p\alpha). \qquad \blacksquare
$$

11.5 Solutions

Let us return to Monge–Ampère operators. From Example 11.1.5 and from (11.16) and (11.18) we see that forms ω, $p(\omega)$ and ω_ε generate the same differential operator:

$$
\Delta_\omega = \Delta_{p(\omega)} = \Delta_{\omega_\varepsilon}.
$$

In general, two differential n-forms ω and ϖ generate the same Monge–Ampère operator if and only if $\omega - \varpi \in I^n$. This means that the operator Δ_ω is uniquely determined by the effective part of $p(\omega)$.

So, from now on we shall suppose that operators Δ_ω are generated by effective forms $\omega \in \Omega^n_\varepsilon(J^1M)$.

Any Monge–Ampère differential operator Δ_ω defines a differential equation $\Delta_\omega(v) = 0$ which we call the *Monge–Ampère* equation.

Example 11.5.1 *For differential operator (11.3) the equation has the form*

$$3(v')^2v'' - 1 = 0. \tag{11.22}$$

In the more general case, the effective differential form $A(x,u,p)\,dx + B(x,u,p)\,dp$ generates the following ordinary differential equation:

$$A(x,v,v') + B(x,v,v')v'' = 0.$$

Example 11.5.2 *For differential operators (11.4)–(11.6) the equations have the forms:*

$$\det \operatorname{Hess} v = 0,$$

$$v_{q_1q_2}v_{q_3q_3} - v_{q_1q_3}v_{q_2q_3} = 0$$

and

$$\sum_{i=1}^{n} v_{q_iq_i} = 0$$

respectively.

Example 11.5.3 *The following equation is known as* the Von Karman equation *on a plane [44]. It arises in aerodynamics and it describes the behavior of the velocity potential v in the transonic approximation of gas (see [44]):*

$$v_{q_1}v_{q_1q_1} - v_{q_2q_2} = 0. \tag{11.23}$$

The effective form

$$\omega = p_1\,dq_2 \wedge dp_1 + dq_1 \wedge dp_2 \tag{11.24}$$

corresponds to the equation.

Example 11.5.4 *The form*

$$\omega = u^{\alpha}\, dq_1 \wedge dp_2 + (\alpha u^{\alpha-1} p_2^2 - p_1 + F(u)) dq_1 \wedge dq_2$$

corresponds to the equation

$$v_t = (v^{\alpha} v_x)_x + F(v).$$

Here $q_1 = t$, $q_2 = x$, α *is a real number and* F *is a smooth function. This equation is often called the* reaction–diffusion equation *[6, 87] or* the equation of non-linear thermal conductivity *[20]. Later on it will be considered in more detail.*

Example 11.5.5 *An effective differential 4-form that corresponds to the* Khokhlov–Zabolotskaya equation *[98]*

$$\frac{\partial}{\partial q_1}\left(\frac{\partial v}{\partial q_2} - v\frac{\partial v}{\partial q_1}\right) = \frac{\partial^2 v}{\partial q_3^2} + \frac{\partial^2 v}{\partial q_4^2} \tag{11.25}$$

is

$$\begin{aligned}\omega = & \frac{1}{2} dp_1 \wedge dq_1 \wedge dq_3 \wedge dq_4 - \frac{1}{2} dp_2 \wedge dq_2 \wedge dq_3 \wedge dq_4 \\ & + u dp_1 \wedge dq_2 \wedge dq_3 \wedge dq_4 + p_1^2 dq_1 \wedge dq_2 \wedge dq_3 \wedge dq_4 \\ & + dp_3 \wedge dq_1 \wedge dq_2 \wedge dq_4 - dp_4 \wedge dq_1 \wedge dq_2 \wedge dq_3.\end{aligned}$$

This equation describes the propagation of a bounded three-dimensional sound beam [98] and later on it will be considered in more detail.

Let us write the coordinate representation of the equation $\Delta_\omega = 0$ for the classical case $M = \mathbb{R}^2$.

It is not hard to see that any effective 2-form ω on $J^1\mathbb{R}^2$ has the following form:

$$\begin{aligned}\omega = & E\, dq_1 \wedge dq_2 + B(dq_1 \wedge dp_1 - dq_2 \wedge dp_2) \\ & + C\, dq_1 \wedge dp_2 - A\, dq_2 \wedge dp_1 + D\, dp_1 \wedge dp_2,\end{aligned} \tag{11.26}$$

where A, B, C, D and E and are smooth functions on $J^1\mathbb{R}^2$.

The corresponding equation then takes the form of the classical Monge–Ampère equation:

$$Av_{xx} + 2Bv_{xy} + Cv_{yy} + D(v_{xx}v_{yy} - v_{xy}^2) + E = 0,$$

where $x = q_1$, and $y = q_2$.

In general case we denote the Monge–Ampère differential equation $\Delta_\omega = 0$ by E_ω.

Note that for any non-vanishing smooth function h on J^1M two forms ω and $h\omega$ generate the same differential equations:

$$E_\omega = E_{h\omega}.$$

A *classical* (or *regular*) *solution* of the Monge–Ampère equation E_ω in an open domain $N \subset M$ is a function $v \in C^\infty(N)$ such that

$$\Delta_\omega(v) \equiv 0.$$

A *multivalued,* or *generalized solution* of the Monge–Ampère equation E_ω is an integral submanifold $i : L \hookrightarrow J^1M, \dim L = \dim M$ of the curtan distribution, such that $\omega|_L = 0$.

As we have seen a function $v \in C^\infty(M)$ is a classic solution of the Monge–Ampère equation if and only if

$$= L_v \overset{\text{def}}{=} j_1(v)(M)$$

is a multivalued solution.

Example 11.5.6 *The curve*

$$L : \begin{cases} x = t^3, \\ u = \dfrac{3}{4}t^4, \\ p = t, \end{cases}$$

$t \in \mathbb{R}$, is a multivalued solution of the ordinary differential equation (11.22). Indeed, the map

$$i : L \hookrightarrow J^1M = \mathbb{R}^3, \quad t \longmapsto \left(t^3, \frac{3}{4}t^4, t\right)$$

is an immersion, and

$$\omega_0|_L = 0, \quad \omega|_L = 0.$$

So, L is a multivalued solution of Monge–Ampère equations (11.22).

Example 11.5.7 *The Lissajou figure $u = \cos(at)$, $x = \sin(bt)$, corresponds to a multivalued solution of the differential equation (see Figure 11.1)*

$$(1 - x^2)y'' + xy' + \lambda y = 0$$

with $\lambda = a^2/b^2$.

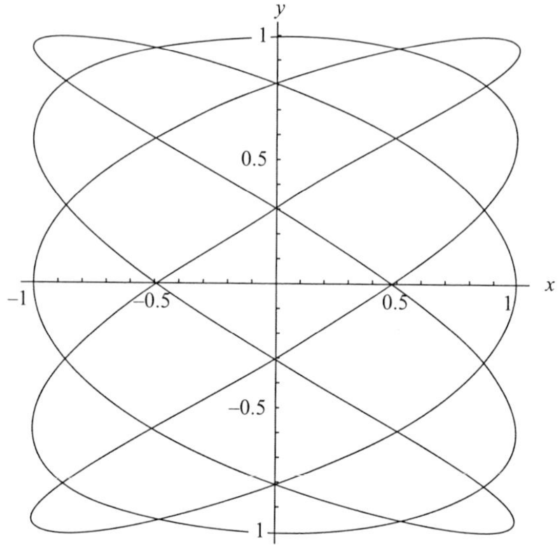

Figure 11.1. The Lissajous figure.

Example 11.5.8 *The surface*

$$L : \begin{cases} q_1 = \dfrac{1}{2}p_1^2 + C_1, \\ q_2 = p_2 + C_2, \\ u = \dfrac{1}{3}p_1^3 + \dfrac{1}{2}p_2^2, \end{cases}$$

$L \subset J^1\mathbb{R}^2$, is a multivalued solution of (11.23). Here C_1, C_2 are constants. The projection of L to $J^0\mathbb{R}^2$ is a two-valued solution.

Similarly,

$$L : \begin{cases} q_1 = \dfrac{1}{2}p_1^2 + C_1, \\ q_2 = p_2 + C_2, \\ u = \dfrac{1}{3}p_1^3 + \dfrac{1}{2}p_2, \end{cases}$$

is a multivalued solution of the non-homogeneous Von Karman equation (see Figure 11.2)

$$v_x v_{xx} - v_{yy} = 1.$$

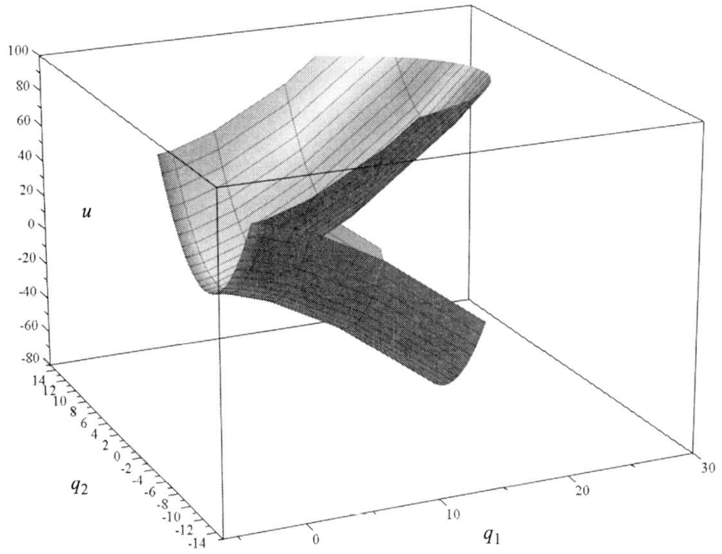

Figure 11.2. The multivalued solution of the non-homogeneous Von Karman equation.

Note that both of these solutions represented by smooth manifolds L are not smooth in the projection on the space $J^0\mathbb{R}^2$.

11.6 Monge–Ampère equations of divergent type

We call elements of $\ker \mathcal{E}$ *divergent forms*, and the corresponding operators Δ_ω and equations E_ω we call *divergent type operators* and *equations* respectively.

The next examples show the relations between this definition and commonly used notions.

Example 11.6.1 *The classical quasi-linear differential equation of divergent type has the form*

$$\sum_{i=1}^{n} \frac{\partial}{\partial q_i} \left(A_i \left(q, v, \frac{\partial v}{\partial q} \right) \right) = 0. \tag{11.27}$$

Consider the following effective form:

$$\theta = \sum_{i=1}^{n} (-1)^{i+1} A_i(q, u, p) dq_1 \wedge \cdots \wedge \widehat{dq_i} \wedge \cdots \wedge dq_n.$$

Then the n-form $\omega = d_p\theta$ defines the equation, and therefore $\mathcal{E}(\omega) = 0$, because the class of ω contains the exact form.

Example 11.6.2 *The Von Karman equation (see [44])*

$$v_x v_{xx} - v_{yy} - v_{zz} = 0$$

is of divergent type, because it can be written as

$$\frac{1}{2}\frac{\partial}{\partial x}(v_x^2) + \frac{\partial}{\partial y}(-v_y) + \frac{\partial}{\partial z}(-v_z) = 0.$$

For the corresponding form we have

$$\omega = p_1\, dp_1 \wedge dq_2 \wedge dq_3 - dq_1 \wedge dp_2 \wedge dq_3 - dq_1 \wedge dq_2 \wedge dp_3$$

and $\omega = d_p\theta$, where

$$\theta = \frac{1}{2}p_1^2\, dq_2 \wedge dq_3 + p_2\, dq_1 \wedge dq_3 - p_3\, dq_1 \wedge dq_2.$$

Therefore $\mathcal{E}(\omega) = 0$.

Example 11.6.3 *The classical Monge–Ampère equation* $\det \operatorname{Hess} v = 1$ *is of divergent type. Indeed, one can take*

$$\theta = p_1\, dp_2 \wedge \cdots \wedge dp_n - q_1\, dq_2 \wedge \cdots \wedge dq_n.$$

Then $\omega = d_p\theta$ is the effective form of divergent type that corresponds to the equation.

12

Symmetries and contact transformations of Monge–Ampère equations

12.1 Contact transformations

Let us define an action of the group of contact diffeomorphisms $\mathrm{Ct}(J^1M)$ on Monge–Ampère operators.

Let F be a contact transformation, $F \in \mathrm{Ct}(J^1M)$, then F^* preserves the Cartan ideal I, $F^*(I^s) = I^s$, and therefore determines a map of effective forms:

$$F^* : \omega + I^s \longmapsto F^*(\omega) + I^s.$$

But F, in general, does not preserve the Cartan form ω_0, and therefore, does not act directly on representatives in $\Omega^s(C^*)$ of the effective forms. Thus we shall define the action F on $\omega = \omega \in \Omega^s_\varepsilon(C^*)$ by taking the effective part $F^*(\omega)_\varepsilon$.

In a similar way, we define an action of the contact diffeomorphism $F \in \mathrm{Ct}(J^1M)$ on the Monge–Ampère operator Δ_ω, where $\omega \in \Omega^n_\varepsilon(J^1M)$, by

$$F(\Delta_\omega) \stackrel{\text{def}}{=} \Delta_{F^*(\omega)_\varepsilon}.$$

Note that, for any $F \in \mathrm{Ct}(J^1M)$ and any integral manifold $L \subset J^1M$ we have

$$F^*(\omega)_\varepsilon|_L = \omega|_{F(L)}.$$

Hence, if L is a multivalued solution of the equation $E_{F^*(\omega)_\varepsilon}$, then $F(L)$ is a solution of the equation E_ω.

We say that two Monge–Ampère operators Δ_{ω_1} and Δ_{ω_2}, where $\omega_1, \omega_2 \in \Omega^n_\varepsilon(J^1M)$, are *contact equivalent* if there exists a contact transformation F, such that

$$\omega_2 = F^*(\omega_1)_\varepsilon,$$

and that two Monge–Ampère equations E_ω and E_ϖ are *contact equivalent* if there exists a contact transformation F and a non-vanishing function $\lambda_F \in C^\infty(J^1M)$, such that

$$\omega_2 = \lambda_F F^*(\omega_1)_\varepsilon.$$

As a corollary of our interpretation of Monge–Ampère equations we obtain the following theorem.

Theorem 12.1.1 (Sophus Lie) *The class of Monge–Ampère equations is closed with respect to contact transformations.*

Example 12.1.1 *The Legendre transformation*

$$\varphi : (x,\ u,\ p) \mapsto (p,\ u - px,\ -x)$$

translates the form (11.2) to the form $\varphi^*(\omega) = 3x^2\, dx - dp$. *Then the non-linear equation (11.22) turns into the following linear equation:*

$$v'' = 3x^2.$$

Example 12.1.2 *(Quasilinear equations) Let us consider a quasi-linear equation of the form*

$$A(v_x, v_y)v_{xx} + 2B(v_x, v_y)v_{xy} + C(v_x, v_y)v_{yy} = 0.$$

This equation is represented by the following effective form:

$$\begin{aligned}\omega = {} & B(p_1, p_2)(dq_1 \wedge dp_1 - dq_2 \wedge dp_2) \\ & + C(p_1, p_2)\, dq_1 \wedge dp_2 - A(p_1, p_2)\, dq_2 \wedge dp_1.\end{aligned}$$

After the Legendre transformation

$$\varphi : (q_1,\ q_2,\ u,\ p_1,\ p_2) \mapsto (p_1,\ p_2,\ u - p_1q_1 - p_2q_2,\ -q_1,\ -q_2) \qquad (12.1)$$

we obtain the following effective form:

$$\begin{aligned}\varphi^*(\omega) = {} & B(-q_1, -q_2)(dq_1 \wedge dp_1 - dq_2 \wedge dp_2) \\ & - A(-q_1, -q_2)dq_1 \wedge dp_2 + C(-q_1, -q_2)dq_2 \wedge dp_1,\end{aligned}$$

which corresponds to the linear equation

$$-A(-q_1, -q_2)v_{q_2q_2} + 2B(-q_1, -q_2)v_{q_1q_2} - C(-q_1, -q_2)v_{q_1q_1} = 0.$$

Example 12.1.3 *The Von Karman equation*

$$v_{q_1} v_{q_1 q_1} - v_{q_2 q_2} = 0 \tag{12.2}$$

becomes the linear equation

$$q_1 v_{q_2 q_2} + v_{q_1 q_1} = 0 \tag{12.3}$$

after a Legendre transformation (12.1). The last equation is known as the Triccomi equation.

Example 12.1.4 *(The Monge–Ampère equation) This equation has the form*

$$\det \operatorname{Hess} v = 1$$

and is generated by the effective differential 2-form

$$\omega = dp_1 \wedge dp_2 - dq_1 \wedge dq_2.$$

After the Euler transformation

$$\varphi : (q_1, q_2, u, p_1, p_2) \mapsto (p_1, q_2, u - p_1 q_1, -q_1, p_2)$$

it becomes

$$\omega = dq_2 \wedge dp_1 - dq_1 \wedge dp_2$$

and corresponds to the Laplace equation

$$v_{q_1 q_1} + v_{q_2 q_2} = 0. \tag{12.4}$$

Example 12.1.5 *(TM equation [84, 21, 110]) The Titeica–Morimoto equation (the TM equation) has the following form:*

$$\det \operatorname{Hess} v = (v - q_1 v_{q_1} - q_2 v_{q_2})^4. \tag{12.5}$$

The effective differential 2-form corresponding to (12.5) is

$$\omega = (u - q_1 p_1 - q_2 p_2)^4 \, dq_1 \wedge dq_2 - dp_1 \wedge dp_2.$$

This equation can be simplified by using the Legendre transformation (12.1) into the form

$$F^*(\omega) = u^4 \, dp_1 \wedge dp_2 - dq_1 \wedge dq_2.$$

The corresponding equation has a simpler form than (12.5):

$$v^4 \det \operatorname{Hess} v = 1. \tag{12.6}$$

Example 12.1.6 *The following equation*

$$v_y^2 v_{xx} - 2v_x v_y v_{xy} + v_x^2 v_{yy} + R(v_x, v_y) = 0$$

where $R(v_x, v_y)$ is a homogeneous polynomial of degree 3 describes linear planar webs and their generalizations. This equation transforms into itself under any diffeomorphism $\phi : \mathbb{R} \to \mathbb{R}$, $u \to \phi(u)$.

Let us explain how contact transformations can be used in finding solutions. Let us consider, for example, the following non-linear wave equation [55]

$$v_{yy} + (f(v_x))_x = 0. \tag{12.7}$$

It corresponds to the form

$$\omega = dq_1 \wedge dp_2 - f'(p_1)\, dq_2 \wedge dp_1.$$

The contact transformation

$$\varphi : (q_1, q_2, u, p_1, p_2) \longmapsto (p_1, q_2, u - p_1 q_1, -q_1, p_2)$$

takes ω to

$$\varphi^*(\omega) = dp_1 \wedge dp_2 + f'(q_1)\, dq_1 \wedge dq_2. \tag{12.8}$$

We obtain the equation

$$\det \operatorname{Hess} v + f'(q_1) = 0. \tag{12.9}$$

Consider, for example, solutions v of the last equation such that $v_{q_1 q_2} = 0$.

Then we obtain

$$v(q_1, q_2) = c_0 + c_1 q_1 + c_2 q_2 + \frac{q_2^2}{2c_3} - c_3 \int f(q_1)\, dq_1, \tag{12.10}$$

where c_0, c_1, c_2, c_3 are constants.

Let us write down the integral manifold L that corresponds to the solution:

$$u = c_0 + c_1 q_1 + c_2 q_2 + \frac{q_2^2}{2c_3} - c_3 \int f(q_1)\, dq_1,$$

$$p_1 = c_1 - c_3 f(q_1),$$

$$p_2 = c_2 + \frac{q_2}{c_3}.$$

Then $\varphi(L)$ gives a solution of the original equation. Thus we obtain

$$u = c_0 + c_1 p_1 + c_2 q_2 + \frac{q_2^2}{2c_3} + p_1 q_1 + c_3 \int f(p_1)\, dp_1,$$

$$q_1 = -c_3 f(p_1) - c_1,$$

$$p_2 = \frac{q_2}{c_3} + c_2.$$

Note that these formulae represent a multivalued solution and (p_1, q_2) are coordinates on $\varphi(L)$.

Let us take, for example, the function

$$f(p_1) = \frac{1}{2} p_1^2,$$

which corresponds to the equation

$$v_{q_1} v_{q_1 q_1} + v_{q_2 q_2} = 0. \tag{12.11}$$

In this case the corresponding multivalued solution takes the form (see Figure 12.1)

$$u = c_0 + c_1 p_1 + c_2 q_2 + \frac{q_2^2}{2c_3} + p_1 q_1 + c_3 \frac{p_1^3}{3},$$

$$q_1 = -c_3 \frac{p_1^2}{2} - c_1,$$

$$p_2 = \frac{q_2}{c_3} + c_2.$$

We obtain another type of solution for (12.9) from the assumption that $v_{q_2 q_2} = 0$.

In this case we find the following class of solutions:

$$v(q_1, q_2) = A(q_1) + q_2 \int (f'(q_1))^{1/2}\, dq_1.$$

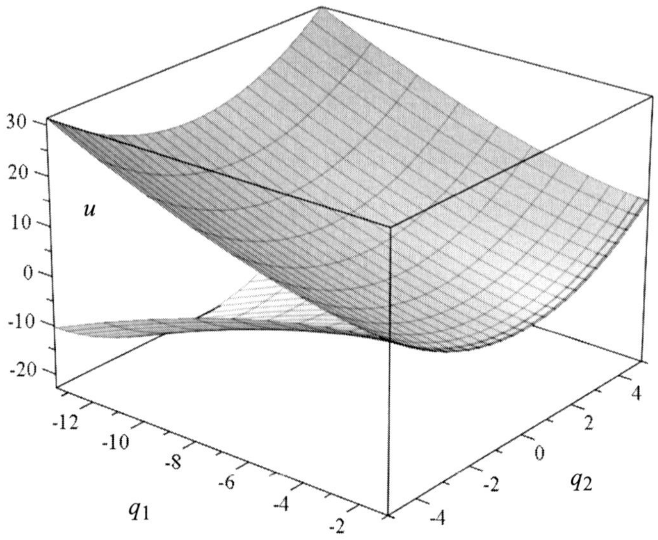

Figure 12.1. The multivalued solution of the Von Karman equation.

The corresponding integral manifold L is

$$u = A(q_1) + q_2 \int (f'(q_1))^{1/2}\, dq_1,$$
$$p_1 = A'(q_1) + q_2\,(f'(q_1))^{1/2},$$
$$p_2 = \int (f'(q_1))^{1/2}\, dq_1$$

and the image $\varphi(L)$ takes the form

$$u = A(p_1) + q_2 \int (f'(p_1))^{1/2}\, dp_1 - p_1 A'(p_1) - p_1 q_2 (f'(p_1))^{1/2},$$
$$q_1 = -A'(p_1) - q_2 (f'(p_1))^{1/2},$$
$$p_2 = \int (f'(p_1))^{1/2}\, dp_1.$$

For example, for the Von Karman equation, $f(p_1) = \frac{1}{2}p_1^2$, taking $A = 0$, we obtain a multivalued solution of the type

$$u = \frac{1}{2}ts^3, \quad q_1 = -ts, \quad q_2 = t, \quad p_1 = s^2, \quad p_2 = \frac{2}{3}s^3$$

(see Figure 12.2).

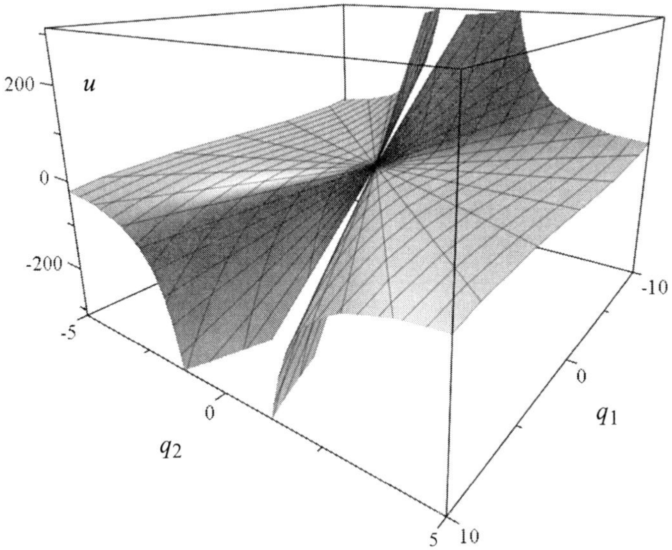

Figure 12.2. The multivalued solution of the Von Karman equation.

Theorem 12.1.2 (Jörgens [69]) *Any smooth solution $v(q_1, q_2)$ of the classical Monge–Ampère equation*

$$\det \operatorname{Hess} v = 1 \tag{12.12}$$

is a polynomial of second degree.

Proof First of all, let us note that the Euler transformation φ transforms our equation into the Laplace equation.

Moreover, if $v_0 = v_0(q_1, q_2)$ is a smooth solution of (12.12) and $L \stackrel{\text{def}}{=} L_{v_0} = j_1(v_0)(\mathbb{R}^2)$ is the corresponding integral manifold, then the image $\varphi(L)$ is a solution of the Laplace equation, and the integral manifold $\varphi(L)$ is projected without singularities onto the (q_1, q_2)-plane.

Indeed, if it not so, a non-zero vector

$$X_a = \lambda_1 \frac{\partial}{\partial p_1} + \lambda_2 \frac{\partial}{\partial p_2}$$

should be tangent to $\varphi(L)$ at some point $a \in \varphi(L)$.

Therefore, the vector

$$d\varphi_a^{-1}(X_a) = -\lambda_1 \left(\frac{\partial}{\partial q_1} + p_1(a)\frac{\partial}{\partial u} \right) + \lambda_2 \frac{\partial}{\partial p_2}$$

should be tangent to L at $\varphi^{-1}(a)$.

But this is impossible, since at the point of contact $\varphi^{-1}(a)$ we should have

$$\left\| \begin{matrix} \dfrac{\partial^2 v_0}{\partial q_1^2} & \dfrac{\partial^2 v_0}{\partial q_1 \partial q_2} \\ \dfrac{\partial^2 v_0}{\partial q_1 \partial q_2} & \dfrac{\partial^2 v_0}{\partial q_2^2} \end{matrix} \right\| \left\| \begin{matrix} -\lambda_1 \\ 0 \end{matrix} \right\| = \left\| \begin{matrix} 0 \\ \lambda_2 \end{matrix} \right\|,$$

for the solution $v_0 = v_0(q_1, q_2)$ of (12.12) and

$$\frac{\partial^2 v_0}{\partial q_1^2} = 0, \qquad \frac{\partial^2 v_0}{\partial q_1 \partial q_2} = -\frac{\lambda_2}{\lambda_1}$$

at the point.

Therefore,

$$\det \left\| \begin{matrix} \dfrac{\partial^2 v_0}{\partial q_1^2} & \dfrac{\partial^2 v_0}{\partial q_1 \partial q_2} \\ \dfrac{\partial^2 v_0}{\partial q_1 \partial q_2} & \dfrac{\partial^2 v_0}{\partial q_2^2} \end{matrix} \right\| = -\left(-\frac{\lambda_2}{\lambda_1} \right)^2 \pm 1.$$

Moreover, observe that the second derivatives $\partial^2 v/\partial q_1^2$ and $\partial^2 v/\partial q_2^2$ of the function v are non-zero.

In fact, if at a some point

$$\frac{\partial^2 v}{\partial q_1^2} = \frac{\partial^2 v}{\partial q_2^2} = 0 \quad \text{and} \quad \frac{\partial^2 v}{\partial q_1 \partial q_2} = b,$$

then at this point the vector

$$\left(\frac{\partial}{\partial q_1} + p_1 \frac{\partial}{\partial u} \right) + b \frac{\partial}{\partial p_2}$$

is tangent to $\varphi(L)$, and so the vector

$$-\frac{\partial}{\partial p_1} + b \frac{\partial}{\partial p_2}$$

is tangent to L, which is impossible.

But the functions $\partial^2 v/\partial q_1^2$ and $\partial^2 v/\partial q_2^2$ are harmonic. Hence, by the Liouville theorem

$$\frac{\partial^2 v}{\partial q_1^2} = -\frac{\partial^2 v}{\partial q_2^2} = \text{constant} \neq 0.$$

Thus, $\partial^2 v/\partial q_1 \partial q_2 = \text{constant}$ and $v(q_1, q_2)$, like $v_0(q_1, q_2)$, is a polynomial of second degree. ■

12.2 Lie equations for contact symmetries

We say that a contact transformation $\varphi \in \mathrm{Ct}(J^1M)$ is a *contact symmetry* or simply *symmetry* of a Monge–Ampère equation E_ω if the form $\varphi^*(\omega)$ generates the same equation E_ω.

It follows from the Lepage theorem that in this case

$$\varphi(\Delta_\omega) = h_\varphi \Delta_\omega$$

for some smooth not vanishing function h_φ on J^1M.

In terms of effective forms this means that the contact transformation φ is a symmetry if the effective part of the form $\varphi^*(\omega)$ is proportional to the effective form ω, i.e.,

$$\varphi^*(\omega)_\varepsilon = h_\varphi \omega,$$

for some smooth function $h_\varphi \in C^\infty(J^1M)$.

In other words,

$$\varphi^*(\omega) = h_\varphi \omega + \theta$$

for some differential form $\theta \in I^n$.

We consider now infinitesimal symmetries of Monge–Ampère equations.

Let $X = X_f$ be a contact vector field. Then the Lie derivative $L_f \stackrel{\text{def}}{=} L_{X_f}$ preserves the Cartan ideal I^* because $L_f(\omega_0) = X_1(f)\omega_0$ and $L_f(d\omega_0) = dL_f(\omega_0) = d(X_1 f) \wedge \omega_0 + X_1(f)\, d\omega_0$.

We say that a contact vector field X_f on the manifold J^1M is an infinitesimal symmetry of the Monge–Ampère equation E_ω if

$$L_f(\omega) = h\omega \bmod I^n \tag{12.13}$$

for some smooth function h on J^1M.

Obviously, the set of all infinitesimal symmetries forms a Lie algebra with respect to the commutator of vector fields.

We denote this algebra by $\mathrm{Sym}(E_\omega)$.

Example 12.2.1 *The Legendre transformation (12.1) is a symmetry of the wave equation*

$$v_{q_1 q_2} = 0.$$

Indeed, the corresponding effective differential 2-form is

$$\omega = dq_1 \wedge dp_1 - dq_2 \wedge dp_2$$

and

$$\varphi^*(\omega) = dp_1 \wedge (-dq_1) - dp_2 \wedge (-dq_2) = \omega.$$

Example 12.2.2 *The translations*

$$T_c : (q, u, p) \longmapsto (q + c, u, p),$$

where $c \in \mathbb{R}^n$ are symmetries of equations $E_\omega \subset J^1\mathbb{R}^n$ if their coefficients do not depend on q. In this case, $T_c^*(\omega) = \omega$.

Example 12.2.3 *The contact vector field*

$$X_f = q_4 \frac{\partial}{\partial q_3} - q_3 \frac{\partial}{\partial q_4} - p_4 \frac{\partial}{\partial p_3} + p_3 \frac{\partial}{\partial p_4},$$

with generating function $f = p_4 q_3 - p_3 q_4$, is an infinitesimal symmetry of the Khokhlov–Zabolotskaya equation (11.25).

Theorem 12.2.1 *A contact vector field X_f is a symmetry of the Monge–Ampère equation E_ω, defined by an effective n-form ω, if and only if*

$$d_p(\iota_f\, \omega) + \iota_f\, d_p\omega + fL_{\iota_1}(\omega) = h\omega \tag{12.14}$$

for some smooth function $h \in C^\infty(J^1 M)$.

Proof Note that

$$\iota_f\, (\omega_0 \wedge \iota_1\, d\omega) = f\iota_1\, d\,\omega - \omega_0 \wedge (\iota_f \iota_1\, d\omega).$$

Therefore,

$$
\begin{aligned}
p(\iota_f\, d\omega) &= \iota_f\, d\omega - \omega_0 \wedge (\iota_1 \iota_f\, d\omega) = \iota_f\, d\omega + \omega_0 \wedge (\iota_f \iota_1\, \omega) \\
&= \iota_f\, d\omega - \iota_f(\omega_0 \wedge \iota_1\, d\omega) + f\iota_1\, d\omega = \iota_f\, d_p\omega + f\iota_1\, d\omega.
\end{aligned}
$$

Using this relation we obtain

$$
p(L_{_f}(\omega)) = p(\iota_f\, d\omega + d(\iota_f\omega)) = d_p(\iota_f\omega) + \iota_f\, d_p\omega + fL_{_1}(\omega).
$$

Note now that X_f is a symmetry if $p(L_{_f}(\omega))$ is proportional to ω. ■

We call (12.14) the *Lie equation* for contact symmetries of E_ω. We also say that a contact vector field X_f is a *symmetry of the Monge–Ampère differential operator* if the generating function f satisfies the Lie equation with $h = 0$.

Example 12.2.4 (The Laplace equation) *As a first example, let us calculate infinitesimal contact symmetries of the Laplace operator on* $\mathbb{R}^2$. *As we have seen the effective differential 2-form corresponding to the Laplace operator is*

$$
\omega = dq_1 \wedge dp_2 - dq_2 \wedge dp_1.
$$

Therefore,

$$
\iota_f\omega = -f_{p_1}\, dp_2 + f_{p_2}\, dp_1 + (f_{q_1} + p_1 f_u)\, dq_2 - (f_{q_2} + p_2 f_u)\, dq_1
$$

and

$$
\begin{aligned}
d_p(\iota_f\omega) &= (f_{q_1q_1} + f_{q_2q_2} + 2p_1 f_{uq_1} + 2p_2 f_{q_2u} + f_{uu}(p_1^2 + p_2^2))\, dq_1 \wedge dq_2 \\
&\quad + (f_{p_2q_1} + f_{q_2p_1} + p_1 f_{p_2u} + p_2 f_{p_1u})(dq_1 \wedge dp_1 - dq_2 \wedge dp_2) \\
&\quad + (f_u - f_{p_1q_1} + f_{q_2p_2} - p_1 f_{p_1u} + p_2 f_{up_2})\, dq_1 \wedge dp_2 \\
&\quad + (-f_{p_1p_1} - f_{p_2p_2})\, dp_1 \wedge dp_2.
\end{aligned}
$$

Note that $d\omega = 0$, $L_{X_1}(\omega) = 0$, *and hence*

$$
\iota_f\, d_p\omega + fL_{_1}(\omega) = 0.
$$

The Lie equation for the generating function f takes the following form:

$$
\begin{aligned}
&f_{q_1q_1} + f_{q_2q_2} + 2p_1 f_{uq_1} + 2p_2 f_{q_2u} + f_{uu}(p_1^2 + p_2^2) = 0, \\
&f_{q_1p_1} - f_{q_2p_2} + p_1 f_{up_1} - p_2 f_{up2} = 0, \\
&f_{p_2q_1} + f_{q_2p_1} + p_1 f_{p_2u} + p_2 f_{p_1u} = 0, \\
&f_{p_1p_1} + f_{p_2p_2} = 0.
\end{aligned}
$$

Differentiating the second equation in p_1 and the third one in p_2 and summing obtained equalities, we obtain

$$f_{up_1} = 0.$$

In the similar way, differentiating the second equation in p_2 and the third one in p_1, and subtracting obtained equalities, we obtain

$$f_{up_2} = 0.$$

So, we see that function f has the following form:

$$f = a(q_1, q_2, u) + b(q_1, q_2, p_1, p_2)$$

for some functions a and b.

Rewriting the Lie equation for these functions we get the following system:

$$a_{q_1q_1} + a_{q_2q_2} + b_{q_1q_1} + b_{q_2q_2} + 2(p_1 a_{q_1u} + p_2 a_{q_2u}) + (p_1^2 + p_2^2)a_{uu} = 0,$$
$$b_{q_1p_1} - b_{q_2p_2} = 0,$$
$$b_{q_1p_2} + b_{q_2p_1} = 0,$$
$$b_{p_1p_1} + b_{p_2p_2} = 0.$$

Differentiating the first equation in p_1 and in p_2, and using the last three equations, we obtain

$$a_{q_1u} + p_1 a_{uu} = 0,$$
$$a_{q_2u} + p_2 a_{uu} = 0.$$

From these two equations it follows that the function a has the form

$$a = Cu + g(q_1, q_2),$$

where C is an arbitrary constant and g is an arbitrary function in q_1, q_2.

Let $w = g + b$. Then the generating functions f has the form

$$f = Cu + w,$$

where the function w satisfies the following system of differential equations:

$$w_{q_1q_1} + w_{q_2q_2} = 0,$$
$$w_{q_1p_2} + w_{q_2p_1} = 0,$$
$$w_{q_1p_1} - w_{q_2p_2} = 0,$$
$$w_{p_1p_1} + w_{p_2p_2} = 0.$$

So, we see that the Lie algebra of infinitesimal symmetries of the Laplace operator is infinite dimensional.

Example 12.2.5 (The TM equation) *As a second example we consider (12.5). Recall that after the Legendre transformation this equation takes the form*

$$v^4 \det \operatorname{Hess} v = 1. \tag{12.15}$$

Solving the Lie equation we find that a basis of generating functions of the Lie algebra symmetries gives the following functions (see [21]):

$$f_1 = up_1, \quad f_2 = up_2, \quad f_3 = u + 2p_1q_1 - p_2q_2, \quad f_4 = u + 2p_2q_2 - p_1q_1,$$
$$f_5 = p_2q_1, \quad f_6 = p_1q_2, \quad f_7 = q_1, \quad f_8 = q_2.$$

The corresponding contact vector fields are:

$$Y_1 = -u\frac{\partial}{\partial q_1} + p_1^2\frac{\partial}{\partial p_1} + p_1p_2\frac{\partial}{\partial p_2},$$

$$Y_2 = -u\frac{\partial}{\partial q_2} + p_1p_2\frac{\partial}{\partial p_1} + p_2^2\frac{\partial}{\partial p_2},$$

$$Y_3 = -2q_1\frac{\partial}{\partial q_1} + q_2\frac{\partial}{\partial q_2} + u\frac{\partial}{\partial u} + 3p_1\frac{\partial}{\partial p_1},$$

$$Y_4 = q_1\frac{\partial}{\partial q_1} - 2q_2\frac{\partial}{\partial p_2} + u\frac{\partial}{\partial u} + 3p_2\frac{\partial}{\partial p_2},$$

$$Y_5 = -q_1\frac{\partial}{\partial q_2} + p_2\frac{\partial}{\partial p_1},$$

$$Y_6 = -q_2\frac{\partial}{\partial q_1} + p_1\frac{\partial}{\partial p_1},$$

$$Y_7 = q_1\frac{\partial}{\partial u} + \frac{\partial}{\partial p_1},$$

$$Y_8 = q_2\frac{\partial}{\partial u} + \frac{\partial}{\partial p_2}.$$

The commutator table is shown on Table 12.1.

Table 12.1. *Commutator table*

	Y_1	Y_2	Y_3	Y_4	Y_5	Y_6	Y_7	Y_8
Y_1	0	0	$3Y_1$	0	Y_2	0	$\frac{2Y_3+Y_4}{3}$	0
Y_2	0	0	0	$3Y_2$	0	Y_1	Y_5	$\frac{Y_3+2Y_4}{3}$
Y_3	$-3Y_1$	0	0	0	$3Y_5$	$-3Y_6$	$3Y_7$	0
Y_4	0	$-3Y_2$	0	0	$-3Y_5$	$3Y_6$	0	$3Y_8$
Y_5	Y_2	0	$-3Y_5$	$3Y_5$	0	$\frac{Y_3-Y_4}{3}$	0	Y_7
Y_6	0	$-Y_1$	$3Y_6$	$-3Y_6$	$\frac{Y_4-Y_3}{3}$	0	Y_8	0
Y_7	$-\frac{2Y_3+Y_4}{3}$	$-Y_5$	$-Y_7$	0	0	$-Y_8$	0	0
Y_8	0	$-\frac{Y_3+2Y_4}{3}$	0	$-3Y_8$	$-Y_7$	0	0	0

Note that the Lie algebra $\mathrm{Sym}(E_\omega)$ *in isomorphic to the Lie algebra* $\mathfrak{sl}(3,\mathbb{R})$ *by the mapping* $\pi : \mathrm{Sym}(E_\omega) \to \mathfrak{sl}(3,\mathbb{R})$, *where*

$$Y_1 \longmapsto -E_{31}, \quad Y_5 \longmapsto E_{23}, \quad Y_2 \longmapsto E_{21}, \quad Y_6 \longmapsto E_{32},$$
$$Y_3 \longmapsto E_{11} + E_{22} - 2E_{33},$$
$$Y_7 \longmapsto E_{13}, \quad Y_4 \longmapsto E_{11} - 2E_{22} + E_{33}, \quad Y_8 \longmapsto -E_{12},$$

here E_{ij} *are the* 3×3 *elementary matrices with the only non-zero element being equal to 1 at* (i,j).

12.3 Reduction

Let X_f be an infinitesimal symmetry of a Monge–Ampère equation E_ω and let $A_t : J^1M \to J^1M$ be the corresponding one-parameter group of contact transformations. It follows from the definition that $A_t(L)$ is a multivalued solution of E_ω whenever it is the integral manifold $L \subset J^1M$.

Indeed, $X_f \in \mathrm{Sym}(E_\omega)$ means that

$$A_t^*(\omega) = h_t\omega + I^n$$

and therefore

$$\omega|_{A_t(L)} = A_t^*(\omega)|_L = 0$$

if $\omega|_L = 0$.

Therefore, from a known solution L and a symmetry X_f one can construct a one-parameter family of solutions $L_t = A_t(L)$.

Consider the special case when $L_t = L$ for all t; that is, the case when the integral manifold L is invariant with respect to the contact vector field X_f. Such solutions are said to be *f-automodel solutions*.

Note that the function f vanishes on the f-automodel solutions. Indeed, by definition of a multivalued solution, the restriction of the Cartan form to L equals zero: $\omega_0|_L = 0$. But $\omega_0(X_f) = f$ and the vector field X_f is tangent to L, hence $f|_L = 0$.

So, any f-automodel solution of the Monge–Ampère equation is also a solution of the first-order partial differential equation $f^{-1}(0) \subset J^1M$.

Now let us consider a set of contact symmetries $X_{f_1}, \ldots, X_{f_k}$ instead of one symmetry X_f. Suppose that these vector fields are linear independent at each point of J^1M and that they span a Lie algebra

$$\mathfrak{g} \stackrel{\text{def}}{=} \langle X_{f_1}, \ldots, X_{f_k} \rangle.$$

In other words, assume that

$$[X_{f_i}, X_{f_j}] = \sum_{l=1}^{k} c_{ij}^l X_{f_l},$$

where c_{ij}^l are the structure constants of the Lie algebra.

First of all, let us note that these vector fields are tangent to the submanifold

$$E_{\mathfrak{g}} \stackrel{\text{def}}{=} \{f_1 = 0, \ \ldots, f_k = 0\} \subset J^1M.$$

Indeed, by the properties of Lagrange brackets, we obtain

$$[f_i, f_j] = \sum_{l=1}^{k} c_{ij}^l f_l = X_{f_i}(f_j) - X_1(f_i) f_j.$$

This implies that the functions

$$X_{f_i}(f_j) = \sum_{l=1}^{k} c_{ij}^l f_l + X_1(f_i) f_i$$

vanish on the submanifold $E_{\mathfrak{g}}$.

We call solutions L of E_ω $\mathfrak{g}$-*invariant* if they are invariant with respect to Lie algebra symmetries $\mathfrak{g}$.

The same reasons as above show that $L \subset E_{\mathfrak{g}}$, or that any $\mathfrak{g}$-invariant solution is a solution of the following system of first-order partial differential equations: $f_1 = 0, \ldots, f_k = 0$.

To find $\mathfrak{g}$-invariant solutions we consider the following construction.

Assume that the space of $\mathfrak{g}$-orbits on $E_{\mathfrak{g}}$ form a smooth manifold $B_{\mathfrak{g}}$, and let $\pi : E_{\mathfrak{g}} \to B_{\mathfrak{g}}$ be the natural projection. Contact vector fields $X_f \in \mathfrak{g}$ are tangent to fibres of the projection and belong to the restriction of the Cartan distribution C on $E_{\mathfrak{g}}$. Therefore, one could define a projection $C_{\mathfrak{g}}$ of the restriction of the Cartan distribution on $B_{\mathfrak{g}}$ as follows:

$$C_{\mathfrak{g}}([a]) = \pi_*(C_a \cap T_a E_{\mathfrak{g}})$$

where $[a]$ is the orbit of $a \in E_{\mathfrak{g}}$.

Let

$$\omega_{\mathfrak{g}} = \iota_{f_1} \circ \cdots \circ \iota_{f_k}(\omega),$$

and consider the restrictions θ_b of $\omega_{\mathfrak{g}}$ on $C_b \cap T_b E_{\mathfrak{g}}$ for $b \in [a]$. If $X_f \in \mathfrak{g}$ and A_t is the corresponding flow then $A_t^*(\theta_b)$ is proportional to $\theta_{A_{-t}(b)}$. Therefore $\omega_{\mathfrak{g}}$ determines a one-dimensional subbundle $l_{\mathfrak{g}}$ in the bundle of $(n-k)$-forms on $C_{\mathfrak{g}}$.

Note also that $C_{\mathfrak{g}}$ is a contact distribution on $B_{\mathfrak{g}}$. Indeed, if $L' \subset B_{\mathfrak{g}}$ is an s-dimensional integral manifold of $C_{\mathfrak{g}}$ then by construction $\pi^{-1}(L') \subset E_{\mathfrak{g}}$ is an integral manifold of the Cartan distribution, and $s + k \leq n$ where $n = \dim M$.

Therefore, the maximum dimension of the integral manifolds is $n - k$, and $\dim B_{\mathfrak{g}} = 2(n-k) + 1$, $\dim C_{\mathfrak{g}} = 2(n-k)$.

Let us now define a *Monge–Ampère differential equation on a contact manifold* (N^{2r+1}, θ) as a one-dimensional subbundle l of the bundle effective r-forms on the distribution $\ker \theta$, and call a solution of the equation any r-dimensional integral manifold L such that $l|_L = 0$.

Returning to the above construction we shall call $l_{\mathfrak{g}}$ the *reduction of the Monge–Ampère equation* E_ω *by the contact symmetry algebra* $\mathfrak{g}$.

Summarizing we obtain the following.

Theorem 12.3.1 *Let $\mathfrak{g}$ be an algebra Lie of contact symmetries, $\mathfrak{g} \subset \mathrm{Sym}(E_\omega)$, such that the orbit space $B_{\mathfrak{g}} = E_{\mathfrak{g}}/\mathfrak{g}$ is smooth. Then $(B_{\mathfrak{g}}, C_{\mathfrak{g}})$ is a contact manifold and $\omega_{\mathfrak{g}}$ determines a Monge–Ampère equation $l_{\mathfrak{g}}$ in such a way that projection of any $\mathfrak{g}$-invariant solution L of E_ω with smooth orbit space $L/\mathfrak{g}$ is a solution of $l_{\mathfrak{g}}$, and the preimage $\pi^{-1}(L')$ of any solution L' of $l_{\mathfrak{g}}$ is a $\mathfrak{g}$-invariant solution of E_ω.*

We now consider the special case of the construction. Let us assume that there is a smooth submanifold $\Gamma \subset E_{\mathfrak{g}}$ such that each $\mathfrak{g}$-orbit of $\mathfrak{g}$ on $E_{\mathfrak{g}}$ intersects Γ at only one point and that Γ is transversal to the orbits.

Then the restriction of the Cartan distribution on Γ gives us a contact distribution C_Γ, and the restriction of the form $\omega_{\mathfrak{g}}$ on Γ gives the reduced Monge–Ampère equation for $\mathfrak{g}$-invariant solutions of E_ω.

As we have seen, for any symmetry X_f and a multivalued solution L of a differential equation the one-parameter family $L_t = A_t(L)$ consists of solutions of the equation. One could use this observation to reproduce new solutions from known ones.

12.4 Examples

12.4.1 The boundary layer equation

The differential equation which describes the flow in a boundary layer in aerodynamics (see [64]) has the form:

$$v_{xx} = v v_t. \tag{12.16}$$

With the notation $q_1 = t, q_2 = x$ the corresponding effective differential 2-form is as follows:

$$\omega = u p_1\, dq_1 \wedge dq_2 - dq_1 \wedge dp_2.$$

Since the coefficients of this form do not depend on q_1 and q_2, it has symmetries of the form

$$f = a_1 p_1 + a_2 p_2.$$

The corresponding f-invariant solution takes the form of the *traveling wave*:

$$v = y(a_2 t - a_1 x), \tag{12.17}$$

for some function y.

By substituting this expression in (12.16), we obtain the following equation:

$$a_1^2 y'' = a_2 y y', \tag{12.18}$$

which is the reduced Monge–Ampère equation for given translation symmetry.

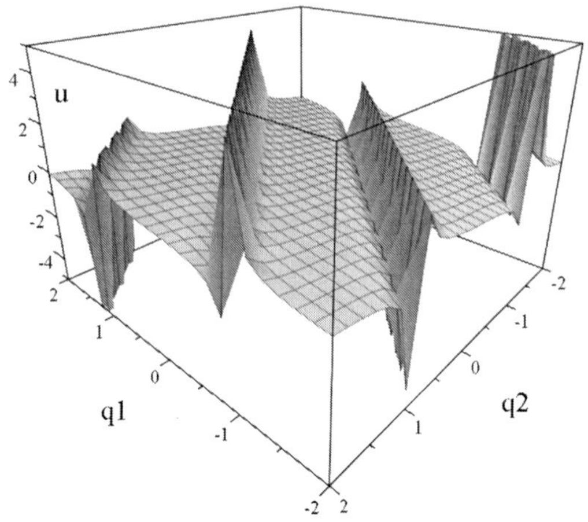

Figure 12.3. The traveling wave solution of the boundary layer equation.

If $a_1 \neq 0$ and $a_2 \neq 0$, then the general solution of this ordinary equation is

$$y(\tau) = \frac{C_1 a_1^2 \sqrt{2}}{a_2} \tan\left(\frac{C_1}{\sqrt{2}}(\tau + C_2)\right),$$

where C_1 and C_2 are constants.

The traveling wave solution is

$$v(t,x) = \frac{C_1 b^2 \sqrt{2}}{a} \tan\left(\frac{C_1}{\sqrt{2}}(at + bx + C_2)\right). \qquad (12.19)$$

(see Figure 12.3).

Besides translations the boundary layer equation also has scale symmetry with the generation function

$$f = q_2 p_2 + 2q_1 p_1$$

which corresponds to the following scale transformations:

$$u \longmapsto u, \quad q_1 \longmapsto \alpha^2 q_1, \quad q_2 \longmapsto \alpha q_2.$$

The corresponding equation for f-invariant solutions is

$$2q_1 \frac{\partial v}{\partial q_1} + q_2 \frac{\partial v}{\partial q_2} = 0.$$

The solutions of the last equation can be represented in the form $v(q_1, q_2) = y(\tau)$, where $\tau = q_1/q_2^2$, and y satisfies the following ordinary differential equation:

$$4\tau^2 y'' - yy' + 6\tau y' = 0.$$

The last equation is the reduced Monge–Ampère equation for a given scale symmetry.

One could also use the symmetry $f = q_2 p_2 + 2q_1 p_1$ to obtain new solutions from (12.19).

The flow for X_f consists of point transformations

$$A_\tau : (q_1, q_2, u) \mapsto (q_1 \exp(-2\tau), q_2 \exp(-\tau), u).$$

Therefore if $v = w(q_1, q_2)$ is a solution, then the functions

$$v_\tau = w(\exp(-2\tau), q_2 \exp(-\tau))$$

are also solutions. In particular, the functions

$$v_\tau(t, x) = \frac{C_1 b^2 \sqrt{2}}{a} \left(\tan \left(\frac{C_1}{\sqrt{2}} (a \exp(-2\tau)t + b \exp(-\tau)x + C_2) \right) \right)$$

are solutions.

12.4.2 The thermal conductivity equation

The non-linear thermal conductivity equation has the form

$$v_t = (v^{-4/3} v_x)_x + \gamma v, \tag{12.20}$$

where $\alpha \neq 0$ and γ are constants [112, 99], $q_1 = t$ and $q_2 = x$.

Later on we shall investigate this equation in more detail, and now we consider the case of symmetry with a generating function

$$f = q_2^2 p_2 + 3q_2 u.$$

To find f-invariant solutions we obtain the system of two partial differential equations:

$$v_{q_1} = (v^{-4/3} v_{q_2})_{q_2} + \gamma v,$$
$$q_2 v_{q_2} = -3v.$$

Hence, f-invariant solutions have the form

$$v(q_1, q_2) = c\frac{\exp(\gamma q_1)}{q_2^3}$$

where c is a constant.

12.4.3 The Petrovsky–Kolmogorov–Piskunov equation

This equation [48]

$$v_t = Dv_{xx} + F(v), \tag{12.21}$$

where F is a smooth function and $D \neq 0$ is a constant, can be defined by the following effective 2-form ω:

$$\omega = (F(u) - p_1)\, dq_1 \wedge dq_2 + D\, dq_1 \wedge dp_2,$$

where $q_1 = t$ and $q_2 = x$.

This equation is invariant with respect to translations in x and t, and one can consider the traveling wave solutions in the form

$$v = y(at + bx).$$

The reduced equation is as follows

$$y'' = sy' + rF(y), \tag{12.22}$$

where $y = y(\tau)$, $\tau = at + bx$, $s = a/Db^2$ and $r = -1/Db^2$.

Let us choose $s = 1$, i.e., $a = Db^2$.

The last equation has been considered in Part I. Using the results, we can construct f-invariant solutions of (12.21).

We consider here the case of (12.21):

$$F(v) = kv(1 - v).$$

This corresponds to the so-called *Fisher equation* (see [87]) .

If we take $a = \frac{6}{5}$ and $b = \sqrt{5k/6D}$, then the reduced equation has the form

$$y'' = y' + \frac{6}{25}y^2 - \frac{6}{25}y. \tag{12.23}$$

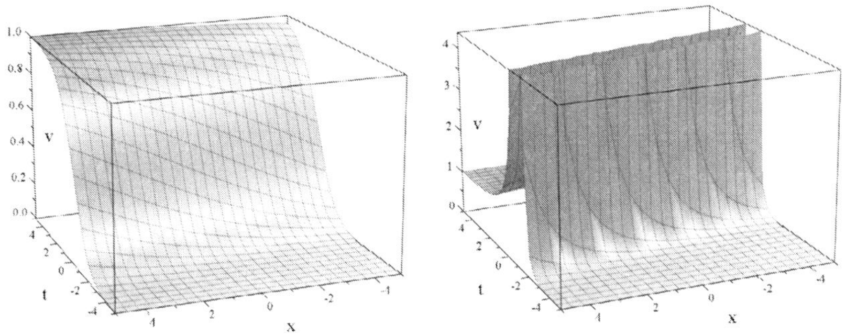

Figure 12.4. The f-invariant solutions of the Fisher equation.

The last equation itself has symmetries with generation functions:

$$f_1 = p \quad \text{and} \quad f_2 = \left(p - \frac{2}{5}\right) \exp\left(-\frac{\tau}{5}\right).$$

Using these symmetries we find a first integral:

$$\exp\left(-\frac{6}{5}\tau\right) (25p^2 + 4y^2 - 20py - 4y^3) = C_1.$$

Therefore

$$p = \frac{2}{5}y \pm \frac{1}{5}\sqrt{\left(C_1 \exp\left(\frac{6}{5}\tau\right) + 4y^3\right)}.$$

If $C_1 = 0$, then we obtain the equation

$$y' - \frac{2}{5}y(1 \pm \sqrt{y}) = 0$$

and solutions

$$y(x) = \left(C \exp\left(-\frac{\tau}{5}\right) \pm 1\right)^{-2}$$

where $C > 0$ is a constant (see Figure 12.4).

The corresponding f-invariant solution of (12.21) is

$$v(t,x) = \left(C \exp\left(-\frac{6t}{25} - \frac{1}{5}\sqrt{\frac{5k}{6D}}x\right) \pm 1\right)^{-2}.$$

Table 12.2. *Generators of the infinite-dimensional Lie algebra* Sym (E_ω) *and their action.*

Symmetry	The action A_t on (q, u, p)
$f_1 = q_2p_2 + q_3p_3 + 2u$	$(q_1, e^{-t}q_2, e^{-t}q_3, e^{2t}u, e^{2t}p_1, e^{3t}p_2, e^{3t}p_3)$
$f_2 = 2q_1p_1 + 3(p_2q_2 + p_3q_3)$	$(e^{-2t}q_1, e^{-3t}q_2, e^{-3t}q_3, u, e^{2t}p_1, e^{3t}p_2, e^{3t}p_3)$
$f_3 = q_3p_2 - q_2p_3$	$(q_1, q_2 \cos t - q_3 \sin t, q_2 \sin t + q_3 \cos t, u, p_1,$ $p_2 \cos t - p_3 \sin t, p_2 \sin t + p_3 \cos t)$
$f_4 = p_1$	$(q_1 - t, q_2, q_3, u, p_1, p_2, p_3)$
$f_5 = p_2$	$(q_1, q_2 - t, q_3, u, p_1, p_2, p_3)$
$f_6 = p_3$	$(q_1, q_2, q_3 - t, u, p_1, p_2, p_3)$
$f_7 = H$	$(q_1, q_2, q_3, u + tH, p_1, p_2 + tH_{q_2}, p_3 + tH_{q_3})$

12.4.4 The Von Karman equation

This equation describes a behavior of the velocity potential v in the transonic approximation of gas motion. It also defines a spreading of rays in the neighborhood of a caustic in the homogeneous atmosphere [44]:

$$v_x v_{xx} - v_{yy} - v_{zz} = 0. \tag{12.24}$$

The corresponding effective differential 3-form is

$$\omega = p_1\, dp_1 \wedge dq_2 \wedge dq_3 + dp_2 \wedge dq_1 \wedge dq_3 + dp_3 \wedge dq_2 \wedge dq_1,$$

where $q_1 = x, q_2 = y$ and $q_3 = z$.

Table 12.2 contains generators of the infinite-dimensional Lie algebra $\mathrm{Sym}(E_\omega)$ [113] and their action.

In the table $H = H(q_2, q_3)$ is a harmonic function:

$$\frac{\partial^2 H}{\partial q_2^2} + \frac{\partial^2 H}{\partial q_3^2} = 0.$$

If we restrict ourselves to contact symmetries that preserve the class of solutions vanishing at infinity then one obtains a six-dimensional Lie subalgebra

$$\mathfrak{g}_6 = \langle f_1, \ldots, f_6 \rangle.$$

Table 12.3. *Table of commutators for* $\mathfrak{g}_6$.

	f_1	f_2	f_3	f_4	f_5	f_6
f_1	0	0	0	0	$-f_5$	$-f_6$
f_2	0	0	0	$-2f_4$	$-3f_5$	$-3f_6$
f_3	0	0	0	0	f_6	$-f_5$
f_4	0	$2f_4$	0	0	0	0
f_5	f_5	$3f_5$	$-f_6$	0	0	0
f_6	f_6	$3f_6$	f_5	0	0	0

The structure of this algebra can be seen from the commutators given in Table 12.3.

Selecting a two-dimensional subalgebra $\mathfrak{g}_2 \subset \mathrm{Sym}(E_\omega)$, we shall reduce the problem of finding the $\mathfrak{g}$-invariant solutions to integration of an ordinary differential equation.

Let us consider, for example, the subalgebra $\mathfrak{g}_2$ generated by the spiral symmetry $f_4 + \gamma f_3$, $\gamma \in \mathbb{R}$ and the scale symmetry f_5.

Thus,

$$\mathfrak{g}_2 = \langle f_1, f_4 + \gamma f_3 \rangle.$$

To find the $\mathfrak{g}_2$-invariant solutions we should integrate the following system of differential equations:

$$\begin{aligned} &v_{q_1} v_{q_1 q_1} - v_{q_2 q_2} - v_{q_3 q_3} = 0 \\ &v_{q_1} + \gamma(q_3 v_{q_2} - q_2 v_{q_3}) = 0 \\ &q_2 v_{q_2} + q_3 v_{q_3} + 2v = 0. \end{aligned}$$

The last two equations give

$$v = (q_2^2 + q_3^2)^{-1} g(\eta),$$

where

$$\eta = q_1 - \frac{1}{\gamma} \arctan \frac{q_2}{q_3},$$

and the reduced equation takes the following form:

$$(\gamma^2 g'^2 - 1) g'' - 4\gamma^2 g = 0.$$

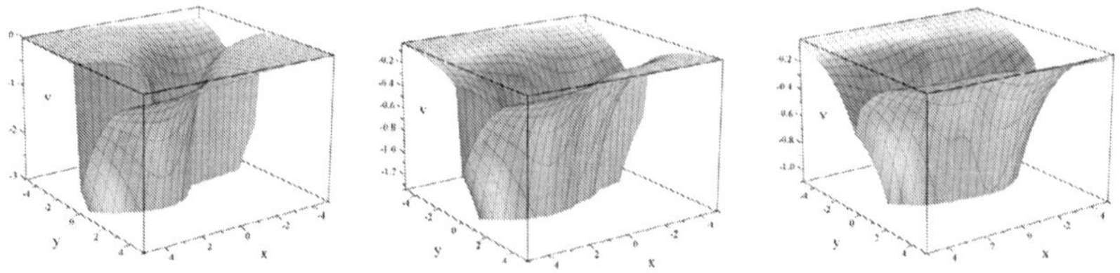

Figure 12.5. The solution of von Karman equation at $t = -1, 0$ and 1.

or

$$\frac{1}{2}(g'^2)' = \frac{2\gamma^2(g^2)'}{(\gamma^2 g'^2 - 1)}.$$

Then

$$(\gamma^2 g'^2 - 1)' = 4\gamma^4(g^2)'$$

and

$$g'^2 - 4\gamma^2 g^2 = c_1$$

for some constant c.

This equation has two families of solutions

$$g_-(\tau) = -\frac{c_1}{2\gamma}\exp(2\gamma(\tau - c_2)) + \frac{1}{8\gamma}\exp(-2\gamma(\tau - c_2))$$

and

$$g_+(\tau) = -\frac{c_1}{2\gamma}\exp(-2\gamma(\tau + c_2)) + \frac{1}{8\gamma}\exp(2\gamma(\tau + c_2)).$$

Finally, $\mathfrak{g}_2$-invariant solutions take one of the following two forms:

$$\begin{aligned} v_+(q_1, q_2, q_3) &= -\frac{c_1}{2\gamma(q_2^2 + q_3^2)}\exp(-2\gamma(\eta + c_2)) \\ &\quad + \frac{1}{8\gamma(q_2^2 + q_3^2)}\exp(2\gamma(\eta + c_2)), \\ v_-(q_1, q_2, q_3) &= -\frac{c_1}{2\gamma(q_2^2 + q_3^2)}\exp(2\gamma(\eta - c_2)) \\ &\quad + \frac{1}{8\gamma(q_2^2 + q_3^2)}\exp(-2\gamma(\eta - c_2)). \end{aligned}$$

12.5 Symmetries of the reduction

As we have seen, the reduction of Monge–Ampère equations by an effective and contact action of a Lie algebra $\mathfrak{g}$ decreases the number of independent variables of a Monge–Ampère equation on dim $\mathfrak{g}$.

In particular, if for an equation with n independent variables we know an $(n-1)$-dimensional subalgebra with effective contact action, then we can reduce this equation to the ordinary one. To solve the last equation we also need symmetries.

This leads us to finding subalgebras $\mathfrak{g} \subset \mathrm{Sym}(E_\omega)$ such that the reduction $l_\mathfrak{g}$ has symmetries inherited from $\mathrm{Sym}(E_\omega)$.

For example, it is easy to see that the function $f = p_2 q_1 - q_2$ is a generating function of a symmetry for the Titeica–Morimoto equation (12.5). To find f-invariant solutions one should solve the system

$$\det \mathrm{Hess}\, v = (v - q_1 v_{q_1} - q_2 v_{q_2})^4,$$

$$q_1 v_{q_2} - v = 0.$$

From the second equation we get

$$v(q_1, q_2) = \frac{q_2^2}{2q_1} + w(q_1),$$

and the reduced equation takes the form

$$w'' - q_1 w^4 + 4w^3 w' q_1^2 - 6w^2 (w')^2 q_1^3 + 4w(w')^3 q_1^4 - (w')^4 q_1^5 = 0.$$

To solve this equation we need two symmetries.

In order to find them we should know which elements of $\mathrm{Sym}(E_\omega) = \mathfrak{sl}_3(\mathbb{R})$ give us symmetries of the reduced equation.

To this end let us consider the reduction $\pi : E_\mathfrak{g} \to B_\mathfrak{g}$ and let $X_f \in \mathrm{Sym}(E_\omega)$ be a contact symmetry. If we expect that this symmetry produces a symmetry for the reduction we should assume first of all that X_f preserves the bundle π. It certainly happens when $[f, \mathfrak{g}] \subset \mathfrak{g}$, that is, when f belongs to the normalizer

$$N_\mathfrak{g} = \{f \in \mathrm{Sym}(E_\omega), [f, \mathfrak{g}] \subset \mathfrak{g}\}$$

of the subalgebra $\mathfrak{g}$ into the algebra $\mathrm{Sym}(E_\omega)$.

Indeed, in this case the flow A_t of X_f transforms $\mathfrak{g}$-orbits to $\mathfrak{g}$-orbits, and the relation

$$X_f(g) = [f, g] + X_1(f)g = 0 \text{ on } E_{\mathfrak{g}}$$

for $g \in \mathfrak{g}$ shows that the vector field X_f is tangent to $E_{\mathfrak{g}}$.

Thus the flow A_t preserves the restriction of the Cartan distribution on $E_{\mathfrak{g}}$. Denote by $\overline{A_t} : B_{\mathfrak{g}} \to B_{\mathfrak{g}}$ the induced transformations, and by $\overline{X_f}$ the corresponding vector field, that is,

$$\pi_* : X_f \longmapsto \overline{X_f}.$$

Note that transformations $\overline{A_t}$ preserve the distribution $C_{\mathfrak{g}}$ and therefore $\overline{X_f}$ is a contact vector field on the contact manifold $(B_{\mathfrak{g}}, C_{\mathfrak{g}})$.

Moreover, the formula

$$L_f \circ \iota_g = \iota_g \circ L_f + \iota_{[f,g]}$$

shows that the effective part of $L_f(\omega_{\mathfrak{g}})$ is proportional to $\omega_{\mathfrak{g}}$ and therefore $\overline{A_t}$ preserves the bundle $l_{\mathfrak{g}}$.

Summarizing we find the following.

Theorem 12.5.1 *The projections $\overline{X_f} = \pi_*(X_f)$ of contact vector fields X_f, $f \in N_{\mathfrak{g}}$, are contact symmetries of the reduced Monge–Ampère equation $l_{\mathfrak{g}}$.*

Corollary 12.5.1 $\pi_* : N_{\mathfrak{g}}/\mathfrak{g} \to \operatorname{Sym} l_{\mathfrak{g}}$ *is a Lie algebra monomorphism.*

Let us continue with the above example for the TM equation (see [21]). One can easily check that functions f_7 and $2f_3 + f_4$ belong to $N_{f_5 - f_8}$, and $f = f_5 - f_8$ is the symmetry we discussed above.

We identify $B_{\mathfrak{g}}$ with a hypersurface Γ:

$$\Gamma = \{p_2 = 0,\ q_2 = 0\}.$$

Then the restriction of the forms ω_0 and $\iota_f\, \omega$ on Γ are

$$\omega_0' = du - p_1\, dq_1$$

and

$$\omega' = -q_1(u - p_1 q_1)^4\, dq_1 + dp_1.$$

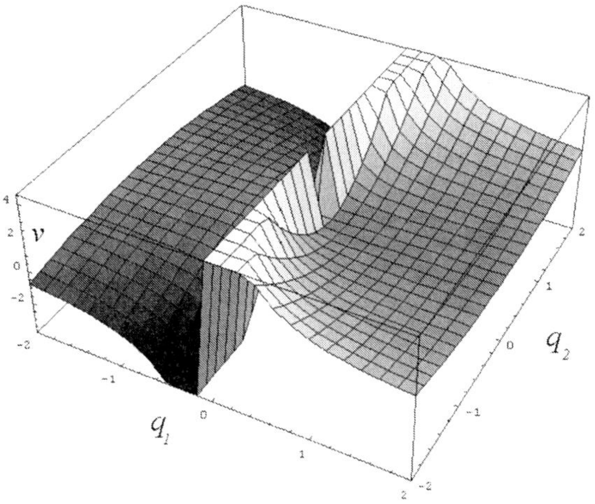

Figure 12.6. The f-invariant solution of the TM equation.

Denote q_1 by x and p_1 by p, then

$$\omega_0' = du - p\,dx,$$
$$\omega' = dp - x(u - px)^4\,dx.$$

The corresponding vectors fields $\overline{X}_{f_7}$ and $\frac{1}{3}\overline{X}_{2f_1+f_4}$ are

$$\frac{1}{3}\overline{X}_{2f_1+f_4} = -x\frac{\partial}{\partial x} + u\frac{\partial}{\partial u} + 2p\frac{\partial}{\partial p},$$
$$\overline{X}_{f_7} = x\frac{\partial}{\partial u} + \frac{\partial}{\partial p}.$$

The distribution $\langle \omega_0', \omega' \rangle$ corresponds to the reduction of the TM equation. This distribution is a completely integrable distribution with two shuffle symmetries, and therefore can be integrated.

The corresponding f-invariant solutions have the following form:

$$v(q_1, q_2) = \frac{q_2^2}{2q_1} + \frac{(q_1^3 + a^3)^{2/3}}{a^3} - \frac{q_1^3}{2a^4}\,{}_2F_1\left(\frac{2}{3}, \frac{1}{3}, \frac{5}{3}, -\frac{q_1^3}{a^3}\right) + bq_2$$

where ${}_2F_1$ is the hypergeometric function and a, b are constants (see Figure 12.6).

12.6 Monge–Ampère equations in symplectic geometry

Let us begin with the following observation. If coefficients of the Monge–Ampère equation E_ω do not depend on the variable v, then coefficients of the effective form ω do not depend on u and therefore $X_1 \in \mathrm{Sym}(E_\omega)$. This allows us to use the symplectic manifold T^*M instead of the contact J^1M.

Namely, in this case

$$\iota_1(\omega) = 0 \quad \text{and} \quad L_1(\omega) = 0$$

and therefore

$$\omega = \pi^*(\omega')$$

for some form $\omega' \in \Omega^n(T^*M)$. Here $\pi : J^1M \to T^*M$ is the natural projection.

The form ω' is clearly effective, i.e., $\perp(\omega') = 0$, where

$$\perp : \Omega^s(T^*M) \to \Omega^{s-2}(T^*M)$$

is the operator of interior multiplication by the bivector field X_Ω dual to the structure form $\Omega = -d\rho$ on T^*M.

Let us denote the $C^\infty(T^*M)$-modules of effective s-forms on T^*M by $\Omega^s_\varepsilon(T^*M)$.

The projection $\pi(L)$ of any multivalued solution L of E_ω is an embedded Lagrange manifold such that the restriction $\omega'|_{\pi(L)} = 0$. Conversely, if L' is the Lagrange manifold such that $\omega'|_{L'} = 0$, then there is an integral manifold $L \subset J^1M$, such that $L' = \pi(L)$ and L' is a multivalued solution of E_ω.

For this reason Lagrange manifolds $L' \subset T^*M$ such that $\omega'|_{L'} = 0$; we also call these *multivalued solutions*.

Example 12.6.1 *The effective form*

$$\omega' = p_1\, dq_2 \wedge dp_1 + dq_1 \wedge dp_2,$$

on T^*M corresponds to the Von Karman equation on a plane. The following two-parameter family of Lagrangian manifolds with constants c_1 and c_2

$$L'_c : \begin{cases} q_1 = \dfrac{1}{2}p_1^2 - c_1, \\ q_2 = p_2 - c_2 \end{cases} \tag{12.25}$$

satisfies the relation $\omega'|_{L'_c} = 0$.

To find L_c we note that $\omega_0|_{L_c} = 0$ and

$$du = p_1\,dq_1 + p_2\,dq_2 = \pi^*(\rho) \tag{12.26}$$

on L_c. Let γ be some path on L'_c, then the integral $\int_\gamma \rho$ depends on the ends of γ because $\Omega|_{L'_c} = 0$. If we choose p_1 and p_2 to be coordinates on L'_c. Then

$$u(a,b) = \int_\gamma \rho = \int_{c_1}^{a} p_1^2\,dp_1 + \int_{c_2}^{b} p_2\,dp_2 = \frac{1}{3}a^3 + \frac{1}{2}b^2 + c_3.$$

Finally, we obtain a three-parameter family

$$u = \frac{1}{3}p_1^3 + \frac{1}{2}p_2^2 + c_3,$$

$$q_1 = \frac{1}{2}p_1^2 - c_1,$$

$$q_2 = p_2 - c_2$$

of solutions of the Von Karman equation. The projection of the family into $J^0\mathbb{R}^2$ gives the following family of two-valued solutions:

$$u(q_1, q_2) = \pm\frac{1}{3}(2q_1 + c_1)^{3/2} + \frac{1}{2}(q_2 + c_2)^2 + c_3.$$

Proposition 12.6.1 *Let functions $f_0, f_1 \in C^\infty(J^1M)$ be such that $f_0(a) = f_1(a) \neq 0$. Then there exists a local contact diffeomorphism φ such that $\varphi_*(X_{f_0}) = X_{f_1}$ and $\varphi(a) = a$.*

Proof We use the path lifting method. Without loss of generality we can assume that $X_{f_1,a} = X_{f_2,a}$. Consider the path

$$f_t = (1-t)f_0 + tf_1.$$

Let us show that there exists a path of contact diffeomorphisms φ_t, such that $\varphi_0 = \mathrm{id}$, and

$$\varphi_{t*}(X_{f_t}) = X_{f_0}.$$

Let Y_t be the path of vector fields which corresponds to φ_t. Since φ_t is contact we have $Y_t = X_{H_t}$ for some path H_t of functions.

Differentiating in t, we obtain

$$\frac{d}{dt}\varphi_{t*}(X_{f_t}) = 0 \tag{12.27}$$

or

$$[X_{f_t}, X_{H_t}] + X_{\dot{f}_t} = 0,$$

where $\dot{f}_t = df_t/dt = f_1 - f_0$.

In terms of generating functions this equation takes the form

$$[f_t, H_t] + \dot{f}_t = 0$$

or

$$X_{f_t}(H_t) - X_1(f_t)H_t + f_1 - f_0 = 0. \tag{12.28}$$

Since the vectors $X_{f_0,a} = X_{f_1,a} \neq 0$, the vector field

$$X_{f_t} = (1-t)X_{f_0} + tX_{f_1}$$

is regular in some neighborhood of the point a, for all $t \in [0, 1]$.

Therefore (12.28) has a smooth solution H_t such that $H_t(a) = 0$ and $d_a H_t = 0$. ■

Corollary 12.6.1 *Let f be a contact symmetry of the equation E_ω such that $f(a) \neq 0$. Then there exists a local contact diffeomorphism φ such $\varphi(a) = a$, and the equation $E_{\varphi^*(\omega)}$ has the contact symmetry X_1.*

In other words, for such a type of equation one could use the symplectic geometry on T^*M instead of the contact one on J^1M.

13

Conservation laws

13.1 Definition and examples

We say that a differential $(n-1)$-form θ on J^1M is a *conservation law* for the Monge–Ampère equation E_ω, if

$$d\theta = g\omega \bmod I^n \tag{13.1}$$

for some smooth function g.

In this case the restriction $\theta|_L$ to an arbitrary multivalued solution L of E_ω is a closed form:

$$d\theta|_L = 0. \tag{13.2}$$

We denote by $\mathrm{CL}(\omega)$ the space of all conservation laws for the Monge–Ampère equation E_ω.

Example 13.1.1 *Let us consider a case when $n = 1$. Then a conservation law is a 0-form, or a first integral. Indeed, the second-order differential equation*

$$v'' = F(x, v, v') \tag{13.3}$$

can be represented by the following effective differential form:

$$\omega = dp - F(x, u, p)\, dx.$$

Let $H \in C^\infty(J^1M)$ be a conservation law, and let L be a solution. Then the vector field

$$D = \frac{\partial}{\partial x} + p\frac{\partial}{\partial u} + F\frac{\partial}{\partial p}$$

is tangent to L.

The definition of a conservation law means that

$$dH = 0 \bmod \langle \omega_0, \omega \rangle$$

or

$$D(H) = H_x + pH_u + FH_p = 0. \tag{13.4}$$

Example 13.1.2 *Consider the model equation of one-dimensional gas dynamics [111, 70]:*

$$\frac{\partial \rho}{\partial t} + c(\rho)\frac{\partial \rho}{\partial x} = 0. \tag{13.5}$$

The following effective 2-form

$$\omega = (p_1 + c(u)p_2)\, dq_1 \wedge dq_2.$$

corresponds to this equation, $q_1 = t$, $q_2 = x$, $u = \rho$. *Then the differential 1-form*

$$\theta = u\, dq_2 - C(u)\, dq_1 \tag{13.6}$$

is a conservation low of (13.5). Here $C'(u) = c(u)$.

Example 13.1.3 *A divergent-type Monge–Ampère equation can be considered as a differential equation defined by a conservation law.*

13.2 Calculus for conservation laws

Here we describe a method for computation of conservation laws.

Let θ be a conservation law. Then $d\theta - g\omega \in I^n$ for some function $g \in C^\infty(J^1M)$. In other words, the operator $\Delta_{g\omega}$ is of divergent type, and therefore

$$\mathcal{E}(g\omega) = 0.$$

Conversely, if $\mathcal{E}(g\omega) = 0$, then from Corollary 11.4.1 if follows that the n-form

$$g\omega - \omega_0 \wedge \perp (d_p(g\omega))$$

is closed.

Therefore, if there are no topological obstructions (for example, if $H^n(M,\mathbb{R}) = 0$), then the form $g\omega - \omega_0 \wedge \perp d_p(g\omega)$ is exact and defines a conservation law θ:

$$d\theta = g\omega - \omega_0 \wedge \perp d_p(g\omega).$$

If M is a compact orientable manifold then the condition for the n-form $g\omega - \omega_0 \wedge \perp d_p(g\omega)$ to be exact, is equivalent to the fact that

$$\int_M (g\omega)|_{j_1(h)} = 0$$

for some smooth function $h \in C^\infty(J^1M)$.

Note that this condition is always satisfied if there exists at least one smooth solution of the Monge–Ampère equation.

Thus, we obtain the following theorem.

Theorem 13.2.1 *The conservation laws for the Monge–Ampère equation E_ω are in one-to-one correspondence with solutions of the equation*

$$\mathcal{E}(g\omega) = 0 \tag{13.7}$$

if either $H^n(M,\mathbb{R}) = 0$ or if M is a compact orientable manifold and there exists at least one smooth solution of the equation E_ω.

To each function $g \in C^\infty(J^1M)$ such that $\mathcal{E}(g\omega) = 0$ there is a corresponding conservation law θ_g that satisfies the relation

$$d\theta_g = g\omega - \omega_0 \wedge \perp (d_p(g\omega)).$$

The function g is called a *generating function* of the conservation law θ_g.

Corollary 13.2.1 *If the operator Δ_ω is of divergent type then the generating functions of conservations laws satisfy the following equation:*

$$d_p(\iota_g\omega) + d_pg \wedge \perp (d_p\omega) + X_1(g)\omega = 0.$$

Now we give an explicit formula for conservation laws.

Let us fix a solution $h_0 \in C^\infty(M)$ of the Monge–Ampère equation, $\Delta_\omega(h_0) = 0$, and let $u \in C^\infty(J^1M)$ be the function such that $u([h]^1_a) = h(a)$.

Denote by $A_t : J^1M \to J^1M$ the flow along the contact vector field X_{u-h_0}.

Note that

$$A_t : (q,\ u,\ p) \longmapsto \left(q,\ (u - h_0)e^t + h_0,\ \left(p - \frac{\partial h_0}{dq}\right)e^t + \frac{\partial h_0}{\partial q}\right).$$

Let now θ_g be a conservation law with a generating function g.

Then from

$$\frac{d}{dt}(A_t^*(\theta_g)) = A_t^*(L_{u-h_0}(\theta_g))$$

and

$$\begin{aligned} L_{u-h_0}(\theta_g) &= \iota_{u-h_0}\, d\gamma_g + d(\iota_{u-h_0}\,\theta_g) \\ &= g\iota_{u-h_0}\omega - (u-h_0) \perp (d_p(g\omega)) \\ &\quad + \omega_0 \wedge \iota_{u-h_0}\,(\perp (d_p(g\omega))) + d(\iota_{u-h_0}\,\theta_g), \end{aligned}$$

we obtain

$$\begin{aligned} \theta_g - A_{-\infty}^*(\theta_g) &= \int_{-\infty}^{0} \frac{d}{dt} A_t^*(\theta_g)\, dt \\ &= \int_{-\infty}^{0} A_t^*(g\iota_{u-h_0}\omega - (u-h_0) \perp d_p(g\omega))\, dt \\ &\quad + \omega_0 \wedge \int_{-\infty}^{0} e^t A_t^*(\iota_{u-h_0}(\perp (d_p(g\omega)))\, dt \\ &\quad + d\left(\int_{-\infty}^{0} A_t^*(\iota_{u-h_0}\,\omega)\, dt\right). \end{aligned}$$

Put

$$\tilde{\theta}_g = \int_{-\infty}^{0} A_t^*[g\iota_{u-h_0}\omega - (u-h_0) \perp d_p(g\omega)]\, dt.$$

Then the forms $\tilde{\theta}_g$ and θ_g determine the same conservation law.

Theorem 13.2.2 *Let h_0 be a smooth solution of a Monge–Ampère equation E_ω, and let g be a generating function of a conservation low, $\mathcal{E}(g\omega) = 0$. The conservation low corresponding to g is given by the formula*

$$\tilde{\theta}_g = \int_{-\infty}^{0} A_t^*(g\iota_{u-h_0}\omega - (u-h_0) \perp d_p(g\omega))\, dt. \tag{13.8}$$

Example 13.2.1 (The SG equation) *The Sine-Gordon equation*

$$v_{xy} = \sin v$$

can be defined by the effective 2-form $\omega = \omega_{SG}$

$$\omega = \frac{1}{2} dq_1 \wedge dp_1 - \frac{1}{2} dq_2 \wedge dp_2 - \sin u \, dq_1 \wedge dq_2$$

where $q_1 = x$, $q_2 = y$.

Let us find the generating functions g for conservation laws.

In our case

$$d\omega = -\cos u \, du \wedge dq_1 \wedge dq_2$$

and $d_p\omega = 0$. *The action of the Euler operator on* $g\omega$ *can be written as*

$$\mathcal{E}(g\omega) = gL_1(\omega) + d_p(\iota_g\omega) + L_1(g)\omega. \tag{13.9}$$

Computing the last three terms in this last formula, we obtain

$$gL_1(\omega) = -g \cos u \, dq_1 \wedge dq_2,$$

$$d_p(\iota_g\omega) = d_p - \left(\frac{1}{2}(g_{q_1} + p_1 g_u) + g_{p_2} \sin u\right) dq_1$$
$$+ \left(\frac{1}{2}(g_{q_2} + p_2 g_u) + g_{p_2} \sin u\right) dq_2 - \frac{1}{2} g_{p_1} \, dp_1 + \frac{1}{2} g_{p_2} \, dp_2,$$

$$L_1(g)\omega = g_u\omega.$$

Then straightforward calculations give the following expression for g:

$$g = (aq_1 + b)p_1 + (-aq_2 + c)p_2,$$

where a, b and c are constants.

After substituting it in the formula

$$d\theta_g = g\omega - \omega_0 \wedge (\iota_g\omega)$$

and taking the integral (13.8), we find conservation laws:

$$\theta_g = \left(\frac{1}{4} u g_{q_1} - g_{p_2} \cos u - \frac{1}{2} p_1 p_2 g_{p_2}\right) dq_1$$
$$+ \left(-\frac{1}{4} u g_{q_2} + g_{p_1} \cos u + \frac{1}{2} p_1 p_2 g_{p_1}\right) dq_2$$
$$+ \frac{1}{4}(p_2 g_{p_2} - p_1 g_{p_1}) \, du + \frac{1}{4} u g_{p_1} dp_1 - \frac{1}{4} u g_{p_2} \, dp_2.$$

Thus, the space of conservation laws of the Sine-Gordon equation is three-dimensional. The following conservation laws

$$\theta_1 = \left(-\frac{1}{4}u_{q_1} - u_{q_2} - \cos u - \frac{1}{4}\sin u\right) dq_1 + \left(-\frac{1}{4}uu_{q_2q_2} + \frac{1}{4}u_{q_2}^2\right) dq_2,$$

$$\theta_2 = \left(\frac{1}{4}u_{q_1} + u_{q_2} + \cos u + \frac{1}{4}\sin u\right) dq_2 + \left(\frac{1}{4}uu_{q_1q_1} - \frac{1}{4}u_{q_1}^2\right) dq_1,$$

$$\theta_3 = \left(\frac{1}{4}uu_{q_1} + q_2\cos u + \frac{1}{4}q_2u_{q_1}u_{q_2} + \frac{1}{4}uq_1u_{q_1q_1} + \frac{1}{4}uq_2\sin u - \frac{1}{4}q_1u_{q_1}^2\right) dq_1$$

$$+ \left(\frac{1}{4}uu_{q_2} + q_1\cos u + \frac{1}{4}q_1u_{q_1}u_{q_2} + \frac{1}{4}uq_2u_{q_2q_2} + \frac{1}{4}uq_1\sin u - \frac{1}{4}q_2u_{q_2}^2\right) dq_2$$

give a basis in $\mathrm{CL}(\omega_{SG})$ *(see [69]).*

Example 13.2.2 (The Von Karman equation) *Generating functions for conservation laws of the Von Karman equation in the three-dimensional space have the following form:*

$$g = g_0(q_1, q_2) + c_1p_1 + c_2p_2 + c_3p_3 + c_4(q_3p_2 - q_2p_3)$$

where $g_0(q_1, q_2)$ *is an arbitrary harmonic function and* c_1, c_2, c_3, c_4 *are constants. The corresponding to the function g conservation law* θ_g *is*

$$\theta_g = f_1\, dq_2 \wedge dq_3 - f_2\, dq_1 \wedge dq_3 + f_3\, dq_1 \wedge dq_2,$$

where

$$f_1 = \frac{1}{3}c_1p_1^3 + \frac{1}{2}(c_2p_2 + c_3p_3) + \frac{1}{2}c_4p_1^2(q_3p_2 - q_2p_3) + \frac{1}{2}c_1(p_2^2 + p_3^2),$$

$$f_2 = \frac{1}{6}(p_1^3 + (p_2^2 - p_3^2))(c_2 + c_4q_3) + p_2p_3(c_3 - c_4q_2) + c_1p_1p_2$$

$$+ p_2g_0 + q_1p_1\frac{\partial g_0}{\partial q_2},$$

$$f_3 = \left(\frac{1}{6}p_1^3 + (p_2^2 - p_3^2)\right)(c_3 - c_4q_3) - p_2p_3(c_2 + c_4q_2) - c_1p_1p_2$$

$$- p_3g_0 - q_1p_1\frac{\partial g_0}{\partial q_3}.$$

13.3 Symmetries and conservations laws

The famous Noether theorem establishes the relationship between symmetries and conservation lows in the calculus of variations. In this section we show the similar relations for the divergent-type Monge–Ampère equations.

Theorem 13.3.1 *Let ω be an effective n-form of divergent type. Then for any contact symmetry $X_f \in \mathrm{Sym}(E_\omega)$ the following differential $(n-1)$-form*

$$\theta = \iota_f\, \omega - f \perp (d_p \omega)$$

is a conservation law.

Proof By definition of symmetry, we have

$$(p \circ L_f)\omega = h_f \omega,$$

for some function $h_f \in C^\infty(J^1 M)$. Using the formula

$$(p \circ L_f)\omega = f\mathcal{E}(\omega) + d_p(\iota_f \omega - f \perp (d_p\omega)) + d\omega_0 \wedge (\iota_f \perp (d_p \omega))$$

and the condition $\mathcal{E}(\omega) = 0$, we obtain

$$d_p(\iota_f \omega - f \perp (d_p\omega)) = h_f \omega + d\omega_0 \wedge (\iota_f \perp (d_p\omega)) = h_f \omega \bmod I^n. \qquad \blacksquare$$

Theorem 13.3.2 *Let ω be an effective n-form of divergent type such that $d_p\omega = 0$ and $p(L_1(\omega)) = h\omega$ for some function $h \in C^\infty(J^1M)$. Then for any conservation law θ_g with the generating function g the contact vector field X_g is a symmetry of the equation E_ω.*

Proof We have

$$\mathcal{E}(g\omega) = g\mathcal{E}(\omega) + X_1(g)\omega + d_p(\iota_g \omega) = 0.$$

Hence,

$$d_p(\iota_g \omega) = -X_1(g)\omega$$

and

$$p(L_{X_g}(\omega)) = d_p(\iota_g \omega) + gL_{X_1}(\omega) = (-X_1(g) + h)\omega. \qquad \blacksquare$$

13.4 Shock waves and the Hugoniot–Rankine condition

13.4.1 Shock Waves for ODEs

In this section we consider discontinuous solutions of Monge–Ampère equations on $\mathbb{R}$.

Proposition 13.4.1 *Let $E_\omega = \{y'' = F(x, y, y')\}$ be an ordinary differential equation. A smooth function $H \in C^\infty(J^1\mathbb{R})$ is a first integral for E_ω if and only if*

$$\int_{\mathbb{R}} j_1^*(h)(H)\, ds = 0, \tag{13.10}$$

for any test function $s \in C_0^\infty(\mathbb{R})$ and any smooth solution $h \in C^\infty(\mathbb{R})$ of E_ω.

Proof Indeed (13.10) gives

$$\int_{\mathbb{R}} j_1^*(h)(H)\, ds = \int_{\mathbb{R}} (j_1^*(h)(H))' s\, dx = 0$$

for all $s \in C_0^\infty(\mathbb{R})$, and therefore, $j_1^*(h)(H)$ is a constant for any solution h. ■

This proposition shows us the way to extend a concept of solution for ODEs (see Figure 13.1). Let us define a discontinuous solutions of the equation E_ω. To this end we fix a point $a \in \mathbb{R}$. This point divides $\mathbb{R}$ into two parts: $\mathbb{R}^- = \{q | \, q < a\}$ and $\mathbb{R}^+ = \{q | \, q > a\}$. Consider a discontinuous function $v = v^+ \cup v^-$, where $v^+ \in C^\infty(\overline{\mathbb{R}^+})$ and $v^- \in C^\infty(\overline{\mathbb{R}^-})$. Here $\overline{\mathbb{R}^\pm} = \mathbb{R} \backslash \mathbb{R}^\mp$.

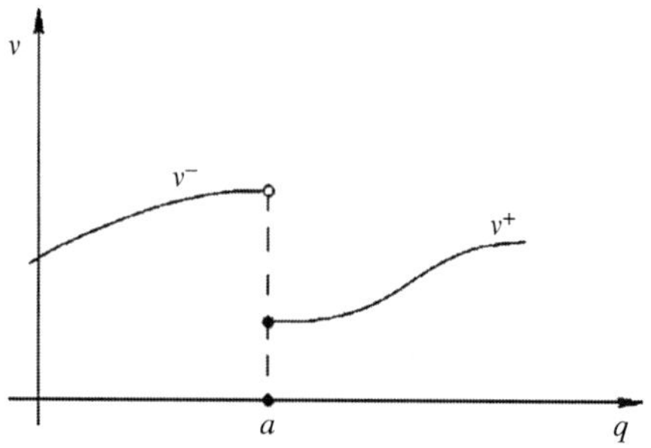

Figure 13.1. A discontinuous solution of an ODE.

The function v is called a *discontinuous solution corresponding to the integral H* if

$$\int_{\mathbb{R}} H|_{j_1(v)}\, ds = 0$$

for any test function $s \in C_0^\infty(\mathbb{R})$.

Theorem 13.4.1 *A function v is a discontinuous solution corresponding to an integral H if and only if* $\Delta_\omega(v) = 0$ *in* $\mathbb{R}\backslash\{a\}$ *and at the point a the following condition*

$$H([v^+]_a^1) = H([v^-]_a^1) \tag{13.11}$$

holds.

Proof Let $\operatorname{supp} s$ be the support of s. If $a \notin \operatorname{supp} s$, say $\operatorname{supp} s \subset \mathbb{R}^-$, then

$$\int_{\mathbb{R}} j_1^*(v^-)(H)\, ds = \int_{\mathbb{R}} (j_1^*(v^-)(H))' s\, dx = 0$$

and therefore $j_1^*(v^-)(H) = \text{constant}$.

If $a \in \operatorname{supp} s$, then

$$\int_{\mathbb{R}} H|_{j_1(v)}\, ds = H([v^+]_a^1)\ s(a) - H([v^-]_a^1)\ s(a)$$

and therefore v is a solution if and only if $H([v^+]_a^1) = H([v^-]_a^1)$. ■

Condition (13.11) is called the *Hugoniot–Rankine condition.*

13.4.2 Discontinuous solutions

Let us extend the notion of discontinuous solutions for Monge–Ampère equations in arbitrary dimension.

Suppose that the smooth manifold M is divided into two domains M^+ and M^- by a submanifold M^0, $\operatorname{codim} M^0 = 1$:

$$M = M^+ \cup M^0 \cup M^-.$$

Consider a discontinuous function $v = v^+ \cup v^-$. We assume that v^+ and v^- are smooth functions in the closures M^+ and M^-, $v^+ \in C^\infty(\overline{M^+})$ and $v^- \in C^\infty(\overline{M^-})$.

The function v is called a *discontinuous solution corresponding to a conservative law* θ if

$$\int_M ds \wedge \theta|_{j_1(v)} = 0$$

for any test function $s \in C_0^\infty(M)$.

Let us denote by $\theta^+ \overset{\text{def}}{=} \theta|_{j_1(v^+)}$ and $\theta^- \overset{\text{def}}{=} \theta|_{j_1(v^-)}$ the form θ evaluated on the solutions v^+ and v^- respectively.

Theorem 13.4.2 *A function v is a discontinuous solution corresponding to a conservatively low θ if and only if $\Delta_\omega(v) = 0$ in the domain $M \backslash M^0$ and at points of M^0 the Hugoniot–Rankine condition*

$$\theta^+|_{M^0} = \theta^-|_{M^0} \tag{13.12}$$

holds.

Proof If $\operatorname{supp} s \subset M^+$ then

$$\int_M ds \wedge \theta|_{j_1(v)} = \int_M d(s\theta^+) - \int_M s\theta^+ = -\int_M s\theta^+ = 0$$

for all such test functions, and therefore $\theta^+ = 0$.

If $M_s = \operatorname{supp} s \cap M^0 \neq \varnothing$, then

$$\int_M ds \wedge \theta|_{j_1(v)} = \int_{M_s} s \left(\theta^+ - \theta^-\right)|_{M_0} = 0$$

implies the Hugoniot–Rankine condition (see [70]). ■

Remark 13.4.1 *In the case of submanifold $M^0 = S^{-1}(0)$ defined by a function $S \in C^\infty(M)$, where $dS \neq 0$ at the points of M^0, the Hugoniot–Rankine condition is equivalent to the following one:*

$$[\theta^+|_{M^0} - \theta^-|_{M^0}] \wedge dS = 0$$

at points of M^0.

Example 13.4.1 *Suppose that the discontinuity boundary (transfer curve) M^0 for the model equation (13.5) is given by*

$$M^0 = \{q_2 = \varphi(q_1)\}.$$

Then the restriction of θ to M^0 is

$$\theta|_{M^0} = u\,d\varphi - C(u)\,dq_1 = (u\varphi' - C(u))dq_1$$

and (13.12) takes the form of the classical Hugoniot–Rankine condition:

$$(v^+ - v^-)\varphi' = C(v^+) - C(v^-).$$

13.4.3 Shock waves

The Hugoniot–Rankine condition allows us to describe the rise of shock waves [70] as some kind of symbiosis of geometrical and analytical approaches. We will illustrate this on multivalued solutions of (13.5). Suppose that the multivalued solution describes a turning-over wave. The corresponding multivalued solution L is shown in Figure 13.2.

In the hatched region the solution is three-valued and we must select along the three branches.

First, we divide the solution at time $t = t_2$ into three parts: I, II and III (see Figure 13.3). These parts can be thought of as an incident, a reflected (or self-reflected) and again an incident (double-reflected) wave respectively.

Consider a transfer curve M^0 and lift it to branches I and II; we obtain two curves, M^1 and M^3. The Hugoniot–Rankine condition says that $\theta|_{M^1}$ and $\theta|_{M^3}$, viewed as 1-forms on M^0, must coincide.

If we unfold surface L the region bounded by M^1, M^3 and the section $q_1 = t_2$ becomes a curvilinear triangle C_1C_3P (see Figure 13.4).

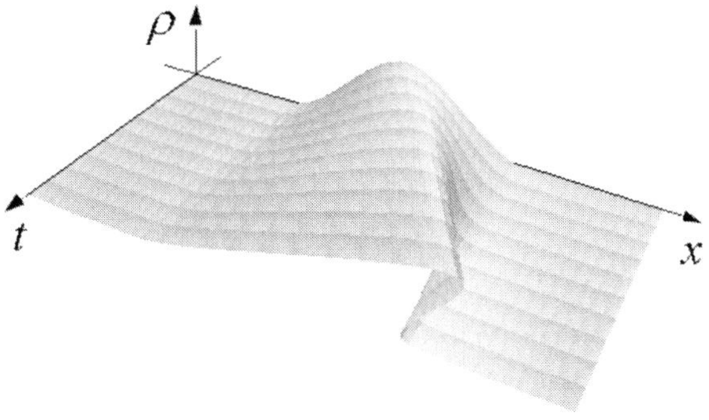

Figure 13.2. The multivalued solution of $\partial\rho/\partial t + c(\rho)\partial\rho/\partial x = 0$.

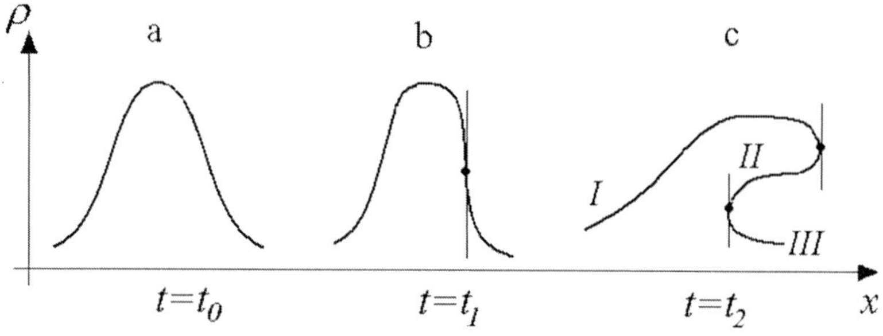

Figure 13.3. Formation of a turned-over wave.

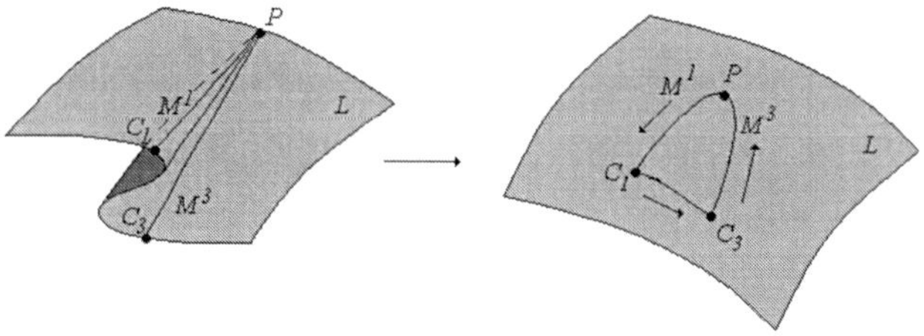

Figure 13.4. Unfolding of a multivalued solution.

Next, integrating the form θ along this triangle, we obtain

$$\int_{M^1} \theta + \int_{q_1=t_2} \theta + \int_{M^3} \theta = 0.$$

The Hugoniot–Rankine condition gives

$$\int_{q_1=t_2} \theta = \int_{q_1=t_2} u\, dq_2 = 0$$

or that areas S_1 and S_2 (see Figure 13.5) coincide.

This is the so-called *Maxwell rule*, i.e., *the areas cut off by the line $x = c$ are equal:* $S_1 = S_2$. This rule lets us construct discontinuous solutions (shock waves) from multivalued ones.

An application of the Hugoniot–Rankine condition to the description of singularities in non-linear acoustics will be considered later on.

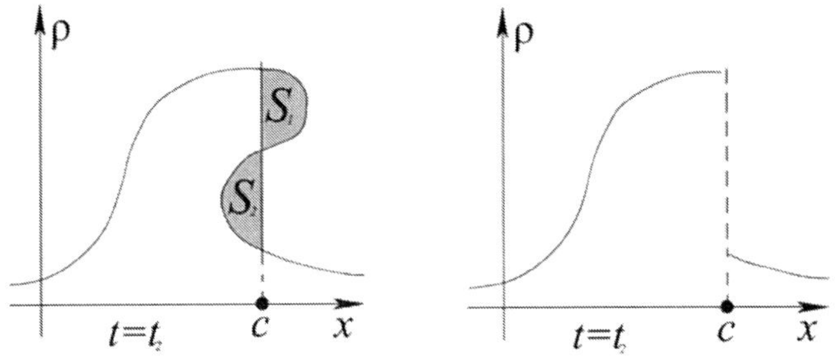

Figure 13.5. The Maxwell rule.

13.5 Calculus of variations and the Monge–Ampère equation

In practice, the most interesting differential equations of the second order arise as Euler–Lagrange equations of variational problems. Here we give an overview of the calculus of variations from the point of view of contact geometry.

We consider the class of variational problems associated with the Monge–Ampère equations. More exactly, we consider the functionals associated with effective n-forms. We will justify the notion of the Euler operator given above and we will show that the Euler–Lagrange equations for this class of variational problems are Monge–Ampère equations corresponding to the forms $\mathcal{E}(\omega)$, where ω is an effective form which defines the initial variational problem.

We also describe the symmetries and conservation laws for such variational problems and give a proof of a contact analog of Noether's theorem.

Finally, we discuss the Euler operator from the cohomology viewpoint. We use a filtration in the exterior form algebra of any contact manifold (a "nonholonomic" filtration introduced and studied in [75]) and show that the Euler operator is the second differential in the spectral sequence based on this filtration.

13.5.1 The Euler operator

Below we give a sketch of an invariant exposition of variational calculus, using the machinery of effective forms developed in the previous chapters.

We associate to each differential n-form $\omega \in \Omega_\varepsilon^n(J^1M)$ the functional $\check{\omega}$, acting as follows:

$$\check{\omega}(h) = \int_M j_1(h)^*(\omega),$$

where $h \in C_0^\infty(M)$.

Note that the Lagrangian of the functional $\check{\omega}$, written in the canonical local coordinates, contains a non-linearity of the same type as the Monge–Ampère operators.

The next theorem shows why the operator $\mathcal{E}$ was called the Euler operator.

Theorem 13.5.1 *Extremals of the functional* $\check{\omega}$ *are solutions of the Monge–Ampère equation associated with the n-form* $\mathcal{E}(\omega)$.

Proof Let $h_0 \in C_0^\infty(M)$ be an extremal of $\check{\omega}$. Fix a test function $s \in C_0^\infty(M)$ and let

$$\varphi(t) = \int_M j_1(h_0 + ts)^*(\omega).$$

Since h_0 is an extremal, $\varphi'(0) = 0$.

On the other hand,

$$\varphi'(0) = \int_M j_1(h_0)^* L_s \omega.$$

It follows from Theorem 11.4.2 that

$$\varphi'(0) = \int_M j_1(h_0)^*(s\,\mathcal{E}(\omega)) = \int_M s\, j_1(h_0)^*(\mathcal{E}(\omega)) = 0$$

for all test functions s, and therefore

$$j_1(h_0)^*(\mathcal{E}(\omega)) = \Delta_{\mathcal{E}(\omega)}(h_0) = 0.$$

■

13.5.2 Symmetries and conservation laws in variational problems

We call a contact vector field $X_f \in \mathrm{ct}(J^1M)$ a *symmetry* of the functional $\check{\omega}$ if $L_f(\omega) \in I^n$.

To give a motivation for this definition, let us suppose that the manifold M is compact. Let $A_t : J^1M \to J^1M$ be a one-parameter group of shifts along the contact vector field X_f. Then for an arbitrary function $h \in C^\infty(M)$ and sufficiently small t, the integral manifolds $L_t = A_t(L)$, $L = j_1(h)(M)$, can be projected diffeomorphically on M and, therefore, $L_t = j_1(h_t)(M)$ for some family of smooth functions $h_t \in C^\infty(M)$.

Moreover, from the relation $L_f(\omega) \in I^n$ it follows that $A_t^*(\omega) - \omega \in I^n$ and

$$j_1(h)(\omega) = j_1(h)(A_t^*\omega) = j_1(h_t)(\omega),$$

$\check{\omega}(h_t) = \check{\omega}(h)$.

Using Theorem 11.4.2 we obtain the following proposition.

Proposition 13.5.1 *A contact vector field X_f is a symmetry of the variational problem if and only if the generating function f satisfies the following equation:*

$$d_p \iota_f \omega + \iota_f d_p \omega + f L_1 \omega = 0.$$

Theorem 13.5.2 (Noether) *Let a contact vector field X_f be a symmetry of the functional $\breve{\omega}$. Then the differential $(n-1)$-form $\iota_f \omega - f \perp d_p \omega$ represents the conservation law for the Euler equation $E_{\mathcal{E}(\omega)}$.*

Proof Using 11.4.2 we get

$$0 = p(L_f \omega) = f\mathcal{E}(\omega) + d_p(\iota_f \omega - f \perp d_p \omega) + \top(\iota_f \perp d_p \omega), \tag{13.13}$$

i.e.,

$$f\mathcal{E}(\omega) + d(\iota_f \omega - f \perp d_p \omega) \in I^n. \tag{13.14}$$

■

Theorem 13.5.3 *Suppose that $L_1(\mathcal{E}(\omega)) = 0$. Then every conservation law for the Euler equation $E_{\mathcal{E}(\omega)}$ is determined by some contact symmetry of the equation.*

Proof Note that the Monge–Ampère operator $\Delta_{\mathcal{E}(\omega)}$ is of divergent type. Indeed,

$$\mathcal{E}^2(\omega) = L_1 \mathcal{E}(\omega) = 0.$$

The statement of the theorem follows now from Theorem 13.3.2 and the relation $d_p \circ \mathcal{E} = 0$. ■

13.5.3 Classical variational problems

Let $\omega_0 \in \Omega^n(M)$ be a volume form. Then any function $L \in C^\infty(J^1 M)$ defines an effective form $\omega = L\, \pi_1^*(\omega_0)$ and the corresponding variational problem. This is the classical case, where L is the Lagrangian.

In this case

$$\mathcal{E}(\omega) = d_p \iota_L(\pi_1^* \omega_0) + X_1(L)\pi_1^*(\omega_0).$$

The correspondence between contact symmetries and conservation laws has the form

$$f \longmapsto \theta = k\iota_f(\pi_1^*\omega_0) - f\iota_L(\pi_1^*\omega_0).$$

The contact symmetries of the variational problem can be found from the following Lie equation:

$$d_p(L\iota_f\pi_1^*\omega_0) + X_f(L)\pi_1^*(\omega_0) = 0.$$

The equation for generating functions of conservation laws of the Euler differential equation takes the form

$$d_p\iota_g\mathcal{E}(\omega) + L_1(g\mathcal{E}(\omega)) = 0.$$

13.6 Effective cohomology and the Euler operator

Let M be a contact manifold of dimension $2n+1$ and C be the contact distribution. Denote by J the ideal in the algebra of differential forms $\Omega^*(M)$ generated by the module $C_0 = \operatorname{Ann}(C) \subset \Omega^1(M)$, that is,

$$J = C_0 \wedge \Omega^*(M).$$

We consider the following filtration in the algebra of differential forms $\Omega^*(M)$ associated with the ideal:

$$F_{p,q} = C_0^{-q} \wedge \Omega^{p+2q}(M), \qquad q \leq 0,\ p \geq 0.$$

This is a decreasing filtration of $\Omega^*(M)$,

$$F_{p,q} \subset \Omega^{p+q}(M),\ \ \Omega^k(M) = \bigcup_{p+q=k} F_{p,q}$$

and

$$F_{p,q} \supset F_{p+1,q-1},$$

and this filtration is compatible with the de Rham differential d:

$$dF_{p,q} \subset F_{p,q+1}.$$

Consider now a spectral sequence $\{E_r^{p,q}, d_r^{p,q} : E_r^{p,q} \to E_r^{p+r,q-r+1}\}$ associated with this filtration. This sequence is converged to the de Rham cohomology of the contact manifold.

The term s of $\{E_r^{p,q}, d_r^{p,q}\}$ are located in the fourth quadrant:

$$0 \leq p + 2q \leq 2n + 1, \quad p \geq 0, \, q \leq 0.$$

The multiplicative structure in the spectral sequence is inherited from the structure in the algebra of differential forms.

The E_0-term is

$$E_0^{p,q} = F_{p,q}/F_{p+1,q-1}.$$

Note that $E_0^{p,q}$ are non-trivial only for $q = 0$ and $q = -1$, because C_0 has dimension 1.

In this case

$$E_0^{p,0} = \frac{F_{p,0}}{F_{p+1,-1}} = \frac{\Omega^p(M)}{C_0 \wedge \Omega^{p-1}(M)} \simeq \Lambda^p(C^*)$$

is the module of p-forms on distribution C, and

$$E_0^{p,-1} = \frac{F_{p,-1}}{F_{p+1,-2}} = C_0 \wedge \Omega^{p-2}(M) \approx \Lambda^{p-2}(C^*)$$

where we have used the identification

$$C^* = \Omega^1(M)/C_0.$$

The complexes $d_0^{p,q} : E_0^{p,q} \to E_0^{p,q+1}$ are two-term sequences

$$0 \to \Lambda^{p-2}(C^*) \approx C_0 \wedge \Omega^{p-2}(M) \overset{d_0^{p,-1}}{\to} \Lambda^p(C^*) \to 0$$

where the differential acts as follows:

$$d_0^{p,-1} : \vartheta \wedge \omega \longmapsto \text{class}(d(\vartheta \wedge \omega)) = d\vartheta|_C \wedge \omega|_C$$

for any $\vartheta \in C_0$.

Note that any local basis 1-form ϑ in C_0 defines a symplectic structure $d\vartheta|_C \in \Lambda^2(C^*)$ on the contact distribution. If $\vartheta' = h\,\vartheta$ is another basis in C_0 then $d\vartheta'\big|_C = h\, d\vartheta|_C$ and therefore any two of these symplectic structures are proportional.

Thus

$$E_1^{p,-1} = \ker d_0^{p,-1} = \ker((\top : \Lambda^{p-2}(C^*) \to \Lambda^p(C^*)))$$

and

$$E_1^{p,0} = \operatorname{coker} d_0^{p,-1} = \operatorname{coker}(\top : \Lambda^{p-2}(C^*) \to \Lambda^p(C^*))$$

where $\top$ is the operator defined by one of the symplectic structures.

Using the properties of operator $\top$ and the Hodge–Lepage decomposition we get the following description:

$$E_1^{p,-1} = \begin{cases} 0, & \text{if } p \le n+1, \\ \top^{n-p+2}(\Lambda_\varepsilon^{p-2}(C^*)) & \text{if } p \ge n+2, \end{cases}$$

and

$$E_1^{p,0} = \begin{cases} 0, & \text{if } p \ge n+1, \\ \Lambda_\varepsilon^p(C^*) & \text{if } p \le n. \end{cases}$$

Elements of

$$\operatorname{coker}(\top : \Lambda^{p-2}(C^*) \to \Lambda^p(C^*)) = \ker(\bot\colon \Lambda^p(C^*) \to \Lambda^{p-2}(C^*))$$

are effective p-forms, and the elements of

$$\ker((\top : \Lambda^p(C^*) \to \Lambda^{p+2}(C^*)))$$

we shall call *co-effective p-forms*.

As we have seen co-effective $(n+k)$-forms are exactly the forms in $\top^k(\Lambda_\varepsilon^{n-k}(C^*))$. We denote the module of co-effective p-forms we denote by $\Lambda_\varkappa^p(C^*)$.

Note that we have $\Lambda_\varepsilon^p(C^*) = 0$ if $p \ge n+1$ and $\Lambda_\varkappa^p(C^*) = 0$ if $p \le n-1$, in the middle dimension $\Lambda_\varkappa^n(C^*) = \Lambda_\varepsilon^n(C^*)$.

The differentials $d_1^{p,q} : E_1^{p,q} \to E_1^{p+1,q}$ give us two complexes for calculation of $E_2^{p,q}$; the first one for effective forms when $q = 0$, is

$$0 \to \Lambda_\varepsilon^0(C^*) \xrightarrow{d_\varepsilon} \Lambda_\varepsilon^1(C^*) \xrightarrow{d_\varepsilon} \Lambda_\varepsilon^0(C^*) \xrightarrow{d_\varepsilon} \cdots \xrightarrow{d_\varepsilon} \Lambda_\varepsilon^n(C^*) \to 0$$

and the second one for co-effective forms when $q = -1$, is

$$0 \to \Lambda_\varepsilon^n(C^*) = \Lambda_\varkappa^n(C^*) \xrightarrow{d_\varkappa} \Lambda_\varkappa^{n+1}(C^*) \xrightarrow{d_\varkappa} \Lambda_\varkappa^{n+2}(C^*) \xrightarrow{d_\varkappa} \cdots$$
$$\xrightarrow{d_\varkappa} \Lambda_\varkappa^{2n}(C^*) \to 0.$$

We have denoted by d_ε and $d_\varkappa$ the corresponding differentials $d_1^{p,0}$ and $d_1^{p,-1}$. The cohomology of the complexes we call the *effective* and the *co-effective cohomology* of the contact manifold and denote these by $H_\varepsilon^p(M)$, $0 \le p \le n$, and $H_\varkappa^p(M)$, $n \le p \le 2n$ respectively.

Example 13.6.1 *If $M = J^1N$ is the contact manifold, then $H_\varkappa^p(M) = 0$ when $p \ge n+1$.*

Let us look at the term $E_2^{p,q}$. First of all the above description shows that

$$E_2^{p,-1} = \begin{cases} 0, & \text{if } p \le n+1, \\ H_\varkappa^{p-2}(M), & \text{if } p \ge n+2, \end{cases}$$

and

$$E_2^{p,0} = \begin{cases} 0, & \text{if } p \ge n+1, \\ H_\varepsilon^{p}(M), & \text{if } p \le n. \end{cases}$$

The differential $d_2^{p,q} : E_2^{p,q} \to E_2^{p+2,q-1}$ is non-trivial only for $p = n$, $q = 0$, and acts as follows:

$$d_2^{n,0} : E_2^{p,0} = \Lambda_\varepsilon^n(C^*) \to \Lambda_\varepsilon^n(C^*) = E_2^{n+2,-1}.$$

Therefore we obtain

$$E_3^{p,-1} = \begin{cases} 0, & \text{if } p \le n+1, \\ \operatorname{coker} d_2^{n,0}, & \text{if } p = n+2, \\ H_x^{p-2}(M), & \text{if } p \ge n+3, \end{cases}$$

and

$$E_3^{p,0} = \begin{cases} 0, & \text{if } p \ge n+1, \\ \ker d_2^{n,0}, & \text{if } p = n, \\ H_\varepsilon^{p}(M), & \text{if } p \le n-1. \end{cases}$$

The next differentials $d_3^{p,q} : E_3^{p,q} \to E_3^{p+3,q-2}$ are obviously trivial. Therefore,

$$E_\infty^{p,q} = E_3^{p,q}.$$

Proposition 13.6.1 *Let $M = J^1N$, $n = \dim N$, and let $\Lambda^n_\varepsilon(C^*) \to H^n_\varepsilon(M) = \Lambda^n_\varepsilon(C^*)\backslash \operatorname{Im} d_\varepsilon$ be the natural projection and $H^n_\varkappa(M) = \ker d_\varkappa \to \Lambda^n_\varepsilon(C^*) = \Lambda^n_\varkappa(C^*)$ be the embedding. Then the following diagram commutes.*

$$
\begin{array}{ccc}
\Lambda^n_\varepsilon(C^*) & \xrightarrow{\ \mathcal{E}\ } & \Lambda^n_\varepsilon(C^*) \\
\Big\downarrow & & \Big\uparrow \\
H^n_\varepsilon(M) & \xrightarrow{\ d_2^{n,0}\ } & H^n_\kappa(M)
\end{array}
$$

Proof It follows from the original construction of the Euler operator. ■

Owing to this proposition we call differential $d_2^{n,0}$ the *Euler operator* on the contact manifold M and denote it by $\mathcal{E}$.

Summing up we arrive at the following result.

Theorem 13.6.1 *The spectral sequence $(E_r^{p,q}, d_r^{p,q})$ on a contact manifold M collapses in the term E_3 and has the following description.*

1. The term $E_0^{p,q}$

$$
E_0^{p,q} = \begin{cases} \Lambda^p(C^*), & \text{if } q = 0, \\ \Lambda^{p-2}(C^*), & \text{if } q = -1, \\ 0, & \text{otherwise.} \end{cases}
$$

2. The term $E_1^{p,q}$

$$
E_1^{p,q} = \begin{cases} \Lambda^p_\varepsilon(C^*), & \text{if } 0 \le p \le n,\ q = 0, \\ \top^{n-p+2}(\Lambda^{p-2}_\varepsilon(C^*)), & \text{if } n+2 \le p,\ q = -1, \\ 0, & \text{otherwise.} \end{cases}
$$

3. The term $E_2^{p,q}$

$$
E_2^{p,q} = \begin{cases} H^p_\varepsilon, & \text{if } 0 \le p \le n,\ q = 0, \\ H^{p-2}_\varkappa(M), & \text{if } n+2 \le p,\ q = -1, \\ 0, & \text{otherwise.} \end{cases}
$$

4. *The term* $E_3^{p,q} = E_\infty^{p,q}$

$$E_3^{p,q} = \begin{cases} H_\varepsilon^p(M), \text{ if } p \leq n-1; \quad \ker \mathcal{E}, \quad \text{if } p = n, \quad q = 0, \\ \operatorname{coker} \mathcal{E}, \text{ if } p = n+2; \quad H_\varkappa^{p-2}(M), \quad \text{if } p \geq n+3, \quad q = -1, \\ 0, \quad \text{otherwise.} \end{cases}$$

5. *The de Rham cohomology* $H^k(M, \mathbb{R})$ *of the contact manifold coincides with the effective cohomology* $H_\varepsilon^k(M)$ *if* $k \leq n-1$, *or with the co-effective cohomology* $H_\varkappa^{k-1}(M)$ *if* $k \geq n+3$, $\dim M = 2n+1$. *In the middle dimensions* $H^n(M, \mathbb{R}) = \ker \mathcal{E}$ *and* $H^{n+1}(M, \mathbb{R}) = \operatorname{coker} \mathcal{E}$.

14
Monge–Ampère equations on two-dimensional manifolds and geometric structures

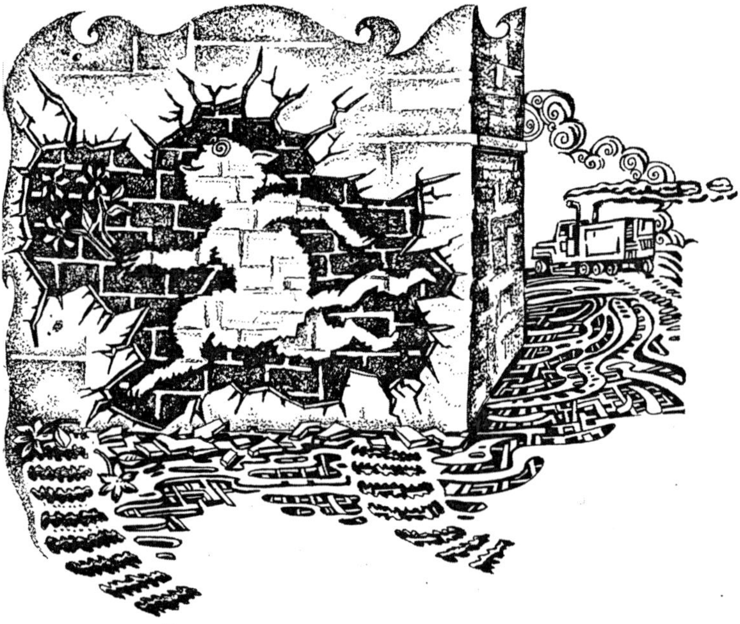

In this chapter we consider Monge–Ampère equations on a two-dimensional manifold M. In this case Monge–Ampère equations endow the space J^1M with structures that are very important in differential geometry: the non-holonomic almost product structure, the non-holonomic almost complex structure (a partial case of the Cauchy–Riemann (CR) structure) and the non-holonomic almost tangent structure. We use this link in order to find the geometrical nature for the differential equations.

We also discuss some methods for integration of the Cauchy problem for Monge–Ampère equations.

14.1 Non-holonomic geometric structures associated with Monge–Ampère equations

14.1.1 Non-holonomic structures on contact manifolds

Let N be a $2n$-dimensional smooth manifold and let A be a smooth field of endomorphisms on N:

$$A_x : T_xN \to T_xN$$

for $x \in N$.

Recall that A is called a (classial) *almost product structure*, an *almost complex structure* or an *almost tangent structure* if $A^2 = 1, -1$ or 0, respectively.

Let Φ be a $(2n+1)$-dimensional contact manifold with a contact structure

$$P : \Phi \ni a \longmapsto P(a) \subset T_a\Phi.$$

Suppose that on each subspace $P(a)$ $(a \in \Phi)$ of the tangent space $T_a\Phi$ an endomorphism $A_a : P(a) \to P(a)$ is given and the map $a \longmapsto A_a$ is smooth. Then we call the field $A : \Phi \ni a \longmapsto A_a \in \mathrm{End}(P(a))$ a *non-holonomic field of endomorphisms* on the contact manifold Φ.

If for a non-holonomic field of endomorphisms A one has $A^2 = 1, -1$ or 0, then the field A is called a *non-holonomic almost product structure* , a *non-holonomic almost complex structure* or a *non-holonomic almost tangent structure*, respectively.

A non-holonomic almost complex structure is a partial case of so-called *Cauchy–Riemann structure* or simply *CR-structure*.

Let A be a non-holonomic almost complex structure. A two-dimensional integral submanifold $L \hookrightarrow J^1M$ is called a *complex curve* of the non-holonomic almost complex structure A if $A_x : T_xL \to T_xL$ for all $x \in L$.

14.1.2 Non-holonomic fields of endomorphisms generated by Monge–Ampère equations

As we have seen any effective differential form $\omega \in \Omega^2_\varepsilon(J^1M)$ generates the non-holonomic field of endomorphisms

$$A_\omega : J^1M \ni a \longmapsto A_{\omega_a} \in \mathrm{End}(C_a), \tag{14.1}$$

where

$$X \rfloor \omega = A_\omega X \rfloor \Omega.$$

for any vector field $X \in D(C)$.

Here

$$\Omega_a \stackrel{\text{def}}{=} d\omega_0|_{C(a)}$$

is the standard symplectic structure on $C(a)$ and $A_\omega(X)_a = A_{\omega,a}(X_a)$ for all $a \in J^1M$.

A function $\text{Pf}(\omega) \in C^\infty(J^1M)$ defined pointwise by the formula

$$\text{Pf}(\omega)(a) \stackrel{\text{def}}{=} \text{Pf}(\omega_a)$$

is called a *Pfaffian* of the form ω.

The following properties of the operator field A_ω follow from the properties of the linear operator $A_{\omega,a}$.

Proposition 14.1.1

1. *A_ω is symmetric with respect to Ω, i.e.,*

$$\Omega(A_\omega X, Y) = \Omega(X, A_\omega Y)$$

for any vector field $X, Y \in D(C)$.

2. *Vector fields $X, A_\omega X \in D(C)$ are skew-orthogonal, i.e.,*

$$\Omega(A_\omega X, X) = 0.$$

3. *For any multivalued solution L of the equation E_ω one has*

$$A_{\omega_a} : T_aL \to T_aL$$

for all $a \in L$.

4. *The square of A_ω is scalar:*

$$A_\omega^2 + \text{Pf}(\omega) = 0. \tag{14.2}$$

5. *Let q_1, q_2, u, p_1, p_2 be the canonical local coordinates in J^1M. The vector fields*

$$\frac{d}{dq_1} \stackrel{\text{def}}{=} \frac{\partial}{\partial q_1} + p_1\frac{\partial}{\partial u}, \quad \frac{d}{dq_2} \stackrel{\text{def}}{=} \frac{\partial}{\partial q_2} + p_2\frac{\partial}{\partial u}, \quad \frac{\partial}{\partial p_1}, \quad \frac{\partial}{\partial p_2} \tag{14.3}$$

form a local basis in the module $D(C)$. Then

$$A_\omega = \left(B\frac{d}{dq_1} + C\frac{d}{dq_2} - E\frac{\partial}{\partial p_2}\right)\otimes dq_1 + \left(B\frac{\partial}{\partial p_1} + D\frac{d}{dq_2} - A\frac{\partial}{\partial p_2}\right)\otimes dp_1$$
$$\left(-A\frac{d}{dq_1} + E\frac{\partial}{\partial p_1} - B\frac{d}{dq_2}\right)\otimes dq_2$$
$$+\left(-D\frac{d}{dq_1} + C\frac{\partial}{\partial p_1} - B\frac{d}{dq_2}\right)\otimes dp_2$$

or

$$A_\omega = \left\| \begin{matrix} B & -A & 0 & -D \\ C & -B & D & 0 \\ 0 & E & B & C \\ -E & 0 & -A & -B \end{matrix} \right\| \qquad (14.4)$$

if

$$\omega = A\,dp_1 \wedge dq_2 + B(dq_1 \wedge dp_1 + dp_2 \wedge dq_2) + C\,dq_1 \wedge dp_2$$
$$+ D\,dp_1 \wedge dp_2 + E\,dq_1 \wedge dq_2$$

and

$$\mathrm{Pf}(\omega) = DE - AC + B^2.$$

We say that a Monge–Ampère equation E_ω is *non-degenerate* at a point $a \in J^1M$ if $\mathrm{Pf}(\omega)(a) \neq 0$.

If $\mathrm{Pf}(\omega)(a) = 0$, then an equation E_ω is said to be *degenerate* at the point a.

Using classification of effective forms we say that a Monge–Ampère equation E_ω is *hyperbolic, elliptic* or *parabolic* at the point $a \in J^1M$ if $\mathrm{Pf}(\omega)(a)$ is negative, positive or zero at the point.

An equation E_ω is called a *mixed type* (or *variable type*) at a point a if $\mathrm{Pf}(\omega)$ changes sign at the point.

Example 14.1.1 *The Von Karman equation*

$$v_x v_{xx} - v_{yy} = 0$$

has the Pfaffian $-v_x$, and therefore, it is hyperbolic in the domain $v_x > 0$, elliptic in the domain $v_x < 0$ and of mixed type at points of the hypersurface $v_x = 0$.

Note that a Monge–Ampère equation is determined by an effective form ω up to multiplication on a non-vanishing smooth function h, i.e., $E_\omega = E_{h\omega}$. Since the multiplication of ω by h does not change the sign of the Pfaffian:

$$\operatorname{Pf}(h\omega) = h^2 \operatorname{Pf}(\omega),$$

we see that the above definitions do not depend on the representative ω.

If $\operatorname{Pf}(\omega) \neq 0$, we can normalize the form ω:

$$\omega \mapsto \frac{1}{\sqrt{|\operatorname{Pf}(\omega)|}}\omega.$$

If $|\operatorname{Pf}(\omega)| = 1$, then the form ω is called *normed*. For a normed form ω its Pfaffian $\operatorname{Pf}(\omega)$ is equal to -1 for a hyperbolic equation and $+1$ for an elliptic equation. The operator A_ω corresponding to the normed form ω is also called *normed* and is denoted by A.

It is clear that $A^2 = 1$ for hyperbolic equations and $A^2 = -1$ for elliptic equations. Therefore, we obtain a non-holonomic almost product structure on the manifold J^1M for hyperbolic equations and a non-holonomic almost complex structure for elliptic equations.

For parabolic equations we have $A_\omega^2 = 0$. This is a non-holonomic almost tangent structure on the manifold J^1M.

14.1.3 Non-degenerate equations

In this section we consider hyperbolic and elliptic equations together. Let E_ω be a non-degenerate Monge–Ampère equation, where ω is a normed effective form. Then $\operatorname{Pf}(\omega) = \varepsilon$ and $\varepsilon = -1$ for a hyperbolic equation and $\varepsilon = 1$ for elliptic ones. Moreover $A^2 = -\varepsilon$.

Denote by $C_a^{\mathbb{C}}$ the complexification of the Cartan subspace C_a and by $A_a^{\mathbb{C}}$ the complexification of the operator A_a. Then $C_a^{\mathbb{C}}$ splits into the direct sum

$$C_a^{\mathbb{C}} = C_+(a) \oplus C_-(a),$$

where

$$C_\pm(a) \stackrel{\text{def}}{=} \{X \in C_a^{\mathbb{C}} \mid A_a^{\mathbb{C}} X = \mp\iota\sqrt{\varepsilon}X\}$$

are complex eigensubspaces of the operator $A_a^{\mathbb{C}}$. We will call these distributions *characteristic distributions*.

The subspaces $C_+(a)$ and $C_-(a)$ are skew-orthogonal and since the complexification $\Omega_a^{\mathbb{C}}$ of the form Ω_a is non-degenerate, each of them is a complex

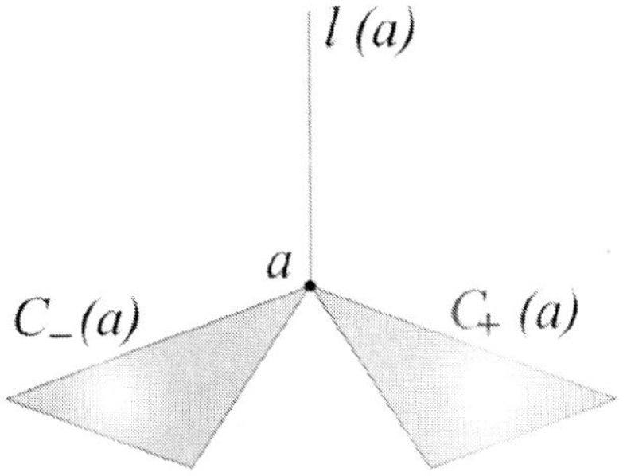

m

m,keepaspectratio]Fig8-1

Figure 14.1. A decomposition of $T_a^{\mathbb{C}}(J^1M)$ into a direct summ.

symplectic plane. So, we obtain two two-dimensional (of complex dimension) complex distributions on J^1M:

$$C_\pm : J^1M \ni a \mapsto C_\pm(a) \subset T_a^{\mathbb{C}}(J^1M).$$

Let us consider the derived distributions $C_+^{(1)}$ and $C_-^{(1)}$. Let X and Y be basic vector fields in the distribution C_- or in the distribution C_+. From the non-degeneracy of the form $\Omega|_{C_\pm}$ it follows that

$$\begin{aligned}\omega_0^{\mathbb{C}}([X,Y]) &= X(\omega_0^{\mathbb{C}}(Y)) - Y(\omega_0^{\mathbb{C}}(X)) - d\omega_0^{\mathbb{C}}(X,Y) = -d\omega_0^{\mathbb{C}}(X,Y)\\ &= -\Omega^{\mathbb{C}}(X,Y) \neq 0.\end{aligned}$$

Therefore, the derived distributions $C_+^{(1)}$ and $C_-^{(1)}$ are three-dimensional and at each point $a \in J^1M$ the intersection

$$l(a) \stackrel{\text{def}}{=} C_+^{(1)}(a) \cap C_-^{(1)}(a)$$

is a complex line. We obtain a one-dimensional complex distribution

$$l : J^1M \ni a \mapsto l(a) \subset T_a^{\mathbb{C}}(J^1M).$$

The complexified tangent space to J^1M at any point a splits into the direct sum:

$$T_a^{\mathbb{C}}(J^1M) = C_+(a) \oplus l(a) \oplus C_-(a)$$

(see Figure 14.1). It is clear that in the hyperbolic case the line $l(a)$ is a real line. But in the elliptic case it is also real. Indeed, since operator A is real, the subspaces $C_+(a)$ and $C_-(a)$ are complex conjugate: $C_+(a) = \overline{C_-(a)}$. Then the subspaces $C_+^{(1)}(a)$ and $C_-^{(1)}(a)$ are also complex conjugate and its intersection is complex conjugate to itself: $l(a) = \overline{l(a)}$. Therefore this complex line is generated by a real vector Z: $l(a) = \mathbb{C}Z, Z \in T_a(J^1M)$.

We see that non-degenerate Monge–Ampère operators generate $\mathcal{AP}$-structures on J^1M, where $n = 5, P_1 = C_+, P_2 = l, P_3 = C_-$.

Note that the non-degenerate Monge–Ampère equation, in contrast to the Monge–Ampère operator, generates an $\mathcal{AP}$-structure (C_+, l, C_-) up to the change C_+ and C_-. Indeed, effective 2-forms $\omega_1 = \omega$ and $\omega_2 = -\omega$ generate the same equation, but $C_-^1 = C_+^2$ and $C_+^1 = C_-^2$. Here $C_\pm^i$ are eigensubspaces of the operator $A_{\omega_i}^{\mathbb{C}}$ $(i = 1, 2)$.

So, for a non-degenerate Monge–Ampère equation on J^1M there exists an invariant defined distribution l of real lines. The lines of this distribution are transversal to the Cartan space at each point. We will use this fact for contact classification of Monge–Ampère equations.

Example 14.1.2 *Let us consider the equation*

$$v_{xx} + \varepsilon v_{yy} = f(x, y, v, v_x, v_y),$$

where $\varepsilon = -1$ *or* $\varepsilon = 1$. *In the case where* $\varepsilon = -1$ *we have the nonlinear wave equation and in the case where* $\varepsilon = 1$ *we have the nonlinear Laplace equation.*

The effective form is

$$\omega = f\, dq_1 \wedge dq_2 - \varepsilon\, dq_1 \wedge dp_2 + dq_2 \wedge dp_1.$$

The Pfaffian $\mathrm{Pf}(\omega) = \varepsilon$. *In basis (14.3) of the module* $D(C)$ *the operator* A *has the form*

$$A_\omega = \left\| \begin{array}{cccc} 0 & 1 & 0 & 0 \\ -\varepsilon & 0 & 0 & 0 \\ 0 & f & 0 & -\varepsilon \\ -f & 0 & 1 & 0 \end{array} \right\|.$$

Then

$$C_+ = \left\langle \varepsilon\frac{d}{dq_2} + f\frac{\partial}{\partial p_2} - \iota\sqrt{\varepsilon}\frac{d}{dq_1},\ \frac{d}{dq_1} + f\frac{\partial}{\partial p_1} + \iota\sqrt{\varepsilon}\frac{d}{dq_2} \right\rangle,$$

$$C_- = \left\langle \varepsilon\frac{d}{dq_2} + f\frac{\partial}{\partial p_2} + \iota\sqrt{\varepsilon}\frac{d}{dq_1},\ \frac{d}{dq_1} + f\frac{\partial}{\partial p_1} - \iota\sqrt{\varepsilon}\frac{d}{dq_2} \right\rangle$$

and the distribution l *is generated by the vector field*

$$Z = 2\frac{\partial}{\partial u} + f_{p_1}\frac{\partial}{\partial p_1} + \varepsilon f_{p_2}\frac{\partial}{\partial p_2}.$$

So, a hyperbolic Monge–Ampère equation defines a pair of distributions of real planes on J^1M. On the other hand, any unordered pair of arbitrary real distributions $C_{1,0}$ and $C_{0,1}$ on J^1M such that

1. $\dim C_{1,0} = \dim C_{0,1} = 2$;
2. at each point $a \in J^1M$ $C(a) = C_{1,0}(a) \oplus C_{0,1}(a)$;
3. at each point $a \in J^1M$ the subspaces $C_{1,0}(a)$ and $C_{0,1}(a)$ are skew-orthogonal with respect to the symplectic structure Ω;

determines the operator A up to the sign. Therefore a hyperbolic Monge–Ampère equation can be regarded as such an unordered pair $\{C_{1,0}, C_{0,1}\}$.

In a similar way, an elliptic Monge–Ampère equation can be viewed as an unordered pair $\{C_{1,0}, C_{0,1}\}$ of complex conjugate distributions on J^1M such that:

1. $\dim_{\mathbb{C}} C_{1,0} = \dim_{\mathbb{C}} C_{0,1} = 2$;
2. at each point $a \in J^1M$ $C_a^{\mathbb{C}} = C_{1,0}(a) \oplus C_{0,1}(a)$;
3. at each point $a \in J^1M$ the subspaces $C_{1,0}(a)$ and $C_{0,1}(a)$ are skew-orthogonal with respect to the complexification $\Omega_a^{\mathbb{C}}$ of the symplectic structure Ω_a.

Let L be a multivalued solution of a non-degenerate Monge–Ampère equation E_ω and let $a \in L$. Due to Proposition 14.1.1, the tangent space T_aL is invariant under operator A_ω and the restriction of the symplectic form Ω on T_aL is zero. Therefore the complexification of T_aL is intersected with the complex planes $C_+(a)$ and $C_-(a)$. By dimension reason, the intersection $(T_aL)^{\mathbb{C}}$ with each of $C_\pm(a)$ is a complex line and $(T_aL)^{\mathbb{C}}$ splits into the direct sum

$$(T_aL)^{\mathbb{C}} = h_+(a) \oplus h_-(a),$$

where

$$h_\pm(a) \stackrel{\text{def}}{=} (T_aL)^{\mathbb{C}} \cap C_\pm(a).$$

It is clear that for hyperbolic equations $h_+(a)$ and $h_-(a)$ are real lines. For elliptic equations these lines are complex. These are called *characteristic directions* of L at the point a and the corresponding distributions h_+ and h_- are called *characteristic distributions* of the multivalued solution L (see break Figure 14.2).

Remark 14.1.1 *For linear hyperbolic Monge–Ampère equations of the form*

$$Av_{xx} + 2Bv_{xy} + Cv_{yy} + E = 0,$$

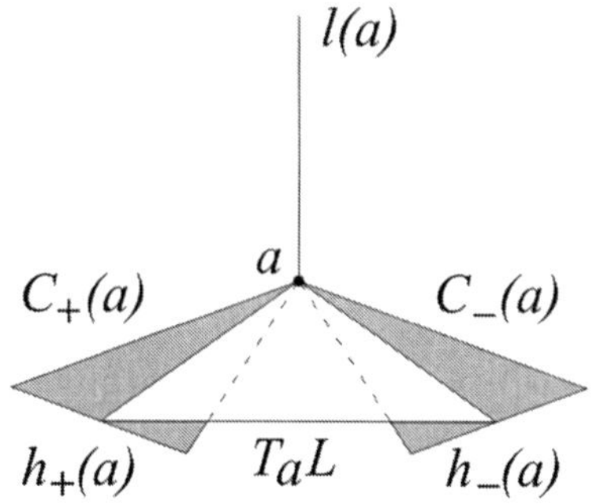

Figure 14.2. Characteristic directions.

where $A, B, C, E \in C^{\infty}(M)$, the constructed characteristic distributions can be projected to the manifold M and we get the classical characteristics.

Theorem 14.1.1 *A two-dimensional manifold $L \subset J^1M$ is a multivalued solution of a hyperbolic Monge–Ampère equation E_ω if and only if at any point $a \in L$ the tangent plane T_aL has one-dimensional intersections with subspaces $C_+(a)$ and $C_-(a)$.*

For a multivalued solution of the elliptic Monge–Ampère equations we have the following result.

Theorem 14.1.2 *A two-dimensional manifold L is a multivalued solution of an elliptic Monge–Ampère equation E_ω if and only if L is a complex curve of the non-holonomic complex structure A.*

14.1.4 Parabolic equations

Now we consider a parabolic Monge–Ampère equation E_ω. We assume that $\omega_a \neq 0$ for all $a \in J^1M$.

In this case the operator A_ω is nilpotent ($A_\omega^2 = 0$) and $\omega \wedge \omega = 0$.

This means that the 2-form ω_a on the Cartan subspace C_a is decomposable, i.e., $\omega_a = \alpha_a \wedge \beta_a$ for some 1-forms α and β on C. The intersection of the kernels of the 1-forms α_a and β_a generates a two-dimensional subspace $P(a) \subset C(a)$ (see Figure 14.3):

$$P_\omega(a) \stackrel{\text{def}}{=} \mathrm{Ann}(\alpha_a) \cap \mathrm{Ann}(\beta_a).$$

Moreover, the 2-form ω_a is effective if and only if the plane $P_\omega(a)$ is Lagrangian.

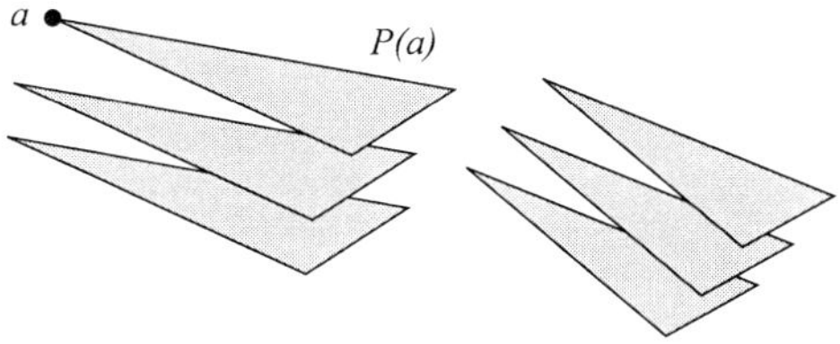

Figure 14.3. An almost tangent structure for a parabolic equation.

Indeed, let $X_a, Y_a \in P(a)$ be two vectors, then $X_a \rfloor \omega_a = Y_a \rfloor \omega_a = 0$, and

$$(X_a \wedge Y_a) \rfloor (\omega_a \wedge \Omega_a) = \Omega_a(X_a, Y_a)\, \omega_a = 0.$$

Therefore, $\Omega_a(X_a, Y_a) = 0$.

In other words, a parabolic effective differential form ω generates a two-dimensional distribution

$$P_\omega : J^1M \ni a \mapsto P_\omega(a) \subset C(a)$$

of Lagrangian planes (see Figure 14.3), and any such distribution defines a parabolic Monge–Ampère equation.

Let L be a multivalued solution of a parabolic Monge–Ampère equation E_ω. Then $\omega|_L = 0$ and therefore 1-forms $\alpha|_L$ and $\beta|_L$ are linear dependent, and

$$P_\omega(a) \cap T_a L \neq 0$$

for all $a \in L$.

Theorem 14.1.3 *An integral two-dimensional manifold L is a multivalued solution of a parabolic Monge–Ampère equation E_ω if and only if the tangent bundle of L has a non-trivial intersection with the distribution P_ω.*

If L is a multivalued solution of a parabolic Monge–Ampère equation E_ω, then the line

$$h(a) \stackrel{\text{def}}{=} L \cap P_\omega(a)$$

is called the *characteristic direction* of L at the point a and the corresponding distribution h is called the *characteristic distribution* of the multivalued solution L.

14.2 Intermediate integrals

14.2.1 Classical and non-holonomic intermediate integrals

Let us consider a second-order partial differential equation

$$F(x, y, v, v_x, v_y, v_{xx}, v_{xy}, v_{yy}) = 0. \tag{14.5}$$

A function $H \in C^\infty(J^1 M)$ is called a (classical) *intermediate integral* of this equation if all functions $v = v(x, y)$ satisfying $H(x, y, v, v_x, v_y) = \text{constant}$ are solutions of (14.5).

Let us formulate the definition of an intermediate integral for Monge–Ampère equations in terms of effective differential 2-forms on $J^1 M$.

Let ω be an effective differential 2-form on $J^1 M$. A function $H \in C^\infty(J^1 M)$ is called an *intermediate integral* of the Monge–Ampère equation E_ω if

$$\omega = (dH \wedge \theta)_\varepsilon$$

for some differential 1-form θ or, equivalently,

$$\omega = dH \wedge \theta + \lambda_0 \, d\omega_0 + \lambda_1 \wedge \omega_0 \tag{14.6}$$

for some $\lambda_k \in \Omega^k(J^1 M)$.

Example 14.2.1 *The function* $H = e^{q_1}(p_2 + u)$ *is an intermediate integral of the equation*

$$v_{xy} + v_x + v_y + v = 0. \tag{14.7}$$

Here $q_1 = x, q_2 = y, p_1 = v_x, p_2 = v_y, u = v$. *Indeed, if* v *is a solution of the family of equations* $e^x(v_y + v) = \text{constant}$, *then*

$$e^x(v_{xy} + v_x + v_y + v) = \frac{\partial}{\partial x}(e^x(v_y + v)) = 0$$

and therefore the function v *is a solution of (14.7).*

The effective differential 2-form corresponding to (14.7)

$$\omega = dq_1 \wedge dp_1 - dq_2 \wedge dp_2 + 2(p_1 + p_2 + u)\, dq_1 \wedge dq_2$$

can be represented in the form $\omega = (dH \wedge \theta)_\varepsilon$ *with*

$$\begin{aligned}\theta = {} & ((p_1 + p_2 + u)\gamma_1 - p_1\gamma_2)\, dq_1 + (2e^{-q_1} + p_2(\gamma_1 - \gamma_2))\, dq_2 \\ & + \gamma_1 \, dp_2 + \gamma_2 \, du,\end{aligned}$$

where γ_1 *and* γ_2 *are smooth functions on* $J^1 M$.

Let us introduce a notion of a non-holonomic intermediate integral which is a natural generalization of the classical one mentioned above (see [76]).

A three-dimensional distribution I on J^1M is called a *non-holonomic intermediate integral* of the Monge–Ampère equation E_ω if the following conditions hold:

1. $I(a)$ is a subspace of the Cartan subspace at any point $a \in J^1M$, i.e., $I(a) \subset C(a)$;
2. any two-dimensional integral submanifold L of I is a multivalued solution of E_ω;
3. I has a characteristic symmetry.

Any classical intermediate integral H generates a non-holonomic intermediate integral I according to the rule

$$I = \mathcal{F}\langle \omega_0, dH \rangle. \tag{14.8}$$

The contact vector field X_H is a characteristic symmetry of the distribution I.

In the general case a non-holonomic intermediate integral is locally generated by the Cartan form ω_0 and some 1-form ϑ:

$$I = \mathcal{F}\langle \omega_0, \vartheta \rangle.$$

Since the Cartan distribution on five-dimensional manifolds has no three-dimensional integral submanifolds, we see that the distribution I is not completely integrable and therefore $\dim I^{(1)} \geq 4$.

It is not difficult to verify that the distribution I has a characteristic symmetry if and only if $\dim I^{(1)} = 4$. In this case for $\omega_0 = du - p\,dq$ and $\vartheta = \vartheta_1\,dq_1 + \vartheta_2\,dq_2 + \vartheta_3\,du + \vartheta_4\,dp_1 + \vartheta_5\,dp_2$ the vector field

$$V = -\vartheta_4 \frac{\partial}{\partial q_1} - \vartheta_5 \frac{\partial}{\partial q_2} - (p_1\vartheta_4 + p_2\vartheta_5)\frac{\partial}{\partial u} + (\vartheta_1 + p_1\vartheta_3)\frac{\partial}{\partial p_1} + (\vartheta_2 + p_2\vartheta_3)\frac{\partial}{\partial p_2}$$

is a characteristic symmetry of I.

Moreover, the distribution I is generated by a certain function H according to the rule (14.8) if and only if $\dim I^{(1)} = 4$ and $I^{(2)} = I^{(1)}$, i.e., the distribution $I^{(1)}$ is completely integrable.

Example 14.2.2 *Let us consider a hyperbolic Monge–Ampère equation E_ω. It generates the real $\mathcal{AP}$-structure (C_+, l, C_-). Suppose that the distribution*

$C_+^{(1)}$ is not completely integrable and $\dim C_+^{(2)} = 4$. *Then the intersection $C_+^{(2)}(a)$ and $C_-(a)$ is a line for any point $a \in J^1M$. We get a one-dimensional distribution*

$$l_1 : a \mapsto l_1(a) \stackrel{\text{def}}{=} C_+^{(2)}(a) \cap C_-(a).$$

Construct a three-dimensional distribution $I \stackrel{\text{def}}{=} C_+ \oplus l_1$. If $\dim I^{(1)} = 4$, *then I is a non-holonomic intermediate integral of E_ω. Indeed, $I(a)$ is a subspace of the Cartan space $C(a)$ at any point $a \in J^1M$. Moreover, since the distribution C_+ is not completely integrable, any two-dimensional integral manifold L of I intersects the planes of the distributions C_+ and C_-. Therefore $\omega|_L = 0$ and we see that L is a multivalued solution of E_ω.*

Example 14.2.3 *The Martin potential of one-dimensional isentropic flows of a politropic gas satisfies the following Monge–Ampère equation:*

$$\text{Hess}\, v + q_2^{-\alpha} = 0 \tag{14.9}$$

with $\alpha = 1 + 1/\gamma$, where γ is the adiabatic exponent (see [93]). We consider the case where $\gamma = 3$. The corresponding effective differential 2-form is

$$\omega = q_2^{-4/3}\, dq_1 \wedge dq_2 + dp_1 \wedge dp_2.$$

The Pfaffian of ω is $\text{Pf}(\omega) = -q_2^{-4/3} < 0$, *therefore the equation is hyperbolic. The corresponding normed operator in the basis (14.3) is*

$$A = \frac{1}{q_2^{2/3}} \left\| \begin{matrix} 0 & 0 & 0 & -1 \\ 0 & 0 & 1 & 0 \\ 0 & 1 & 0 & 0 \\ -1 & 0 & 0 & 0 \end{matrix} \right\|.$$

The distributions C_+ and C_- are generated by the vector fields

$$X_+ = \frac{\partial}{\partial p_2} - q_2^{2/3}\frac{d}{dq_1}, \quad Y_+ = \frac{\partial}{\partial p_1} + q_2^{2/3}\frac{d}{dq_2}$$

and

$$X_- = \frac{\partial}{\partial p_2} + q_2^{2/3}\frac{d}{dq_1}, \quad Y_- = \frac{\partial}{\partial p_1} - q_2^{2/3}\frac{d}{dq_2}$$

respectively. The distribution l is generated by the vector field

$$Z = \frac{\partial}{\partial p_2} + 3q_2 \frac{\partial}{\partial u}$$

and $l_1 = \mathcal{F}\langle X_- \rangle$. Therefore $I = \mathcal{F}\langle \omega_0, \theta \rangle$, where $\theta = dp_1 - q_2^{-2/3}\, dq_2$. We see that $\theta = dH$, where $H = p_1 - 3q_2^{1/3}$. The corresponding first-order differential equation is $v_{q_1} = 3q_2^{1/3} + c_1$, where c_1 is an arbitrary constant. Any solution of this equation has the form

$$v(q_1, q_2) = 2q_1 q_2^{1/3} + c_1 q_1 + h(q_2)$$

and is a solution of (14.9). Here h is an arbitrary smooth function. The considered equation has also non-holonomic intermediate integrals $I_\varphi = \mathcal{F}\langle \omega_0, \theta_\varphi \rangle$, where

$$\theta_\varphi = q_2^{-2/3}(dq_1 - k_\varphi\, dq_2) + k_\varphi\, dp_1 + dp_2,$$

$k_\varphi = \frac{1}{3} q_2^{-1/3}(\varphi(H) - p_2)$ and φ is an arbitrary smooth function on $\mathbb{R}$. The characteristic symmetry of I_φ is

$$V_\varphi = -k_\varphi \frac{\partial}{\partial q_1} - \frac{\partial}{\partial q_2} - (p_1 k_\varphi + p_2)\frac{\partial}{\partial u} + q_2^{-2/3}\left(\frac{\partial}{\partial p_1} - k_\varphi \frac{\partial}{\partial p_2}\right).$$

More details on the method of finding classical and non-holonomic intermediate integrals can be found in [76, 79, 93].

14.2.2 Cauchy problem and non-holonomic intermediate integrals

Non-holonomic intermediate integrals can be used for solving the Cauchy problem for Monge–Ampère equations.

Let E_ω be a Monge–Ampère equation and let I be its non-holonomic intermediate integral with a characteristic symmetry V. A smooth curve $\mathcal{K} \subset J^1 M$ is called *corresponding to I Cauchy data* for E_ω if the following two conditions hold:

1. $\mathcal{K}$ is an integral curve of the distribution I;
2. the tangent line to $\mathcal{K}$ is transversal to the vector field V at each point $a \in \mathcal{K}$, i.e., $V_a \not\subset T_a\mathcal{K}$.

A multivalued solution L of E_ω is called a *solution of the Cauchy problem* if $L \supset \mathcal{K}$.

Let $\mathcal{K}$ be corresponding to I Cauchy data and A_t be a translations group along the vector field V. Then the surface $L = \cup_t A_t(\mathcal{K})$ is a multivalued solution of the Cauchy problem.

14.3 Symplectic Monge–Ampère equations

Let ω be an effective differential 2-form on the smooth manifold J^1M which is invariant with respect to the contact vector field X_1. Then ω is a lift of differential form ω' on T^*M with respect to the projection $\pi : J^1M \to T^*M, \omega = \pi^*(\omega')$. If $L \subset J^1M$ is a solution of E_ω, then the projection $L' = \pi(L)$ is a submersed Lagrangian manifold, and $\omega'|_{L'} = 0$.

This observation allows us to consider a special class of Monge–Ampère equations defined by X_1-invariant effective forms.

We call such equations *symplectic*, because they can be studied in terms of the symplectic geometry of T^*M.

14.3.1 A field of endomorphisms A_ω on T^*M

The cotangent bundle T^*M is a four-dimensional symplectic manifold with a structure differential 2-form $\Omega = d\rho$. A differential 2-form $\omega \in \Omega^2(T^*M)$ is effective if and only if

$$\omega \wedge \Omega = 0. \tag{14.10}$$

We define a Pfaffian of the differential 2-form ω on T^*M as a function $\mathrm{Pf}(\omega) \in C^\infty(T^*M)$ such that

$$\mathrm{Pf}(\omega)(a) \stackrel{\mathrm{def}}{=} \mathrm{Pf}(\omega_a).$$

The operator field associated with ω generates the field of endomorphisms A_ω on T^*M:

$$A_\omega : T^*M \ni a \longmapsto A_{\omega_a} \in \mathrm{End}(T_a(T^*M))$$

by the rule

$$X \rfloor \omega = A_\omega X \rfloor \Omega.$$

The field of endomorphisms A_ω on T^*M inherits all properties of the non-holonomic linear operator A_ω on J^1M (see Proposition 14.1.1).

Proposition 14.3.1

1. *Vector fields* $X, A_\omega X$ *are skew-orthogonal, i.e.,*

$$\Omega(A_\omega X, X) = 0.$$

2. *Any multivalued solution* L *of the equation* E_ω, *i.e., a Lagrangian submanifold such that* $\omega|_L = 0$, *is invariant with respect to* A_ω:

$$A_{\omega_a} : T_a L \to T_a L$$

for any point $a \in L$.

3. *The square of* A_ω *is scalar and*

$$A_\omega^2 + \mathrm{Pf}(\omega) = 0.$$

4. *Let*

$$\omega = E\, dq_1 \wedge dq_2 + B(dq_1 \wedge dp_1 - dq_2 \wedge dp_2)$$
$$+\, C\, dq_1 \wedge dp_2 - A\, dq_2 \wedge dp_1 + D\, dp_1 \wedge dp_2,$$

in canonical coordinates, where A, B, C, D, E *are smooth functions on* T^*M. *Then in the basis*

$$\frac{\partial}{\partial q_1}, \frac{\partial}{\partial p_1}, \frac{\partial}{\partial q_2}, \frac{\partial}{\partial p_2} \tag{14.11}$$

of the module $D(T^*M)$ *we obtain the following representation of* A_ω:

$$A_\omega = \left(B\frac{\partial}{\partial q_1} + C\frac{\partial}{\partial q_2} - E\frac{\partial}{\partial p_2}\right) \otimes dq_1 + \left(B\frac{\partial}{\partial p_1} + D\frac{\partial}{\partial q_2} - A\frac{\partial}{\partial p_2}\right) \otimes dp_1$$
$$+ \left(-A\frac{\partial}{\partial q_1} + E\frac{\partial}{\partial p_1} - B\frac{\partial}{\partial q_2}\right) \otimes dq_2$$
$$+ \left(-D\frac{\partial}{\partial q_1} + C\frac{\partial}{\partial p_1} - B\frac{\partial}{\partial q_2}\right) \otimes dp_2.$$

Equation E_ω is called *non-degenerate* at a point a if $\mathrm{Pf}(\omega)(a) \neq 0$ and *degenerate* if $\mathrm{Pf}(\omega)(a) = 0$.

We say that the symplectic Monge–Ampère equation E_ω with $\omega \in \Omega^2_\varepsilon(T^*M)$ is *hyperbolic*, *elliptic* or *parabolic* at a point $a \in T^*M$ if $\mathrm{Pf}(\omega)(a)$ is negative, positive or zero, and $\omega_a \neq 0$.

The equation E_ω is called *mixed-type* (or *variable type*) at a point a, if $\omega_a \neq 0$ and $\mathrm{Pf}(\omega)$ changes sign at this point.

Example 14.3.1 *The Triccomi equation*

$$xv_{xx} + v_{yy} = 0$$

has $\mathrm{Pf}(\omega) = x$*, and therefore is hyperbolic in the domain* $x < 0$*, elliptic in the domain* $x > 0$ *and of mixed type on the hypersurface* $x = 0$.

As above, in the case where $\mathrm{Pf}(\omega) \neq 0$, we can normalize the form ω in such a way that $|\mathrm{Pf}(\omega)| = 1$. We call such forms ω *normed*. For a normed form ω its Pfaffian $\mathrm{Pf}(\omega)$ is -1 for a hyperbolic equation and is $+1$ for an elliptic one.

The operator A_ω corresponding to the normed form ω is also called *normed* and is denoted by A.

Then $A^2 = 1$ for hyperbolic equations and $A^2 = -1$ for elliptic equations. For parabolic equations we have $A_\omega^2 = 0$.

14.3.2 Non-degenerate symplectic equations

Let A be the normed operator that corresponds to a non-degenerate Monge–Ampère equation E_ω.

It follows from $A^2 = \varepsilon$, where $\varepsilon = \mathrm{Pf}(\omega) = \pm 1$ that the operator A is an almost product structure for the hyperbolic equation and an almost complex structure for elliptic equations.

The complexification of the tangent space $T_a(T^*M)$ at $a \in T^*M$ splits into the direct sum of two skew-orthogonal complex symplectic planes:

$$T_a^{\mathbb{C}}(T^*M) = V_+(a) \oplus V_-(a), \tag{14.12}$$

where

$$V_+(a) \overset{\text{def}}{=} \{X \in T_a(T^*M)^{\mathbb{C}} \mid A_a^{\mathbb{C}} X = -\iota\sqrt{\varepsilon} X\},$$

$$V_-(a) \overset{\text{def}}{=} \{X \in T_a(T^*M)^{\mathbb{C}} \mid A_a^{\mathbb{C}} X = \iota\sqrt{\varepsilon} X\}.$$

The corresponding distributions V_+ and V_- we will call *characteristic distributions* (as in the contact situation).

Example 14.3.2 *Let us consider the equation*

$$v_{xx} + \varepsilon v_{yy} = f(x, y, v_x, v_y),$$

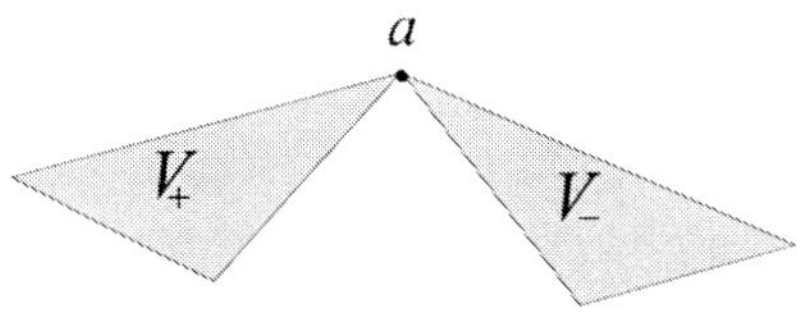

Figure 14.4. A decomposition of $T_a(T^*M)$ into a direct sum.

where $\varepsilon = -1$ *or* $\varepsilon = 1$. *The effective form is*

$$\omega = f\, dq_1 \wedge dq_2 - \varepsilon\, dq_1 \wedge dp_2 + dq_2 \wedge dp_1.$$

The Pfaffian $\mathrm{Pf}(\omega) = \varepsilon$. *In basis (14.11) of the module* $D(T^*M)$ *the operator* A *has the form*

$$A_\omega = \begin{Vmatrix} 0 & 1 & 0 & 0 \\ -\varepsilon & 0 & 0 & 0 \\ 0 & f & 0 & -\varepsilon \\ -f & 0 & 1 & 0 \end{Vmatrix}.$$

Then,

$$V_+ = \left\langle \varepsilon\frac{\partial}{\partial q_2} + f\frac{\partial}{\partial p_2} - \iota\sqrt{\varepsilon}\frac{\partial}{\partial q_1}, \frac{\partial}{\partial q_1} + f\frac{\partial}{\partial p_1} + \iota\sqrt{\varepsilon}\frac{\partial}{\partial q_2} \right\rangle,$$

$$V_- = \left\langle \varepsilon\frac{\partial}{\partial q_2} + f\frac{\partial}{\partial p_2} + \iota\sqrt{\varepsilon}\frac{\partial}{\partial q_1}, \frac{\partial}{\partial q_1} + f\frac{\partial}{\partial p_1} - \iota\sqrt{\varepsilon}\frac{\partial}{\partial q_2} \right\rangle.$$

For hyperbolic equations $V_+(a)$ and $V_-(a)$ can be considered as real planes. On the other hand, any unordered pair of arbitrary real distributions $V_{1,0}$ and $V_{0,1}$ on T^*M such that:

1. $\dim V_{1,0} = \dim V_{0,1} = 2$;
2. at each point $a \in T^*M$ $T_a(T^*M) = V_{1,0}(a) \oplus V_{0,1}(a)$;
3. at each point $a \in T^*M$ the subspaces $V_{1,0}(a)$ and $V_{0,1}(a)$ are skew-orthogonal with respect to the symplectic structure Ω_a,

determines a hyperbolic operator A up to sign.

In the elliptic case the operator A is defined (up to sign) by any unordered pair $\{V_{1,0}, V_{0,1}\}$ of complex conjugate distributions on T^*M such that:

1. $\dim_{\mathbb{C}} V_{1,0} = \dim_{\mathbb{C}} V_{0,1} = 2$;
2. at each point $a \in T^*M$ $T_a(T^*M)^{\mathbb{C}} = V_{1,0}(a) \oplus V_{0,1}(a)$;

3. at each point $a \in T^*M$ the subspaces $V_{1,0}(a)$ and $V_{0,1}(a)$ are skew-orthogonal with respect to the complexification $\Omega_a^{\mathbb{C}}$ of the symplectic structure Ω_a.

Therefore a non-degenerate symplectic Monge–Ampère equation can be viewed as such unordered pairs of distributions $\{V_{1,0}, V_{0,1}\}$.

We see that hyperbolic and elliptic symplectic Monge–Ampère equations generate $\mathcal{AP}$-structures on T^*M with $n = 4, P_1 = V_+, P_2 = V_-$.

The following two theorems explain the geometrical sense of multivalued solutions for non-degenerate symplectic Monge–Ampère equations.

Theorem 14.3.1 *A Lagrangian two-dimensional manifold $L \subset T^*M$ is a solution of a symplectic hyperbolic equation $E_\omega, \omega \in \Omega_\varepsilon^2(T^*M)$, if and only if the tangent spaces T_aL for all $a \in L$ have one-dimensional intersections $h_+(a)$ and $h_-(a)$ with real symplectic planes $V_+(a)$ and $V_-(a)$:*

$$T_aL = h_+(a) \oplus h_-(a).$$

The lines $h_\pm(a) \stackrel{\text{def}}{=} T_aL \cap V_\pm(a)$ are called the *characteristic directions* of the multivalued solution L at the point a and the corresponding distributions $h_\pm$ are called *characteristic distributions* of the multivalued solution L.

Theorem 14.3.2 *A Lagrangian surface $L \subset T^*M$ is a multivalued solution of a symplectic elliptic Monge–Ampère equation E_ω if and only if L is a complex curve of the almost complex structure A.*

14.3.3 Symplectic parabolic equations

Now we consider a symplectic parabolic Monge–Ampère equation E_ω, where $\omega \in \Omega_\varepsilon^2(T^*M)$.

Suppose that $\omega_a \neq 0$ at all points $a \in T^*M$. In this case $\omega \wedge \omega = 0$ and the form ω is decomposable, i.e., $\omega = \alpha \wedge \beta$ for some differential 1-forms α and β on T^*M.

Therefore, the equation E_ω determines the two-dimensional Lagrangian distribution $P_\omega = \ker \alpha \cap \ker \beta$ on T^*M.

Theorem 14.3.3 *A Lagrangian two-dimensional manifold $L \subset T^*M$ is a solution of a symplectic parabolic equation E_ω, if and only if tangent spaces T_aL have non-trivial intersections with the Lagrangian distribution P_ω for all $a \in L$.*

The intersection $h(a) \stackrel{\text{def}}{=} L \cap P_{\omega,a}$ mentioned in this theorem is called the *characteristic direction* of L at the point a and the corresponding distribution h is called the *characteristic distribution* of the multivalued solution L.

14.3.4 Intermediate integrals

Let us discuss the definition of intermediate integral for symplectic Monge–Ampère equations in terms of effective differential 2-forms on T^*M.

Let ω be an effective differential 2-form on T^*M. A function $H \in C^\infty(T^*M)$ is called an *intermediate integral* of the Monge–Ampère equation E_ω if

$$\omega = (dH \wedge \theta)_\varepsilon \tag{14.13}$$

for some differential 1-form θ on T^*M or, equivalently, $\omega = dH \wedge \theta + \lambda\Omega$ for some function λ and some 1-form θ.

Example 14.3.3 *For the wave equation*

$$v_{xy} = 0$$

the function $H = q_1 + p_1$ is an intermediate integral. Indeed, if v is a solution of the equation $x + v_x = 0$, then v is also a solution of the wave equation. On the other hand, the effective 2-form corresponding to the wave equation $\omega = dq_1 \wedge dp_1 - dq_2 \wedge dp_2$ can be represented in the form (14.13) with $\theta = (2 - \gamma)dq_1 + \gamma\, dp_1$. Here γ is an arbitrary smooth function on T^*M.

If a Monge–Ampère equation E_ω admits two intermediate integrals H_1 and H_2 then $\omega = h(dH_1 \wedge dH_2)_\varepsilon$, i.e.,

$$\omega = \lambda_1\, dH_1 \wedge dH_2 + \lambda_2\Omega$$

for some smooth functions h, λ_1 and λ_2.

14.4 Cauchy problem for hyperbolic Monge–Ampère equations

We call the form ω $(+)$-*integrable* or $(-)$-*integrable* if the distribution $C_+^{(1)}$ or $C_-^{(1)}$ is completely integrable respectively (see [72]).

If the form ω is $(+)$-integrable and $(-)$-integrable simultaneously, then the form ω is called $(\pm)$-*integrable*.

Let E_ω be a hyperbolic Monge–Ampère equation. A smooth curve $\mathcal{K} \subset J^1M$ is called a *Cauchy data* for E_ω if the following two conditions hold:

1. $\mathcal{K}$ is an integral curve of the Cartan distribution, i.e., $\omega_0|_\mathcal{K} = 0$;
2. the tangent line to $\mathcal{K}$ at each point $a \in \mathcal{K}$ is not characteristic, i.e., $T_a\mathcal{K} \not\subset C_-(a)$ and $T_a\mathcal{K} \not\subset C_+(a)$.

A multivalued solution L of E_ω is called a *solution of the Cauchy problem* if $L \supset \mathcal{K}$.

14.4.1 Constructive methods for integration of Cauchy problem

A-method

Suppose that ω is $(+)$-integrable. Let $H \in C^\infty(J^1M)$ be an integral of the distribution $C_+^{(1)}$ and let $M_c = H^{-1}(c)$. Since, $l \subset C_+^{(1)}$, we see that $l(a) \subset T_a(M_c)$ for any point $a \in M_c$.

Hence, the restriction of the Cartan distribution on M_c is a three-dimensional distribution. Denote this restriction by C_c.

The restriction of the distribution C_- on M_c is a one-dimensional distribution, denote it by l_-.

Therefore,

$$C_c(a) = C_+(a) \oplus l_-(a).$$

Moreover, the skew-orthogonality of distributions C_+ and C_- implies that the distribution l_- is the characteristic distribution for C_c.

Now we can solve the Cauchy problem in the following way.

1. Consider two independent integrals $H_1, H_2 \in C^\infty(J^1M)$ of the distribution $C_+^{(1)}$. Functions $H_1|_\mathcal{K}, H_2|_\mathcal{K}$ are functionally dependent. Therefore, we can find a smooth function $F(x, y) \in C^\infty(\mathbb{R}^2)$ such that $F(H_1|_\mathcal{K}, H_2|_\mathcal{K}) = 0$.
2. Let $H = F(H_1, H_2)$ and let $M_0 = H^{-1}(0) \subset J^1M$ be a smooth hypersurface. Denote by $L(a) \ni a$ the integral curves of the distribution l_-. Then $\mathcal{K} \subset M_0$, and

$$L = \bigcup_{a \in \mathcal{K}} L(a)$$

is a smooth surface in some neighborhood of $\mathcal{K}$.

This surface is integral because l_- is a characteristic distribution for C_c, and by construction it has one-dimensional intersections with the characteristic distributions $C_\pm$. Therefore L is a solution of the Cauchy problem.

B-method

Let ω be a $(\pm)$-integrable effective form, and let H and G be integrals of the distributions $C_+^{(1)}$ and $C_-^{(1)}$ respectively.

Consider a submanifold

$$N^c = \{H = c_1, G = c_2\} \subset J^1 M,$$

where $c = (c_1, c_2) \in \mathbb{R}^2$.

Let C_c be the restriction of the Cartan distribution on N^c. This is a two-dimensional distribution and

$$C_c(a) = l_+(a) \oplus l_-(a)$$

where $l_\pm$ are the restrictions of the characteristic distributions $C_\pm$ on N^c.

Moreover,

$$T_a N^c = l(a) \oplus l_-(a) \oplus l_+(a).$$

If $X \in D(l_+)$, $Y \in D(l_-)$, then

$$\omega_0([X, Y]) = X(\omega_0(Y)) - Y(\omega_0(X)) - d\omega_0(X, Y) = 0.$$

Therefore, C_c is a completely integrable distribution and its two-dimensional integral manifolds are solutions of the Monge–Ampère equation.

This observation gives the following method of solution to the Cauchy problem.

1. As above we find integrals H (for the distribution C_+) and G (for the distribution C_-) such that $H|_{\mathcal{K}} = 0$, $G|_{\mathcal{K}} = 0$.
2. Find an integral $F \in C^\infty(N^0)$ of the distribution C_0 on $N^0 = \{F = G = 0\}$.
3. On the curve $\mathcal{K}$ we have $F|_{\mathcal{K}} = c$. Then the two-dimensional manifold

$$L = \{H = 0, G = 0, F = c\}$$

is the solution of the Cauchy problem.

Example

Consider the non-linear wave equation

$$v_{xy} = f(x, y, v, v_x, v_y).$$

This equation is represented by the following effective form:

$$\omega = -2f\, dq_1 \wedge dq_2 + dq_1 \wedge dp_1 - dq_2 \wedge dp_2.$$

Then Pfaffian $\mathrm{Pf}(\omega) = -1$ and operator A_ω has the form

$$A_\omega = \left\| \begin{array}{cccc} 1 & 0 & 0 & 0 \\ 0 & -1 & 0 & 0 \\ 0 & -2f & 1 & 0 \\ 2f & 0 & 0 & -1 \end{array} \right\|$$

in basis (14.3).

The characteristic distributions are

$$C_+ = \left\langle P_1 = \frac{d}{dq_1} + f\frac{\partial}{\partial p_2},\ P_2 = \frac{\partial}{\partial p_1} \right\rangle,$$

$$C_- = \left\langle Q_1 = \frac{d}{dq_2} + f\frac{\partial}{\partial p_1},\ Q_2 = \frac{\partial}{\partial p_2} \right\rangle.$$

The distribution l is generated by the vector field

$$Z = \frac{\partial}{\partial u} + f_{p_2}\frac{\partial}{\partial p_1} + f_{p_1}\frac{\partial}{\partial p_2}.$$

Moreover,

$$[Q_1, Z] = (-f_u + p_2 f_{p_2 u} - 2f_{p_1} f_{p_2} + ff_{p_1 p_2} + f_{q_2 p_2})\frac{\partial}{\partial p_1}$$

$$+ (p_2 f_{p_1 u} + ff_{p_1 p_1} + f_{q_2 p_1})\frac{\partial}{\partial p_2} - f_{p_1}\frac{\partial}{\partial u},$$

$$[Q_2, Z] = f_{p_2 p_2}\frac{\partial}{\partial p_1} + f_{p_1 p_2}\frac{\partial}{\partial p_2}.$$

Therefore the distribution $C_-^{(1)}$ is completely integrable if and only if the following conditions hold:

$$f_{p_2 p_2} = 0,$$

$$-f_u + p_2 f_{p_2 u} - f_{p_1} f_{p_2} + ff_{p_1 p_2} + f_{q_2 p_2} = 0.$$

From the first equation it follows that

$$f = s(q, u, p_1)p_2 + r(q, u, p_1),$$

and from second one that

$$r_{q_2} - s_u + sr_{p_1} - rs_{p_1} = 0.$$

For example, differential equations of the form

$$v_{xy} = \alpha v_x + \beta v_y - \alpha\beta u + \gamma,$$

where $\alpha, \beta, \gamma \in \mathbb{R}$ are $(\pm)$-integrable and hence can be solved by the B-method.

15
Systems of first-order partial differential equations on two-dimensional manifolds

In this chapter we consider Jacobi differential equations on two-dimensional manifolds. They are systems of two non-linear first-order partial differential equations of the following form:

$$A_1 + B_1\frac{\partial v_1}{\partial x} - C_1\frac{\partial v_1}{\partial y} - D_1\frac{\partial v_2}{\partial y} + E_1\frac{\partial v_2}{\partial x} + F_1 \det J_v = 0,$$

$$A_2 + B_2\frac{\partial v_1}{\partial x} - C_2\frac{\partial v_1}{\partial y} - D_2\frac{\partial v_2}{\partial y} + E_2\frac{\partial v_2}{\partial x} + F_2 \det J_v = 0,$$

where

$$\det J_v = \begin{vmatrix} \dfrac{\partial v_1}{\partial x} & \dfrac{\partial v_1}{\partial y} \\ \dfrac{\partial v_2}{\partial y} & \dfrac{\partial v_2}{\partial x} \end{vmatrix}$$

is the Jacobian and A_i, B_i, C_i, D_i, E_i and F_i $(i = 1, 2)$ are functions in x, y and v. For the first time the geometry of such systems of first-order differential equations on two-dimensional manifolds was discussed in [71]. Classification problems for Jacobi equation was studied in [25].

Although the coordinate representation of Monge–Ampère equations and Jacobi equations looks different, their geometies are very closed.

We associate the same geometrical structures with Jacobi equations: almost product structures for hyperbolic equations, almost complex structures for elliptic equations and almost tangent structures for parabolic ones.

15.1 Non-linear differential operators of first order on two-dimensional manifolds

We start with a simple observation. Let $M = N \times \mathbb{R}^m$ and N be a two-dimensional manifold. Any differential 2-form $\omega \in \Omega^2(M)$ generates a differential operator

$$\Delta_\omega : C^\infty(N, \mathbb{R}^m) \to \Omega^2(N)$$

in the following way:

$$\Delta_\omega(v) \overset{\text{def}}{=} \omega|_{L_v},$$

where $L_v \subset N \times \mathbb{R}^m$ is a graph of the vector-function v.

Suppose that $(x, u) = (x_1, x_2, u_1, \ldots, u_m)$ are local coordinates on M such that x_1, x_2 are local coordinates on the manifold N and $u_1, \ldots, u_m$ are coordinates in $\mathbb{R}^m$. Let

$$\omega = \sum_{i,j=1(i<j)}^{m} F_{ij}(x, u)du_i \wedge du_j + \sum_{i=1}^{m} B_i(x, u)du_i \wedge dx_1$$
$$+ \sum_{i=1}^{m} C_i(x, u)du_i \wedge dx_2 + A(x, u)dx_1 \wedge dx_2$$

be a coordinate representation of the 2-form ω and let $v = (v_1, \ldots, v_m) \in C^\infty(N, \mathbb{R}^m)$ be a vector-function. Then

$$L_v = \{u_1 = v_1(x_1, x_2), \ldots, u_m = v_m(x_1, x_2)\} \subset M$$

and, therefore,

$$\Delta_\omega(v) = \left(\sum_{i,j=1(i<j)}^{m} F_{ij}(x, v) \det J_{i,j} + \sum_{i=1}^{m} \left(C_i(x, v)\frac{\partial v_i}{\partial x_1} - B_i(x, v)\frac{\partial v_i}{\partial x_2} \right) \right.$$
$$\left. + A(x, v) \right) dx_1 \wedge dx_2,$$

where

$$\det J_{i,j} = \det \begin{Vmatrix} \dfrac{\partial v_i}{\partial x_1} & \dfrac{\partial v_j}{\partial x_1} \\ \dfrac{\partial v_i}{\partial x_2} & \dfrac{\partial v_j}{\partial x_2} \end{Vmatrix}.$$

If we consider k differential 2-forms $\omega_1, \ldots, \omega_k$ on M, we obtain a system of k partial differential equations of first order:

$$\begin{cases} \Delta_{\omega_1}(v) = 0, \\ \vdots \quad \vdots \quad \vdots \quad \vdots \quad \vdots \\ \Delta_{\omega_k}(v) = 0. \end{cases} \tag{15.1}$$

It is clear, that if a_{ij} are smooth functions on the manifold M and matrix $\|a_{ij}\|_{i,j=1,\ldots,k}$ is non-degenerate at each point of M, then the system

$$\begin{cases} a_{11}\Delta_{\omega_1}(v) + \cdots + a_{1k}\Delta_{\omega_k}(v) = 0, \\ \vdots \qquad \vdots \qquad \vdots \qquad \vdots \\ a_{1k}\Delta_{\omega_1}(v) + \ldots + a_{kk}\Delta_{\omega_k}(v) = 0 \end{cases}$$

is equivalent to (15.1). Therefore, we can identify the system (15.1) with a smooth field of k-dimensional subspaces

$$E : M \ni a \longmapsto E(a) \subset \Lambda^2(T_a^*M),$$

A two-dimensional submanifold $L \subset M$ is a *multivalued solution* of the system of first-order partial differential equations E if L is integral for each $\omega_i \in \Lambda(E) (i = 1, \ldots, k)$.

Example 15.1.1 (Variational calculus) *Any differential 2-form $\theta \in \Omega^2(M)$ also defines a functional $\hat{\Omega}$ on orientable two-dimensional submanifolds:*

$$\hat{\Omega} : L \to \int_L \Omega.$$

Let X be a vector field on M with the one-parameter group $\varphi_t : M \to M$. Then for variation of $\hat{\Omega}$ we find

$$\hat{\Omega}(\varphi_t(L)) - \hat{\Omega}(L) = \int_{\varphi_t(L)} \Omega - \int_L \Omega = \int_L (\varphi_t^*(\Omega) - \Omega)$$

$$= t \int L_X(\Omega) + o(t) = t \int X \rfloor d\Omega + o(t).$$

Hence, a submanifold $L \subset M$ is an extremal of $\hat{\Omega}$ if the restriction $\omega|_L = 0$ for all 2-forms $\omega \overset{\text{def}}{=} X \rfloor d\Omega$ and for all vector field X on M. The system of *differential equations* E_Ω, *where*

$$E_\Omega(a) = \{X_a \rfloor d_a\Omega \in \Lambda^2(T_a^*M) | \forall X_a \in T_aM\}$$

is called the Euler equation *for the functional* $\hat{\Omega}$.

15.2 Jacobi equations

If we suppose that $m = k = 2$ we obtain a system of two partial differential equations. In order to find its coordinate representation we consider two differential 2-forms

$$\begin{aligned}\omega_i = A_i\, dx_1 \wedge dx_2 + B_i\, du_1 \wedge dx_1 + C_i\, du_1 \wedge dx_2 \\ + D_i\, du_2 \wedge dx_1 + E_i\, du_2 \wedge dx_2 + F_i\, du_1 \wedge du_2\end{aligned} \tag{15.2}$$

$(i = 1, 2)$ on the manifold M. The corresponding system of differential equations is

$$\begin{aligned} A_1 - B_1\frac{\partial v_1}{\partial x_2} + C_1\frac{\partial v_1}{\partial x_1} - D_1\frac{\partial v_2}{\partial x_2} + E_1\frac{\partial v_2}{\partial x_1} + F_1 \det J_v = 0, \\ A_2 - B_2\frac{\partial v_1}{\partial x_2} + C_2\frac{\partial v_1}{\partial x_1} - D_2\frac{\partial v_2}{\partial x_2} + E_2\frac{\partial v_2}{\partial x_1} + F_2 \det J_v = 0, \end{aligned} \tag{15.3}$$

where $\det J_v$ is the determinant of the Jacobi matrix

$$J_v = \begin{Vmatrix} \dfrac{\partial v_1}{\partial x_1} & \dfrac{\partial v_1}{\partial x_2} \\ \dfrac{\partial v_2}{\partial x_1} & \dfrac{\partial v_2}{\partial x_2} \end{Vmatrix}.$$

This gives rise to the following definitions.

Any smooth field of planes

$$E : M \ni a \longmapsto E(a) \in \Lambda^2(T_a^*M)$$

is called a *Jacobi system* of partial differential equations on manifold M $\dim M = Y$ or a *Jacobi equation*.

Planes $E(a)$ are called Jacobi planes.

A two-dimensional submanifold $L \subset M$ is called a *multivalued solution* of the Jacobi equation $E = \mathcal{F}\langle\omega_1, \omega_2\rangle$ if L is an integral manifold for each ω, i.e., $\omega_1|_L = \omega_2|_L = 0$.

As we have seen in 6.3 each non-degenerate equation E generates a field of endomorphisms A_E on the smooth manifold M by the following rule:

$$A_E : M \ni a \longmapsto A_{E(a)} \in \mathrm{End}(T_a M),$$

where the endomorphism $A_{E(a)}$ is defined by the formula

$$X \rfloor \omega_{2,a} = A_{E(a)} X \rfloor \omega_{1,a},$$

$X \in T_a M$ for an orthogonal basis ω_1, ω_2. The endomorphism $A_{E(a)}$ does not depend on the choice of (oriented) orthogonal basis of the plane $E(a)$.

Example 15.2.1 (Cauchy–Riemann systems) *The Cauchy–Riemann system*

$$\begin{aligned} \frac{\partial v_1}{\partial x_2} - \frac{\partial v_2}{\partial x_1} &= 0, \\ \frac{\partial v_1}{\partial x_1} + \frac{\partial v_2}{\partial x_2} &= 0 \end{aligned} \tag{15.4}$$

is a Jacobi equation which is generated by the 2-forms

$$\omega_1 = du_1 \wedge dx_1 + du_2 \wedge dx_2,$$

$$\omega_2 = du_1 \wedge dx_2 - du_2 \wedge dx_1.$$

The corresponding field of endomorphisms in the basis

$$\frac{\partial}{\partial x_1}, \frac{\partial}{\partial u_1}, \frac{\partial}{\partial x_2}, \frac{\partial}{\partial u_2} \tag{15.5}$$

of the module $D(M)$ is

$$A_E = \begin{Vmatrix} 0 & 0 & 1 & 0 \\ 0 & 0 & 0 & -1 \\ -1 & 0 & 0 & 0 \\ 0 & 1 & 0 & 0 \end{Vmatrix}.$$

Example 15.2.2 (Wave equation) *The wave equations*

$$\begin{aligned} \frac{\partial v_1}{\partial x_2} - \frac{\partial v_2}{\partial x_1} &= 0, \\ \frac{\partial v_1}{\partial x_1} - \frac{\partial v_2}{\partial x_2} &= 0 \end{aligned} \tag{15.6}$$

are generated by the forms

$$\omega_1 = du_1 \wedge dx_1 + du_2 \wedge dx_2,$$
$$\omega_2 = du_1 \wedge dx_2 + du_2 \wedge dx_1.$$

The field of endomorphisms in the basis (15.5) is

$$A_E = \left\| \begin{matrix} 0 & 0 & 1 & 0 \\ 0 & 0 & 0 & 1 \\ 1 & 0 & 0 & 0 \\ 0 & 1 & 0 & 0 \end{matrix} \right\|.$$

Note that in these examples ω_1 is a standard symplectic structure on M.

Example 15.2.3 (Equation of the Chaplygin gas) *The following system describes the non-stationary flow of the Chaplygin gas [14, 109]:*

$$\begin{aligned} &\frac{\partial \rho}{\partial t} + \frac{\partial}{\partial x}(\rho v) = 0, \\ &\frac{\partial v}{\partial t} + v\frac{\partial v}{\partial x} + \frac{c^2}{2}\frac{\partial}{\partial x}(\rho^{-2}) = 0. \end{aligned} \tag{15.7}$$

Here v and ρ are the velocity and the normalized density of a gas, t and x are the time and a space coordinate, while c is a constant. This equation also describe processes in unstable plasmas. The corresponding 2-forms ω_1, ω_2 and the field of endomorphisms are the following

$$\omega_1 = -u_2\, du_1 \wedge dx_1 + du_1 \wedge dx_2 - u_1\, du_2 \wedge dx_1,$$
$$\omega_2 = c^2 u_1^{-3}\, du_1 \wedge dx_1 - u_2\, du_2 \wedge dx_1 + du_2 \wedge dx_2$$

and

$$A_E = \left\| \begin{matrix} \dfrac{u_2}{u_1} & 0 & -\dfrac{1}{u_1} & 0 \\ 0 & 0 & 0 & 1 \\ \dfrac{c^2 + u_1^2 u_2^2}{u_1^3} & 0 & -\dfrac{u_2}{u_1} & 0 \\ 0 & -\dfrac{c^2}{u_1^4} & 0 & 0 \end{matrix} \right\|.$$

Here $x_1 = t$, $x_2 = x$, $u_1 = \rho$ and $u_2 = v$.

Example 15.2.4 (The Gibbons–Tsarev system) *The system (see [27])*

$$
\begin{aligned}
&v_2 \frac{\partial v_1}{\partial x_1} - \frac{\partial v_1}{\partial x_2} + \frac{1}{v_1 - v_2} = 0, \\
&v_1 \frac{\partial v_2}{\partial x_1} - \frac{\partial v_2}{\partial x_2} + \frac{1}{v_2 - v_1} = 0.
\end{aligned}
\tag{15.8}
$$

is generated by the differential 2-forms

$$\omega_1 = u_2\, du_1 \wedge dx_2 + du_1 \wedge dx_1 + \frac{1}{u_1 - u_2}\, dx_1 \wedge dx_2,$$

$$\omega_2 = u_1\, du_2 \wedge dx_2 + du_2 \wedge dx_1 + \frac{1}{u_2 - u_1}\, dx_1 \wedge dx_2.$$

The following example establishes a connection between Jacobi equations and Monge–Ampère equations.

Example 15.2.5 (Equation of one-dimensional gas flow) *A one-dimensional non-stationary flow of barotropic non-viscous gas is described by the following system:*

$$
\begin{aligned}
&\frac{\partial v}{\partial t} + \frac{\partial u}{\partial x} = 0, \\
&\frac{\partial u}{\partial t} + \frac{\partial v}{\partial x}(p(v)) = 0,
\end{aligned}
\tag{15.9}
$$

where u is the velocity of the gas, $v = 1/\rho$, ρ and p are the density and the pressure of the gas respectively. The function $p = p(v)$ is the equation of state of a gas [93]. Then,

$$\omega_1 = du_1 \wedge dx_1 + du_2 \wedge dx_2,$$

$$\omega_2 = du_1 \wedge dx_2 - p'(u_2)\, du_2 \wedge dx_1$$

and

$$A_E = \begin{Vmatrix} 0 & 0 & -1 & 0 \\ 0 & 0 & 0 & p'(u_2) \\ p'(u_2) & 0 & 0 & 0 \\ 0 & -1 & 0 & 0 \end{Vmatrix}.$$

Example 15.2.6 (Monge–Ampère equations) *Any Monge–Ampère equation*

$$Av_{xx} + 2Bv_{xy} + Cv_{yy} + D(v_{xx}v_{yy} - v_{xy}^2) + E = 0$$

on T^*M can be considered as a Jacobi equation as follows

$$\begin{aligned} &\frac{\partial v_2}{\partial x_1} - \frac{\partial v_1}{\partial x_2} = 0, \\ &A\frac{\partial v_1}{\partial x_1} + 2B\frac{\partial v_1}{\partial x_2} + C\frac{\partial v_2}{\partial x_2} + D\left(\frac{\partial v_1}{\partial x_1}\frac{\partial v_2}{\partial x_2} - \frac{\partial v_1}{\partial x_2}\frac{\partial v_2}{\partial x_1}\right) + E = 0. \end{aligned} \tag{15.10}$$

Here $x_1 = x$, $x_2 = y$, $v_1 = v_x$ and $v_2 = v_y$.

Let $E_1 = \mathcal{F}\langle\omega_1^1, \omega_2^1\rangle$ and $E_2 = \mathcal{F}\langle\omega_1^2, \omega_2^2\rangle$ be two Jacobi equations. We will say that equations E_1 and E_2 are *equivalent* if there exists a diffeomorphism F of the smooth manifold M such that $F^*(E_1) = E_2$, i.e.,

$$F^*(\omega_1^1) = a_{11}\omega_1^2 + a_{12}\omega_2^2,$$

$$F^*(\omega_2^1) = a_{21}\omega_1^2 + a_{22}\omega_2^2.$$

Note that since F is a diffeomorphism we have $\det \|a_{ij}\| \neq 0$.

One can write the last formulae without the functions a_{ij}:

$$F^*(\omega_i^1) \wedge \omega_1^2 \wedge \omega_2^2 = 0 \qquad (i = 1, 2).$$

Theorem 15.2.1 *The class of Jacobi equations is invariant with respect to diffeomorphisms of the smooth manifold M.*

This theorem is also true for general systems (15.1).

The equation E is called *elliptic, hyperbolic* or *parabolic* at a point $a \in M$ if the Jacobi plane $E(a)$ is so. The equation E is called *non-degenerate* at a point $a \in M$ if the Jacobi plane $E(a)$ is elliptic or hyperbolic.

Example 15.2.7 *Cauchy–Riemann equations (15.4) and the wave equation (15.6) are elliptic and hyperbolic respectively.*

Example 15.2.8 *Equation (15.7) is elliptic. Indeed, if we choose the volume form*

$$\mu = dx_1 \wedge dx_2 \wedge du_1 \wedge du_2, \tag{15.11}$$

then the matrix $Q = \|q(\omega_i, \omega_j)\|$ *has the form*

$$Q = \begin{Vmatrix} -2u_1 & 0 \\ 0 & -\dfrac{2c^2}{u_1^3} \end{Vmatrix}$$

and

$$\varepsilon(E) = \operatorname{sign}\det Q = 1.$$

Example 15.2.9 *Equation (15.8) is hyperbolic. Indeed, for the volume form (15.11) the matrix* Q *is*

$$Q = \begin{Vmatrix} 0 & u_2 - u_1 \\ u_2 - u_1 & 0 \end{Vmatrix}.$$

Hence,

$$\varepsilon(E) = \operatorname{sign}(-(u_2 - u_1)^2) = -1.$$

Note that $\omega_1 \wedge \omega_1 = \omega_2 \wedge \omega_2 = 0$ *and* $\omega_1 \wedge \omega_2 = h\mu \neq 0$. *Let us introduce the orthogonal basis* ω_1', ω_2' *of* E:

$$\omega_1' = \omega_1 + \omega_2,$$

$$\omega_2' = \omega_1 - \omega_2.$$

Then we have $\omega_1' \wedge \omega_2' = 0$ *and* $\omega_1' \wedge \omega_1' = -\omega_2' \wedge \omega_2' \neq 0$. *The matrix* $Q' = \|q(\omega_i', \omega_j')\|$ *is diagonal:*

$$Q' = \begin{Vmatrix} 2(u_2 - u_1) & 0 \\ 0 & -2(u_2 - u_1) \end{Vmatrix}.$$

We assume $u_2 \neq u_1$, *then*

$$A_E = \begin{Vmatrix} \dfrac{u_1 + u_2}{u_1 - u_2} & 0 & \dfrac{2u_1u_2}{u_1 - u_2} & 0 \\ -\dfrac{2}{(u_1 - u_2)^2} & 1 & -\dfrac{2u_1}{(u_1 - u_2)^2} & 0 \\ -\dfrac{2}{u_1 - u_2} & 0 & -\dfrac{u_1 + u_2}{u_1 - u_2} & 0 \\ \dfrac{2}{(u_1 - u_2)^2} & 0 & \dfrac{2u_2}{(u_1 - u_2)^2} & -1 \end{Vmatrix}. \tag{15.12}$$

We see that $A_\Pi^2 = 1$ *and* $\det A_\Pi = 1$. *Note that the form* ω_1' *is closed, i.e.,* $d\omega_1' = 0$, *hence* ω_1' *is a symplectic structure on* M. *To represent this symplectic structure in canonical coordinates let us construct a diffeomorphism* $F : \mathbb{R}^2 \times \mathcal{D} \to M = \mathbb{R}^4$,*:*

$$F : \begin{cases} x_1 = X_1, \\ x_2 = X_2, \\ u_1 = \dfrac{1}{2}(U_1 - \sqrt{U_1^2 - 4U_2}), \\ u_2 = \dfrac{1}{2}(U_1 + \sqrt{U_1^2 - 4U_2}). \end{cases}$$

where X_1, X_2 *are coordinates on* $\mathbb{R}^2$, $\mathcal{D} = \{(U_1, U_2) \in \mathbb{R}^2 \,|\, U_1^2 - 4U_2 > 0\}$. *The inverse transformation is*

$$F^{-1} : \begin{cases} X_1 = x_1, \\ X_2 = x_2, \\ U_1 = u_1 + u_2, \\ U_2 = u_1 u_2. \end{cases}$$

Then,

$$F^*(\omega_1') = dU_1 \wedge dX_1 + dU_2 \wedge dX_2,$$

$$F^*(\omega_2') = \frac{1}{\sqrt{U_1^2 - 4U_2}}(-U_1\, dU_1 \wedge dX_1 - 2U_2\, dU_1 \wedge dX_2 + 2\, dU_2 \wedge dX_1$$
$$+ U_1\, dU_2 \wedge dX_2 - 2\, dX_1 \wedge dX_2).$$

The 2-forms $\omega_1'' \stackrel{\text{def}}{=} F^*(\omega_1')$ *and* $\omega_2'' \stackrel{\text{def}}{=} F^*(\omega_2')$ *form a basis of* E. *Note that* ω_1'' *is a canonical symplectic structure on* $\mathbb{R}^2 \times D$. *Therefore, (15.8) is equivalent to the following system:*

$$\begin{aligned} &\frac{\partial v_2}{\partial x_1} - \frac{\partial v_1}{\partial x_2} = 0, \\ &2v_2\frac{\partial v_1}{\partial x_1} - v_1\frac{\partial v_1}{\partial x_2} - v_1\frac{\partial v_2}{\partial x_1} + 2\frac{\partial v_2}{\partial x_2} + 2 = 0, \end{aligned} \tag{15.13}$$

if $v_1^2 - 4v_2 > 0$.

For this equation we have

$$A_E = \frac{1}{\sqrt{(u_1^2 - 4u_2)}} \begin{Vmatrix} u_1 & 0 & 2u_2 & 0 \\ 0 & u_1 & -2 & -2 \\ -2 & 0 & -u_1 & 0 \\ 2 & 2u_2 & 0 & -u_1 \end{Vmatrix}.$$

(We denoted $u_i = U_i$, $i = 1, 2$).

Note that $A_\Pi^2 = 1$ and $\det A_\Pi = 1$.

The corresponding Monge–Ampère equation is (see (15.10)):

$$v_y v_{xx} - \frac{1}{2} v_x v_{xy} + v_{yy} + 1 = 0.$$

15.3 Symmetries of Jacobi equations

A diffeomorphism F of the manifold M is called a *symmetry* of the Jacobi equation E if it takes E to itself.

If $E = \mathcal{F}\langle\omega_1, \omega_2\rangle$, this means that

$$F^*(\omega_i) \wedge \omega_1 \wedge \omega_2 = 0$$

or, equivalently,

$$F^*(\omega_1) = a_{11}\omega_1 + a_{12}\omega_2,$$
$$F^*(\omega_2) = a_{21}\omega_1 + a_{22}\omega_2,$$

for some $a_{ij} \in C^\infty(M)$ such that $\det \|a_{ij}\| \neq 0$, $(i, j = 1, 2)$.

If F is a symmetry of E, and L is a multivalued solution of E, then $F(L)$ is also a multivalued solution of E.

An *infinitesimal symmetry* of the Jacobi equation $E = \mathcal{F}\langle\omega_1, \omega_2\rangle$ is a vector field X on M such that

$$L_X(\omega_i) \wedge \omega_1 \wedge \omega_2 = 0 \qquad (i = 1, 2)$$

or, equivalently,

$$L_X(\omega_1) = a_{11}\omega_1 + a_{12}\omega_2,$$
$$L_X(\omega_2) = a_{21}\omega_1 + a_{22}\omega_2,$$

for some $a_{ij} \in C^\infty(M)$ such that $\det \|a_{ij}\| \neq 0$.

Note that the functions a_{ij} in the last formula are not arbitrary. Indeed, if the 2-forms ω_1 and ω_2 are orthogonal, then

$$L_X(\omega_1 \wedge \omega_2) = L_X(\omega_1) \wedge \omega_2 + \omega_1 \wedge L_X(\omega_2) = 0,$$

$$L_X(\omega_1 \wedge \omega_1) = \varepsilon(\Pi) L_X(\omega_2 \wedge \omega_2).$$

Hence,

$$a_{12} + \varepsilon(\Pi) a_{21} = 0,$$

$$a_{11} = a_{22}.$$

Example 15.3.1 *The translation*

$$F : (x_1, x_2, u_1, u_2) \to (x_1 + c_1, x_2 + c_2, u_1 + c_3, u_2 + c_4),$$

where $c_1, \ldots, c_4 \in \mathbb{R}$, is a symmetry of the Cauchy–Riemann system (15.4) and of the wave equation (15.6). The translation

$$F : (x_1, x_2, u_1, u_2) \to (x_1 + c_1, x_2 + c_2, u_1, u_2)$$

is a symmetry of the Jacobi system (15.3), where the coefficients A_i, B_i, C_i, D_i, E_i and F_i $(i = 1, 2)$ do not depend on x_1 and x_2.

Example 15.3.2 *Note that*

$$\omega_1 = du_1 \wedge dx_1 + du_2 \wedge dx_2$$

in Examples 15.2.1 and 15.2.2 is a symplectic structure on $\mathbb{R}^4$. Therefore a Hamiltonian vector field X_H, which is defined by the equality

$$X_H \rfloor \omega_1 = -dH,$$

is a symmetry of the Cauchy–Riemann system and the wave equation if

$$L_{X_H}(\omega_2) = a\omega_1 + b\omega_2$$

for some smooth functions a and b on $\mathbb{R}^4$.

Theorem 15.3.1 *Each element A_t of the translation group $\{A_t\}_{t \in I}$ along trajectories of an infinitesimal symmetry X of E is a symmetry of the Jacobi equation E.*

Proof From the formula

$$\frac{d}{dt}A_t^*(\omega)\bigg|_{t=t_0} = A_{t_0}^*(L_X(\omega))$$

it follows that

$$\frac{d\omega_i}{dt} = A_t^*(L_X(\omega_i)) = A_{i1}(t)\omega_1(t) + A_{i2}(t)\omega_2(t), \tag{15.14}$$

where

$$\omega_j(t) \stackrel{\text{det}}{=} A_t^*(\omega_j), \qquad A_{ij}(t) \stackrel{\text{det}}{=} A_t^*(a_{ij}) \qquad (i,j=1,2).$$

Note that

$$\omega_i(0) = \omega_i \qquad (i=1,2). \tag{15.15}$$

Therefore, we can solve the Cauchy problem (15.15) for the ordinary differential equation (15.14):

$$\omega_i(t) = B_{i1}(t)\omega_1 + B_{i2}(t)\omega_2$$

$(i = 1,2)$. Here $\|B_{ij}(t)\|_{i,j=1,2}$ is a fundamental matrix of the linear system (15.14). So, A_t is a symmetry of Jacobi equation E for each t. ■

15.4 Geometric structures associated with Jacobi's equations

Let $E = \mathcal{F}\langle\omega_1, \omega_2\rangle$ be a non-degenerate Jacobi equation. We can choose a basis of E such that $\omega_1 \wedge \omega_2 = 0$ and $\omega_1 \wedge \omega_1 = \varepsilon\omega_2 \wedge \omega_2$, where $\varepsilon = -1$ for the elliptic equation and $\varepsilon = 1$ for the hyperbolic one. Then $A_E^2 = \varepsilon$. We get an almost complex structure in the elliptic case and almost product structure in the hyperbolic case.

The complexification of the tangent space T_aM at $a \in M$ splits in to the direct sum of two skew-orthogonal complex symplectic planes:

$$(T_aM)^{\mathbb{C}} = V_+(a) \oplus V_-(a), \tag{15.16}$$

where

$$V_+(a) \stackrel{\text{def}}{=} \{X \in (T_aM)^{\mathbb{C}} \mid A^{\mathbb{C}}_{E,a}X = -\iota\sqrt{\varepsilon}X\},$$

$$V_-(a) \stackrel{\text{def}}{=} \{X \in (T_aM)^{\mathbb{C}} \mid A^{\mathbb{C}}_{E,a}X = \iota\sqrt{\varepsilon}X\}.$$

We call the corresponding distributions V_+ and V_- *characteristic distributions*. For hyperbolic equations these distributions can be viewed as being real.

Therefore the non-degenerate Jacobi equation generate a $\mathcal{AP}$-structure on M with $n = 4$, $P_1 = V_+$, $P_2 = V_-$.

Theorem 15.4.1 *A two-dimensional manifold $L \subset M$ is a multivalued solution of a hyperbolic Jacobi equation E, if and only if the tangent spaces T_aL for all $a \in L$ have one-dimensional intersections $h_+(a)$ and $h_-(a)$ with real planes $V_+(a)$ and $V_-(a)$: $T_aL = h_+(a) \oplus h_-(a)$.*

The lines $h_\pm(a) \stackrel{\text{def}}{=} T_aL \cap V_\pm(a)(i,j = 0,1; i \neq j)$ are called the *characteristic directions* of the multivalued solution L at the point a and the corresponding distributions $h_\pm$ are called *characteristic distributions* of the multivalued solution L.

Theorem 15.4.2 *A surface $L \subset M$ is a solution of an elliptic Jacobi equation E if and only if L is a complex curve of the almost-complex structure A_E.*

Let E be a non-degenerate Jacobi equation. Due to (15.16) (see Section 3.5), the modules $\Omega^s(M)^{\mathbb{C}}$ fall into the direct sum

$$\Omega^s(M)^{\mathbb{C}} = \bigoplus_{p+q=s} \Omega^{p,q}(M)^{\mathbb{C}} \tag{15.17}$$

and the de Rham differential $d : \Omega^s(M)^{\mathbb{C}} \to \Omega^{s+1}(M)^{\mathbb{C}}$ splits in the direct sum

$$d = d_{1,0} \oplus d_{0,1} \oplus d_{2,-1} \oplus d_{-1,2},$$

where $d_{2,-1}$ and $d_{-1,2}$ are tensor invariants of the Jacobi equation.

15.5 Conservation laws of Jacobi equations

A differential 1-form θ is called a *conservation law* of the Jacobi equation $E = \mathcal{F}\langle\omega_1, \omega_2\rangle$ if its differential $d\theta$ vanishes on each solution of E:

$$d\theta|_L = 0.$$

This means that

$$d\theta = a\omega_1 + b\omega_2$$

for smooth functions a, b on M.

Let L be a solution of E and let θ be a conservation law. Assume that $\mathcal{D} \subset M$ is a domain in L, and $\partial\mathcal{D}$ is a boundary of $\mathcal{D}$. From the Stokes theorem it follows that

$$\int_{\partial\mathcal{D}} \theta = \int_{\mathcal{D}} d\theta = 0.$$

Example 15.5.1 (Cauchy–Riemann systems) *The differential 1-forms*

$$\theta_1 = u_1\,dx_1 + u_2\,dx_2$$

and

$$\theta_2 = u_1\,dx_2 - u_2\,dx_1$$

are the conservation law of (15.4). Indeed, $d\theta_1 = -\omega_1$ *and* $d\theta_2 = \omega_2$*.*

Theorem 15.5.1 *If a Jacobi equation E has a conservation law θ such that the form $d\theta$ is non-degenerate, then locally E can be written as a Monge–Ampère equation on T^*M.*

Proof The form $d\theta$ determines a symplectic structure on M. Due to the Darboux theorem, locally there exist canonical coordinates x_1, x_2, u_1, u_2 on M such that $d\theta$ has the form

$$d\theta = du_1 \wedge dx_1 + du_2 \wedge dx_2.$$

Let ω_1 be a 2-form, such that $E = \mathcal{F}\langle\omega_1, d\theta\rangle$. If we write E in local coordinates, we get Monge–Ampère equations (15.10). ■

Let us describe a method of calculation of conservation laws for hyperbolic Jacobi equations.

The next theorem follows from the decompositions (15.17).

Theorem 15.5.2 *A differential 1-form $\theta \in \Omega^1(M)$ is a conservation law of the hyperbolic Jacobi equation E if and only if*

$$d\theta \in \Omega^2(V_+) \oplus \Omega^2(V_-), \tag{15.18}$$

or using decomposition $\theta = \theta_+ + \theta_-$, where $\theta_{i,j} \in \Omega^1(V_\pm)$, if and only if

$$d_{0,1}\theta_+ + d_{1,0}\theta_- = 0.$$

Here $P_{1,0} = V_+$ and $P_{0,1} = V_-$.

Example 15.5.2 *Let us find conservation laws for (15.13). The distributions V_+ and V_- are defined as*

$$V_+ = \left\langle \left(\sqrt{u_1^2 - 4u_2} + u_1\right)\frac{\partial}{\partial x_1} - 2\frac{\partial}{\partial x_2} + 2\frac{\partial}{\partial u_2}, \right.$$
$$\left. \left(\sqrt{u_1^2 - 4u_2} + u_1\right)\frac{\partial}{\partial u_1} + 2u_2\frac{\partial}{\partial u_2}\right\rangle,$$
$$V_- = \left\langle \left(\sqrt{u_1^2 - 4u_2} - u_1\right)\frac{\partial}{\partial x_1} + 2\frac{\partial}{\partial x_2} - 2\frac{\partial}{\partial u_2}, \right.$$
$$\left. \left(\sqrt{u_1^2 - 4u_2} - u_1\right)\frac{\partial}{\partial u_1} - 2u_2\frac{\partial}{\partial u_2}\right\rangle.$$

Moreover, the modules $\Omega^1(V_+)$ and $\Omega^1(V_-)$ are generated by the forms

$$2\,dx_1 + \left(u_1 - \sqrt{u_1^2 - 4u_2}\right)du_1,$$
$$2u_2\,dx_2 + \left(\sqrt{u_1^2 - 4u_2} - u_1\right)du_1 + \left(\sqrt{u_1^2 - 4u_2} - u_1\right)du_2,$$

and

$$2\,dx_1 + \left(u_1 + \sqrt{u_1^2 - 4u_2}\right)du_1,$$
$$-2u_2\,dx_2 + \left(\sqrt{u_1^2 - 4u_2} + u_1\right)du_1 + \left(\sqrt{u_1^2 - 4u_2} + u_1\right)du_2$$

respectively.

Therefore, the module $\Omega^{1,1}(M)$ *is generated by:*

$$dx_1 \wedge du_1;$$

$$dx_2 \wedge du_1 + dx_2 \wedge du_2;$$

$$du_2 \wedge dx_1 - u_2\, du_1 \wedge dx_2;$$

$$2u_2\, dx_1 \wedge dx_2 - 2u_2\, du_1 \wedge du_2 + u_1\, du_2 \wedge dx_1 + u_1 u_2\, du_1 \wedge dx_2.$$

Let

$$\theta = a\, dx_1 + b\, dx_2 + c\, du_1 + h\, du_2$$

be a conservation law of (15.13). Then formula (15.18) implies that the projection of the differential $d\theta$ *to the free module* $\Omega^{1,1}(M)$ *is zero. It is easy to see that*

$$\frac{\partial a}{\partial u_1} = \frac{\partial c}{\partial x_1},$$

$$\frac{\partial b}{\partial u_1} = \frac{\partial c}{\partial x_2},$$

$$\frac{\partial a}{\partial x_2} = \frac{\partial b}{\partial x_1},$$

$$\frac{\partial a}{\partial u_2} = \frac{\partial h}{\partial x_1}.$$

15.6 Cauchy problem for hyperbolic Jacobi equations

Let E be a hyperbolic Jacobi equation.

A smooth curve $\mathcal{K} \subset M$ is called a *Cauchy data* if the tangent to $\mathcal{K}$ at each point $a \in \mathcal{K}$ is contained neither in $V_+(a)$ nor in $V_-(a)$. A multivalued solution L of the hyperbolic Jacobi equation E is called the *solution of the Cauchy problem* if $L \supset \mathcal{K}$.

A hyperbolic Jacobi equation E is called a *(+)-type equation* (or *(−)-type equation*) if the distribution V_+ (or V_-) is completely integrable.

Example 15.6.1 *Suppose that in (15.9) we have* $p'(u_2) < 0$. *Then this equation is hyperbolic. Let* $f(u_2) \stackrel{\text{def}}{=} \sqrt{-p'(u_2)}$. *The corresponding distributions* $V_\pm$ *for*

(15.9) are generated by the vector fields

$$X_{\pm} = \frac{\partial}{\partial x_1} \mp f(u_2)\frac{\partial}{\partial x_2}, \quad Y_{\pm} = f(u_2)\frac{\partial}{\partial u_1} \mp \frac{\partial}{\partial u_2}.$$

Here $X_{\pm}, Y_{\pm} \in D(V_{\pm})$. If $f f' = 0$ (i.e., $p'' = 0$), then $Z_{\pm} \stackrel{\text{def}}{=} [X_{\pm}, Y_{\pm}] = 0$ and distributions $V_{\pm}$ are completely integrable.

Let E be a $(+)$-type equation and let $H_1, H_2 \in C^{\infty}(M)$ be functionally independent 1st integrals of the distribution V_+. Then for any multivalued solution L of E we can find a function $H = F(H_1, H_2)$ such that the restriction H to L is a constant: $H|_L = c =$ constant. Let us consider a three-dimensional submanifold

$$M_c \stackrel{\text{def}}{=} \{H = c\} \subset M.$$

Then,

1. the distribution V_- is tangent to the submanifold M_c;
2. the restriction l_- of the distribution V_- on M_c is a one-dimensional distribution:

$$l_- : M_c \ni a \longmapsto l_-(a) \stackrel{\text{def}}{=} V_-(a) \cap T_a L.$$

Hence, any invariant with respect to the l_- two-dimensional submanifold $L' \subset M_c$ is a multivalued solution of E.

From the above observation we obtain the following method for the Cauchy problem.

Let $\mathcal{K}$ be a Cauchy data. Consider two independent integrals $H_1, H_2 \in C^{\infty}(M)$ of the distribution V_+. For the functions $H_1|_{\mathcal{K}}, H_2|_{\mathcal{K}}$ one can find a smooth function $F(x, y) \in C^{\infty}(\mathbb{R}^2)$ such that $F(H_1|_{\mathcal{K}}, H_2|_{\mathcal{K}}) = 0$. By setting $H \stackrel{\text{def}}{=} F(H_1, H_2)$ and $M_0 \stackrel{\text{def}}{=} \{H = 0\} \subset M$ one obtains on the manifold M_0 a characteristic distribution l_-. But $\mathcal{K} \subset M_0$, hence the manifold

$$L \stackrel{\text{def}}{=} \bigcup_{a \in \mathcal{K}} L(a)$$

is a solution of the Cauchy problem. Here we denote by $L(a)$ the integral curve of the distribution l_- passing through the point a.

PART IV

Applications

16
Non-linear acoustics

In this chapter an application of symmetries and conservation laws to non-linear acoustics are given. This is based on the Khokhlov–Zabolotskaya (KZ) equation, which is the main equation of three-dimensional non-linear acoustics. We investigate the structure of the Lie algebra of contact symmetries of this equation, conservation laws and invariant solutions. This leads to some non-linear effects: the rise of the gradient catastrophy and self-diffraction.

Here we follow results obtained in [41, 42, 69, 70].

16.1 Symmetries and conservation laws of the KZ equation

16.1.1 KZ equation and its contact symmetries

The propagation of a bounded three-dimensional sound beam is described, taking into account the non-linear properties of medium, by the *Khokhlov–Zabolotskaya equation [98]*:

$$\frac{\partial}{\partial q_1}\left(\frac{\partial \rho}{\partial q_2} - \rho \frac{\partial \rho}{\partial q_1}\right) = \frac{\partial^2 \rho}{\partial q_3^2} + \frac{\partial^2 \rho}{\partial q_4^2}. \tag{16.1}$$

Here ρ is a deviation of the density from an equilibrium state,

$$q_1 = \frac{c}{\varepsilon}\left(t - \frac{x}{c}\right), \quad q_2 = x, \quad q_3 = \sqrt{\frac{2}{\varepsilon}}y, \quad q_4 = \sqrt{\frac{2}{\varepsilon}}z,$$

and x is the coordinate in the direction of the propagation, y, z are transverse coordinates, c is the velocity of sound in the medium, $\varepsilon = \frac{1}{2}(\gamma + 2)$ and γ is the adiabatic exponent.

The effective differential 4-form, which represents the KZ equation is

$$\begin{aligned}\omega =& \frac{1}{2}dp_1 \wedge dq_1 \wedge dq_3 \wedge dq_4 - \frac{1}{2}dp_2 \wedge dq_2 \wedge dq_3 \wedge dq_4 \\ &+ u\,dp_1 \wedge dq_2 \wedge dq_3 \wedge dq_4 + p_1^2\,dq_1 \wedge dq_2 \wedge dq_3 \wedge dq_4 \\ &+ dp_3 \wedge dq_1 \wedge dq_2 \wedge dq_4 - dp_4 \wedge dq_1 \wedge dq_2 \wedge dq_3.\end{aligned}$$

Straightforward calculations show that the Lie algebra $\mathfrak{g}_{KZ}$ of symmetries for the KZ equation E_ω is generated by the contact vector fields with the following generating functions:

$$A = \left(q_1 q_2 + \frac{3(q_3^2 + q_4^2)}{4}\right) p_1 + \frac{5}{2}q_2^2 p_2 + 3q_2(q_3 p_3 + q_4 p_4) + 4q_2 u + q_1,$$

$$S_1 = q_2 p_2 + \frac{q_3 p_3 + q_4 p_4}{2} + u,$$

$$S_2 = q_1 p_1 + \frac{q_3 p_3 + q_4 p_4}{2} - u,$$

$$R = q_4 p_3 - q_3 p_4,$$

$$T = p_2,$$

$$T_1(k) = kp_1 + k',$$

$$T_3(l) = 2lp_3 + l'q_3p_1 + q_3l'',$$

$$T_4(m) = 2mp_4 + m'q_4p_1 + q_4m'',$$

where $k = k(q_2)$, $l = l(q_2)$ and $m = m(q_2)$ are arbitrary smooth functions.

Therefore, the Lie algebra of contact symmetries of the KZ equation is infinite dimensional and consists of point symmetries only.

Note that the symmetries S_1 and S_2 correspond to scale transformations and R and T correspond to rotation in the plane (q_3, q_4) and translation along q_2 respectively.

Below we listed the transformations on the solutions which correspond to the above symmetries:

A

$$u(q) \mapsto \frac{u(\overline{q})}{(1 + (5/2)tq_2)^{8/5}} - \frac{t\overline{q}_1}{(1 + (5/2)tq_2)^2}$$

where

$$\overline{q}_1 = \frac{q_1(1 + (5/2)tq_2) - 3t/4(q_3^2 + q_4^2)}{(1 + (5/2)tq_2)^{7/5}},$$

$$\overline{q}_2 = \frac{q_2}{1 + (5/2)tq_2},$$

$$\overline{q}_3 = \frac{q_3}{(1 + (5/2)tq_2)^{6/5}},$$

$$\overline{q}_4 = \frac{q_4}{(1 + (5/2)tq_2)^{6/5}}.$$

$\mathbf{S_1}$

$$u(q) \mapsto e^t u(q_1, e^t q_2, e^{t/2} q_3, e^{t/2} q_4).$$

$\mathbf{S_2}$

$$u(q) \mapsto e^{-t} u(e^t q_1, q_2, e^{t/2} q_3, e^{t/2} q_4).$$

R

$$u(q) \mapsto u(q_1, q_2, q_3 \cos t - q_4 \sin t, q_4 \cos t + q_3 \sin t).$$

T

$$u(q) \mapsto u(q_1, q_2 + t, q_3, q_4)$$

$\mathbf{T}_1(k)$

$$u(q) \mapsto u(q_1 + tk(q_2), q_2, q_3, q_4) + tk'(q_2).$$

$\mathbf{T}_3(k)$

$$u(q) \mapsto u(q_1 + tk'(q_3 + kt), q_2, q_3 + 2kt, q_4) + tk''(q_3 + kt).$$

$\mathbf{T}_4(k)$

$$u(q) \mapsto u(q_1 + tk'(q_4 + kt), q_2, q_3, q_4 + 2kt) + tk''(q_4 + kt).$$

16.1.2 The structure of the symmetry algebra

As a linear space, the symmetry algebra $\mathfrak{g}_{KZ}$ is the direct sum

$$\mathfrak{g}_{KZ} = \mathfrak{g}_5 \oplus \mathfrak{g}_\infty$$

of the five-dimensional subspace

$$\mathfrak{g}_5 \stackrel{\text{def}}{=} \langle A, S_1, S_2, R, T \rangle$$

and the infinite-dimensional subspace

$$\mathfrak{g}_\infty \stackrel{\text{def}}{=} \langle T_1(k), T_3(l), T_4(m) \rangle.$$

In this sense, the KZ equation is similar to linear equations, which also have ∞-dimensional symmetry algebras. Roughly speaking, the KZ equation can be thought of as an intermediary between linear equations and truly non-linear equations, which have a few symmetries if any.

Structure of $\mathfrak{g}_5$

First of all we remark that $\mathfrak{g}_5$ is closed with respect to the Lagrange bracket and therefore forms a five-dimensional Lie algebra. Moreover all commutators with S_2 and R are vanishing and therefore symmetries S_2 and R belong to the centre of the Lie algebra.

The commutation relations between generators A, S_1 and T are given by the table

	A	S_1	T
A	0	A	$5S_1+S_2$
S_1	$-A$	0	T
T	$-5S_1-S_2$	$-T$	0

Let

$$b_+ = -\frac{2T}{5}, \quad b_- = A, \quad h = 2S_1 + \frac{2S_2}{5},$$

be the new generators instead of A, S_1 and T, then commutation relations between them take the form

$$[h, b_+] = 2b_+, \quad [h, b_-] = -2b_-, \quad [b_+, b_-] = h.$$

In other words, we obtain the following description of the Lie algebra.

Proposition 16.1.1 *The Lie algebra $\mathfrak{g}_5$ is a direct sum of the Lie algebra $\mathfrak{sl}_2$ and the abelian two-dimensional Lie algebra:*

$$\mathfrak{g}_5 = \mathfrak{sl}_2 \oplus \mathbb{R}^2,$$

where

$$\mathfrak{sl}_2 = \langle b_+, b_-, h\rangle \text{ and } \mathbb{R}^2 = \langle S_2, R\rangle.$$

Structure of $\mathfrak{g}_\infty$

First we remark that $\mathfrak{g}_\infty$ is closed with respect to the Lagrange bracket and therefore forms an ∞-dimensional Lie algebra. Moreover $[\mathfrak{g}_5, \mathfrak{g}_\infty] \subset \mathfrak{g}_\infty$, and therefore $\mathfrak{g}_\infty$ is an ideal in the symmetry Lie algebra.

We can represent $\mathfrak{g}_\infty$ as a $\mathfrak{g}_5$-invariant direct sum of three copies of $C^\infty(\mathbb{R})$ under realizations T_1, T_3 and T_4.

Then realization $T_1 : k(x) \longmapsto k(q_2)p_1 + k'(q_2)$ gives us a commutative algebra:

$$[T_1(k), T_1(l)] = 0$$

for all $k, l \in C^\infty(\mathbb{R})$.

Elements of algebra $\mathfrak{g}_5$ under realization T_1 are represented by the following linear differential operators:

$$b_+ \Rightarrow \frac{2}{5}\frac{d}{dx}, \quad b_- \Rightarrow -\frac{5x^2}{2}\frac{d}{dx}+x, \quad h \Rightarrow -2x\frac{d}{dx}+\frac{2}{5},$$

$$S_2 \Rightarrow 1, \quad R \Rightarrow 0.$$

Realization $T_3 : k(x) \longmapsto 2k(q_2)p_3+k'(q_2)q_3p_1+q_3k''(q_2)$ gives the following representation for elements of $\mathfrak{g}_5$:

$$b_+ \Rightarrow \frac{2}{5}\frac{d}{dx}, \quad b_- \Rightarrow -\frac{5x^2}{2}\frac{d}{dx}+3x, \quad h \Rightarrow -2x\frac{d}{dx}+\frac{6}{5},$$

$$S_2 \Rightarrow \frac{1}{2}, \quad R \Rightarrow T_3 \rightarrow -T_4$$

and for $T_4 : k(x) \longmapsto 2k(q_2)p_4 + k'(q_2)q_4p_1 + q_3k''(q_2)$ we obtain

$$b_+ \Rightarrow \frac{2}{5}\frac{d}{dx}, \quad b_- \Rightarrow -\frac{5x^2}{2}\frac{d}{dx}+3x, \quad h \Rightarrow -2x\frac{d}{dx}+\frac{6}{5},$$

$$S_2 \Rightarrow \frac{1}{2}, \quad R \Rightarrow T_4 \rightarrow T_3.$$

We unify the last two into a single realization of $C^\infty(\mathbb{R}, \mathbb{C}) : B : k(x)+\iota l(x) \longmapsto T_3(k) + T_4(l)$.

Then elements $b_\pm, h$ and S_2 will have the same representations as above but with $R = -\iota$.

Denote by $\Omega^\alpha(\mathbb{R}) = \{f(x)(dx)^\alpha\}$ the module of α-forms on the line. Then the Lie derivative gives a representation of the Lie algebra of vector fields on $\mathbb{R}$ by differential operators of the form

$$\Phi_\alpha : k(x)\frac{d}{dx} \rightarrow k(x)\frac{d}{dx} + \alpha k'(x).$$

This observation shows that T_1 corresponds to $\Phi_{-1/5}$ but T_3, and T_4 to $\Phi_{-3/5}$

$$[B_3(k), B_4(l)] = 0$$

and

$$[B_3(k), B_3(l)] = 2B_1(kl' - k'l), \quad [B_3(k), B_3(l)] = 2B_3(kl' - k'l).$$

This shows that

1. $\mathfrak{g}_5$ and $\mathfrak{g}_\infty$ are subalgebras of $\mathfrak{g}_{KZ}$.
2. $\mathfrak{g}_\infty$ is an ideal of $\mathfrak{g}_{KZ}$, that is $[\mathfrak{g}_5, \mathfrak{g}_\infty] \subset \mathfrak{g}_\infty$.
3. $\mathfrak{g}_\infty = \mathfrak{g}^0 \oplus \mathfrak{g}^1$, where

$$\mathfrak{g}^0 \stackrel{\text{def}}{=} \langle T_1(m) \rangle \qquad \text{and} \qquad \mathfrak{g}^1 \stackrel{\text{def}}{=} \langle T_3(k), T_4(l) \rangle.$$

This decomposition is subject to the commutation relations

$$[\mathfrak{g}^1, \mathfrak{g}^1] \subset \mathfrak{g}^0, \qquad [\mathfrak{g}_\infty, \mathfrak{g}^0] = 0.$$

4. $[\mathfrak{g}_5, \mathfrak{g}^0] \subset \mathfrak{g}^0$, $[\mathfrak{g}_5, \mathfrak{g}^1] \subset \mathfrak{g}^1$.
5. $\mathfrak{g}_5 = \mathfrak{g}_2 \oplus \mathfrak{g}_3$, where

$$\mathfrak{g}_2 \stackrel{\text{def}}{=} \langle S_2, R \rangle$$

is the center of $\mathfrak{g}_5$, i.e., $[\mathfrak{g}_2, \mathfrak{g}_5] = 0$, and $\mathfrak{g}_3$ is the commutator subalgebra, i.e.,

$$\mathfrak{g}_3 \stackrel{\text{def}}{=} [\mathfrak{g}_5, \mathfrak{g}_5].$$

We have seen that $\mathfrak{g}_3$ is isomorphic to the Lie algebra $\mathfrak{sl}_2$.

16.1.3 Classification of one-dimensional subalgebras of $\mathfrak{sl}(2, \mathbb{R})$

Here we classify elements of $\mathfrak{g} = \mathfrak{sl}(2, \mathbb{R})$, with respect to the adjoint action. We recall that each element of $\mathfrak{sl}(2, \mathbb{R})$ has the from

$$X = x_0 h + x_+ b_+ + x_- b_-,$$

where $x_0, x_+, x_- \in \mathbb{R}$.

Using the commutation relations, we find

$$\mathrm{ad}_X(h) = [h, X] = 2x_+ b_+ - 2x_- b_-,$$

$$\mathrm{ad}_X(b_+) = [b_+, X] = -2x_0 b_+ + x_- h,$$

$$\mathrm{ad}_X(b_-) = [b_-, X] = 2x_0 b_- - x_+ h.$$

Hence, the matrix representation of the operator ad_X in the basis (h, b_+, b_-) has the form

$$\mathrm{ad}_X = \begin{Vmatrix} 0 & x_- & -x_+ \\ 2x_+ & -2x_0 & 0 \\ -2x_- & 0 & 2x_0 \end{Vmatrix}.$$

We remind that the bilinear form

$$k(Y_1, Y_2) \stackrel{\text{def}}{=} \mathrm{tr}(\mathrm{ad}_{Y_1} \circ \mathrm{ad}_{Y_2}),$$

is called the *Killing form*.

The Killing form has the following important property:

$$k(\mathrm{ad}_X\, Y_1, Y_2) + k(Y_1, \mathrm{ad}_X\, Y_2) = 0,$$

which says that ad_X are skew symmetric with respect to k.

This property is equivalent to the condition

$$\frac{d}{dt} k(\exp(t\, \mathrm{ad}_X) Y_1, \exp(t\, \mathrm{ad}_X) Y_2) = 0,$$

which means that the Killing form is invariant under the adjoint representation.

As a result we see that if Y_1 and Y_2 are on the same orbit, then

$$k(Y_1, Y_1) = k(Y_2, Y_2).$$

Moreover, the orbits of the adjoint representation are level surfaces of the quadratic form $k(Y, Y)$.

A straightforward computation gives

$$k(Y, Y) = 8(y_0^2 + y_- y_+),$$

if $Y = y_0 h + y_+ b_+ + y_- b_-$.

This quadratic form is of signature $(+, +, -)$. Therefore the adjoint representation can be viewed as the group of Lorenz transformations on the three-dimensional Minkowski space $\mathfrak{sl}(2, \mathbb{R})$.

The light cone $k(Y, Y) = 0$ divides this space into disconnected parts. Finally, we obtain the following list of inequivalent one-dimensional subspaces of the algebra $\mathfrak{sl}(2, \mathbb{R})$:

1. $\langle b_+ \rangle$, $k = 0$,
2. $\langle b_+ - b_- \rangle$, $k < 0$,
3. $\langle h \rangle$, $k > 0$.

16.1.4 Classification of symmetries

Now we apply this result to the symmetry algebra $\mathfrak{g}_{KZ}$ of the KZ equation. We recall that

$$\mathfrak{g}_{KZ} = \mathfrak{g}_5 \oplus \mathfrak{g}_\infty \qquad \text{and} \qquad \mathfrak{g}_5 = \mathfrak{g}_2 \oplus \mathfrak{g}_3,$$

where $\mathfrak{g}_\infty$ is an ideal of the algebra $\mathfrak{g}_{KZ}$ and $\mathfrak{g}_2$ is a center of $\mathfrak{g}_5 \cong \mathfrak{g}_{KZ}/\mathfrak{g}_\infty$, and $\mathfrak{g}_3$ is the commutator subalgebra of $\mathfrak{g}_5$, i.e., $\mathfrak{g}_3 = [\mathfrak{g}_5, \mathfrak{g}_5]$.

Since $\mathfrak{g}_3 \cong \mathfrak{sl}(2, \mathbb{R})$, our previous result can be viewed as the list of inequivalent one-dimensional subalgebras of $\mathfrak{g}_3$:

1. $\langle T \rangle$,
2. $\langle 2T + 5A \rangle$,
3. $\langle 5S_1 + S_2 \rangle$

of the Lie algebra $\mathfrak{g}_3$ under the adjoint representation.

Let us classify one-dimensional subalgebras of $\mathfrak{g}_5$. Since $\mathfrak{g}_5$ has a non-trivial center, the corresponding Lie group of transformations $\mathfrak{g}_5$ is not isomorphic to the adjoint representation $\mathrm{Ad}\,\mathfrak{g}_5$. The Lie algebra of $\mathrm{Ad}\,\mathfrak{g}_5$ is isomorphic to the quotient algebra $\mathfrak{g}_5/\mathfrak{g}_2 \cong \mathfrak{g}_3 \cong \mathfrak{sl}(2, \mathbb{R})$. The group $\mathrm{Ad}\,\mathfrak{g}_5$ coincides with the adjoint action of the group $\mathfrak{g}_3$ on $\mathfrak{g}_5$ and leaves invariant each element of the center $\mathfrak{g}_2$. Therefore the list of inequivalent one-dimensional subalgebras of $\mathfrak{g}_5$ is given by the following list:

1. $\langle T + c_1 S_2 + c_2 R \rangle$,
2. $\langle 2T + 5A + c_1 S_2 + c_2 R \rangle$,
3. $\langle 5S_1 + S_2 + c_1 S_2 + c_2 R \rangle$,
4. $\langle c_1 S_2 + c_2 R \rangle$,

where $c_1, c_2 \in \mathbb{R}$.

Let us discuss the classification problem for the full algebra $\mathfrak{g}_{KZ}$. Since $\mathfrak{g}_\infty$ is an ideal if $\mathfrak{g}_{KZ}$ and $\mathfrak{g}_{KZ}/\mathfrak{g}_\infty \cong \mathfrak{g}_5$, the obtained results can be viewed as the list of inequivalent one-dimensional subalgebras of the quotient algebra $\mathfrak{g}_{KZ}/\mathfrak{g}_\infty$:

1. $\langle T + c_1 S_2 + c_2 R + \mathfrak{g}_\infty \rangle$,
2. $\langle 2T + 5A + c_1 S_2 + c_2 R + \mathfrak{g}_\infty \rangle$,
3. $\langle 5S_1 + S_2 + c_1 S_2 + c_2 R + \mathfrak{g}_\infty \rangle$,
4. $\langle c_1 S_2 + c_2 R + \mathfrak{g}_\infty \rangle$.

This gives the following preliminary classification of one-dimensional subalgebras of the full algebra $\mathfrak{g}_{KZ}$:

1. $\langle T + c_1 S_2 + c_2 R + Y \rangle$,
2. $\langle 2T + 5A + c_1 S_2 + c_2 R + Y \rangle$,
3. $\langle 5S_1 + S_2 + c_1 S_2 + c_2 R + Y \rangle$,
4. $\langle c_1 S_2 + c_2 R + Y \rangle$,
5. $\langle Y \rangle$.

Here $c_1, c_2 \in \mathbb{R}$ and $Y \in \mathfrak{g}_\infty$.

In other words, there are five equivalence types, and two subalgebras from two different classes cannot be transformed to each other under the adjoint representation of the group generated by the Lie algebra $\mathfrak{g}_{KZ}$. On the other hand, two subalgebras from the same class can be equivalent to each other.

16.1.5 Conservation laws

To find conservation laws for the KZ equation one should solve the equation $\mathcal{E}(g\omega) = 0$ with respect to the generating function g. Straightforward computations show that

$$g = a(q_2, q_3, q_3) + q_1 b(q_2, q_3, q_4)$$

where $a(q_2, q_3, q_4)$ is biharmonic with respect to q_3, q_4 and

$$b(q_2, q_3, q_4) = w(q_3, q_4) + \int_0^{q_2} \left(\frac{\partial^2 a}{\partial q_3^2} + \frac{\partial^2 a}{\partial q_4^2} \right) dq_2,$$

where $w(q_3, q_4)$ is a harmonic function.

The corresponding conservation laws are

$$\theta_g = \left((up_1 - p_2)g - \frac{1}{2}\frac{\partial g}{\partial q_1} u^2 \right) dq_2 \wedge dq_3 \wedge dq_4 + u \frac{\partial g}{\partial q_1} dq_1 \wedge dq_3 \wedge dq_4$$
$$+ \left(p_3 g - \frac{\partial g}{\partial q_3} u \right) dq_1 \wedge dq_2 \wedge dq_4 + \left(u \frac{\partial g}{\partial q_4} - p_4 g \right) dq_1 \wedge dq_2 \wedge dq_3.$$

Recall that the solution of the KZ equation represents the deviation from the equilibrium density which, for bounded sound beams, is proportional to the velocity of the sound beam along the q_2-axis. Thus, θ_g may be viewed as the "energy" of a medium whose density is proportional to $\partial g/\partial q_1$. Consequently, the conservation law θ_g for $g = q_1$ represents the conservation of energy, while for $g = 1$ it represents the conservation of momentum.

16.2 Singularities of solutions of the KZ equation

16.2.1 Caustics

Here we construct solutions invariant under the abelian subalgebra

$$\mathfrak{g} = \left\langle q_3\frac{\partial}{\partial q_1} + 2q_2\frac{\partial}{\partial q_3},\ q_4\frac{\partial}{\partial q_1} + 2q_2\frac{\partial}{\partial q_4}\right\rangle.$$

The corresponding generating functions of the vector fields are

$$f = p_1q_3 + 2p_3q_2 \qquad \text{and} \qquad g = p_1q_4 + 2p_4q_2$$

respectively.

Each $\mathfrak{g}$-invariant solution $\rho(q_1,\dots,q_4)$ has the form

$$\rho = v(x, y),$$

where

$$x = q_2, \qquad y = 4q_1q_2 - (q_3^2 + q_4^2)$$

are invariants for $\mathfrak{g}$.

For such type of solutions the KZ equation has the form

$$2\frac{\partial v}{\partial y} - 4x^2\left(v\frac{\partial^2 v}{\partial y^2} + \left(\frac{\partial v}{\partial y}\right)^2\right) + x\frac{\partial^2 v}{\partial x\partial y} + y\frac{\partial^2 v}{\partial y^2} = 0.$$

The order of this equation can be reduced by integrating with respect to y:

$$x\frac{\partial v}{\partial x} + (y - 4x^2v)\frac{\partial v}{\partial y} + v = h(x),$$

where $h = h(x)$ is an arbitrary function.

We shall consider solutions $v(x, y)$ such that $v(x, y) \to 0$ when $(x, y) \to \infty$. For these solutions $h(x) \equiv 0$.

This leads us to the following first-order equation

$$x\frac{\partial v}{\partial x} + (y - 4x^2 v)\frac{\partial v}{\partial y} + v = 0.$$

After substitution

$$v = \frac{1}{x}w(\eta, \theta), \qquad \eta = \frac{y}{x}, \qquad \theta = \ln x$$

we obtain the classical Euler equation

$$\frac{\partial w}{\partial \theta} = 4w\frac{\partial w}{\partial \eta}.$$

Solutions of this equation can be described as follows: for a given initial density

$$w|_{\theta=0} = w_0(\eta)$$

one can find $w = w(\eta, \theta)$ from the equation

$$w = w_0(\eta + 4w\theta)$$

or from the system

$$\begin{aligned} \eta + 4\theta w &= \varsigma, \\ w &= w_0(\varsigma). \end{aligned} \tag{16.2}$$

Geometrically it can represented as follows. Let us fix η and θ and let us plot on the plane (ς, w) the graph of the line $l = l(\eta, \theta)$ given by the first equation in (16.2) and the graph K of the function $w = w_0(\varsigma)$.

If the line l and the curve K have only one common point, then the system (16.2) has only one solution (ζ_0, w_0). This means that w (and, therefore, v also) is a one-valued function. We can see this situation if the initial curve $w_0(\eta)$ is convex or concave.

If the line l intersects the curve K at several points then w is a multivalued function. In this case caustics arise. Physically the appearance of a caustic means additional acoustic waves in a domain that is bounded by this caustic (see Figure 16.1).

Let us describe the process causing the appearance of caustics in geometrical terms.

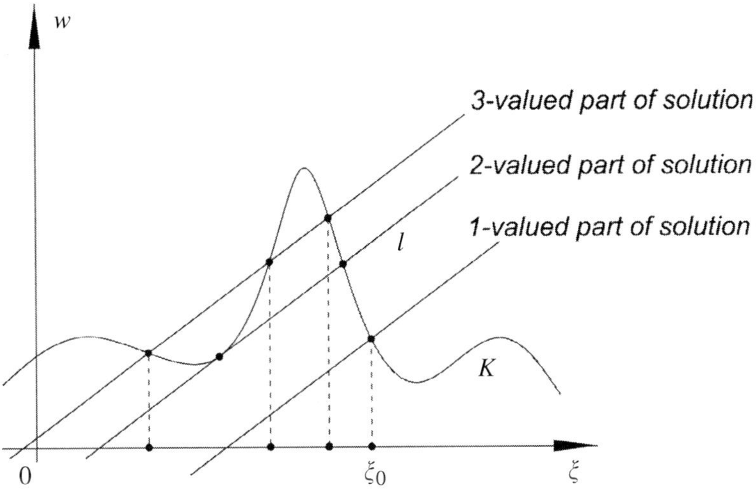

Figure 16.1. How multivalued solutions arise.

Let us now fix θ only. It is equivalent to the corresponding cross-section of the sound beam by the plane q_2 = constant and η shall increase monotonically in time. From the geometrical point of view this means that we fix the slope of the line l and then translate it parallel to itself by increasing the ordinate w. Then, up to a certain moment, K and l will intersect at only one point: as η increases further, the solution becomes multivalued. The corresponding times when "gradient" catastrophes occur can be computed using the given system (16.2).

16.2.2 Contact shock waves

Here we use the conservation laws to describe the propagation of perturbations in non-linear acoustics.

Let $M_0 \subset \mathbb{R}^4$ be a three-dimensional submanifold that divide $\mathbb{R}^4$ into two domains M^+ and M^-:

$$\mathbb{R}^4 = M^+ \cup M^0 \cup M^-.$$

Let us consider so-called *contact shock waves.*

They are continuous solutions of the KZ equation. That is, we consider solutions of the form $v = v^+ \cup v^-$, where v^+ is defined on M^+ and v^- on M^- respectively.

Assume that the boundary of the beam is given by the function $S(q)$. In this case the Hugoniot–Rankine condition

$$[\theta^+|_{M^0} - \theta^-|_{M^0}] \wedge dS = 0$$

yields the Hamilton–Jacobi equation

$$v^+(q)\left(\frac{\partial S}{\partial q_1}\right)^2 - \frac{\partial S}{\partial q_1}\frac{\partial S}{\partial q_2} + \left(\frac{\partial S}{\partial q_3}\right)^2 + \left(\frac{\partial S}{\partial q_4}\right)^2 = 0 \tag{16.3}$$

independent of the choice of conservation law.

Given an arbitrary background solution $v^+(q)$, the solution $S(q)$ of this equation can only be found by numerical methods.

We shall examine in more detail the case of a zero background $v^+ = 0$, assuming that the initial perturbation at $q_2 = 0$ is localized inside a disk whose radius changes according to a given law $R(q_1)$, that is,

$$S|_{q_2=0} = S_0(q_1, q_3, q_4) = q_3^2 + q_4^2 - R^2(q_1).$$

Solving the Hamilton–Jacobi equation with this initial value, we find that at the cross-section $q_2 = c =$ constant the perturbation is localized in a disk of radius

$$R_1(q_1, q_2) = \left| R(z) + \frac{2q_2}{R'(z)} \right| \tag{16.4}$$

where $z, 0 \leqslant z \leqslant q_1$ is a solution of the equation:

$$\frac{q_1 - z}{q_2} = \left(\frac{1}{R'(z)}\right)^2. \tag{16.5}$$

Consider now several particular cases of $R(q_1)$.

1. $R(q_1) = R_0 + \varepsilon q_1$, the radius of the perturbation domain grows with linear speed ε for the cross-section $q_2 = 0$. In this case, at the cross-section $q_2 =$ constant > 0 the radius of the perturbation disc also grows with speed ε in time and ε^{-1} in q_2:

$$R_1(q_1, q_2) = R_0 + \varepsilon q_1 + \frac{1}{\varepsilon} q_2.$$

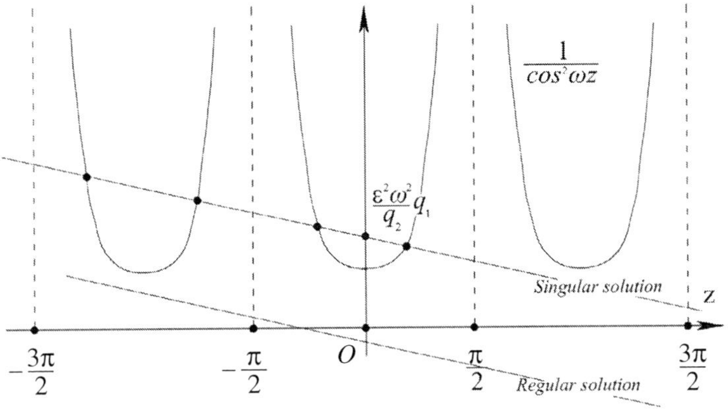

Figure 16.2. How self-diffraction arises.

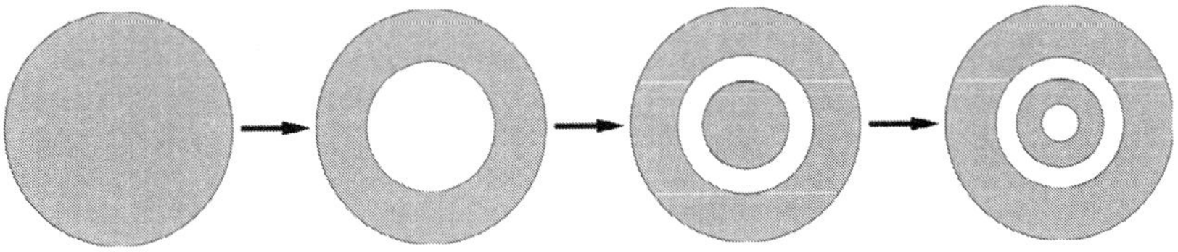

Figure 16.3. Self-diffraction.

It follows that, in order to get a stable beam, it is necessary that for the boundary of the initial perturbation to perform self-sustained oscillations.

2. $R(q_1) = R_0 + \varepsilon \sin(\omega q_1)$. In this case (16.5) for z takes the form

$$\varepsilon^2 \omega^2 \frac{q_1 - z}{q_2} = \frac{1}{\cos^2(\omega z)}.$$

Depending on

$$\gamma \stackrel{\text{def}}{=} \varepsilon^2 \omega^2 \frac{q_1}{q_2}$$

the equation admits a distinct number of solutions on the interval $0 \leqslant z \leqslant q_1$. Fixing q_2 we see that as q_1 grows, the sound beam necessarily separates into layers – the phenomenon of "self-diffraction" of sound beams (see Figure 16.2).

The time at which separation occurs and the radius of the rings that are thus formed can be computed using relation (16.4) (see Figure 16.3).

3. $S_0(q_1, q_3, q_4) = (Aq, q) - \eta(q_1)$, where $q = (q_3, q_4)$, A is a symmetric positive matrix, $(\cdot, \cdot)$ is the scalar product and $\eta(q_1)$ is the given law governing the evolution of the boundary. Then

$$S_0(q_1, q_2, q) = S_0(q_1^*, q^*),$$

where q_1^*, q^* and q_1, q_2, q are connected by

$$q_1 - q_1^* = \frac{4q_2}{(\eta'(q_1^*))^2}(Aq^*, q^*),$$

$$q = \left(1 + \frac{4q_2}{\eta'(q_1^*)}A\right)q^*.$$

Let us consider in more detail the case

$$\eta(q_1) = \eta_0 + \varepsilon q_1,$$

where $\eta_0 \in \mathbb{R}$. Then we have

$$S(q_1, q_2, q) = (ABq, q) + \eta_0 - \varepsilon q_1$$

with

$$B = \left(1 + \frac{4q_2}{\varepsilon}A\right)^{-1}.$$

From this relation it follows that at the cross-section $q_2 =$ constant the perturbation is localized in the domain obtained from the section $q_2 = 0$ by applying the affine transformation with the matrix

$$\left(1 + \frac{4q_2}{\varepsilon}A\right)^{-1/2}.$$

This means that if we look, say, at the section

$$q_2 = -\frac{4\lambda}{\varepsilon},$$

where λ is an eigenvalue of A, then the shock wave shall be focused in the corresponding eigendirection (see Figure 16.4).

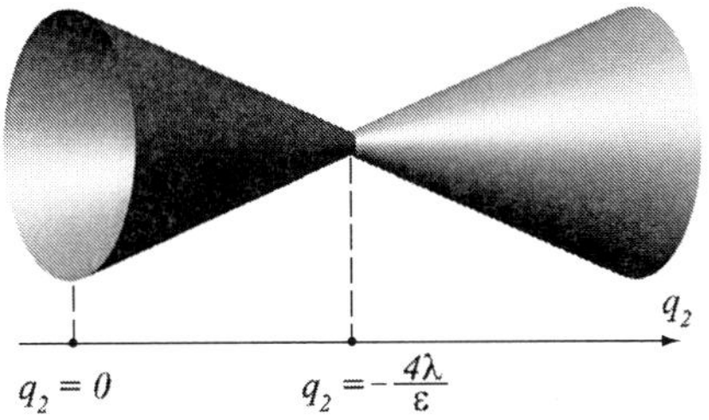

Figure 16.4. The shock wave collapses.

For example, for a circular perturbation $A = a\mathbf{1}$ the shock wave collapses at the section

$$q_2 = -\frac{4a}{\varepsilon}.$$

17

Non-linear thermal conductivity

Here we consider a problem of *group classification* for the non-linear thermal conductivity equation. This is the model equation for many processes: heat and mass transfer, theory of combustion, biology and ecology. It describes species propagation in natural habitat, membrane ion transport, nerve impulse propagation, the spread of chemical concentration waves, self-organization in biochemical systems, the formation of apex zones in plants, non-linear effects in plasma and so on (see [6, 20, 87, 99]).

17.1 Symmetries of the TC equation

17.1.1 TC equation

The equation

$$v_t = (v^\alpha v_x)_x + F(v) \tag{17.1}$$

is often called the *reaction–diffusion equation* or the *equation of non-linear thermal conductivity*.

We will call it the *thermal conductivity equation* (*TC equation*). This is the Kolmogorov–Petrovsky–Piskunov equation [48] with non-linear diffusion coefficient.

Here α is an arbitrary real number, v^{α} is a diffusion coefficient and $F(v)$ is the so-called source function. We consider only the case of non-linear diffusion, i.e.,

$$\alpha \neq 0.$$

For the differential TC equation we have the following effective differential form:

$$\omega = u^{\alpha}\, dq_1 \wedge dp_2 + (\alpha u^{\alpha-1} p_2^2 - p_1 + F(u)) dq_1 \wedge dq_2.$$

Here $q_1 = t$ and $q_2 = x$.

17.1.2 Group classification of TC equation

Here we follow [63] and [49]. Note that (17.1) has infinitesimal symmetries with generation functions $f_1 = p_1$ and $f_2 = p_2$ for all α and $F(v)$.

Below we list equations with non-trivial algebras of symmetries. There are ten classes of the equtations, all others have the symmetries p_1 and p_2 only.

Case 1

$$F(v) = 0, \quad \alpha \neq 0, \quad \alpha \neq -\frac{4}{3}.$$

The TC equation has the form

$$v_t = (v^{\alpha} v_x)_x.$$

The basis of $\mathfrak{g}_{TC}$ is

$$f_1 = p_1, \quad f_2 = p_2, \quad f_3 = u + \alpha q_1 p_1, \quad f_4 = 2u - \alpha q_2 p_2,$$

with the following Lie algebra structure:

	X_1	X_2	X_3	X_4
X_1	0	0	$-\alpha X_1$	0
X_2	0	0	0	αX_2
X_3	αX_1	0	0	0
X_4	0	$-\alpha X_2$	0	0

Here $X_i \stackrel{\text{def}}{=} X_{f_i}$ are contact vector fields with the generating functions f_i, $i = 1, 2, 3, 4$.

Case 2

$$F(v) = 0, \quad \alpha = -\frac{4}{3}.$$

The TC equation has the form

$$v_t = (v^{-4/3} v_x)_x.$$

The basis of $\mathfrak{g}_{TC}$ is

$$f_1 = p_1, \ f_2 = p_2, \ f_3 = 3u - 4q_1p_1, \ f_4 = 3u + 2q_2p_2, \ f_5 = q_2^2p_2 + 3q_2u,$$

with the following Lie algebra structure:

	X_1	X_2	X_3	X_4	X_5
X_1	0	0	$4X_1$	0	0
X_2	0	0	0	$-2X_2$	$-X_4$
X_3	$-4X_1$	0	0	0	0
X_4	0	$2X_2$	0	0	$-2X_5$
X_5	0	X_4	0	$2X_5$	0

Case 3

$$F(v) = \gamma v, \quad \alpha \neq 0, \quad \alpha \neq -\frac{4}{3}, \quad \gamma \neq 0.$$

The TC equation has the form

$$v_t = (v^\alpha v_x)_x + \gamma v.$$

The basis of $\mathfrak{g}_{TC}$ is

$$f_1 = p_1, \quad f_2 = p_2, \quad f_3 = \alpha q_2p_2 - 2u, \quad f_4 = (p_1 - \gamma u)\exp(-\alpha\gamma q_1),$$

with the following Lie algebra structure:

	X_1	X_2	X_3	X_4
X_1	0	0	0	$\alpha\gamma X_4$
X_2	0	0	$-\alpha X_2$	0
X_3	0	αX_2	0	0
X_4	$-\alpha\gamma X_4$	0	0	0

Case 4

$$F(v) = \gamma v, \quad \alpha = -\frac{4}{3}, \quad \gamma \neq 0.$$

The TC equation has the form

$$v_t = (v^{-4/3} v_x)_x + \gamma v.$$

The basis of $\mathfrak{g}_{TC}$ is

$$f_1 = p_1, \quad f_2 = p_2, \quad f_3 = 3u + 2q_2 p_2, \quad f_4 = q_2^2 p_2 + 3q_2 u,$$

with the following Lie algebra structure:

	X_1	X_2	X_3	X_4
X_1	0	0	0	0
X_2	0	0	$-2X_2$	$-X_3$
X_3	0	$2X_2$	0	$-2X_4$
X_4	0	X_3	$2X_4$	0

Case 5

$$F(v) = \mu v^\beta, \quad \alpha \neq 0, \quad \beta \neq \alpha + 1, \quad \beta \neq -\frac{1}{3}, \quad \beta \neq 1.$$

The TC equation has the form

$$v_t = (v^\alpha v_x)_x + \mu v^\beta. \tag{17.2}$$

The basis of $\mathfrak{g}_{TC}$ is

$$f_1 = p_1, \qquad f_2 = p_2, \quad f_3 = 2\frac{\beta - 1}{\beta - \alpha - 1} q_1 p_1 + q_2 p_2 + \frac{2}{\beta - \alpha - 1} u,$$

with the following Lie algebra structure:

	X_1	X_2	X_3
X_1	0	0	$-2\frac{\beta - 1}{\beta - \alpha - 1} X_1$
X_2	0	0	$-X_2$
X_3	$2\frac{\beta - 1}{\beta - \alpha - 1} X_1$	X_2	0

(17.3)

Case 6

$$F(v) = \mu v^{-1/3}, \quad \alpha = -\frac{4}{3}, \quad \mu > 0.$$

The TC equation has the form

$$v_t = (v^{-4/3} v_x)_x + \mu v^{-1/3}.$$

The basis of $\mathfrak{g}_{TC}$ is

$$f_1 = p_1, \qquad f_2 = p_2, \qquad f_3 = 3u - 4q_1 p_1,$$

$$f_4 = \left(u + \frac{1}{\sqrt{3\mu}} p_2\right) \exp\left(2\sqrt{\frac{\mu}{3}} q_2\right),$$

$$f_5 = \left(u - \frac{1}{\sqrt{3\mu}} p_2\right) \exp\left(-2\sqrt{\frac{\mu}{3}} q_2\right),$$

with the following Lie algebra structure:

	X_1	X_2	X_3	X_4	X_5
X_1	0	0	$4X_1$	0	0
X_2	0	0	0	$-\frac{2}{3}\sqrt{3\mu}X_4$	$\frac{2}{3}\sqrt{3\mu}X_5$
X_3	$-4X_1$	0	0	0	0
X_4	0	$\frac{2}{3}\sqrt{3\mu}X_4$	0	0	$-\frac{4}{3\sqrt{3\mu}}X_2$
X_5	0	$-\frac{2}{3}\sqrt{3\mu}X_5$	0	$\frac{4}{3\sqrt{3\mu}}X_2$	0

Case 7

$$F(v) = \mu v^{-1/3}, \quad \alpha = -\frac{4}{3}, \quad \mu < 0.$$

The TC equation has the form

$$v_t = (v^{-4/3} v_x)_x + \mu v^{-1/3}.$$

The basis of $\mathfrak{g}_{TC}$ is

$$f_1 = p_1, \qquad f_2 = p_2, \qquad f_3 = 3u - 4q_1p_1,$$
$$f_4 = u\cos\left(2\sqrt{-\frac{\mu}{3}}q_2\right) + p_2\frac{1}{\sqrt{-3\mu}}\sin\left(2\sqrt{-\frac{\mu}{3}}q_2\right),$$
$$f_5 = u\sin\left(2\sqrt{-\frac{\mu}{3}}q_2\right) - p_2\frac{1}{\sqrt{-3\mu}}\cos\left(2\sqrt{-\frac{\mu}{3}}q_2\right),$$

with the following Lie algebra structure:

	X_1	X_2	X_3	X_4	X_5
X_1	0	0	$4X_1$	0	0
X_2	0	0	0	$\frac{2}{3}\sqrt{-3\mu}X_5$	$-\frac{2}{3}\sqrt{-3\mu}X_4$
X_3	$-4X_1$	0	0	0	0
X_4	0	$-\frac{2}{3}\sqrt{-3\mu}X_5$	0	0	$-\frac{2}{3\sqrt{-3\mu}}X_2$
X_5	0	$\frac{2}{3}\sqrt{-3\mu}X_4$	0	$\frac{2}{3\sqrt{-3\mu}}X_2$	0

Case 8

$$F(v) = \gamma v + \mu v^{1+\alpha}, \quad \alpha \neq 0, \quad \alpha \neq -\frac{4}{3}.$$

The TC equation has the form

$$v_t = (v^\alpha v_x)_x + \gamma v + \mu v^{1+\alpha}.$$

The basis of $\mathfrak{g}_{TC}$ is

$$f_1 = p_1, \qquad f_2 = p_2, \quad f_3 = (p_1 - \gamma u)\exp(-\alpha\gamma q_1),$$

with the following Lie algebra structure:

	X_1	X_2	X_3
X_1	0	0	$\alpha\gamma X_3$
X_2	0	0	0
X_3	$-\alpha\gamma X_3$	0	0

Case 9

$$F(v) = \gamma v + \mu v^{-1/3}, \quad \alpha = -\frac{4}{3}, \quad \gamma \neq 0, \quad \mu > 0.$$

The TC equation has the form

$$v_t = (v^{-4/3} v_x)_x + \gamma v + \mu v^{-1/3}.$$

The basis of $\mathfrak{g}_{TC}$ is

$$f_1 = p_1, \quad f_2 = p_2, \quad f_3 = (p_1 - \gamma u) \exp\left(\frac{4}{3}\gamma q_1\right),$$
$$f_4 = \left(u + \frac{1}{\sqrt{3\mu}} p_2\right) \exp\left(2\sqrt{\frac{\mu}{3}} q_2\right),$$
$$f_5 = \left(u - \frac{1}{\sqrt{3\mu}} p_2\right) \exp\left(-2\sqrt{\frac{\mu}{3}} q_2\right),$$

with the following Lie algebra structure:

	X_1	X_2	X_3	X_4	X_5
X_1	0	0	$-\frac{4}{3}\gamma X_3$	0	0
X_2	0	0	0	$-\frac{2}{3}\sqrt{3\mu} X_4$	$\frac{2}{3}\sqrt{3\mu} X_5$
X_3	$\frac{4}{3}\gamma X_3$	0	0	0	0
X_4	0	$\frac{2}{3}\sqrt{3\mu} X_4$	0	0	$-\frac{4}{3\sqrt{3\mu}} X_2$
X_5	0	$-\frac{2}{3}\sqrt{3\mu} X_5$	0	$\frac{4}{3\sqrt{3\mu}} X_2$	0

Case 10

$$F(v) = \gamma v + \mu v^{-1/3}, \quad \alpha = -\frac{4}{3}, \quad \gamma \neq 0, \quad \mu < 0.$$

The TC equation has the form

$$v_t = (v^{-4/3} v_x)_x + \gamma v + \mu v^{-1/3}.$$

The basis of $\mathfrak{g}_{TC}$ is

$$f_1 = p_1, \qquad f_2 = p_2, \quad f_3 = (p_1 - \gamma u)\exp\left(\frac{4}{3}\gamma q_1\right),$$

$$f_4 = u\cos\left(2\sqrt{-\frac{\mu}{3}}q_2\right) + p_2\frac{1}{\sqrt{-3\mu}}\sin\left(2\sqrt{-\frac{\mu}{3}}q_2\right),$$

$$f_5 = u\sin\left(2\sqrt{-\frac{\mu}{3}}q_2\right) - p_2\frac{1}{\sqrt{-3\mu}}\cos\left(2\sqrt{-\frac{\mu}{3}}q_2\right),$$

with the following Lie algebra structure:

	X_1	X_2	X_3	X_4	X_5
X_1	0	0	$-\frac{4}{3}\gamma X_3$	0	0
X_2	0	0	0	$\frac{2}{3}\sqrt{-3\mu}X_5$	$-\frac{2}{3}\sqrt{-3\mu}X_4$
X_3	$\frac{4}{3}\gamma X_3$	0	0	0	0
X_4	0	$-\frac{2}{3}\sqrt{-3\mu}X_5$	0	0	$-\frac{2}{3\sqrt{-3\mu}}X_2$
X_5	0	$\frac{2}{3}\sqrt{-3\mu}X_4$	0	$\frac{2}{3\sqrt{-3\mu}}X_2$	0

17.2 Invariant solutions

Here we find exact solutions using symmetry algebras $\mathfrak{g}_{TC}$.

We begin with some elementary remarks. Let us consider, for example, the equation

$$v_t = (vv_x)_x + \mu, \tag{17.4}$$

where μ is a constant. This equation is a special case of the classification when $\alpha = 1$ and $\beta = 0$.

First of all solutions which are invariant with respect to the symmetry $f = bp_1 - ap_2$, $a, b \in \mathbb{R}$, are exactly traveling wave solutions, i.e., the solutions of the form

$$v(t,x) = \Phi(at + bx).$$

The structure of the Lie algebra $\mathfrak{g}_{TC}$ of (17.4) is as follows:

	X_1	X_2	X_3
X_1	0	0	$-X_1$
X_2	0	0	$-X_2$
X_3	X_1	X_2	0

Note that

$$X_f = bX_1 - aX_2 = -b\frac{\partial}{\partial q_1} + a\frac{\partial}{\partial q_2}.$$

Since

$$[X_f, X_1] = [X_f, X_2] = 0$$

and

$$[X_f, X_3] = b[X_1, X_3] - a[X_2, X_3] = -bX_1 + aX_2 = -X_f,$$

we see that the normalizer of the subalgebra $\mathfrak{g} = \langle X_f \rangle$ coincides with $\mathfrak{g}_{TC}$: $N_{\mathfrak{g}} = \mathfrak{g}_{TC}$. Let

$$g_1 = f_2 = p_2, \quad g_2 = f_3 = q_1p_1 + q_2p_2 - u$$

and

$$\Gamma = \{f = bp_1 - ap_2 = 0, \quad q_1 = 0\}.$$

Then Γ is invariant under the vector fields

$$X_{g_1} = -q_1\frac{\partial}{\partial q_1} - q_2\frac{\partial}{\partial q_2} - u\frac{\partial}{\partial u}$$

and

$$X_{g_2} = -\frac{\partial}{\partial q_2}.$$

The variables

$$y \stackrel{\text{def}}{=} u, \quad x \stackrel{\text{def}}{=} q_2, \quad p \stackrel{\text{def}}{=} p_2$$

can be choosen as coordinates on Γ. Note that

$$X_f \rfloor \omega = -bu\,dp_2 - b(p_2^2 - p_1 + \mu)\,dq_2 - a(p_2^2 - p_1 + \mu)\,dq_1$$

and

$$\omega_2 = -\frac{1}{b}(X_f \rfloor \omega)|_\Gamma = y\,dp + \left(p^2 - \frac{a}{b}p + \mu\right) dx.$$

Therefore the reduced equation on Γ has the form

$$(yy')' - \frac{a}{b}y' + \frac{\mu}{b} = 0. \tag{17.5}$$

This equation, by construction, has a two-dimensional Lie algebra of contact symmetries. Namely, the Lie algebra generated by vector fields

$$Y_0 \overset{\text{def}}{=} X_{g_1}|_\Gamma = -\frac{\partial}{\partial x}, \quad Y_1 \overset{\text{def}}{=} X_{g_2}|_\Gamma = -x\frac{\partial}{\partial x} - y\frac{\partial}{\partial y}.$$

The algebra can be used for integrating equation (17.5) in quadratures. Using the method described in Part I, we pass to the forms

$$\bar{\omega}_1 = dx - \frac{x}{y}\,dy - b\frac{xp - y}{p^2 b - ap + \mu}\,dp,$$

$$\bar{\omega}_2 = -\frac{1}{y}\,dy - \frac{bp}{p^2 b - ap + \mu}\,dp.$$

Note that $d\bar{\omega}_2 = 0$ and, therefore, there exists a function H_1 such that $\bar{\omega}_2 = dH_1$. If we put $\mu = a^2/4b$, then

$$H_1(x, y, p) = \frac{2ab(1-p)}{(2bp - a)(2b - a)} - \ln(y(2bp - a)).$$

To find the restriction of the form $\bar{\omega}_1$ to the surface $H_1(x, y, p) = C_1$, we note that

$$y = \frac{1}{C_1(2bp - a)} \exp\left(\frac{2ab(1-p)}{(2bp - a)(2b - a)}\right)$$

and

$$\bar{\omega}_1|_{H_1 = C_1} = dx + 4b^2 \exp\left(\frac{2ab(1-p)}{(2bp - a)(2b - a)}\right) [C_1(2bp - a)^3]dp.$$

The last form is closed too, and its integral is

$$H_0(x,p) = x + \frac{4b(bp-a)}{(2bp-a)a^2C_1} \exp\left(\frac{2ab(1-p)}{(2bp-a)(2b-a)}\right).$$

From the equation $H_0(x,p) = C_2$ we get

$$p(x) = \frac{a(2+\mathrm{LambertW}(a^2C_1/2b\exp(2(a-b)/(2b-a))(x-C_2)))}{2b(\mathrm{LambertW}(a^2C_1/2b\exp(2(a-b)/(2b-a))(x-C_2))+1)}. \tag{17.6}$$

Here LambertW is the special function (see Figure 17.1) defined by

$$\mathrm{LambertW}(x)\exp(\mathrm{LambertW}(x)) = x.$$

From (17.6) we obtain the general solution of the differential equation (17.5):

$$y(x) = \frac{a}{2b}(x+c_2)\left(1+\frac{1}{\mathrm{LambertW}(c_1(x+c_2))}\right), \tag{17.7}$$

where

$$c_1 = \frac{1}{2b}a^2C_1\exp\left(-2\frac{b-a}{2b-a}\right), \qquad c_2 = -C_2.$$

So, the curve L_Γ has the form (17.7). We get the corresponding solution of (17.4) as the solution of the equation $f = 0$, i.e.,

$$bu_t - au_x = 0,$$

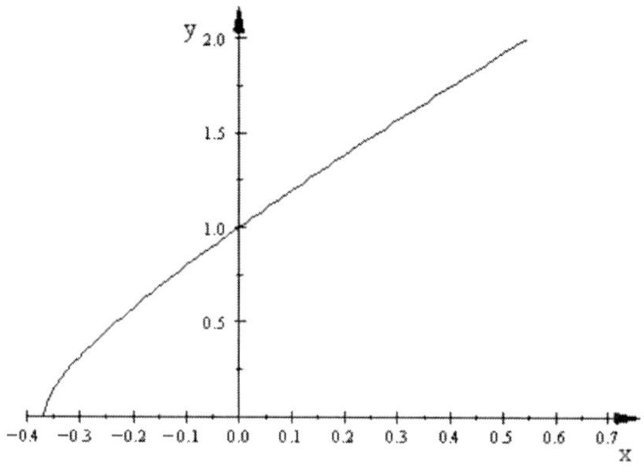

Figure 17.1. The function $y(x) = x(1 + (1/(\mathrm{LambertW}(x))))$.

with the Cauchy data:

$$u(0,x) = \frac{a}{2b}(x + c_2)\left(1 + \frac{1}{\text{LambertW}(c_1(x + c_2))}\right).$$

This solution is

$$u(t,x) = \frac{a}{2b^2}(at + bx + c_2)\left(1 + \frac{1}{\text{LambertW}(c_1(at + bx + c_2))}\right).$$

Below we list some other exact solutions of the non-linear diffusion equation. Here C, C_1 and C_2 are arbitrary constants.

1. The equation:

$$v_t = (v^\alpha v_x)_x. \tag{17.8}$$

The symmetry:

$$f = 2u - \alpha q_2 p_2.$$

The f-invariant solution:

$$v(t,x) = x^{2/\alpha}\left(C - \frac{2(\alpha + 2)}{\alpha}t\right)^{-1/\alpha}.$$

2. The equation:

$$v_t = (v^{-4/3} v_x)_x + \gamma v.$$

The symmetry:

$$f = q_2^2 p_2 + 3q_2 u.$$

The f-invariant solution:

$$v(t,x) = C\frac{\exp(\gamma t)}{x^3}.$$

3. The equation:

$$v_t = (v^\alpha v_x)_x + \gamma v.$$

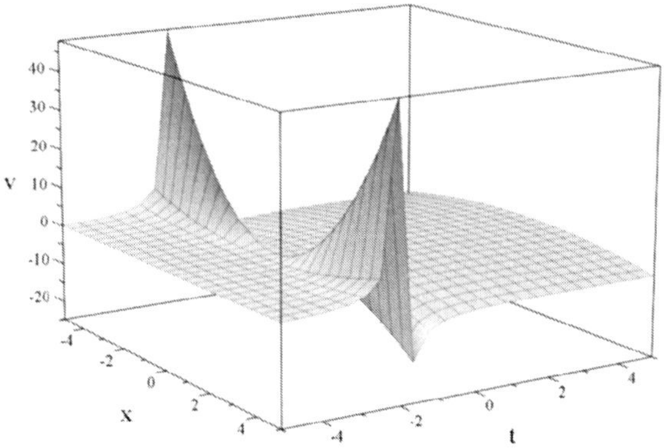

Figure 17.2. The equation $v_t = (vv_x)_x + v$, with the solution $v(t,x) = -\frac{1}{6}x^2 e^{t+2}/(e^{t+2} - 1)$.

The symmetry:

$$f = 2u - \alpha q_2 p_2.$$

The f-invariant solution:

$$v(t,x) = x^{2/\alpha} \exp((t + C)\gamma) \left(-\frac{1}{2} \frac{\alpha^2 \gamma}{(2 + \alpha)(\exp((t + C)\alpha\gamma) - 1)} \right)^{1/\alpha}.$$

Figure 17.2 presents the graph of the f-automodel solution.

4. The equation:

$$v_t = (vv_x)_x + \gamma v + \mu v^2.$$

The generating function of symmetry:

$$f = (p_1 - \gamma u) \exp(-\gamma q_1).$$

If $\mu > 0$, then the f-invariant solution:

$$v(t,x) = \exp(\gamma t) \sqrt[4]{\frac{2C_1}{\mu(\tan^2(C_2 - \sqrt{2}\sqrt{\mu}x) + 1)}}.$$

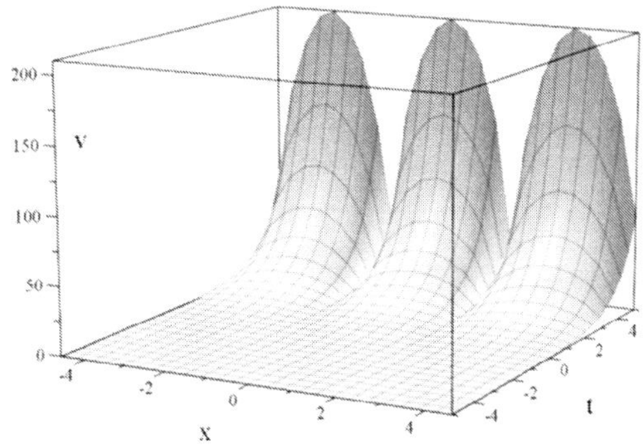

Figure 17.3. The equation $v_t = (vv_x)_x + v + \frac{1}{2}v^2$, with the solution $v(t,x) = \sqrt{2}e^t(\tan^2 x + 1)^{-1/4}$.

If $\mu < 0$, then

$$v(t,x) = \exp(\gamma t) \times \sqrt{\frac{1}{\sqrt{-2\mu}} \exp(-\sqrt{-2\mu}x - C_2)(\exp(2\sqrt{-2\mu}x + 2C_2) + C_1)}.$$

Figure 17.3 shows the solution.
Note that if we put $\gamma = 1$ and $\mu = -1$, we obtain the Fisher equation with a non-linear diffusion coefficient:

$$v_t = (vv_x)_x + v(1 - v).$$

The corresponding solution is

$$v(t,x) = \exp(t)\sqrt{\frac{1}{\sqrt{2}} \exp(-\sqrt{2}x - C_2)(\exp(2\sqrt{2}x + 2C_2) + C_1)}.$$

5. The equation:

$$v_t = (v^2 v_x)_x + \gamma v + \mu v^3.$$

The symmetry:

$$f = (p_1 - \gamma u) \exp(-2\gamma q_1).$$

If $\mu > 0$, then the f-invariant solution:

$$v(t,x) = \exp(\gamma t)(C_1 \cos(\sqrt{3\mu}x) - C_2 \sin(\sqrt{3\mu}x))^{1/3}.$$

If $\mu < 0$, then

$$v(t,x) = \exp\left(\gamma t - \frac{\sqrt{-3\mu}}{3}x\right)(C_1 \exp(2\sqrt{-3\mu}x) + C_2)^{1/3}.$$

6. The reduced equation for traveling wave solutions $v(t,x) = y(at + bx)$ of (17.2) is

$$(y^\alpha y')' - \frac{a}{b^2}y' + \frac{\mu}{b^2}y^\beta = 0.$$

Here $y = y(z)$, $z = at + bx$. This ODE inherits the symmetries (17.2)

$$X_1 = \frac{\partial}{\partial z},$$
$$X_2 = -z\frac{\partial}{\partial z} - \frac{1}{\alpha}y\frac{\partial}{\partial y} + \left(1 - \frac{1}{\alpha}\right)p\frac{\partial}{\partial p}.$$

For $\beta = 1 - \alpha$ and $a = 0$, we get a first integral

$$H = (y')^2 y^{2\alpha} + \frac{\mu}{b^2}y^2$$

of the reduced equation

$$(y^\alpha y')' + \frac{\mu}{b^2}y^{1-\alpha} = 0.$$

For more solutions of (17.1) see [49].

18

Meteorology applications

The results in this section are based on the recent contributions [93, 94] to the ongoing research into balanced models in geophysical fluid dynamics, in which the objective is to understand the properties of the equations that assist solution strategies. The relationships between the ellipticity of the operators, the Hamiltonian structure of the evolution equations and the stability of flows described by such balanced models are subtle and our understanding of these issues is far from complete. This chapter only illustartes how geometry can offer insights into some of these problems. For example, the quaternionic structure associated with a Monge–Ampère equation facilitates an understanding of why complex coordinates arise in a natural way in these models – something that would be difficult to deduce from the physics alone.

18.1 Shallow water theory and balanced dynamics

The motion of a typical particle in shallow water theory can be described by expressing the current Cartesian horizontal coordinates

$$x = x(a,b,t), \quad y = y(a,b,t) \tag{18.1}$$

as functions, on the right, of the Lagrangian particle labels a, b and the time t. The incompressibility hypothesis requires the current depth, or height, h to be a function $h(a,b,t)$ with the property

$$\frac{h(a,b,0)}{h(a,b,t)} = \frac{\partial(x,y)}{\partial(a,b)}, \tag{18.2}$$

where the Jacobian on the right is that of the mapping (18.1). The time derivative of (18.2) following a particle gives the differential continuity equation.

The equations of horizontal momentum balance for flows over a bed which is rotating with position-dependent Coriolis parameter $f(y)$ (where the y-coordinate denotes north) are

$$\begin{cases} \ddot{x} + g\dfrac{\partial h}{\partial x} - \dot{y}f = 0, \\ \ddot{y} + g\dfrac{\partial h}{\partial y} + \dot{x}f = 0. \end{cases} \tag{18.3}$$

Here g is a given constant, representing the combined effect of the acceleration due to gravity and a centrifugal component due to the Earth's rotation and the overdot denotes the time derivative following a particle.

Another important kinematic concept is the so-called potential vorticity defined by

$$\xi = \frac{1}{h}\left(\frac{\partial \dot{y}}{\partial x} - \frac{\partial \dot{x}}{\partial y} + f\right), \tag{18.4}$$

which is conserved on particles. Common choices of the Coriolis parameter are $f =$ constant or $f = \beta y$, ($\beta \in \mathbb{R}$) as approximations to $2\Omega \sin\phi$, depending on the purpose. Here β and Ω are constants[1] (related to the spin of the Earth) and ϕ is latitude.

The semi-geostrophic approximation to (18.3), in the case when f is a constant, is the replacement of the true acceleration by the time derivative of another

[1] Note that to avoid confusion we also use Ω to denote a symplectic structure.

vector

$$u_g = -\frac{g}{f}\frac{\partial h}{\partial y}, \qquad v_g = \frac{g}{f}\frac{\partial h}{\partial x}, \tag{18.5}$$

following the particle. This vector is a notional velocity, called the geostrophic velocity. The semi-geostrophic approximation seeks to find motions satisfying

$$\begin{cases} \dot{u}_g + g\dfrac{\partial h}{\partial x} - \dot{y}f = 0, \\ \dot{v}_g + g\dfrac{\partial h}{\partial y} + \dot{x}f = 0, \end{cases} \tag{18.6}$$

together with the continuity equation obtained from (18.2). Associated with these equations, the potential vorticity

$$\Xi = \frac{1}{h}\left[f + \frac{\partial v_g}{\partial x} - \frac{\partial u_g}{\partial y} + \frac{1}{f}\frac{\partial(u_g, v_g)}{\partial(x, y)}\right] \tag{18.7}$$

is conserved.

We introduce a transformation of coordinates

$$X = x + \frac{g}{f^2}\frac{\partial h}{\partial x}, \qquad Y = y + \frac{g}{f^2}\frac{\partial h}{\partial y}. \tag{18.8}$$

For a thorough discussion of this transformation, including its Legendre and contact properties, see Roulstone and Sewell [97]. The coordinates (X, Y) are called *geostrophic coordinates* because, when f is a constant, $\dot{X} = u_g$, $\dot{Y} = v_g$. Motivated by the issue of the dependence of the Coriolis parameter, f, on latitude, Salmon [101] studied certain *generalized* semi-geostrophic equations with pseudo-Hamiltonian form in (X, Y)-space, namely

$$\dot{X} = -\frac{1}{f(X,Y)}\frac{\partial \Psi}{\partial Y}, \qquad \dot{Y} = \frac{1}{f(X,Y)}\frac{\partial \Psi}{\partial X}, \tag{18.9}$$

where

$$\Psi(X, Y, t) = \frac{1}{2}(u_g^2 + v_g^2) + gh. \tag{18.10}$$

The results of Salmon [102] are also of interest when f is independent of position (in which case (18.9) can easily be written in canonical form) and henceforth, in this chapter, we consider f to be a given constant. In the coordinates (18.8) the potential vorticity takes the Jacobian form (noting (18.5))

$$\Xi = \frac{f}{h}\frac{\partial(X,Y)}{\partial(x,y)} = \frac{1}{h}\left[f + \frac{\partial v_g}{\partial x} - \frac{\partial u_g}{\partial y} + \frac{1}{f}\frac{\partial(u_g, v_g)}{\partial(x,y)}\right]. \tag{18.11}$$

The definition (18.11) can be rewritten as

$$\Xi = \frac{gf}{(H - \frac{1}{2}f(x^2+y^2))}\det \mathrm{Hess}(H), \tag{18.12}$$

where

$$H = \frac{1}{2}f(x^2+y^2) + gh$$

and

$$\mathrm{Hess}(H) = \begin{Vmatrix} \dfrac{\partial^2 H}{\partial x^2} & \dfrac{\partial^2 H}{\partial x \partial y} \\ \dfrac{\partial^2 H}{\partial y \partial x} & \dfrac{\partial^2 H}{\partial y^2} \end{Vmatrix}.$$

This is the starting point for a Monge–Ampère type equation in which $H(x, y, t)$ is the unknown, to be found in conjunction with suitable boundary conditions.

18.2 A geometric approach to semi-geostrophic theory

Consider fluid flow in a region $\mathcal{D} \subset \mathbb{R}^2$. The cotangent bundle $T^*\mathbb{R}^2$ has coordinates $q_1 \stackrel{\text{def}}{=} x$, $q_2 \stackrel{\text{def}}{=} y$, $p_1 = p$, $p_2 = q$ which span the phase space of a fluid obeying Newton's second law (the shallow water equations as above).

The transfromation

$$F : (q_1, q_2, p_1, p_2) \mapsto (Q_1 = q_1 + p_1,\ Q_2 = q_2 + p_2,\ P_1 = p_1,\ P_2 = p_2)$$

is a symplectic, and the function

$$f \stackrel{\text{def}}{=} \frac{1}{2}(p_1^2 + p_2^2)$$

is a generating function for F.

Taking a suitably scaled geopotential function ϕ in most applications, we obtain a natural lift F_c of the transformation F to the contact space of 1-jets:

$$F_c : F : (q_1, q_2, u, p_1, p_2) \mapsto \left(Q_1 = q_1 + p_1,\ Q_2 = q_2 + p_2, V = u + \frac{1}{2}(p_1^2 + p_2^2), \right.$$
$$\left. P_1 = p_1,\ P_2 = p_2 \right).$$

The map F_c is a contact transformation.

The graph of the geopotential ϕ is a Legendrian submanifold

$$L = \left\{ u = \phi, p_1 = \frac{\partial \phi}{\partial q_1}, p_2 = \frac{\partial \phi}{\partial q_2} \right\}$$

in $J^1\mathcal{D}$.

Let ω be an effective 2-form in $J^1\mathcal{D}$ such that being evaluated on the graph of the geopotential ϕ yields a Monge–Ampère equation

$$\Delta_\omega(\phi) = 0.$$

We note that the contact transformation F_c, when restricted to the graph of ϕ, is the *geostrophic momentum transformation*, and in meteorological terminology, the vector field

$$S = -p_2 \frac{\partial}{\partial q_1} + p_1 \frac{\partial}{\partial q_2} = u_g \frac{\partial}{\partial q_1} + v_g \frac{\partial}{\partial q_2}$$

is known as the *geostrophic wind.*

The semi-geostrophic equations can be written in the "pseudo-Hamiltonian" form

$$\dot{Q}_1 = -\frac{\partial \Phi}{\partial Q_2}, \qquad \dot{Q}_2 = \frac{\partial \Phi}{\partial Q_1}, \tag{18.13}$$

where

$$\frac{d}{dt} = \frac{\partial}{\partial t} + \dot{q}_1 \frac{\partial}{\partial q_1} + \dot{q}_2 \frac{\partial}{\partial q_2} = \frac{\partial}{\partial t} + \dot{Q}_1 \frac{\partial}{\partial Q_1} + \dot{Q}_2 \frac{\partial}{\partial Q_2}. \tag{18.14}$$

The mapping $\{q_1, q_2\} \mapsto \{Q_1, Q_2\}$ is a Legendre transformation

$$Q_1 = \psi_{q_1}, \quad Q_2 = \psi_{q_2},$$

where

$$\psi = \phi + \frac{1}{2}(q_1^2 + q_2^2)$$

and

$$q_1 = \Psi_{Q_1}, \quad q_2 = \Psi_{Q_2}$$

with $\Psi = \frac{1}{2}(Q_1^2 + Q_2^2) - \Phi$.

The singularities of this map can be interpreted as atmospheric fronts [15, 16].

The conservation law for the potential vorticity can be expressed (using the Hamiltonian representation of the semi-geostrophic equations and a continuity equation), as

$$\frac{d}{dt}\left(\frac{\partial(q_1, q_2)}{\partial(a, b)}\right) = \frac{d}{dt}(1 + \phi_{q_1q_1} + \phi_{q_2q_2} + \det \operatorname{Hess} \phi) = 0.$$

The Monge–Ampère operator corresponding to the conservation law can be described as a pull-back of 2-form $dQ_1 \wedge dQ_2 \in \Omega^2(J^1(\mathcal{D}))$ under the transformation F:

$$\begin{aligned} F^*(dQ_1 \wedge dQ_2) &= d(q_1 + p_1) \wedge d(q_2 + p_2) \\ &= dq_1 \wedge dq_2 + dq_1 \wedge dp_2 + dp_1 \wedge dq_2 + dp_1 \wedge dp_2 \end{aligned}$$

and the corresponding Monge–Ampère equation is

$$1 + \phi_{q_1q_1} + \phi_{q_2q_2} + \det \operatorname{Hess} \phi = 0.$$

Now we show how this equation relates to a classical geometric structure on the phase space $T^*\mathbb{R}^2 = \mathbb{R}^4$. Namely, it is a part of *hyper-Kähler triplet* of the Monge–Ampère equations introduced in [94, 95] which we will explain in the next section.

18.3 Hyper-Kähler structure and Monge–Ampère operators

Let $\omega \in \Omega^2_\varepsilon(\mathbb{R}^4)$ be an effective 2-form representing an elliptic Monge–Ampère operator $\Delta_\omega : \mathbb{C}^\infty \mapsto \Omega^2(\mathbb{R}^2)$ with constant coefficients.

Positivity of the Pfaffian $\operatorname{Pf}(\omega) > 0$ (in an open domain of a point $x \in \mathbb{R}^4$) provides the ellipticity condition for the corresponding Monge–Ampère operator Δ_ω and as was shown the existence of almost complex structure in $\mathbb{R}^4$.

We will identify the points $(x, p, y, q) \in \mathbb{R}^4$ with quaternions $l \in \mathbb{H}$:

$$l = x + \mathbf{i}p + \mathbf{j}y + \mathbf{k}q$$

and the points $(x, p), (y, q) \in \mathbb{R}^2$ with $z, w \in \mathbb{C} \times \mathbb{C} = \mathbb{C}^2$. Then using the standard multiplication of quaternions we have

$$l = z + \mathbf{j}\bar{w}, \quad z = x + \mathbf{i}p, \quad w = y + \mathbf{i}q.$$

Let

$$\bar{l} = x - \mathbf{i}p - \mathbf{j}y - \mathbf{k}q = \bar{z} - \mathbf{j}\bar{w}$$

be the quaternion conjugated to l. Then there are three Kähler forms $\omega_I, \omega_J, \omega_K \in \Omega^2(\mathbb{R}^4)$ such that

$$-\frac{1}{2} dl \wedge d\bar{l} = \mathbf{i}\omega_I + \mathbf{j}\omega_J + \mathbf{k}\omega_K.$$

These forms are expressed in the coordinates in $\mathbb{R}^4$ as

$$\omega_I = dx \wedge dp + dy \wedge dq, \quad \omega_J = dx \wedge dy + dq \wedge dp, \quad \omega_K = dx \wedge dq + dp \wedge dy.$$

The 2-forms have a many remarkable features, the most important of which is their *self-duality* with respect to the Hodge star-operator $* : \Omega^2(\mathbb{R}^4) \mapsto \Omega^2(\mathbb{R}^4)$ via the Euclidian standard metric on $\mathbb{R}^4$.

Let us consider the corresponding Monge–Ampère operators. It is easy to see that

$$\omega_I = dx \wedge dp + dy \wedge dq = \Omega$$

and hence $\Delta_{\omega_I} \equiv 0$.

The other two operators are elliptic:

$$\Delta_{\omega_K} = \Delta, \quad \Delta_{\omega_J} = 1 - \text{detHess},$$

where Δ is the usual Laplace operator in $\mathbb{R}^2$ and detHess is a determinant of the second derivatives matrix for a "potential" ϕ. It is easy to check that the transposed quaternion wedge product $d\bar{l} \wedge dl$ produces an anti-self-dual triple of 2-forms:

$$-\frac{1}{2}\bar{dl} \wedge dl = \mathbf{i}\tilde{\omega_I} + \mathbf{j}\tilde{\omega_J} + \mathbf{k}\tilde{\omega_K}.$$

$$\tilde{\omega_I} = dx \wedge dp - dy \wedge dq,$$

$$\tilde{\omega_J} = dx \wedge dy - dq \wedge dp,$$

$$\tilde{\omega_K} = dx \wedge dq - dp \wedge dy.$$

The latter triple encodes the hyperbolic Monge–Ampère operator with constant coefficients in $\mathbb{R}^2$. In complex coordinates z, w self-dual forms have the following expressions:

$$\omega_I = \frac{\iota}{2}(dz \wedge d\bar{z} + dw \wedge d\bar{w}),$$

$$\omega_J = \frac{1}{2}(dz \wedge dw + d\bar{z} \wedge d\bar{w}),$$

$$\omega_K = -\frac{\iota}{2}(dz \wedge dw - d\bar{z} \wedge d\bar{w}),$$

and we have that

$$\omega_J + \iota\omega_K = dz \wedge dw \in \Omega^{2,0}(\mathbb{C}^2)$$

is a holomorphic symplectic form on $\mathbb{C}^2$ with respect to the complex structure **i**.

We consider $\mathbb{C}^2$ as a hyper-Kähler manifold endowed with three complex structures I, J, K subjected to quaternionic algebra relations. The twistor space of $\mathbb{C}^2$ is the product $P : \mathbb{C}P^1$. It admits the structure of a complex contact three-dimensional manifold.

As we mentioned above, the set of all elliptic Monge–Ampère operators are in one-to-one correspondence with the set of all (almost) complex structures in $\mathbb{C}^2$ and this correspondence is realized as a homeomorphism between the Grassmanian LG_ω associated with an elliptic Monge–Ampère operator Δ_ω and the projective space $\mathbb{C}P^1$. The precise correspondence goes as follows. Let us parametrize the points of $\mathbb{C}P^1 = \mathbb{S}^2$ by a non-zero complex number ξ such that the Cartesian coordinates $(a, b, c) \in \mathbb{R}^3, a^2 + b^2 + c^2 = 1$ are related to ξ by

$$(a, b, c) = \left(\frac{(1 - \xi\bar{\xi})}{1 + \xi\bar{\xi}}, \frac{\xi + \bar{\xi}}{1 + \xi\bar{\xi}}, \frac{\iota(\xi - \bar{\xi})}{1 + \xi\bar{\xi}}\right).$$

Denote the corresponding complex structure endomorphism by I_ξ:

$$I_\xi : \begin{pmatrix} z \\ w \end{pmatrix} \mapsto \begin{pmatrix} \dfrac{\iota(1-\xi\bar{\xi})}{1+\xi\bar{\xi}} z + \dfrac{2\bar{\xi}}{1+\xi\bar{\xi}} w \\ \dfrac{-2\xi}{1+\xi\bar{\xi}} z - \dfrac{\iota(1-\xi\bar{\xi})}{1+\xi\bar{\xi}} w \end{pmatrix}.$$

It is easy to check that the self-dual triple $\omega_I, \omega_J, \omega_K$ is invariant with respect to the transformation induced by I_ξ:

$$I_\xi^*(\omega_I) = \omega_I, \quad I_\xi^*(\omega_J) = \omega_J, \quad I_\xi^*(\omega_K) = \omega_K.$$

The "general" elliptic Monge–Ampère operator Δ_{ω_ξ} corresponding to the point $\xi \in \mathbb{C}P^1$ is defined by the effective 2-form

$$\omega_\xi = \frac{\xi + \bar{\xi}}{1+\xi\bar{\xi}} \omega_J + \frac{\iota(\xi - \bar{\xi})}{1+\xi\bar{\xi}} \omega_K.$$

Its Pfaffian is

$$\operatorname{Pf}(\omega_\xi) = \frac{4(\alpha^2+\beta^2)}{(1+\alpha^2+\beta^2)^2},$$

where $\xi = \alpha + \iota\beta$.

Introduce a new "complex coordinates" associated with I_ξ:

$$z_\xi = z + \xi\bar{w}, \quad w_\xi = w - \xi\bar{z}.$$

We have

$$\begin{aligned} dz_\xi \wedge dw_\xi &= d(z+\xi\bar{w}) \wedge d(w - \xi\bar{z}) \\ &= dz \wedge dw - \xi(dz \wedge d\bar{z} + dw \wedge d\bar{w}) + \xi^2\, d\bar{z} \wedge d\bar{w}) \\ &= \omega_J + \iota\omega_K + 2\iota\xi\omega_I + \xi^2(\omega_J - \iota\omega_K) \\ &= (1+\xi^2)\omega_J + (1-\xi^2)\iota\omega_K + 2\iota\xi\omega_I. \end{aligned}$$

The corresponding effective 2-form is given by a quadratic pencil of Monge–Ampère operators

$$\omega^\xi = (1+\xi^2)\omega_J + (1-\xi^2)\iota\omega_K$$

within the initial hyper-Kähler triplet of forms.

Now we consider the canonical transformation for "near-local balanced models" (see the next subsection):

$$X = x + \iota c q + a p, \quad Y = y - \iota c p + a q, \tag{18.15}$$

where $a, c \in \mathbb{R}$, $\iota = \sqrt{-1}$ and $(p, q) = (g\,\partial h/\partial x, g\,\partial h/\partial y)$ if we define $\hat{z} = x + \iota c q$, $\hat{w} = y - \iota c p$ on a two-dimensional complex manifold, then there exists a hyper-Kähler triple of 2-forms

$$\hat{\omega}_I = \Re(d\hat{z} \wedge d\hat{w}) = dx \wedge dy - c^2\, dp \wedge dq, \tag{18.16}$$

$$\hat{\omega}_J = \frac{\iota}{2}(d\hat{z} \wedge d\hat{\bar{z}} + d\hat{w} \wedge d\hat{\bar{w}}) = c(dx \wedge dq + dp \wedge dy), \tag{18.17}$$

$$\hat{\omega}_K = \Im(d\hat{z} \wedge \hat{w}) = -c(dx \wedge dp + dy \wedge dq) = -c\Omega, \tag{18.18}$$

where Ω is the canonical 2-form. The first two 2-forms, (18.16) and (18.17), represent Legendre-conjugate elliptic Monge–Ampère operators. The transformation $(\hat{z}, \hat{w}) \mapsto (X, Y)$, implicit in (18.15), has the following action on the hyper-Kähler triple:

$$\hat{\omega}_I \mapsto dx \wedge dy + a(dx \wedge dq + dp \wedge dy) + (a^2 - c^2)\, dp \wedge dq, \tag{18.19}$$

$$\hat{\omega}_J \mapsto c(dx \wedge dq + dp \wedge dy) + 2ac\, dp \wedge dq, \tag{18.20}$$

$$\hat{\omega}_K \mapsto \hat{\omega}_K \;\; (= -c\Omega). \tag{18.21}$$

18.4 Monge–Ampère operators with constant coefficients and plane balanced models

The main meteorological output of this geometric description is that the general family of the (elliptic) Monge–Ampère operators with constant coefficients contains all known plane balance models:

$$1 + \phi_{q_1 q_1} + a\phi_{q_2 q_2} + (a^2 - c^2)\det \mathrm{Hess}\,\phi = \zeta^{\mathbf{C}}/f, \tag{18.22}$$

where $\zeta^{\mathbf{C}} = h\Xi$ is *constrained vorticity* (see the discussion of this notion in [82]).

It also gives a geometric explanation to the rather miraculous appearance of a complex structure in one such model.

The ellipticity condition for this general Monge–Ampère operator is satisfied for the models with

$$(c^2 - 1)\zeta^{\mathbf{C}}/f < c^2.$$

The most popular among them correspond to the following choice of parameters:

(a) the semi-geostrophic model ($a = 1,\ c = 0$ for a positive $\zeta^{\mathbf{C}}/f$);
(b) the *L_1-dynamics of Salmon* [100] with $a = c = 1$;
(c) the *$\sqrt{3}$-dynamics of McIntyre and Roulstone* [81] which corresponds to $a = 1, c = \sqrt{3}$ if $\zeta^{\mathbf{C}}/f < \frac{3}{2}$.

Our classification theorem in $d = 2$ gives in the same time a geometric description of all plane "near-balanced models ($0 < c < \sqrt{3}$) with uniform potential vorticity if $\zeta^{\mathbf{C}}/f < \frac{3}{2}$.

They:

- correspond geometrically to the points of the Monge–Ampère elliptic Lagrangian Grassmanian ($\mathbb{CP}^1$). In other words, all such models correspond to a choice of complex structure on $\mathbb{R}^4$ which is parametrized by $\mathbb{C}P^1$;
- all of them are *linearizable* by the choice of a proper canonical coordinate transformation.

PART V

Classification of Monge–Ampère equations

19

Classification of symplectic MAOs on two-dimensional manifolds

The problem of equivalence and classification of Monge–Ampère equations goes back to Sophus Lie's papers from the 1870s and 1880s (see [65, 66, 67]). Sophus Lie have raised the following problem. *Find equivalence classes of non-linear second-order differential equations with respect to the group of contact transformations.*

Sophus Lie himself had found conditions to transform a Monge–Ampère equation (MAE) to a quasi-linear one and to a linear equation with constant coefficients. The important steps in a solution of this problem were made by

Darboux and Goursat [29], who had basically treated the hyperbolic Monge–Ampère equations.

As far as we know, a complete proof of Lie's theorems had never been published. The first results in this direction were obtained in [73, 74, 77].

In this part we consider the problem of local equivalence for Monge–Ampère equations and Monge–Ampère operators.

In [53, 54, 55, 59, 60, 61] it was shown that for Monge–Ampère equations of general type this problem can be reduced to the equivalence problem for e-structures. This leads to a solution of the equivalence problems and to a classification of Monge–Ampère equations and Monge–Ampère operators.

In this chapter we consider the problem of classification of symplectic Monge–Ampère operators of hyperbolic, elliptic and mixed types.

Let ω_1 and ω_2 be two effective differential 2-forms on the cotangent bundle T^*M.

We say that two symplectic Monge–Ampère operators Δ_ω and $\Delta_{\tilde{\omega}}$ (or two symplectic effective differential 2-forms ω and $\tilde{\omega}$) are equivalent if there exists a symplectic transformation φ of T^*M, such that $\varphi^*(\omega) = \tilde{\omega}$. We say that two Monge–Ampère operators Δ_ω and $\Delta_{\tilde{\omega}}$ (or two effective differential 2-forms ω and $\tilde{\omega}$) are locally symplectic equivalent at a point $a \in T^*M$ if there exists a symplectic transformation φ of some neighborhood $\mathcal{O}_a$ of a such that $\varphi^*(\omega) = \tilde{\omega}$ and $\varphi(a) = a$.

19.1 e-Structures

Recall that an *e-structure* (or an *absolute parallelism*) on a smooth n-dimensional manifold Φ is a collection of n vector fields $X_1, \ldots, X_n$ that are linearly independent at each point. This means that $X_1, \ldots, X_n$ form a global basis of the module $D(\Phi)$.

Let $X_1, \ldots, X_n$ be an e-structure on the n-dimensional smooth manifold Φ. Since the vector fields $X_1, \ldots, X_n$ are linearly independent at each point then commutators $[X_i, X_j]$ are linear combinations of the fields and we obtain *structure functions* of the e-structures. They are functions f_{ij}^k such that

$$[X_i, X_j] = \sum_{k=1}^{n} f_{ij}^k X_k.$$

Here we restrict ourselves to the case when $n = 4$ and Φ is a symplectic manifold with a structure form Ω.

Let us define differential 1-forms on Φ by

$$\theta_1 = -X_2 \rfloor \Omega, \qquad \theta_2 = X_1 \rfloor \Omega, \qquad \theta_3 = -X_4 \rfloor \Omega, \qquad \theta_4 = X_3 \rfloor \Omega.$$

Assume that the vector fields $X_1, \ldots, X_4$ form the four-dimensional Lie algebras over $\mathbb{R}$:

$$\mathfrak{g} = \bigoplus_{i=1}^{4} \mathbb{R}X_i.$$

Then the structure functions f_{ij}^k are constant and we denote f_{ij}^k by c_{ij}^k.

The Maurer–Cartan equations take the form

$$d\theta_k = -\sum_{1 \leqslant i < j \leqslant 4} c_{ij}^k \theta_i \wedge \theta_j.$$

Indeed, since the bases $X_1, \ldots, X_4$ and $\theta_1, \ldots, \theta_4$ are dual, we get

$$d\theta_k(X_i, X_j) = X_i(\theta_k(X_j)) - Y(\theta_k(X_i)) - \theta_k([X_i, X_j]) = -\theta_k([X_i, X_j]) = -c_{ij}^k.$$

Proposition 19.1.1 *Let $X_1^1, \ldots, X_4^1$ and $X_1^2, \ldots, X_4^2$ be two e-structures on a four-dimensional smooth symplectic manifold Φ. Suppose that they form four-dimensional Lie algebras $\mathfrak{g}_1$ and $\mathfrak{g}_2$ over $\mathbb{R}$. Suppose also that the corresponding structure constant of $\mathfrak{g}_1$ and $\mathfrak{g}_2$ coincide. Then for any point $a \in \Phi$ there exists its neighborhood $\mathcal{O}_a$ and a preserving the point a local diffeomorphism $\varphi : \mathcal{O}_a \to \varphi(\mathcal{O}_a)$ such that $\varphi^*(\theta_i^2) = \theta_i^1$ and $\varphi_*(X_i^1) = X_i^2 (i = 1, \ldots, 4)$.*

Proof Instead of the diffeomorphism φ we will construct its graph Γ. Consider the natural projection $\pi : \Phi^2 \to \Phi$ and denote $\bar{\theta}_i^l = \pi^*(\theta_i^l)$. Consider on Φ^2 a four-dimensional distribution $\mathcal{F}\langle \bar{\theta}_1^1 - \bar{\theta}_1^2, \ldots, \bar{\theta}_4^1 - \bar{\theta}_4^2 \rangle$. This distribution is completely integrable. Indeed

$$\begin{aligned} d(\bar{\theta}_k^1 - \bar{\theta}_k^2) &= -\sum_{1 \leqslant i < j \leqslant 4} c_{ij}^k (\bar{\theta}_i^1 \wedge \bar{\theta}_j^1 - \bar{\theta}_i^2 \wedge \bar{\theta}_j^2) \\ &= -\sum_{1 \leqslant i < j \leqslant 4} c_{ij}^k (\bar{\theta}_i^1 \wedge (\bar{\theta}_j^1 - \bar{\theta}_i^2) + (\bar{\theta}_i^1 - \bar{\theta}_i^2) \wedge \bar{\theta}_j^2). \end{aligned}$$

The point (a, a) belongs to one of the maximal integral manifolds of $\mathcal{F}$, say Γ. Let $\gamma : \Gamma \to \Phi$ be the projection on Φ. Then $\gamma^*(\theta_i^2) = \bar{\theta}_i^2|_\Gamma = \bar{\theta}_i^1|_\Gamma$. We have

$$(\gamma^{-1})^* \pi^*(\theta_i^2) = \theta_i^1$$

and the map $\pi_1 \circ \gamma^{-1}$ is required. ■

19.2 Classification of non-degenerate Monge–Ampère operators

19.2.1 Differential invariants of non-degenerate operators

Let ω be an effective differential 2-form such that $\mathrm{Pf}(\omega) = \varepsilon = \pm 1$ and let A be the operator that corresponds to ω. Then $A^2 = \varepsilon$ and the complexification of the tangent space $T_a(T^*M)$ at a point $a \in T^*M$ splits in to the direct sum of two skew-orthogonal complex symplectic planes:

$$T_a(T^*M)^{\mathbb{C}} = V_+(a) \oplus V_-(a), \tag{19.1}$$

where

$$V_\pm(a) = \{X \in T_a(T^*M)^{\mathbb{C}} \mid A_a^{\mathbb{C}} X = \mp\iota\sqrt{\varepsilon} X\}.$$

Let us denote by $V_1 = V_+$ and $V_2 = V_-$ the corresponding distributions, and $V_j^{(i)}$ their derivations. Let us assume that $V_j^{(i)}$ are also distributions . Then for each $j = 1, 2$ one has the following choices:

(0) $V_j^{(1)} = V_j$, that is V_j is a completely integrable distribution;
(1) $V_j \neq V_j^{(1)}$ but $V_j^{(1)} = V_j^{(2)}$, that is V_j is not completely integrable but $V_j^{(1)}$ is a three-dimensional completely integrable distribution; and
(2) $V_j \neq V_j^{(1)} \neq V_j^{(2)}$, distributions $V_j, V_j^{(1)}$ and $V_j^{(2)}$ are not completely integrable.

We say that the symplectic Monge–Ampère operator (or equation) *belongs to class* $H_{k,l}$ at a point $a \in T^*M (k, l = 0, 1, 2; k \leq l)$ if case (k) holds for one of V_j and case (l) holds for the other.

Classes $H_{k,l}$ are obviously invariant under symplectic transformations.

As we have seen in Part I, there are the following decompositions of the de Rham complex:

$$\Omega^s(T^*M)^{\mathbb{C}} = \bigoplus_{p+q=s} \Omega^{p,q}(T^*M),$$

$$d = d_{1,0} \oplus d_{0,1} \oplus d_{2,-1} \oplus d_{-1,2},$$

where $\Omega^{p,q}(T^*M) = \Omega^p(V_+) \otimes \Omega^q(V_-)$ and $d_{i,j} : \Omega^{p,q}(T^*M) \to \Omega^{p+i,q+k}(T^*M)$.

Remark that $d_{-1,2}$ and $d_{2,-1}$ are tensor invariants of the Monge–Ampère equation E_ω.

Proposition 19.2.1

1. *The distribution* V_+ *is completely integrable if and only if* $d_{2,-1} = 0$.
2. *The distribution* V_- *is completely integrable if and only if* $d_{-1,2} = 0$.

Proof The distribution V_+ is generated by a pair of 1-forms $\alpha_-, \beta_- \in \Omega^{0,1}(T^*M)$, with $\alpha_- \wedge \beta_- \neq 0$. This distribution is completely integrable if and only if $d\alpha_- = \alpha_- \wedge \gamma_1 + \beta_- \wedge \delta_1$ and $d\beta_- = \alpha_- \wedge \gamma_2 + \beta_- \wedge \delta_2$ for some 1-forms $\gamma_i, \delta_i \in \Omega^1(T^*M)^{\mathbb{C}}$ $(i = 1, 2)$. The latter means that $d\alpha_-, d\beta_- \notin \Omega^{2,0}(T^*M)$. ■

The formula

$$W_\omega \rfloor \Omega^2 = 2d\omega$$

uniquely determines the real vector field W_ω. Let μ_ω be the the dual (real) differential 1-form

$$\mu_\omega \stackrel{\text{def}}{=} W_\omega \rfloor \Omega.$$

Using decomposition (19.1), we obtain $W_w = W_+ + W_-$, where $W_+ \in D(V_+)$ and $W_- \in D(V_-)$. Since the distributions V_+ and V_- are skew-orthogonal, 1-forms $W_+ \rfloor \Omega^{\mathbb{C}}$ and $W_- \rfloor \Omega^{\mathbb{C}}$ belong to $\Omega^{1,0}(T^*M)$ and $\Omega^{0,1}(T^*M)$ correspondingly. Denote them by

$$\mu_+ \stackrel{\text{def}}{=} W_+ \rfloor \Omega, \qquad \mu_- \stackrel{\text{def}}{=} W_- \rfloor \Omega.$$

Two three-dimensional distributions $\ker \mu_+$ and $\ker \mu_-$ define the two-dimensional distribution $\ker \mu_+ \cap \ker \mu_-$. Note that

$$\mu_+(W_+) = \mu_+(W_-) = \mu_-(W_+) = \mu_-(W_-) = 0,$$

therefore the distribution $\ker \mu_+ \cap \ker \mu_-$ is generated by the vector fields W_+ and W_-: $\ker \mu_+ \cap \ker \mu_- = \mathcal{F}\langle W_+, W_- \rangle$.

Let Q be their commutator

$$Q \stackrel{\text{def}}{=} [W_+, W_-].$$

Then using decomposition (19.1) again, we get two vector fields $Q_+ \in D(V_+)^{\mathbb{C}}$ and $Q_- \in D(V_-)^{\mathbb{C}}$ such that $Q = Q_+ + Q_-$.

Suppose now that the three-dimensional distributions $\ker \mu_+$ and $\ker \mu_-$ are both integrable, i.e., in a some neighborhood $\mathcal{O}_a$ of a point a the equation belongs to the class $H_{1,1}$.

Since this distribution $\mathcal{F}\langle W_+, W_-\rangle$ is also completely integrable, one can define two functional invariants g_+ and g_- of the form ω by the following formula:

$$[W_+, W_-] = g_+ W_+ + g_- W_-.$$

Distributions $\ker \mu_+$ and $\ker \mu_-$ are completely integrable, therefore

$$\mu_+ \wedge d\mu_+ = \mu_- \wedge d\mu_- = 0.$$

Then

$$\mu_+ \wedge (W_+ \rfloor d\mu_+) = W_+ \rfloor (\mu_+ \wedge d\mu_+) = 0,$$

i.e. the 1-forms μ_+ and $W_+ \rfloor d\mu_+$ are linear dependent. Therefore

$$W_+ \rfloor d\mu_+ = g_0 \mu_+$$

for some function g_0.

This function is an invariant of ω. Note also that $W_- \rfloor d\mu_- = -g_0 \mu_-$.

Since $\mu_+ \in \Omega^{1,0}$, we have

$$d\mu_+ = d_{1,0}\mu_+ + d_{0,1}\mu_+ + d_{-1,2}\mu_+.$$

and $\mu_+ \wedge d_{1,0}\mu_+ = 0$.

Then,

$$\mu_+ \wedge d\mu_+ = \mu_+ \wedge d_{0,1}\mu_+ + \mu_+ \wedge d_{-1,2}\mu_+.$$

Since $\mu_+ \wedge d\mu_+ = 0$, $\mu_+ \wedge d_{0,1}\mu_+ \in \Omega^{2,1}$ and $\mu_+ \wedge d_{-1,2}\mu_+ \in \Omega^{1,2}$, we see that

$$\mu_+ \wedge d_{0,1}\mu_+ = 0 \tag{19.2}$$

and $\mu_+ \wedge d_{-1,2}\mu_+ = 0$.

Since $d_{-1,2}\mu_+ \in \Omega^{0,2}$ and $\mu_+ \in \Omega^{1,0}$, the last equality is realized if and only if $d_{-1,2}\mu_+ = 0$.

Then,

$$d\mu_+ = d_{1,0}\mu_+ + d_{0,1}\mu_+. \tag{19.3}$$

In the similar way we obtain

$$d\mu_- = d_{1,0}\mu_- + d_{0,1}\mu_-. \tag{19.4}$$

From (19.2) it follows that

$$d_{0,1}\mu_+ = \mu_+ \wedge \gamma_-,$$

for some uniquely determined differential 1-form $\gamma_- \in \Omega^{0,1}$. In a similar way we obtain a uniquely determined differential 1-form $\gamma_+ \in \Omega^{1,0}$ such that

$$d_{1,0}\mu_- = \mu_- \wedge \gamma_+.$$

We denote by X_+ and X_- the dual vector fields:

$$X_\pm \rfloor \Omega = \gamma_\pm. \tag{19.5}$$

Lemma 19.2.1 *$\gamma_-(W_-) = g_+$ and $\gamma_+(W_+) = -g_-$.*

Proof Since (19.3) and the fact that

$$W_- \rfloor d_{1,0}\mu_+ = W_+ \rfloor d_{0,1}\mu_- = 0,$$

we obtain

$$W_- \rfloor d\mu_+ = W_- \rfloor d_{0,1}\mu_+ = W_- \rfloor (\mu_+ \wedge \gamma_-) = -\gamma_-(W_-)\mu_+$$

and

$$W_+ \rfloor d\mu_- = W_+ \rfloor d_{1,0}\mu_- = W_+ \rfloor (\mu_- \wedge \gamma_+) = -\gamma_+(W_+)\mu_-.$$

Moreover

$$\begin{aligned} [W_+, W_-] \rfloor \Omega &= [L_{W_+}, \iota_{W_-}](\Omega) \\ &= L_{W_+}(W_- \rfloor \Omega) - W_- \rfloor L_{W_+}(\Omega) \\ &= W_+ \rfloor d\mu_- - W_- \rfloor d\mu_+ \\ &= -\gamma_+(W_+)\mu_- + \gamma_-(W_-)\mu_+. \end{aligned}$$

On the other hand,

$$[W_+, W_-] \rfloor \Omega = g_+\mu_+ + g_-\mu_-.$$

Therefore, $g_+ = \gamma_-(W_-)$ and $g_- = -\gamma_+(W_+)$. ■

19.2.2 Hyperbolic operators

In the hyperbolic case we can assume that above functions, vector fields, differential forms and tensor fields are real.

Consider two cases.

Case 1

Vectors $W_{+,a}$, $W_{-,a}$, $Q_{+,a}$, and $Q_{-,a}$ are linear independent at a point $a \in T^*M$.

In this case the vector fields W_+, W_-, Q_+, and Q_- are linear independent in some neighborhood $\mathcal{O}_a$ of a.

We define two functional invariants of the form ω:

$$r_+ \stackrel{\text{def}}{=} \Omega(W_+, Q) \qquad \text{and} \qquad r_- \stackrel{\text{def}}{=} \Omega(W_-, Q).$$

From non-degeneracy of the form Ω on the distributions V_+ and V_- it follows that the functions $r_\pm$ do not vanish in some neighborhood of the point a. Therefore one can normalize the vector fields:

$$X_1 \stackrel{\text{def}}{=} W_+, \qquad X_2 \stackrel{\text{def}}{=} \frac{1}{r_+} Q_+$$

$$X_3 \stackrel{\text{def}}{=} W_-, \qquad X_4 \stackrel{\text{def}}{=} \frac{1}{r_-} Q_-.$$

We have

$$\Omega(X_1, X_2) = \Omega(X_3, X_4) = 1$$

and since the distributions V_+ and V_- are skew-orthogonal,

$$\Omega(X_1, X_3) = \Omega(X_1, X_4) = \Omega(X_2, X_3) = \Omega(X_2, X_4) = 0.$$

Define differential 1-forms

$$\theta_1 \stackrel{\text{def}}{=} -X_2 \rfloor \Omega, \qquad \theta_3 \stackrel{\text{def}}{=} -X_4 \rfloor \Omega,$$

$$\theta_2 \stackrel{\text{def}}{=} X_1 \rfloor \Omega, \qquad \theta_4 \stackrel{\text{def}}{=} X_3 \rfloor \Omega$$

$(i = 1, \dots, 4)$.

The local bases X_1, X_2, X_3, X_4 and $\theta_1, \theta_2, \theta_3, \theta_4$ of $D(T^*M)$ and $\Omega^1(T^*M)$ are dual and we have the following representation of the symplectic form:

$$\Omega = \theta_1 \wedge \theta_2 + \theta_3 \wedge \theta_4.$$

Let us now find values of ω on the vector fields X_1, X_2, X_3, X_4. Since $X_1, X_2 \in D(V_+)$ and $X_3, X_4 \in D(V_-)$, we see that $AX_1 = X_1$, $AX_2 = X_2$, $AX_3 = -X_3$ and $AX_4 = -X_4$, where $A = A_\omega$.

Therefore

$$\omega(X_1, X_2) = \Omega(AX_1, X_2) = \Omega(X_1, X_2) = 1.$$

In the similar way we find values of the form ω on the rest of the pairs of the vector fields:

$\omega(\uparrow, \leftarrow)$	X_1	X_2	X_3	X_4
X_1	0	1	0	0
X_2	-1	0	0	0
X_3	0	0	0	-1
X_4	0	0	-1	0

This gives us the following representation of the form ω:

$$\omega = \theta_1 \wedge \theta_2 - \theta_3 \wedge \theta_4.$$

The following theorem provides the canonical representation of hyperbolic effective 2-forms.

Theorem 19.2.1 *Let ω be a hyperbolic normed effective differential 2-form on T^*M and $a \in T^*M$. Suppose that the vectors $W_{+,a}$, $W_{-,a}$, $Q_{+,a}$ and $Q_{-,a}$ are linear independent at the point a. Then in a some neighborhood of a there exists an e-structure $X_1, \ldots, X_4$ such that we have the following representation of the forms Ω and ω:*

$$\Omega = \theta_1 \wedge \theta_2 + \theta_3 \wedge \theta_4, \qquad (19.6)$$

$$\omega = \theta_1 \wedge \theta_2 - \theta_3 \wedge \theta_4.$$

Case 2

In a neighborhood $\mathcal{O}_a \ni a$ the Monge–Ampère operator belongs to the class $H_{1,1}$. In this case the three-dimensional distributions $\ker \mu_+$ and $\ker \mu_-$ are integrable in $\mathcal{O}_a$.

As a corollary of Lemma 19.2.1 one get a table of values for the 1-forms $\mu_+, \mu_-, \gamma_+, \gamma_-$ on the vector fields W_+, W_-, X_+, X_-:

	W_+	X_+	W_-	X_-
μ_+	0	g_-	0	0
γ_+	$-g_-$	0	0	0
μ_-	0	0	0	$-g_+$
γ_-	0	0	g_+	0

Lemma 19.2.2 $\mu_+ \wedge \gamma_+ \wedge \mu_- \wedge \gamma_- = g_+ g_- \Omega^2$.

Proof We have:

$W_+ \rfloor \Omega^2 = 2\mu_+ \wedge \Omega;$

$W_- \rfloor (W_+ \rfloor \Omega^2) = 2W_- \rfloor (\mu_+ \wedge \Omega) = 2\mu_- \wedge \mu_+;$

$X_+ \rfloor (W_- \rfloor (W_+ \rfloor \Omega^2)) = 2X_+ \rfloor (\mu_- \wedge \mu_+) = 2g_- \mu_-;$

$\Omega^2(W_+, W_-, X_+, X_-) = 2g_- X_- \rfloor \mu_- = -2g_+ g_-.$ ■

Corollary 19.2.1 *The vector fields W_+, W_-, X_+, X_- are linear independent in a neighborhood of the point $a \in T^*M$ if and only if $g_+ g_- \neq 0$ at the point.*

Suppose now that $(g_+ g_-)(a) \neq 0$. Due to Lemma 19.2.2, the vector fields W_+, W_-, X_+, X_- form a basis of the module $D(T^*M)$ in some neighborhood of the point a.

Let us consider the normed basis of $D(T^*M)$:

$$X_1 \stackrel{\text{def}}{=} W_+, \quad X_2 \stackrel{\text{def}}{=} \frac{1}{g_-} X_+,$$

$$X_3 \stackrel{\text{def}}{=} W_-, \quad X_4 \stackrel{\text{def}}{=} -\frac{1}{g_+} X_-$$

and a basis of the module $\Omega^1(T^*M)$ of differentia 1-forms on T^*M:

$$\theta_1 \stackrel{\text{def}}{=} -X_2 \rfloor \Omega, \quad \theta_2 \stackrel{\text{def}}{=} X_1 \rfloor \Omega,$$

$$\theta_3 \stackrel{\text{def}}{=} -X_4 \rfloor \Omega, \quad \theta_4 \stackrel{\text{def}}{=} X_3 \rfloor \Omega.$$

These bases are dual and the symplectic structure Ω and the effective form ω have the following canonical forms:

$$\Omega = \theta_1 \wedge \theta_2 + \theta_3 \wedge \theta_4,$$

$$\omega = \theta_1 \wedge \theta_2 - \theta_3 \wedge \theta_4.$$

Local coordinates

Let us find a coordinate representation of hyperbolic Monge–Ampère operators of $H_{1,1}$ class.

Note that for an effective differential 2-form

$$\omega = E\, dq_1 \wedge dq_2 + B(dq_1 \wedge dp_1 - dq_2 \wedge dp_2) + C\, dq_1 \wedge dp_2 - A dq_2 \wedge dp_1$$

we have

$$W_\omega = a\frac{\partial}{\partial q_1} + b\frac{\partial}{\partial q_2} + c\frac{\partial}{\partial p_1} + e\frac{\partial}{\partial p_2},$$

where

$$a = A_{p_2} - B_{p_1} - D_{q_2}, \qquad b = B_{p_2} - C_{p_1} + D_{q_1},$$

$$c = B_{q_1} + C_{q_2} - E_{p_2}, \qquad e = E_{p_1} - A_{q_1} - B_{q_2}.$$

Theorem 19.2.2 *Any normed hyperbolic 2-form ω of $H_{1,1}$ class is (locally) symplectic equivalent to the differential 2-form*

$$-2f\, dq_1 \wedge dq_2 + dq_1 \wedge dp_1 - dq_2 \wedge dp_2,$$

where $f = f(q,p)$ is a smooth function.

Proof Let H_+ and H_- be integrals of the distributions $\mathcal{F}\langle\mu_+\rangle$ and $\mathcal{F}\langle\mu_-\rangle$ respectively. Without loss generality, we can suppose that $H_+ = q_1, H_- = q_2$. Then $\mu_+ = h_+\, dq_1$ and $\mu_- = h_-\, dq_2$ for some functions $h_\pm$, and the vector fields W_+ and W_- have the forms $W_+ = -h_+\partial/\partial p_1$ and $W_- = -h_-\partial/\partial p_2$.

Note that $\omega(W_+, W_-) = 0$, and therefore the form ω does not contain the term $dp_1 \wedge dp_2$. Then the distributions V_+ and V_- have the following forms:

$$V_+ = \mathcal{F}\left\langle X_+ = (1+B)\frac{\partial}{\partial q_1} + C\frac{\partial}{\partial q_2} - E\frac{\partial}{\partial p_2}, \quad Y_+ = (1+B)\frac{\partial}{\partial p_1} - A\frac{\partial}{\partial p_2}\right\rangle,$$

$$V_- = \mathcal{F}\left\langle X_- = (1-B)\frac{\partial}{\partial q_1} - C\frac{\partial}{\partial q_2} + E\frac{\partial}{\partial p_2},\right.$$
$$\left. Y_- = A\frac{\partial}{\partial q_1} + (1+B)\frac{\partial}{\partial q_2} - E\frac{\partial}{\partial p_1}\right\rangle.$$

Since the distributions V_+ and V_- are skew-orthogonal, we have $\ker \mu_+ \supset V_-$ and $\ker \mu_- \supset V_+$. Therefore, $\mu_+(X_-) = h_+(1-B) = 0 \Rightarrow B = 1$ and $\mu_-(X_+) = C = 0$. Then,

$$W_+ = A_{p_2}\frac{\partial}{\partial q_1} - E_{p_2}\frac{\partial}{\partial p_1} + \frac{1}{2}(E_{p_2}A - EA_{p_2})\frac{\partial}{\partial p_2} = -h_+\frac{\partial}{\partial p_1}.$$

Therefore $E_{p_2} = h_+, A = 0$. We see that

$$W_- = E_{p_1}\frac{\partial}{\partial p_2} = -h_-\frac{\partial}{\partial p_2},$$

so

$$E_{p_1} = -h_-,$$

and

$$\omega = -2f dq_1 \wedge dq_2 + dq_1 \wedge dp_1 - dq_2 \wedge dp_2, \tag{19.7}$$

where $2f = -E$. ■

For this form we obtain the following coordinate representation of the operator A_w and the projection operators Pr_+ and Pr_- corresponding to the distributions V_+ and V_- :

$$A_w = \begin{Vmatrix} 1 & 0 & 0 & 0 \\ 0 & -1 & 0 & 0 \\ 0 & -2f & 1 & 0 \\ 2f & 0 & 0 & -1 \end{Vmatrix},$$

$$\mathrm{Pr}_+ = \frac{1}{2}\begin{Vmatrix} 1 & 0 & 0 & 0 \\ 0 & 0 & 0 & 0 \\ 0 & -f & 1 & 0 \\ f & 0 & 0 & 0 \end{Vmatrix},$$

$$\mathrm{Pr}_- = \frac{1}{2}\begin{Vmatrix} 0 & 0 & 0 & 0 \\ 0 & 1 & 0 & 0 \\ 0 & f & 0 & 0 \\ -f & 0 & 0 & 1 \end{Vmatrix}.$$

in the basis $\partial/\partial q_1, \partial/\partial q_2, \partial/\partial p_1, \partial/\partial p_2$

The differential forms μ_+ and μ_- are

$$\mu_+ = -2f_{p_2}\, dq_1,$$
$$\mu_- = 2f_{p_1}\, dq_2$$

and

$$V_+ = \mathcal{F}\left\langle \frac{\partial}{\partial q_1} + f\frac{\partial}{\partial p_2}, \frac{\partial}{\partial p_1} \right\rangle,$$
$$V_- = \mathcal{F}\left\langle \frac{\partial}{\partial q_2} + f\frac{\partial}{\partial p_1}, \frac{\partial}{\partial p_2} \right\rangle.$$

The vector field W has the form

$$W = 2f_{p_2}\frac{\partial}{\partial p_1} - 2f_{p_1}\frac{\partial}{\partial p_2}$$

and its projections to the distributions V_+ and V_- are

$$W_+ = 2f_{p_2}\frac{\partial}{\partial p_1}, \quad W_- = -2f_{p_1}\frac{\partial}{\partial p_2}$$

respectively.

Then

$$[W_+, W_-] = 4f_{p_1} f_{p_2p_2} \frac{\partial}{\partial p_1} - 4f_{p_2} f_{p_1p_1} \frac{\partial}{\partial p_2}$$

and we have the following functional invariants:

$$g_+ = \frac{2f_{p_1} f_{p_2p_2}}{f_{p_2}}, \quad g_- = \frac{2f_{p_2} f_{p_1p_1}}{f_{p_1}}, \quad g_0 = 2f_{p_1p_2}.$$

Bases for modules $\Omega^{p,q}$ are the following:

$$\begin{aligned}
\Omega^{1,0} &= \langle dq_1, dp_1 - f\, dq_2 \rangle, \\
\Omega^{0,1} &= \langle dq_2, dp_2 - f\, dq_1 \rangle, \\
\Omega^{1,1} &= \langle dq_1 \wedge dq_2, dq_1 \wedge dp_2, dq_2 \wedge dp_1, f^2\, dq_1 \wedge dq_2 - f\, dq_1 \wedge dp_1 \\
&\qquad + f\, dq_2 \wedge dp_2 - dp_1 \wedge dp_2 \rangle, \\
\Omega^{2,0} &= \langle dq_1 \wedge dp_1 - f\, dq_1 \wedge dq_2 \rangle, \\
\Omega^{0,2} &= \langle dq_2 \wedge dp_2 + f\, dq_1 \wedge dq_2 \rangle, \\
\Omega^{1,2} &= \langle dq_1 \wedge dq_2 \wedge dp_2, dq_2 \wedge dp_1 \wedge dp_2 - f\, dq_1 \wedge dq_2 \wedge dp_1 \rangle, \\
\Omega^{2,1} &= \langle dq_1 \wedge dq_2 \wedge dp_1, dq_1 \wedge dp_1 \wedge dp_2 - f\, dq_1 \wedge dq_2 \wedge dp_2 \rangle, \\
\Omega^{2,2} &= \langle dq_1 \wedge dq_2 \wedge dp_1 \wedge dp_2 \rangle,
\end{aligned}$$

and

$$\begin{aligned}
d_{0,1}\mu_+ &= 2(ff_{p_1p_2} + f_{q_2p_2})\, dq_1 \wedge dq_2 + 2f_{p_2p_2}\, dq_1 \wedge dp_2, \\
d_{1,0}\mu_- &= 2(ff_{p_1p_2} + f_{q_1p_1}) dq_1 \wedge dq_2 - 2f_{p_1p_1}\, dq_2 \wedge dp_1.
\end{aligned}$$

Therefore we have

$$\gamma_+ = -\frac{f_{p_1p_1}}{f_{p_1}}(dp_1 - f\, dq_2) - \frac{ff_{p_1p_2} + f_{q_1p_1}}{f_{p_1}}\, dq_1,$$

$$\gamma_- = -\frac{f_{p_2p_2}}{f_{p_2}}(dp_2 - f\, dq_1) - \frac{ff_{p_1p_2} + ff_{q_2p_2}}{f_{p_2}}\, dq_2$$

and the corresponding vector fields:

$$X_+ = -\frac{f_{p_1p_1}}{f_{p_1}}\left(\frac{\partial}{\partial q_1} + f\frac{\partial}{\partial p_2}\right) + \frac{f f_{p_1p_2} + f_{q_1p_1}}{f_{p_1}}\frac{\partial}{\partial p_1},$$

$$X_- = -\frac{f_{p_2p_2}}{f_{p_2}}\left(\frac{\partial}{\partial q_2} + f\frac{\partial}{\partial p_1}\right) + \frac{f f_{p_1p_2} + f_{q_2p_2}}{f_{p_1}}\frac{\partial}{\partial p_2},$$

and

$$X_1 = 2f_{p_2}\frac{\partial}{\partial p_1},$$

$$X_2 = \frac{1}{2f_{p_2}}\frac{\partial}{\partial q_1} - \frac{f f_{p_1p_2} + f_{q_1p_1}}{2f_{p_2}f_{p_1p_1}}\frac{\partial}{\partial p_1} + \frac{f}{2f_{p_2}}\frac{\partial}{\partial p_2},$$

$$X_3 = -2f_{p_1}\frac{\partial}{\partial p_2},$$

$$X_4 = \frac{1}{2f_{p_1}}\frac{\partial}{\partial q_2} + \frac{f}{2f_{p_1}}\frac{\partial}{\partial p_1} - \frac{f f_{p_1p_2} + f_{q_2p_2}}{2f_{p_1}f_{p_2p_2}}\frac{\partial}{\partial p_2}.$$

The coordinate representations of the tensors $d_{2,-1}$ and $d_{-1,2}$ are:

$$d_{2,-1} = f_{p_1}(dq_1 \wedge dp_1 - f\, dq_1 \wedge dq_2) \otimes \frac{\partial}{\partial p_2},$$

$$d_{-1,2} = f_{p_2}(dq_2 \wedge dp_2 + f\, dq_1 \wedge dq_2) \otimes \frac{\partial}{\partial p_1}.$$

Theorem 19.2.3 *Let ω be an effective symplectic differential 2-form of class $H_{1,1}$ and let $g_0 = g_+ = g_- = 0$ in a neighborhood $\mathcal{O}_a \ni a$. Then the form ω is locally equivalent to the form*

$$\tilde{\omega} = -2(A(q)p_1 + B(q)p_2 + C(q))\, dq_1 \wedge dq_2 + dq_1 \wedge dp_1 - dq_2 \wedge dp_2 \tag{19.8}$$

for some functions A, B and C.

Proof From the coordinate representations of the functions $g_\pm$ and g_0 it follows that $f_{p_1p_1} = f_{p_2p_2} = f_{p_1p_2} = 0$. Therefore,

$$f = A(q)p_1 + B(q)p_2 + C(q).$$

■

Now consider the case when the vector fields X_1, X_2, X_3, X_4 form a four-dimensional real Lie algebra $\mathfrak{g}$, i.e.,

$$[X_i, X_j] = \sum_{k=1}^{4} c_{ij}^k X_k,$$

where $i, j = 1, \ldots, 4$ and $c_{ij}^k \in \mathbb{R}$ are structure constants of the Lie algebra $\mathfrak{g}$.

Note that $c_{ij}^k = \theta_k([X_i, X_j])$ and from the formula

$$[X_1, X_3] = [W_+, W_-] = g_+ W_+ + g_- W_- = g_+ X_1 + g_- X_3$$

it follows that $g_+ = c_{13}^1$ and $g_- = c_{13}^3$. Using the definition of the functional invariant g_0, we obtain $X_1 \rfloor d\theta_2 = g_0 \theta_2$.

Then $\theta_2([X_1, X_2]) = -d\theta_2(X_1, X_2) = g_0$ and we see that $g_0 = c_{12}^2$.

Thus, if the vector fields X_1, X_2, X_3, X_4 form a four-dimensional real Lie algebra, then the functional invariants g_0, g_+, g_- are constants.

One can check that under the condition $g_+ g_- \neq 0$ there exists only the Lie algebra and with the structure presented in the following table.

	X_1	X_2	X_3	X_4
X_1	0	$\frac{1}{2}X_1 - 4\kappa X_2 + \frac{1}{2}X_3$	$4\kappa(X_1 + X_3)$	$\frac{1}{2}X_1 + \frac{1}{2}X_3 - 4\kappa X_4$
X_2	$-\frac{1}{2}X_1 + 4\kappa X_2 - \frac{1}{2}X_3$	0	$\frac{1}{2}X_1 - 4\kappa X_2 + \frac{1}{2}X_3$	0
X_3	$-4\kappa(X_1 + X_3)$	$-\frac{1}{2}X_1 + 4\kappa X_2 - \frac{1}{2}X_3$	0	$-\frac{1}{2}X_1 - \frac{1}{2}X_3 + 4\kappa X_4$
X_4	$-\frac{1}{2}X_1 - \frac{1}{2}X_3 + 4\kappa X_4$	0	$\frac{1}{2}X_1 + \frac{1}{2}X_3 - 4\kappa X_4$	0

The function f which generates this Lie algebra is:

$$f(q, p) = (p_1 + \kappa p_2)^2.$$

Here κ is a non-zero real number, and $g_0 = g_- = g_+ = 2\kappa$.

Theorem 19.2.4 *Suppose that ω is an effective symplectic differential 2-form of class $H_{1,1}$ in a neighborhood $\mathcal{O}_a$ of a point a, with $g_+ g_-(a) \neq 0$, and the vector fields X_1, X_2, X_3, X_4 form a Lie algebra.*

Then the form ω *is a locally symplectic equivalent to the differential 2-form*

$$\tilde{\omega} = -2(p_1 + \kappa p_2)^2 dq_1 \wedge dq_2 + dq_1 \wedge dp_1 - dq_2 \wedge dp_2,$$

where κ *is a non-zero real number.*

The corresponding Monge–Ampère operator has the form

$$\Delta_{\tilde{\omega}}(v) = (v_{q_1 q_2} - (v_{q_1} + \kappa v_{q_2})^2) dq_1 \wedge dq_2.$$

19.2.3 Elliptic operators

In this case the vector fields W_+, W_-, Q_+ and Q_- are complex. Moreover, W_+ and W_- (Q_+ and Q_-) are complex conjugate, i.e., $W_+ = \overline{W_-}$ and $Q_+ = \overline{Q_-}$. Note that W and $V \stackrel{\text{def}}{=} AW$ are linear independent at each point of real vector fields.

Define a real vector field X: if $\operatorname{Re} Q_+ \neq 0$ we put $X \stackrel{\text{def}}{=} \operatorname{Re} Q_+$ and $X \stackrel{\text{def}}{=} \operatorname{Im} Q_+$ otherwise. Moreover, set

$$Z \stackrel{\text{def}}{=} AX, \quad \eta \stackrel{\text{def}}{=} V \rfloor \Omega, \quad \xi \stackrel{\text{def}}{=} X \rfloor \Omega, \quad \tau \stackrel{\text{def}}{=} Z \rfloor \Omega.$$

The table below indicates values of the 1-forms ξ, τ, μ, η on the vector fields X, Z, W, V:

	X	Z	W	V
ξ	0	0	w	v
τ	0	0	v	$-w$
μ	$-w$	$-v$	0	0
η	$-v$	w	0	0

(19.9)

where $v \stackrel{\text{def}}{=} \Omega(X, V)$, $w \stackrel{\text{def}}{=} \Omega(X, W)$.

Note that

$$\Omega^2(W, V, X, Z) = 2(v^2 + w^2)$$

and therefore the vector fields W, V, X, Z (and the differential 1-forms ξ, τ, μ, η) are linear independent if and only if $v^2 + w^2 \neq 0$.

As above we will consider two cases.

Case 1

$v^2 + w^2 \neq 0$.

In this case the vector fields W, V, X, Z form a basis of the module $D(\mathcal{O}_a)$ in a some neighborhood $\mathcal{O}_a$ of a.

Let

$$
\begin{aligned}
X_1 &\overset{\text{def}}{=} -\frac{1}{v^2+w^2}(\upsilon V + wW),\\
X_2 &\overset{\text{def}}{=} X,\\
X_3 &\overset{\text{def}}{=} \frac{1}{v^2+w^2}(\upsilon W - wV),\\
X_4 &\overset{\text{def}}{=} -Z.
\end{aligned}
\tag{19.10}
$$

Then $X_3 = AX_1$, $X_4 = -AX_2$, and

$$
\begin{aligned}
&\Omega(X_1, X_2) = \Omega(X_3, X_4) = 1,\\
&\Omega(X_1, X_3) = \Omega(X_1, X_4) = \Omega(X_2, X_3) = \Omega(X_2, X_4) = 0.
\end{aligned}
$$

Let $(\theta_1, \ldots, \theta_4)$ be the dual basis for $(X_1, \ldots, X_4)$. Then

$$\Omega = \theta_1 \wedge \theta_2 + \theta_3 \wedge \theta_4.$$

Calculating values of ω on the vector fields X_1, X_2, X_3, X_4:

$\omega(\uparrow, \leftarrow)$	X_1	X_2	X_3	X_4
X_1	0	0	0	1
X_2	0	0	1	0
X_3	0	-1	0	0
X_4	-1	0	0	0

we get the following representation of the form ω:

$$\omega = \theta_1 \wedge \theta_4 - \theta_2 \wedge \theta_3$$

and the following result.

Theorem 19.2.5 *Let ω be an elliptic normed effective differential 2-form on T^*M and $a \in T^*M$. Suppose that $v^2(a) + w^2(a) \neq 0$. Then in a some neighborhood of a there exists an e-structure $X_1, \ldots, X_4$ such that we have the following* *representation of the forms* Ω *and* ω*:*

$$\Omega = \theta_1 \wedge \theta_2 + \theta_3 \wedge \theta_4,$$
$$\omega = \theta_1 \wedge \theta_4 - \theta_2 \wedge \theta_3. \tag{19.11}$$

Remark 19.2.1 *Similar e-structures for hyperbolic and elliptic equations were obtained by Kruglikov in [53, 54, 55].*

Case 2

The Monge–Ampère operator belongs to the class$H_{1,1}$.

This means that the three-dimensional complex distributions $\ker \mu_+$ and $\ker \mu_-$ are both integrable in $\mathcal{O}_a$.

Let $X \overset{\text{def}}{=} \operatorname{Re} X_-$ and $Z \overset{\text{def}}{=} AX$. These are real vector fields. As above we set

$$\eta \overset{\text{def}}{=} V \rfloor \Omega, \quad \xi \overset{\text{def}}{=} X \rfloor \Omega, \quad \tau \overset{\text{def}}{=} Z \rfloor \Omega.$$

Then we have the same table (19.9) of values of the 1-forms on the vector fields, where as above $v \overset{\text{def}}{=} \Omega(P, V)$, $w \overset{\text{def}}{=} \Omega(P, W)$.

We define the new basis of $D(O_a)$ by (19.10) and get the same canonical representation (19.11) of the form Ω and ω.

Theorem 19.2.6 *Let* ω *be an elliptic normed effective differential 2-form on* T^*M *and* $a \in T^*M$. *Suppose that at each point of some neighborhood of* a *the Monge–Ampère operator belongs to the class* $H_{1,1}$. *Then in a some neighborhood of* a *there exists an e-structure* $X_1, \ldots, X_4$ *such that we have the following representation of the forms* Ω *and* ω*:*

$$\Omega = \theta_1 \wedge \theta_2 + \theta_3 \wedge \theta_4,$$
$$\omega = \theta_1 \wedge \theta_4 + \theta_2 \wedge \theta_3.$$

Example 19.2.1 *As an example we consider the following non-linear Laplace operator:*

$$\Delta_\omega(v) = (v_{q_1q_1} + v_{q_2q_2} - f(q,p))dq_1 \wedge dq_2,$$

which corresponds to the non-linear Laplace equation

$$v_{xx} + v_{yy} = f(x, y, v_x, v_y). \tag{19.12}$$

For this operator the corresponding effective differential 2-form is

$$\omega = -f(q,p)\,dq_1 \wedge dq_2 + dq_1 \wedge dp_2 + dq_2 \wedge dp_1.$$

In the basis $\partial/\partial q_1, \partial/\partial q_2, \partial/\partial p_1, \partial/\partial p_2$ *the operator* A *has the following matrix representation:*

$$A = \begin{Vmatrix} 0 & 1 & 0 & 0 \\ -1 & 0 & 0 & 0 \\ 0 & f & 0 & -1 \\ -f & 0 & 1 & 0 \end{Vmatrix}.$$

The complex distributions V_+ *and* V_- *are*

$$V_+ = \left\langle \frac{\partial}{\partial q_2} + f\frac{\partial}{\partial p_2} + \iota\frac{\partial}{\partial q_1},\ \frac{\partial}{\partial q_1} + f\frac{\partial}{\partial p_1} - \iota\frac{\partial}{\partial q_2} \right\rangle,$$

$$V_- = \left\langle \frac{\partial}{\partial q_2} + f\frac{\partial}{\partial p_2} - \iota\frac{\partial}{\partial q_1},\ \frac{\partial}{\partial q_1} + f\frac{\partial}{\partial p_1} + \iota\frac{\partial}{\partial q_2} \right\rangle.$$

The vector field

$$W = -f_{p_2}\frac{\partial}{\partial p_1} + f_{p_1}\frac{\partial}{\partial p_2}$$

falls into two components

$$W_\pm = \frac{1}{2}\left(-f_{p_2}\frac{\partial}{\partial p_1} + f_{p_1}\frac{\partial}{\partial p_2} \mp \iota\left(f_{p_1}\frac{\partial}{\partial p_1} + f_{p_2}\frac{\partial}{\partial p_2}\right)\right).$$

Therefore,

$$\mu_\pm = \tfrac{1}{2}(f_{p_2}\,dq_1 - f_{p_1}\,dq_2 \pm \iota(f_{p_1}\,dq_1 + f_{p_2}\,dq_2)).$$

We see that $\mu_+ \wedge d\mu_+ = 0$ *and* $\mu_- \wedge d\mu_- = 0$*, therefore the distributions* $\ker \mu_+$ *and* $\ker \mu_-$ *are completely integrable and (19.12) belongs to the class* $H_{2,2}$*. The vector field*

$$Q = \iota\Big((2f_{p_1}f_{p_1p_2} + f_{p_2}(f_{p_2p_2} - f_{p_1p_1}))\frac{\partial}{\partial p_1}$$
$$+(-2f_{p_2}f_{p_1p_2} + f_{p_1}(f_{p_2p_2} - f_{p_1p_1}))\frac{\partial}{\partial p_2}\Big)$$

is a linear combination of the vector fields W_+ and W_- with coefficients

$$g_+ = \frac{(f_{p_2} - \iota f_{p_1})(2f_{p_1p_2} + \iota(f_{p_1p_1} - f_{p_2p_2}))}{2(f_{p_1} - \iota f_{p_2})},$$

$$g_- = \frac{(f_{p_1} - \iota f_{p_2})(f_{p_2p_2} - f_{p_1p_1} - \iota 2f_{p_1p_2})}{2(f_{p_2} - \iota f_{p_1})}$$

respectively.

Example 19.2.2 *Let us construct an e-structure for the following case of the non-linear Laplace operator:*

$$\Delta_\omega(v) = (v_{q_1q_1} + v_{q_2q_2} - v_{q_1}^2)dq_1 \wedge dq_2.$$

For this operator

$$W = 2p_1 \frac{\partial}{\partial p_2},$$

$$V = -2p_1 \frac{\partial}{\partial p_1},$$

$$W_\pm = p_1 \frac{\partial}{\partial p_2} \mp \iota p_1 \frac{\partial}{\partial p_1},$$

$$\gamma_\pm = \frac{1}{2p_1}(-dp_1 \pm \iota(p_1^2\, dq_2 - dp_2)),$$

$$v = 1,$$

$$w = 0.$$

Therefore we have the following e-structure:

$$X_1 = 2p_1 \frac{\partial}{\partial p_1},$$

$$X_2 = -\frac{1}{2p_1}\frac{\partial}{\partial q_1},$$

$$X_3 = 2p_1 \frac{\partial}{\partial p_2},$$

$$X_4 = -\frac{1}{2p_1}\frac{\partial}{\partial q_2} - \frac{p_1}{2}\frac{\partial}{\partial p_2}.$$

The dual basis is

$$\theta_1 = \frac{1}{2p_1}\, dp_1,$$
$$\theta_2 = -p_1\, dq_1,$$
$$\theta_3 = \frac{1}{2p_1}\, dp_2 - \frac{p_1}{2}\, dq_2,$$
$$\theta_4 = -2p_1\, dq_2$$

and

$$\theta_1 \wedge \theta_2 + \theta_3 \wedge \theta_4 = dq_1 \wedge dp_1 + dq_2 \wedge dp_2 = \Omega,$$
$$\theta_1 \wedge \theta_4 + \theta_2 \wedge \theta_3 = -p_1^2\, dq_1 \wedge dq_2 + dq_1 \wedge dp_2 + dq_2 \wedge dp_1 = \omega.$$

19.3 Classification of degenerate Monge–Ampère operators

19.3.1 Non-linear mixed-type operators

Let ω be an effective differential 2-form such that at a point $a \in T^*M$ the Pfaffian $\mathrm{Pf}(\omega)$ vanishes. Assume also that

$$\mathrm{Pf}(\omega) = f^n,$$

where n is a natural number and f is a smooth function such that its differential does not vanish at the point a: $df_a \neq 0$.

Then in a neighborhood of the point a the surface $\{f = 0\}$ is a smooth manifold. Let $X \overset{\text{def}}{=} X_f$ be a Hamiltonian vector field with a Hamiltonian f.

Define the vector field

$$Z \overset{\text{def}}{=} A_\omega X,$$

the differential 1-form

$$\tau \overset{\text{def}}{=} Z \rfloor \Omega,$$

and two smooth functions

$$\upsilon \overset{\text{def}}{=} V(f), \quad w \overset{\text{def}}{=} W(f).$$

Then we have

$$X \rfloor \omega = \tau_\omega, \quad Z \rfloor \omega = f^n\, df, \quad V \rfloor \omega = -f^n \mu.$$

The table below shows the values of the 1-forms df, τ, μ, η on the vector fields X, Z, W, V.

	X	Z	W	V
df	0	0	w	υ
τ	0	0	$-\upsilon$	$F^n w$
μ	w	υ	0	0
η	υ	$-F^n w$	0	0

Note that $d\omega = \mu \wedge \Omega$.

Indeed,

$$2d\omega = W \rfloor \Omega^2 = 2(W \rfloor \Omega) \wedge \Omega = 2\mu \wedge \Omega.$$

Lemma 19.3.1 *For the differential 1-forms df, τ, μ, η the following properties hold:*

1. $\tau \wedge \Omega = df \wedge \omega$;
2. $\eta \wedge \Omega = -\mu \wedge \omega$;
3. $f^n \mu \wedge \Omega = \eta \wedge \omega$;
4. $\tau \wedge \mu \wedge \Omega = \eta \wedge df \wedge \Omega$;
5. $\tau \wedge \eta \wedge \Omega = -f^n \mu \wedge df \wedge \Omega$;
6. $\tau \wedge \mu \wedge df = -w\tau \wedge \Omega - \upsilon\, df \wedge \Omega$.

Proof Since the form ω is effective, $\omega \wedge \Omega = 0$. Then we have

$$0 = X \rfloor (\omega \wedge \Omega) = \tau \wedge \Omega - df \wedge \omega$$

and the first property is proved.

In the similar way one obtains properties 2 and 3.

Properties 4 and 5 follow from properties 1 and 2 by

$$\mu \wedge \tau \wedge \Omega = \mu \wedge df \wedge \omega = -df \wedge \mu \wedge \omega = df \wedge \eta \wedge \Omega,$$

$$-\eta \wedge \tau \wedge \Omega = -\eta \wedge df \wedge \omega = df \wedge \eta \wedge \omega = F^n dF \wedge \mu \wedge \Omega,$$

and property 6 follows from property 4:

$$X \rfloor (\tau \wedge \mu \wedge \Omega) = X \rfloor (\eta \wedge df \wedge \Omega),$$

then

$$\upsilon\, df \wedge \Omega = -w\tau \wedge \Omega - \tau \wedge \mu \wedge df. \qquad \blacksquare$$

Lemma 19.3.2 *One has*

$$2\, df \wedge \tau \wedge \eta \wedge \mu = g\Omega^2,$$

where

$$g \stackrel{\text{def}}{=} F^n w^2 + \upsilon^2.$$

Proof From property 6, we get

$$(f^n w^2 + \upsilon^2)\Omega = \upsilon\, df \wedge \eta + w\tau \wedge \eta - \upsilon\tau \wedge \mu - f^n \mu \wedge df$$

and multiplying both parts by Ω and using properties 7 and 4 of the previous lemma, we find

$$(f^n w^2 + \upsilon^2)\Omega^2 = 2(w\tau \wedge \eta \wedge \Omega - \upsilon\tau \wedge \mu \wedge \Omega) = 2df \wedge \tau \wedge \eta \wedge \mu. \qquad \blacksquare$$

Suppose now that at the point a the function υ does not vanish:

$$\upsilon(a) \neq 0.$$

Then the vector fields X, Z, W, V are linear independent in a neighborhood of the point a and these vector fields form a local basis.

Let

$$X_1 \stackrel{\text{def}}{=} \frac{1}{g}(F^n wW + \upsilon V),$$

$$X_2 \stackrel{\text{def}}{=} X,$$

$$X_3 \stackrel{\text{def}}{=} \frac{1}{g}(\upsilon W - wV),$$

$$X_4 \stackrel{\text{def}}{=} Z$$

and

$$\theta_1 \overset{\text{def}}{=} df,$$
$$\theta_2 \overset{\text{def}}{=} \frac{1}{g}(f^n w\mu + \upsilon\eta),$$
$$\theta_3 \overset{\text{def}}{=} -\tau,$$
$$\theta_4 \overset{\text{def}}{=} \frac{1}{g}(\upsilon\mu - w\eta).$$

Then bases $X_1, \ldots, X_4$ and $\theta_1, \ldots, \theta_4$ are dual and

$$X_1 \rfloor \Omega = \theta_2, \qquad X_2 \rfloor \Omega = -\theta_1, \qquad X_3 \rfloor \Omega = \theta_4, \qquad X_4 \rfloor \Omega = -\theta_3.$$

We see that $A_\omega X_3 = X_1$ and $A_\omega X_2 = X_4$.

In the basis $\theta_1, \ldots, \theta_4$ the symplectic form Ω and the effective form ω have the following canonical representations:

$$\begin{aligned} \Omega &= \theta_1 \wedge \theta_2 + \theta_3 \wedge \theta_4, \\ \omega &= \theta_3 \wedge \theta_2 + f^n \theta_4 \wedge \theta_1. \end{aligned} \tag{19.13}$$

This representation allows us to construct normal forms of mixed-type effective differential 2-forms.

Let ω be a mixed-type effective differential 2-form and (19.13) be a canonical representation, and let vector fields $X_1, \ldots, X_4$ form a four-dimensional Lie algebra $\mathfrak{g}$.

Since $d\Omega = 0$, we obtain the following extra conditions for the structure constants:

$$\begin{aligned} &c_{23}^2 + c_{12}^4 = 0, && c_{24}^2 - c_{12}^3 = 0, && c_{23}^3 + c_{24}^4 = 0, \\ &c_{13}^3 + c_{12}^2 = 0, && c_{34}^3 + c_{24}^2 = -w, && c_{34}^2 - c_{13}^3 - c_{14}^4 = 0, \\ &c_{14}^3 - f_\omega^n c_{24}^4 = \upsilon_\omega, && c_{23}^4 = c_{14}^2 = c_{34}^4 = 0, && c_{ij}^1 = 0 \quad (i,j = 1, \ldots, 4) \end{aligned}$$

(here $w = w_\omega$). Therefore in this case υ_ω = constant. From $\upsilon_\omega(a) \neq 0$ it follows that $c_{14}^3 \neq 0$.

Let

$$\begin{aligned} &a_1 = c_{12}^2, && a_2 = c_{13}^2, && a_3 = c_{23}^2, \\ &a_4 = c_{12}^3, && a_5 = c_{34}^2, && a_6 = c_{23}^3, \\ &a_7 = c_{24}^3, && a_8 = c_{13}^4, && a_9 = c_{14}^3. \end{aligned}$$

Finally we get eight Lie algebras.

(1) Lie algebra $\mathfrak{g}_1$ $(a_1 \neq 0, w_\omega \neq 0)$:

$[\uparrow, \leftarrow]$	X_1	X_2	X_3	X_4
X_1	0	a_1X_2	$-2a_1^2w^{-1}X_2 - a_1X_3$	a_9X_3
X_2	$-a_1X_2$	0	0	0
X_3	$2a_1^2w^{-1}X_2 + a_1X_3$	0	0	$-a_1X_2 - wX_3$
X_4	$-a_9X_3$	0	$a_1X_2 + wX_3$	0

(2) Lie algebra $\mathfrak{g}_2$:

$[\uparrow, \leftarrow]$	X_1	X_2	X_3	X_4
X_1	0	0	0	a_9X_3
X_2	0	0	0	a_7X_3
X_3	0	0	0	$-wX_3$
X_4	$-a_9X_3$	a_7X_3	wX_3	0

(3) Lie algebra $\mathfrak{g}_3$ $(a_4 \neq 0, \beta = 2\lambda^2 a_2 \neq 2a_4)$:

$[\uparrow, \leftarrow]$	X_1	X_2	X_3	X_4
X_1	0	$-\lambda a_2X_2 + a_4X_3$	$a_2X_2 + \lambda a_2X_3$	a_9X_3
X_2	$\lambda a_2X_2 - a_4X_3$	0	0	$a_4X_2 + \lambda a_4X_3$
X_3	$-a_2X_2 - \lambda a_2X_3$	0	0	$\lambda a_2X_2 + (\beta + a_4)X_3$
X_4	$-a_9X_3$	$-a_4X_2 - \lambda a_4X_3$	$-\lambda a_2X_2 - (\beta + a_4)X_3$	0

(4) Lie algebra $\mathfrak{g}_4$ $(a_7 \neq 0, \sigma \neq 0)$:

$[\uparrow, \leftarrow]$	X_1	X_2	X_3	X_4
X_1	0	$\sigma^2a_7X_2 + \sigma a_7X_3$	$-\sigma^3a_7X_2 - \sigma^2a_7X_3$	a_9X_3
X_2	$-\sigma^2a_7X_2 - \sigma a_7X_3$	0	0	$\sigma a_7X_2 + a_7X_3$
X_3	$\sigma^3a_7X_2 + \sigma^2a_7X_3$	0	0	$-\sigma^2a_7X_2 - \sigma a_7X_3$
X_4	$-a_9X_3$	$-\sigma a_7X_2 - a_7X_3$	$\sigma^2a_7X_2 + \sigma a_7X_3$	0

(5) Lie algebra $\mathfrak{g}_5$ ($a_1 \neq 0$):

$[\uparrow, \leftarrow]$	X_1	X_2	X_3	X_4
X_1	0	a_1X_2	$a_2X_2 - a_1X_3 + a_8X_4$	$a_9X_3 + 2a_1X_4$
X_2	$-a_1X_2$	0	0	0
X_3	$-a_2X_2 + a_1X_3 - a_8X_4$	0	0	a_1X_2
X_4	$-a_9X_3 - 2a_1X_4$	0	$-a_1X_2$	0

(6) Lie algebra $\mathfrak{g}_6$:

$[\uparrow, \leftarrow]$	X_1	X_2	X_3	X_4
X_1	0	a_1X_2	$a_2X_2 - a_1X_3 + a_8X_4$	$a_9X_3 + a_1X_4$
X_2	$-a_1X_2$	0	0	0
X_3	$-a_2X_2 + a_1X_3 - a_8X_4$	0	0	0
X_4	$-a_9X_3 - a_1X_4$	0	0	0

(7) Lie algebra $\mathfrak{g}_7$ ($a_4 \neq 0, a_9 \neq 0, \gamma = a_4^2 a_9^{-1}$):

$[\uparrow, \leftarrow]$	X_1	X_2	X_3	X_4
X_1	0	a_4X_3	0	a_9X_3
X_2	$-a_4X_3$	0	γX_3	$a_4X_2 + a_7X_3 - \gamma X_4$
X_3	0	$-\gamma X_3$	0	$-a_4X_3$
X_4	$-a_9X_3$	$-a_4X_2 - a_7X_3 + \gamma X_4$	a_4X_3	0

(8) Lie algebra $\mathfrak{g}_8$ ($a_6 \neq 0$):

$[\uparrow, \leftarrow]$	X_1	X_2	X_3	X_4
X_1	0	0	0	0
X_2	0	0	a_6X_3	$a_7X_3 - a_6X_4$
X_3	0	$-a_6X_3$	0	0
X_4	0	$-a_7X_3 + a_6X_4$	0	0

These Lie algebras admit the following representations by vector fields.

1. Lie algebra $\mathfrak{g}_1$:

$$X_1 = \frac{\partial}{\partial q_1} - (a_1 p_1 + 2a_1^2 w^{-1} p_2)\frac{\partial}{\partial p_1} + (a_1 p_2 + a_9 q_2)\frac{\partial}{\partial p_2},$$

$$X_2 = \frac{\partial}{\partial p_1},$$

$$X_3 = -\frac{\partial}{\partial p_2},$$

$$X_4 = \frac{\partial}{\partial q_2} + (a_1 p_2 + a_9 q_2)\frac{\partial}{\partial p_1} - (w p_2 + a_9 a_1^{-1} w q_2)\frac{\partial}{\partial p_2}.$$

2. Lie algebra $\mathfrak{g}_2$:

$$X_1 = \frac{\partial}{\partial q_1},$$

$$X_2 = \frac{\partial}{\partial p_1},$$

$$X_3 = -\frac{\partial}{\partial p_2},$$

$$X_4 = \frac{\partial}{\partial q_2} - (a_9 q_1 + a_7 p_1 + w p_2)\frac{\partial}{\partial p_2}.$$

3. Lie algebra $\mathfrak{g}_3$:

$$X_1 = \frac{\partial}{\partial q_1} + (\lambda a_2 p_1 + a_2 p_2 - a_2 a_9 a_4^{-1} q_1)\frac{\partial}{\partial p_1} + (a_4 p_1 - \lambda a_2 p_2)\frac{\partial}{\partial p_2},$$

$$X_2 = \frac{\partial}{\partial p_1},$$

$$X_3 = -\frac{\partial}{\partial p_2},$$

$$X_4 = \frac{\partial}{\partial q_2} + (a_4 p_1 - \lambda a_2 p_2)\frac{\partial}{\partial p_1} + (-\lambda a_4 p_1$$

$$+ (2\lambda^2 a_2 + a_4) p_2 - a_9 q_1)\frac{\partial}{\partial p_2}.$$

4. Lie algebra $\mathfrak{g}_4$:

$$X_1 = \frac{\partial}{\partial q_1} + \sigma^2(a_9 q_1 - a_7 p_1 - \sigma a_7 p_2)\frac{\partial}{\partial p_1} + \sigma a_7(p_1 + \sigma p_2)\frac{\partial}{\partial p_2},$$
$$X_2 = \frac{\partial}{\partial p_1},$$
$$X_3 = -\frac{\partial}{\partial p_2},$$
$$X_4 = \frac{\partial}{\partial q_2} + \sigma a_7(p_1 + \sigma p_2)\frac{\partial}{\partial p_1} - (a_7 p_1 + \sigma a_7 p_2 + a_9 q_1)\frac{\partial}{\partial p_2}.$$

5. Lie algebra $\mathfrak{g}_5$:

$$X_1 = \frac{\partial}{\partial q_1} + (a_8 p_2 - 2a_1 q_2)\frac{\partial}{\partial q_2} + (a_2 p_2 - a_1 p_1)\frac{\partial}{\partial p_1} + (a_9 q_2 + a_1 p_2)\frac{\partial}{\partial p_2},$$
$$X_2 = \frac{\partial}{\partial p_1},$$
$$X_3 = (a_8 p_2 - 2a_1 q_2)\frac{\partial}{\partial p_1} - \frac{\partial}{\partial p_2},$$
$$X_4 = \frac{\partial}{\partial q_2} + (a_9 q_2 + a_1 p_2)\frac{\partial}{\partial p_1}.$$

6. Lie algebra $\mathfrak{g}_6$:

$$X_1 = \frac{\partial}{\partial q_1} + (a_8 p_2 - a_1 q_2)\frac{\partial}{\partial q_2} + \left(-a_2 p_2 - a_1 p_1 + \frac{1}{2}a_1 a_9 q_2^2 - \frac{1}{2}a_1 a_8 p_2^2 + a_1^2 p_2 q_2\right)\frac{\partial}{\partial p_1} + (a_9 q_2 + a_1 p_2)\frac{\partial}{\partial p_2},$$
$$X_2 = \frac{\partial}{\partial p_1},$$
$$X_3 = (a_1 q_2 - a_8 p_2)\frac{\partial}{\partial p_1} + \frac{\partial}{\partial p_2},$$
$$X_4 = -\frac{\partial}{\partial q_2} - (a_9 q_2 + a_1 p_2)\frac{\partial}{\partial p_1}.$$

7. Lie algebra $\mathfrak{g}_7$:

$$X_1 = -a_9 \frac{\partial}{\partial q_2} - a_4 a_9 p_2 \frac{\partial}{\partial p_1},$$

$$X_2 = a_4 a_9^{-1} \frac{\partial}{\partial q_1} - a_4^2 (a_9^{-1} q_2 + a_7 p_2) \frac{\partial}{\partial q_2} + a_9^{-1} (a_4^2 p_2 - 1) \frac{\partial}{\partial p_1},$$

$$X_3 = a_4 \frac{\partial}{\partial q_2} + (a_4^2 p_2 - 1) \frac{\partial}{\partial p_1},$$

$$X_4 = \frac{\partial}{\partial q_1} - (a_4 q_2 + a_4 a_7 a_9 p_2) \frac{\partial}{\partial q_2} + (a_7 a_9 p_2 + q_2) \frac{\partial}{\partial p_1} + a_4 p_2 \frac{\partial}{\partial p_2}$$

8. Lie algebra $\mathfrak{g}_8$:

$$X_1 = \frac{\partial}{\partial q_1},$$

$$X_2 = \frac{\partial}{\partial p_1} + (a_7 q_2 - a_6 p_2) \frac{\partial}{\partial p_2} + a_6 q_2 \frac{\partial}{\partial q_2},$$

$$X_3 = -\frac{\partial}{\partial p_2} - a_6 q_2 \frac{\partial}{\partial q_1},$$

$$X_4 = (a_6 p_2 - a_7 q_2) \frac{\partial}{\partial q_1} + \frac{\partial}{\partial q_2}.$$

These representations lead us to the following normal forms of effective differential 2-forms.

1. Lie algebra $\mathfrak{g}_1$:

$$\begin{aligned}\omega =& (-q_1^n + a_1^2 p_2^2 - a_2^2 q_2^2 + w a_1 p_1 p_2 + a_9 w q_2 p_1) dq_1 \wedge dq_2 + (a_1 p_2 \\ &+ a_9 q_2)(dq_1 \wedge dp_1 - dq_2 \wedge dp_2) - (w p_2 + a_9 a_1^{-1} w q_2) dq_2 \wedge dp_1 \\ &+ (a_1 p_1 + 2 a_1^2 w p_2) dq_1 \wedge dp_2 + dp_1 \wedge dp_2.\end{aligned}$$

2. Lie algebra $\mathfrak{g}_2$:

$$\omega = dp_1 \wedge dp_2 - (a_9 q_1 + a_7 p_1 + w p_2) dq_2 \wedge dp_1 - q_1^n dq_1 \wedge dq_2$$

3. Lie algebra $\mathfrak{g}_3$:

$$\begin{aligned}\omega =&(-q_1^n + a_2 a_9^2 a_4^{-1} q_1^2 - \alpha a_4 p_1^2 + \alpha a_2 p_2^2 + 2\alpha\lambda a_2 p_1 p_2\\ &- 2\alpha a_2 a_9 a_4^{-1} q_1 p_2) dq_1 \wedge dq_2 + (a_4 p_1 - \lambda a_2 p_2)(dq_1 \wedge dp_1\\ &- dq_2 \wedge dp_2) + (a_2 a_9 a_4^{-1} q_1 - \lambda a_2 p_1 - a_2 p_2) dq_1 \wedge dp_2\\ &+ ((2\lambda^2 a_2 + a_4) p_2 - \lambda a_4 p_1 - a_9 q_1) dq_2 \wedge dp_1 + dp_1 \wedge dp_2.\end{aligned}$$

Here $\alpha \overset{\text{def}}{=} a_4 + \lambda^2 a_2$.

4. Lie algebra $\mathfrak{g}_4$:

$$\begin{aligned}\omega =& - (q_1^n + \sigma^2 a_9 q_1) dq_1 \wedge dq_2 + \sigma a_7 (p_1 + \sigma p_2)(dq_1 \wedge dp_1 - dq_2 \wedge dp_2)\\ &+ \sigma^2 (a_7 p_1 + \sigma a_7 p_2 - a_9 q_1) dq_1 \wedge dp_2\\ &- (a_7 p_1 + \sigma a_7 p_2 + a_9 q_1) dq_2 \wedge dp_1 + dp_1 \wedge dp_2.\end{aligned}$$

5. Lie algebra $\mathfrak{g}_5$:

$$\begin{aligned}\omega =& - (q_1^n + (a_9 q_2 + a_1 p_2)^2) dq_1 \wedge dq_2 + (a_9 q_2 + a_1 p_2)(dq_1 \wedge dp_1\\ &- dq_2 \wedge dp_2) + (a_1 p_1 - a_2 p_2 + (a_8 a_9 - 2a_1^2) q_2 p_2 - 2 a_1 a_9 q_2^2\\ &+ a_1 a_8 p_2^2) dq_1 \wedge dp_2 + dp_1 \wedge dp_2.\end{aligned}$$

6. Lie algebra $\mathfrak{g}_6$:

$$\begin{aligned}\omega =&(q_1^n + (a_9 q_2 + a_1 p_2)^2) dq_1 \wedge dq_2 - (a_9 q_2 + a_1 p_2)(dq_1 \wedge dp_1\\ &- dq_2 \wedge dp_2) + ((2a_1^2 - a_8 a_9) q_2 p_2 + \tfrac{3}{2} a_1 a_9 q_2^2 - \tfrac{3}{2} a_1 a_8 p_2^2 - a_1 p_1\\ &- a_2 p_2) dq_1 \wedge dp_2 - dp_1 \wedge dp_2.\end{aligned}$$

7. Lie algebra $\mathfrak{g}_7$:

$$\begin{aligned}\omega =&(-a_4^2 a_9 p_2^2 - a_9^{-1}(1 - a_4^2 p_2)^2 f^n) dq_1 \wedge dq_2\\ &+ (a_4 a_9 p_2 - a_4 a_9^{-1}(1 - a_4^2 p_2) f^n)(dq_1 \wedge dp_1 - dq_2 \wedge dp_2)\\ &\times (q_2 + a_7 a_9 p_2)(1 - a_4^2 p_2)(a_4^2 a_9^{-1} f^n - a_9) dq_1 \wedge dp_2\\ &+ (a_9 + a_4^2 a_9^{-1} f^n) dp_1 \wedge dp_2.\end{aligned}$$

Here $f \overset{\text{def}}{=} -a_9^{-1} q_2 + a_4^2 a_9^{-1} p_2 q_2 - a_4 a_9^{-1} p_1 + \frac{1}{2} a_4^2 a_7 p_2^2$.

8. Lie algebra $\mathfrak{g}_7$:

$$\omega = f^n(-dq_1 \wedge dq_2 + a_6 q_2(dq_1 \wedge dp_1 - dq_2 \wedge dp_2) \\ + a_6 q_2(a_7 q_2 - a_6 p_2) dq_2 \wedge dp_1) + (F^n a_6^2 q_2^2 + 1) dp_1 \wedge dp_2.$$

Here $f \stackrel{\text{def}}{=} q_1 + \frac{1}{2} a_7 q_2^2 - a_6 p_2 q_2$.

19.3.2 Linear mixed-type operators

Let us start from the following remark. If an effective differential 2-form ω is non-degenerate, then one can normalize it by multiplication by the function $1/\sqrt{|\operatorname{Pf}(\omega)|}$. If a form ω is of mixed type, then this is not possible. But if the Pfaffian of ω can be written in the form $\operatorname{Pf}(\omega) = f^n$, where $n \in \mathbb{N}$ and f is a smooth function such that $f(a) = 0$ and its differential $df_a \neq 0$ $(a \in T^*M)$, then one can find a symplectic transformation such that $\operatorname{Pf}(\omega)$ transforms to a function of the form q_1^n. This fact is a corollary of the following lemma.

Lemma 19.3.3 *Let F and G be smooth functions on T^*M and let $F(a) = G(a) = 0$ but $d_aF \neq 0$, $d_aG \neq 0$. Then there exists a symplectic diffeomorphism φ of a neighborhood of the point a such that $\varphi(a) = a$ and $\varphi^*(F) = G$.*

Proof Without loss of generality we can assume that $d_aF = d_aG$. Consider a path

$$F_t \stackrel{\text{def}}{=} G + t(F - G),$$

with $F_0 = G$ and $F_1 = F$, and find a path φ_t in the pseudo-group of local symplectic diffeomorphisms such that $\varphi_t(a) = a$ and $\varphi_t^*(F_t) = G$. This is equivalent to

$$\frac{d}{dt}\varphi_t^*(F_t) = 0 \tag{19.14}$$

or

$$X_{H_t}(F_t) + F - G = 0, \tag{19.15}$$

where H_t is the Hamiltonian of the vector field X_t corresponding to the path ϕ_t.

The last equation can be written as

$$X_{F_t}(H_t) = F - G$$

and has a smooth solution H_t with $d_a H_t = 0$, because $X_{F_t} \neq 0$ at the point a. ■

Suppose two differential forms ω_1 and ω_2 have the Pfaffians $f_{\omega_1} = \mathrm{Pf}(\omega_1)$ and $f_{\omega_2} = \mathrm{Pf}(\omega_2)$ respectively, and the point a is the point where $f_{\omega_1}(a) = f_{\omega_2}(a) = 0$, but $df_{\omega_1,a} \neq 0$, $df_{\omega_2,a} \neq 0$. Then from Lemma 19.3.3 there exists a local symplectic diffeomorphism that takes f_{ω_1} and f_{ω_2}, and moreover we can assume that

$$f_{\omega_1} = f_{\omega_2} = q_1$$

in canonical coordinates.

There are two classical equations of mixed-type:

$$v_{xx} + x v_{yy} + \alpha v_x + \beta v_y + \gamma(x, y) = 0 \tag{19.16}$$

and

$$x v_{xx} + v_{yy} + \alpha v_x + \beta v_y + \gamma(x, y) = 0. \tag{19.17}$$

The first is known as the Triccomi equation [108] and the second as the Keldysh one [46]. The effective differential forms corresponding to (19.16) and (19.17) are

$$\omega_T = q_1\, dq_1 \wedge dp_2 - dq_2 \wedge dp_1 + (\alpha p_1 + \beta p_2 + \gamma(q)) dq_1 \wedge dq_2$$

and

$$\omega_K = dq_1 \wedge dp_2 - q_1\, dq_2 \wedge dp_1 + (\alpha p_1 + \beta p_2 + \gamma(q)) dq_1 \wedge dq_2$$

respectively.

The corresponding differential operators $\Delta_T \stackrel{\mathrm{def}}{=} \Delta_{\omega_T}$ and $\Delta_K \stackrel{\mathrm{def}}{=} \Delta_{\omega_K}$ we call *the Triccomi operator* and *the Keldysh operator* respectively.

In the both cases the Pfaffians are the same:

$$\mathrm{Pf}(\omega_T) = \mathrm{Pf}(\omega_K) = q_1.$$

Lemma 19.3.4 *If $\omega \neq 0$, then*

$$Z(f) = 0.$$

Proof Using Lemma 20.4.1, we obtain

$$df \wedge \omega = (Z \rfloor \Omega) \wedge \Omega.$$

Hence,

$$Z \rfloor (df \wedge \omega) = Z(f)\omega - df \wedge (Z \rfloor \omega).$$

But

$$df \wedge (Z \rfloor \omega) = df \wedge (A_\omega Z \rfloor \Omega) = df \wedge (A_\omega^2 X \rfloor \Omega) = f\, df \wedge dfF = 0.$$

Therefore, $Z(f)\omega = 0$. Since $\omega \neq 0$, we find that $Z(f) = 0$. ∎

Theorem 19.3.1 *The mixed-type effective differential 2-form ω is locally symplectic equivalent at the point $a \in \{\mathrm{Pf}(\omega) = 0\}$ to the form*

$$\tilde{\omega}_T = q_1\, dq_1 \wedge dp_2 - dq_2 \wedge dp_1 + h(q,p)\, dq_1 \wedge dq_2 \qquad (19.18)$$

if and only if the vector field Z (see (20.13)) is Hamiltonian and the point a is its regular point, i.e., $Z_a \neq 0$. If, in addition,

$$W + \alpha Z + \beta X = 0, \qquad (19.19)$$

then this form is equivalent to the Triccomi form ω_T.

Proof For the form $\tilde{\omega}_T$ we have $\tilde{X}_T = X_T = \partial/\partial p_1$ and $\tilde{Z}_T = Z_T = -\partial/\partial p_2$. We see that $\tilde{Z}_T$ is a Hamiltonian vector field with Hamiltonian q_2. The path-lifting method allows us to construct a symplectic transformation φ that transforms the Hamiltonian g of the vector field Z to $-q_2$ and preserves the point a and the Pfaffian $f = q_1$. Taking into account coordinate representations of the vector field Z and the Pfaffian f, we obtain $A = 1$, $B = D = 0$, $C = q_1$. Hence, ω has the form

$$\omega = q_1\, dq_1 \wedge dp_2 - dq_2 \wedge dp_1 + E\, dq_1 \wedge dq_2 \qquad (19.20)$$

and the corresponding equation is (19.18), where $E = h$. The first part of the theorem is proved.

Let us prove the second part. For the form ω and for the Triccomi form ω_T we have

$$W = -E_{p_2} \frac{\partial}{\partial p_1} + E_{p_1} \frac{\partial}{\partial p_2},$$

$$W_T = -\beta \frac{\partial}{\partial p_1} + \alpha \frac{\partial}{\partial p_2}$$

respectively. Since (19.19), we obtain $E_{p_1} = \alpha$ and $E_{p_2} = \beta$. Then $E = \alpha p_1 + \beta p_2 + k(q)$. ∎

Remark 19.3.1 *The conditions (19.19) can be written in terms of the form ω:*

$$d\omega = df \wedge (\beta\Omega - \alpha\omega).$$

If $\alpha = \beta = 0$, we obtain

$$d\omega = 0.$$

For the Keldysh form we have

$$X_K = \frac{\partial}{\partial p_1}, \quad Z_K = -q_1\frac{\partial}{\partial p_2},$$
$$W_K = -\beta\frac{\partial}{\partial p_1} + (\alpha - 1)\frac{\partial}{\partial p_2}, \quad V_K = (\alpha - 1)\frac{\partial}{\partial p_1} + \beta q_1\frac{\partial}{\partial p_2}.$$

We see that the vector field Z_K vanishes on the surface $\{q_1 = 0\}$. Therefore, for the form ω we should define other vector fields:

$$Y \stackrel{\text{def}}{=} \frac{1}{F}Z.$$

For the Keldysh form $Y_K = -\partial/\partial p_2$ is a Hamiltonian regular vector fields. Now we can formulate an analog of Theorem 19.3.1.

Theorem 19.3.2 *The mixed-type effective differential 2-form ω is locally symplectic equivalent at the point $a \in \{\mathrm{Pf}(\omega) = 0\}$ to the form*

$$\tilde{\omega}_K = dq_1 \wedge dp_2 - q_1\, dq_2 \wedge dp_1 + f(q,p)\, dq_1 \wedge dq_2$$

if and only if the vector field Y is a Hamiltonian and the point a is its regular point, i.e., $Y_a \neq 0$. If, in addition,

$$W + (\alpha - 1)Y + \beta X = 0, \tag{19.21}$$

then this form is equivalent to the Keldysh form ω_K.

The proof is similar to that for the previous theorem.

Remark 19.3.2 *Received results can be generalized to the case when* $\mathrm{Pf}(\omega) = F^n$, *where $F(a) = 0$ and $dF_a \neq 0$ (see [58]).*

The method, which was used to prove this theorem, gives us an algorithm for constructing symplectic transformations, which send E_ω to the Triccomi equation.

Example 19.3.1 (Flow of a multicomponent gas mixture) *Consider the equation of gas dynamics, which describes two-dimensional transonic stationary flow on a multicomponent gas mixture:*

$$(v_x + b)v_{xx} - v_{yy} = 0, \tag{19.22}$$

where v is the velocity potential. If $b = 0$, we get the Von Karman equation [44]. The corresponding effective form for (19.22) is as follows:

$$\omega = (p_1 + b)dq_2 \wedge dp_1 + dq_1 \wedge dp_2.$$

The Pfaffian of ω is $\mathrm{Pf}(\omega) = -(p_1 + b)$ *and $d\omega = 0$. Simple calculations give*

$$X = \frac{\partial}{\partial q_1}, \qquad Z = \frac{\partial}{\partial q_2}, \qquad W = 0.$$

We see that the field Z is Hamilton with Hamiltonian $g = -p_2$. So, all the conditions of Theorem 19.3.1 are realized. Hence, (19.22) can be reduced to the Triccomi equation (19.16) with $\alpha = \beta = 0$. To construct this transformation we put $Q_1 = -p_1 - b$ and $Q_2 = -g = p_2$. We get the rest of the variables by $P_1 = q_1$ and $P_2 = -q_2$. In new coordinates ω has the form

$$\omega_T = Q_1\, dQ_1 \wedge dP_2 - dQ_2 \wedge dP_1. \tag{19.23}$$

Example 19.3.2 (The Maxwell–Einstein equation) *The following equation arises in the theory of gravitation and describes an axially symmetric stationary gravitation field (see [23, 8])*

$$((x^2 - 1)v_x)_x + ((1 - y^2)v_y)_y = 0. \tag{19.24}$$

The corresponding effective differential 2-form

$$\omega = 2(p_1q_1 - p_2q_2)dq_1 \wedge dq_2 + (1 - q_2^2)dq_1 \wedge dp_2 + (1 - q_1^2)dq_2 \wedge dp_1$$

satisfies Theorem 19.3.2 for $\alpha = 1, \beta = 0$.

Indeed,

$$f = -(1 - q_1^2)(1 - q_2^2)$$

and

$$
\begin{aligned}
X &= -2q_1(1-q_2^2)\frac{\partial}{\partial p_1} - 2q_2(1-q_1^2)\frac{\partial}{\partial p_2}, \\
Y &= 2q_2\frac{\partial}{\partial p_1} - 2q_1\frac{\partial}{\partial p_2}, \\
W &= 0.
\end{aligned}
$$

We see that at points $(\pm 1, \pm 1)$ *and* $(\pm 1, \mp 1)$ *on the plane* q_1, q_2 *the zero level of the Pfaffian is not smooth. Hence, at points a which do not coincide with one of them (19.24) transforms to the equation*

$$
xv_{xx} + v_{yy} - v_x = 0.
$$

20
Classification of symplectic MAEs on two-dimensional manifolds

We say that two symplectic Monge–Ampère equations E_ω and $E_{\widetilde{\omega}}$ are *equivalent* if there exists a symplectic transformation φ of T^*M and a non-vanishing function $h_\varphi \in C^\infty(T^*M)$, such that $\varphi^*(\omega) = h_\varphi \widetilde{\omega}$.

We say that two symplectic Monge–Ampère equations E_ω and $E_{\widetilde{\omega}}$ are *locally equivalent at a point $a \in T^*M$* if there exists a symplectic transformation φ of a neighborhood $\mathcal{O}_a$ of a and a function $h_\varphi \in C^\infty(\mathcal{O}_a)$, $h_\varphi(a) \neq 0$, such that $\varphi^*(\omega) = h_\varphi \widetilde{\omega}$ in a neighborhood.

20.1 Monge–Ampère equations with constant coefficients

The class of Monge–Ampère equations

$$Av_{xx} + 2Bv_{xy} + Cv_{yy} + D(v_{xx}v_{yy} - v_{xy}^2) + E = 0, \tag{20.1}$$

that can be reduced to Monge–Ampère equations with constant coefficients A, B, C, D, E by symplectic transformations is described by the following theorem.

Theorem 20.1.1 *A non-degenerate Monge–Ampère equation E_ω is symplectically equivalent to a Monge–Ampère equation with constant coefficients if and only if*

$$d\omega = \frac{1}{2} d(\ln |\operatorname{Pf}(\omega)|) \wedge \omega. \tag{20.2}$$

In this case a hyperbolic equation is locally equivalent to the wave equation

$$v_{xy} = 0$$

and an elliptic one is equivalent to the Laplace equation

$$v_{xx} + v_{yy} = 0.$$

Remark 20.1.1 *The conditions of the theorem can be formulated in terms of the de Rham complex: a non-degenerate Monge–Ampère equation E_ω is symplectically equivalent to a Monge–Ampère equation with constant coefficients if and only if $d_{-1,2} = d_{2,-1} = 0$.*

Proof First of all note that if a non-degenerate Monge–Ampère equation E_{ω_1} with the normed form ω_1 can be reduced to another Monge–Ampère equation (20.1) with constant coefficients A, B, C, D, E by a symplectic diffeomorphism then $d\omega_1 = 0$.

Secondly, any non-degenerate effective differential 2-form ω can be written as $\omega = h\omega_1$, where $h = \sqrt{|\operatorname{Pf}(\omega_1)|}$, and ω_1 is an effective normed differential 2-form. If the equation E_ω is equivalent to a Monge–Ampère equation with constant coefficients, then $d\omega_1 = 0$. But

$$d\omega_1 = d\left(\frac{1}{h}\omega\right) = -\frac{1}{h^2}dh \wedge \omega + \frac{1}{h}d\omega = 0,$$

and we obtain

$$d\omega - \frac{1}{2}d(\ln|\operatorname{Pf}(\omega)|) \wedge \omega = d\omega - \frac{1}{h}dh \wedge \omega = 0. \qquad \blacksquare$$

Proofs of sufficiency will be given for hyperbolic and elliptic equations separately.

20.1.1 Hyperbolic equations

Let ω be a hyperbolic form such that the condition (20.2) is realized and ω_1 be a normed hyperbolic form: $\omega_1 \overset{\text{def}}{=} 1/\sqrt{-\operatorname{Pf}(\omega)}\omega$. Then $\operatorname{Pf}(\omega_1) = -1$ and the form ω_1 is closed: $d\omega_1 = 0$. From the definition of the Pfaffian it follows that $\omega_1 \wedge \omega_1 + \Omega \wedge \Omega = 0$ and, since ω_1 is effective and take into account the fact that ω_1 is an effective form, i.e., $\omega_1 \wedge \Omega = 0$, the last equality can be rewritten as $(\omega_1 \pm \Omega) \wedge (\omega_1 \pm \Omega) = 0$. Therefore, the 2-forms $\omega_1 + \Omega$ and $\omega_1 - \Omega$ are decomposable and, therefore,

$$\omega_1 \pm \Omega = \gamma_1^\pm \wedge \gamma_2^\pm \tag{20.3}$$

for some differential 1-forms $\gamma_1^\pm$ and $\gamma_2^\pm$.

Denote by $V_{\mp}$ distributions generated by $\gamma_1^{\pm}$ and $\gamma_2^{\pm}$. We have

$$X\rfloor(\omega_1 \pm \Omega) = X\rfloor\omega_1 \pm X\rfloor\Omega = (AX \pm X)\rfloor\Omega$$

for any vector field X. Therefore,

$$X \in \ker(\omega_1 \pm \Omega) \Leftrightarrow AX = \mp X$$

and $V_{\mp} = \ker(A \mp 1)$.

Lemma 20.1.1 *The distributions V_- and V_+ are completely integrable if and only if $d\omega_1 = 0$.*

Proof Since

$$d\omega_1 = d\gamma_1^{\pm} \wedge \gamma_2^{\pm} - \gamma_1^{\pm} \wedge d\gamma_2^{\pm}.$$

Condition $d\omega_1 = 0$ implies

$$d\gamma_i^{\pm} \wedge \gamma_1^{\pm} \wedge \gamma_2^{\pm} = 0, \tag{20.4}$$

for $i = 1, 2$.

On the other hand, if the distributions V_- and V_+ are both completely integrable then $d\gamma_i^{\pm} = f^{\pm}\gamma_1^{\pm} \wedge \gamma_2^{\pm}$ and

$$d\omega_1 = d\gamma_1^{\pm} \wedge \gamma_2^{\pm} - \gamma_1^{\pm} \wedge d\gamma_2^{\pm} = 0.$$

Therefore, if $d\omega_1 = 0$, then distributions V_- and V_+ are completely integrable. Take functionally independent integrals f_1, f_2 and g_1, g_2 for V_- and V_+ respectively. Then

$$\omega_1 + \Omega = df_1 \wedge df_2 \qquad \text{and} \qquad \omega_1 - \Omega = dg_1 \wedge dg_2.$$

Therefore,

$$\Omega = \frac{1}{2}(df_1 \wedge df_2 + dg_1 \wedge dg_2),$$
$$\omega_1 = \frac{1}{2}(df_1 \wedge df_2 - dg_1 \wedge dg_2).$$

If we put

$$q_1 \stackrel{\text{def}}{=} \frac{1}{2}f_1, \quad q_2 \stackrel{\text{def}}{=} \frac{1}{2}g_1, \quad p_1 \stackrel{\text{def}}{=} f_2, \quad p_2 \stackrel{\text{def}}{=} g_2,$$

we get the following representations of ω_1 and Ω:

$$\omega_1 = dq_1 \wedge dp_1 - dq_2 \wedge dp_2,$$
$$\Omega = dq_1 \wedge dp_1 + dq_2 \wedge dp_2. \qquad \blacksquare$$

Note that the form ω_1 corresponds to the wave equation

$$v_{xy} = 0, \tag{20.5}$$

where $x = q_1$, $y = q_2$.

20.1.2 Elliptic equations

First of all we recall the definition of the Nijenhuis tensor (or the so-called *Nijenhuis bracket [26]*).

With each field of endomorphisms A on a smooth manifold Φ one can associate a so-called *Nijenhuis tensor* $N_A \in T\Phi \otimes \Omega^2(\Phi)$ with the following formula:

$$N_A(X,Y) \stackrel{\text{def}}{=} [AX, AY] - A[AX, Y] - A[X, AY] + A^2[X, Y].$$

It is easy to check that N_A is a tensor.

Let A be an *almost complex structure*, $A^2 = -1$, then the Newlander–Nirenberg theorem [88] states that the almost complex structure A is integrable if and only if the Nijenhuis tensor N_A is zero. The integrability of A means there is a complex structure such that

$$A : \begin{cases} \dfrac{\partial}{\partial x_k} \longmapsto \dfrac{\partial}{\partial y_k}, \\ \dfrac{\partial}{\partial y_k} \longmapsto -\dfrac{\partial}{\partial x_k} \end{cases}$$

in local complex coordinates $z_1 = x_1 + \iota y_1, \dots, z_n = x_n + \iota y_n$.

Let ω be an elliptic form such that condition (20.2) is realized and ω_1 be a normed elliptic form: $\omega_1 \stackrel{\text{def}}{=} 1/\sqrt{\operatorname{Pf}(\omega)}\omega$. Then $\operatorname{Pf}(\omega_1) = 1$ and the form ω_1 is closed. The corresponding operator $A \stackrel{\text{def}}{=} A_{\omega_1}$ generates an almost complex structure on T^*M: $A^2 = -1$.

Let us prove that this complex structure is integrable. Let X, Y be arbitrary vector fields. The formulas $\iota_{[X,Y]} = [L_X, \iota_Y]$ and $L_X = \iota_X d + d\iota_X$ imply that

$$\iota_{[X,Y]} = \iota_X\, d\iota_Y + d\iota_X\iota_Y - \iota_Y\iota_X\, d - \iota_Y\, d\iota_X.$$

Let $Z = N_A(X, Y)$. To prove that the vector field Z is zero for any vector fields X and Y it is enough to show that $\iota_Z(\Omega) = 0$. We have:

$$\begin{aligned}
\iota_{[X,Y]}(\Omega) &= \iota_X\, d\iota_Y(\Omega) - d\Omega(X, Y) - \iota_Y\, d\iota_X(\Omega),\\
\iota_{A[X,AY]}(\Omega) &= \iota_{[X,AY]}(\omega_1) = \iota_X\, d\iota_{AY}(\omega_1) - \iota_{AY}\, d\iota_X(\omega_1) - d\omega_1(X, AY)\\
&= -\iota_X\, d\iota_Y(\Omega) + d\Omega(X, Y) - \iota_{AY}\, d\iota_X(\omega_1),\\
\iota_{A[AX,Y]}(\Omega) &= \iota_{AX}\, d\iota_Y(\omega_1) + \iota_Y\, d\iota_X(\Omega) + d\Omega(X, Y),\\
\iota_{[AX,AY]}(\Omega) &= \iota_{AX}\, d\iota_{AY}(\Omega) + d\Omega(AY, AX) - \iota_{AY}\, d\iota_{AX}(\Omega)\\
&= \iota_{AX}\, d\iota_Y(\omega_1) + d\Omega(X, Y) - \iota_{AY}\, d\iota_X(\omega_1).
\end{aligned}$$

Adding these equalities, we find $\iota_Z(\Omega) = 0$ and $Z = 0$, and, therefore, $N_A = 0$.

Due to the theorem of Newlander–Nirenberg, we see that the almost complex structure A is integrable.

Let

$$\theta \stackrel{\text{def}}{=} \Omega^{\mathbb{C}} - \iota\omega_1^{\mathbb{C}},$$

where $\Omega^{\mathbb{C}}$ and $\omega_1^{\mathbb{C}}$ are complexifications of Ω and ω_1 respectively.

Then $\theta(AX, Y) = \theta(X, AY) = \iota\theta(X, Y)$ and $d\theta = 0$. Therefore θ defines a complex symplectic structure on the complex manifold T^*M. Due to the complex version of the Darboux theorem one can find local complex coordinates z_1, z_2 such that $\theta = dz_1 \wedge dz_2$. Then $\Omega = dx_1 \wedge dx_2 - dy_1 \wedge dy_2$ and $\omega_1 = -dx_1 \wedge dy_2 - dy_1 \wedge dx_2$, or

$$\begin{aligned}
\Omega &= dq_1 \wedge dp_1 + dq_2 \wedge dp_2,\\
\omega_1 &= dq_1 \wedge dp_2 + dp_1 \wedge dq_2
\end{aligned}$$

in coordinates $q_1 = x_1, p_1 = x_2, q_2 = y_1, p_2 = -y_2$.

Note that the form ω_1 corresponds to the Laplace equation $v_{xx} + v_{yy} = 0$ where $x = q_1, y = q_2$. Hence, the elliptic equation E_ω can be transformed into the Laplace equation by a symplectic transformation.

20.1.3 Parabolic equations

Now we consider the parabolic Monge–Ampère equation E_ω with the corresponding field of endomorphisms A_ω. Suppose that $\omega \neq 0$, but

$$\mathrm{Pf}(\omega) = 0$$

and

$$A_\omega^2 = 0.$$

Thus, we have $\omega \wedge \omega = 0$, i.e., the form ω is decomposable:

$$\omega = \alpha \wedge \beta$$

for some differential 1-forms α and β. Therefore, the equation E_ω defines a two-dimensional distribution

$$P = \mathcal{F}\langle \alpha, \beta \rangle \tag{20.6}$$

on T^*M. This distribution is Lagrangian because ω is effective.

In terms of the operator A_ω the plane $P(a)$ is a kernel of the linear operator $A_{\omega,a}$.

Indeed, if $X \in P(a)$, then

$$A_{\omega,a}X \rfloor \Omega_a = X \rfloor \omega_a = \alpha(X)\beta - \beta(X)\alpha = 0.$$

On the other hand, if $A_{\omega,a}X = 0$, then $\alpha(X) = \beta(X) = 0$ and, therefore, $X \in P(a)$.

So, we have $P = \ker A_\omega$. Since $A_\omega^2 = 0$, we see that $P_a = \ker A_{\omega,a} \supset \operatorname{Im} A_{\omega,a}$ at any point $a \in T^*M$. Therefore, by reason of a dimensional, at any point $a \in T^*M$ we have

$$\operatorname{Im} A_{\omega,a} = \ker A_{\omega,a} = P(a). \tag{20.7}$$

Theorem 20.1.2 *The distribution P is completely integrable if and only if the Nijenhuis tensor $N_{A_\omega} = 0$. If this condition is realized, then there exists a (local) coordinate system q, p on T^*M such that*

$$\Omega = dq_1 \wedge dp_1 + dq_2 \wedge dp_2,$$
$$\omega = dq_1 \wedge dp_2.$$

In this case the equation E_ω is locally symplectic equivalent to the equation

$$v_{xx} = 0.$$

Proof Assume that

$$N_{A_\omega} = 0. \tag{20.8}$$

From (20.7), we can choose non-zero vector fields $X \in D(P)$ and $Z \notin D(P)$ such that $A_\omega Z \neq X$ at any point of T^*M. Then the vector fields X and $Y = A_\omega Z$ form a basis of the module $D(P)$. Note that $A_\omega X = A_\omega Y = 0$. Therefore we have

$$[X, Y] = N_{A_\omega}(X, Y) - [A_\omega X, A_\omega Y] + A_\omega[A_\omega X, Y] + A_\omega[X, A_\omega Y] = 0.$$

Now suppose P is completely integrable and f_1 and g_2 are its independent integrals, i.e., $\omega = df_1 \wedge dg_2$. Let f_1 and g_1 be some functions on T^*M that are functionally independent on f_1 and g_2. Then we can choose the following (local) coordinates:

$$q_1 \stackrel{\text{def}}{=} f_1, \qquad q_2 \stackrel{\text{def}}{=} f_2, \qquad p_1 \stackrel{\text{def}}{=} g_1, \qquad p_2 \stackrel{\text{def}}{=} g_2.$$

In these coordinates $\omega = dq_1 \wedge dp_2$ and $\Omega = dq_1 \wedge dp_1 + dq_2 \wedge dp_2$. ∎

20.2 Non-degenerate quasilinear equations

The Monge–Ampère equation (20.1) is called *quasilinear* if $D = 0$, i.e., it has the following form:

$$Av_{xx} + 2Bv_{xy} + Cv_{yy} + E = 0. \tag{20.9}$$

The following problem was posed by Sophus Lie: *when is the Monge–Ampère equation (20.1) equivalent to a quasilinear one?*

The following lemma gives necessary and sufficient conditions for such an equivalence.

Lemma 20.2.1 *A Monge–Ampère equation E_ω is locally equivalent to a quasilinear one if and only if there exists a two-dimensional Lagrangian foliation of T^*M by solutions of E_ω.*

Proof Suppose the equation E_ω is quasilinear. It means that the form ω has no term $D\, dp_1 \wedge dp_2$. Hence, ω vanishes on the following two-parameter family of Lagrangian planes

$$\{p_1 = C_1, p_2 = C_2\}. \tag{20.10}$$

Suppose now that there exist a two-dimensional Lagrangian foliation π such that each of its fibres is integral for the form ω. As we have seen all Lagrangian foliations are locally equivalent. Hence, there exists a symplectic diffeomorphism φ which transforms π to the foliation $\varphi(\pi)$ with fibres (20.10). It means that the form $(\varphi^{-1})^*(\omega)$ vanishes on the foliation $\varphi(\pi)$ and, because of this, it has no term $D\,dp_1 \wedge dp$. ■

The following theorem is a simple corollary of the above lemma.

Theorem 20.2.1 *[73] Let M be a real analytic two-dimensional manifold. Any non-degenerate Monge–Ampère equation (20.1) with real analytic coefficients* A, B, C, D, E *is locally equivalent to a quasilinear equation.*

Proof Since the equation is non-degenerate at the point a, the Pfaffian does not vanish at a and therefore, in a neighborhood of a the Monge–Ampère satisfies the Cauchy–Kovalevsky theorem. Therefore, there exists a two-parameter analytic family of solutions, which defines a Lagrangian foliation. Lemma 20.2.1 completes the proof. ■

Note that Lemma 20.2.1 and Theorem 20.2.1 hold not only on T^*M, but also on J^1M. The proof is the same, but one has to consider Legendre foliations instead of Lagrangian ones.

Remark 20.2.1 *Monge–Ampère equations on manifolds of dimension higher than two are, in general, not quasilinearizable.*

20.3 Intermediate integrals and classification

Let ω be an effective differential 2-form on the cotangent bundle T^*M. Assume that the Monge–Ampère equation E_ω is of hyperbolic or parabolic type and admits two independent intermediate integrals $f, g \in C^\infty(T^*M)$. This means that E_ω can be represented by the effective part of the form $df \wedge dg$.

Theorem 20.3.1 *Suppose that a Monge–Ampère equation* E_ω *admits intermediate integrals* f, g *such that the subspace*

$$\mathcal{I} \overset{\text{def}}{=} \{h \in C^\infty(T^*\mathbb{R}^2) \mid h = C_1 + C_2 f + C_3 g,\ (C_1, C_2, C_3 \in \mathbb{R})\}$$

of the vector space $C^\infty(T^*\mathbb{R}^2)$ *generated by a triplet* $(1, f, g)$ *over* $\mathbb{R}$ *is a Poisson subalgebra with respect to the canonical Poisson algebra structure on* $C^\infty(T^*\mathbb{R}^2)$*. Then the equation* E_ω *is symplectically equivalent to one of the following normal forms:*

1. *the wave equation* $v_{xy} = 0$ *for hyperbolic equations;*
2. *the differential equation* $v_{xx} = 0$ *for parabolic equations.*

Proof Since $\mathcal{I}$ is closed with respect to the Poisson brackets, we have

$$\{f,g\} = c_1 + c_2 f + c_3 g$$

for some $c_1, c_2, c_3 \in \mathbb{R}$. Then, up to an isomorphism, there are three possible cases:

1. $\{f,g\} = 1$;
2. $\{f,g\} = 0$;
3. $\{f,g\} = f$.

The functions f and g can be transformed by a symplectic diffeomorphism into p_1 and q_1 in the first case, and into p_1 and q_2 in the second case. The third case is reduced to the first one by an evident re-scale: g is replaced by $g_0 = g/f$. Then we have $\{f, g_0\} = 1$. ω takes the form $\omega = dp_1 \wedge dq_1 + \lambda_2 \Omega$ in the first case, and the form $\omega = dp_1 \wedge dq_2 + \lambda_2 \Omega$ in the second one. The corresponding equations are $v_{xy} = 0$ and $v_{xx} = 0$ respectively. ■

20.4 Classification of generic Monge–Ampère equations

20.4.1 Monge–Ampère equations and e-structures

Here we consider a classification of generic symplectic Monge–Ampère equations.

In order to solve this problem we construct e-structures for Monge–Ampère equations as was done for Monge–Ampère operators above. Our construction does not depend on the type of equation (see [61]).

Thus, let us consider a Monge–Ampère equation E_ω, where ω is an effective 2-form on T^*M. Suppose h is a smooth non-vanishing function on the cotangent bundle T^*M. Then the forms ω and $h\omega$ generate the same Monge–Ampère equation. Because of this, a tensor T_ω is an invariant of the Monge–Ampère equation if it does not change when we multiply the form ω by h, i.e., $T_{h\omega} = T_\omega$, and $T_{\varphi^*\omega} = \varphi^*(T_\omega)$ for all symplectic diffeomorphisms φ.

Remark that the Pfaffian is not an invariant of the Monge–Ampère equations, but a relative one. Indeed,

$$\mathrm{Pf}(h\omega) = h^2\,\mathrm{Pf}(\omega). \tag{20.11}$$

In general, a tensor T_ω is called a *relative invariant* of the Monge–Ampère equation E_ω if $T_{h\omega} = h^k T_\omega$ for some real number k and for any non-vanishing function h, and $T_{\varphi^*(\omega)} = \varphi^*(T_\omega)$. If $k = 0$, then the tensor T_ω is called an *absolute invariant* or *invariant*.

Let us denote the Pfaffian of the form ω by F_ω: $F_\omega = \mathrm{Pf}(\omega)$. Let $X_\omega \stackrel{\text{def}}{=} X_{F_\omega}$ be the Hamiltonian vector field: $X_\omega \rfloor \Omega = -dF_\omega$. Then F_ω and X_ω are invariants of ω with respect to symplectic diffeomorphisms. The same holds for the vector fields $Z_\omega, W_\omega, V_\omega$.

Let us check what happens with them when we multiply the form ω by a non-vanishing function h.

Lemma 20.4.1 *For any smooth function h on T^*M we have:*

$$dh \wedge \omega = (A_\omega X_h \rfloor \Omega) \wedge \Omega,$$

where X_h is the Hamiltonian vector field: $X_h \rfloor \Omega = -dh$.

Proof The form ω is effective, i.e., $\omega \wedge \Omega = 0$. Then

$$0 = X_h \rfloor (\omega \wedge \Omega) = (A_\omega X_h \rfloor \Omega) \wedge \Omega - dh \wedge \omega.$$ ■

Since the vector field X_ω is Hamiltonian and $A_{h\omega} = hA_\omega$, we see that

$$X_{h\omega} = X_{h^2 F_\omega} = h^2 X_\omega + 2hF_\omega X_h \tag{20.12}$$

and

$$Z_{h\omega} = A_{h\omega} X_{h\omega} = h^3 Z_\omega + 2h^2 F_\omega A_\omega X_h. \tag{20.13}$$

Using Lemma 20.4.1, we have

$$W_{h\omega} \rfloor \Omega^2 = 2d(h\omega) = 2(dh \wedge \omega + h d\omega) = 2((A_\omega X_h + hW_\omega) \rfloor \Omega) \wedge \Omega.$$

Therefore

$$W_{h\omega} = hW_\omega + A_\omega X_h. \tag{20.14}$$

For the vector field $V_{h\omega}$ we have

$$V_{h\omega} = h^2 V_\omega - hF_\omega X_h. \tag{20.15}$$

Let us now construct relative invariant vector fields $Y_{1,\omega}, \ldots, Y_{4,\omega}$. We define

$$Y_{1,\omega} \stackrel{\text{def}}{=} X_\omega + 2V_\omega,$$

$$Y_{2,\omega} \stackrel{\text{def}}{=} Z_\omega - 2F_\omega W_\omega.$$

Note that

$$Y_{2,\omega} = A_\omega Y_{1,\omega}.$$

Define two 1-forms σ_1 and σ_2:

$$\sigma_{i,\omega} \stackrel{\text{def}}{=} Y_{i,\omega} \rfloor \Omega \qquad (i = 1, 2)$$

and construct vector fields $Y_{3,\omega}$ and $Y_{4,\omega}$ and differential 1-forms σ_3 and σ_4 using the following formulas:

$$\begin{aligned} Y_{3,\omega} \rfloor \Omega^2 &= 2\sigma_{1,\omega} \wedge d\sigma_{1,\omega}, \\ Y_{4,\omega} &\stackrel{\text{def}}{=} A_\omega Y_{3,\omega}, \\ \sigma_{i,\omega} &\stackrel{\text{def}}{=} Y_{i,\omega} \rfloor \Omega \quad (i = 3, 4). \end{aligned}$$

From (20.12)–(20.15) it follows that

$$\begin{aligned} Y_{1,h\omega} &= X_{h\omega} + 2V_{h\omega} = h^2(X_\omega + 2V_\omega) = h^2 Y_{1,\omega}, \\ Y_{2,h\omega} &= Z_{h\omega} - 2h^2 F_\omega W_{h\omega} = h^3(Z_\omega - 2F_\omega W_\omega) = h^3 Y_{2,\omega}. \end{aligned}$$

Therefore,

$$\sigma_{1,h\omega} \wedge d\sigma_{1,h\omega} = h^2 \sigma_{1,\omega} \wedge d(h^2 \sigma_{1,\omega}) = h^4 \sigma_{1,\omega} \wedge d\sigma_{1,\omega}.$$

Hence, the vector fields $Y_{i,\omega} (i = 1, \dots, 4)$ are relative invariants and

$$Y_{i,h\omega} = h^{i+1} Y_{i,\omega}.$$

From now on we write the vector fieldss $Y_{i,\omega}$ and 1-forms $\sigma_{i,\omega}$ without ω: $Y_i = Y_{i,\omega}$ and $\sigma_i = \sigma_{i,\omega}$.

Let us find values of the forms σ_j on the vector fields Y_i. By definition, put

$$r_\omega \stackrel{\text{def}}{=} \Omega(Y_1, Y_3), \quad s_\omega \stackrel{\text{def}}{=} \Omega(Y_2, Y_4).$$

Then:

$$\sigma_i(Y_i) = 0 \qquad (i = 1, \ldots, 4),$$

$$\sigma_i(Y_{i+1}) = \Omega(Y_i, A_\omega Y_i) = 0 \qquad (i = 1, 3),$$

$$\sigma_{i+1}(Y_i) = \Omega(A_\omega Y_i, Y_i) = 0 \qquad (i = 1, 3),$$

$$\sigma_2(Y_3) = \Omega(A_\omega Y_1, Y_3) = s_\omega,$$

$$\sigma_2(Y_4) = \Omega(A_\omega Y_1, A_\omega Y_3) = -F_\omega r_\omega.$$

We collect this in the following table:

	Y_1	Y_2	Y_3	Y_4
σ_1	0	0	r_ω	s_ω
σ_2	0	0	s_ω	$-F_\omega r_\omega$
σ_3	$-r_\omega$	$-s_\omega$	0	0
σ_4	$-s_\omega$	$F_\omega r_\omega$	0	0

Lemma 20.4.2 *The following relations:*

1. $Y_i \rfloor \omega = \sigma_{i+1}$ $\qquad (i = 1, 3)$,
2. $Y_{i+1} \rfloor \omega = -F_\omega \sigma_i$ $\qquad (i = 1, 3)$,
3. $\sigma_3 \wedge \Omega = \sigma_1 \wedge d\sigma_1$,
4. $\sigma_i \wedge \omega = -\sigma_{i+1} \wedge \Omega$ $\qquad (i = 1, 3)$,
5. $\sigma_{i+1} \wedge \omega = F_\omega \sigma_i \wedge \Omega$ $\qquad (i = 1, 3)$,
6. $\sigma_2 \wedge \sigma_3 \wedge \Omega = \sigma_1 \wedge \sigma_4 \wedge \Omega$,
7. $\sigma_4 \wedge \sigma_2 \wedge \Omega = F_\omega \sigma_1 \wedge \sigma_3 \wedge \Omega$,
8. $\sigma_1 \wedge \sigma_2 \wedge \sigma_3 = s_\omega \sigma_1 \wedge \Omega - r_\omega \sigma_2 \wedge \Omega$

hold.

Proof The proofs of relations 1 and 2 are similar. For example,

$$Y_{i+1} \rfloor \omega = -A_\omega Y_{i+1} \rfloor \Omega = A_\omega^2 Y_i \rfloor \Omega = -F_\omega Y_i \rfloor \Omega = -F_\omega \sigma_i$$

proves relation 2, and

$$2\sigma_1 \wedge d\sigma_1 = Y_3 \rfloor \Omega^2 = 2(Y_3 \rfloor \Omega) \wedge \Omega = 2\sigma_3 \wedge \Omega.$$

proves relation 3.

The proofs of relations 4 and 5 are similar.

Let us prove relation 4. Since $\omega \wedge \Omega = 0$, we have

$$0 = Y_i \rfloor (\omega \wedge \Omega) = \sigma_{i+1} \wedge \Omega + \sigma_i \wedge \omega.$$

To prove relation 6. we note that from relations 4 and 5 it follows that

$$\sigma_3 \wedge \sigma_2 \wedge \Omega = -\sigma_3 \wedge \sigma_1 \wedge \omega = \sigma_1 \wedge \sigma_3 \wedge \omega = -\sigma_1 \wedge \sigma_4 \wedge \Omega.$$

In the similar way relation 7 can be proved. Multiplying both sides of relation 6, by $Y_{1,\omega}$, we get relation 8. ■

Lemma 20.4.3 *One has*

$$2\sigma_1 \wedge \sigma_2 \wedge \sigma_3 \wedge \sigma_4 = g_\omega \Omega^2,$$

where

$$g_\omega \stackrel{\text{def}}{=} F_\omega r_\omega^2 + s_\omega^2.$$

Proof It follows from relation 8 that

$$\left(F_\omega r_\omega^2 + s_\omega^2\right)\Omega = s_\omega \sigma_2 \wedge \sigma_3 + r_\omega F_\omega \sigma_1 \wedge \sigma_3 + r_\omega \sigma_4 \wedge \sigma_2 + s_\omega \sigma_1 \wedge \sigma_4.$$

Multiplying both sides by Ω and using relations 7 and 8, we obtain

$$\left(F_\omega r_\omega^2 + s_\omega^2\right)\Omega^2 = 2(s_\omega \sigma_2 \wedge \sigma_3 + r_\omega \sigma_4 \wedge \sigma_2) \wedge \Omega.$$

Multiplying relation 8 by σ_4 and using relation 6 we obtain from the last formula:

$$\sigma_1 \wedge \sigma_2 \wedge \sigma_3 \wedge \sigma_4 = s_\omega \, \sigma_2 \wedge \sigma_3 \wedge \Omega + r_\omega \sigma_4 \wedge \sigma_2 \wedge \Omega = \frac{1}{2} g_\omega \Omega^2. \quad ■$$

It follows from this lemma that the vectors $Y_{1,a}, \ldots, Y_{4,a}$ are linearly independent if and only if

$$g_\omega(a) \neq 0. \tag{20.16}$$

In what follows we assume that condition (20.16) holds.

Note that

$$r_{h\omega} = h^6 r_\omega, \quad s_{h\omega} = h^7 s_\omega, \quad g_{h\omega} = h^{14} g_\omega.$$

This enables us to construct the following.

1. Functional invariants of the equation E_ω:

$$F \stackrel{\text{def}}{=} F_\omega s_\omega^{-2/7},$$
$$I = r_\omega s_\omega^{-6/7}.$$

2. Invariant vector fields

$$X_1 \stackrel{\text{def}}{=} \frac{s_\omega^{2/7}}{g_\omega}(F_\omega r_\omega Y_3 + s_\omega Y_4),$$
$$X_2 \stackrel{\text{def}}{=} -\frac{1}{s_\omega^{2/7}} Y_1,$$
$$X_3 \stackrel{\text{def}}{=} \frac{s_\omega^{3/7}}{g_\omega}(s_\omega Y_3 - r_\omega Y_4),$$
$$X_4 \stackrel{\text{def}}{=} -\frac{1}{s_\omega^{3/7}} Y_2.$$

3. Invariant differential 1-forms

$$\theta_1 \stackrel{\text{def}}{=} \frac{1}{s_\omega^{2/7}}\sigma_1,$$
$$\theta_2 \stackrel{\text{def}}{=} \frac{s_\omega^{2/7}}{g_\omega}(F_\omega r_\omega \sigma_3 + s_\omega \sigma_4),$$
$$\theta_3 \stackrel{\text{def}}{=} \frac{1}{s_\omega^{3/7}}\sigma_2,$$
$$\theta_4 \stackrel{\text{def}}{=} \frac{s_\omega^{3/7}}{g_\omega}(s_\omega \sigma_3 - r_\omega \sigma_4).$$

Note that for a general Monge–Ampère equation these vector fields (and differential 1-forms) are linearly independent and form bases of the modules $D(\mathcal{O}_a)$ and $\Omega^1(\mathcal{O}_a)$ respectively in a neighborhood $\mathcal{O}_a$ of a point a. In other words they produce an e-structure in $\mathcal{O}_a$ related to the Monge–Ampère equation.

Let us show how to normalize the form ω. We set

$$\widehat{\omega} \stackrel{\text{def}}{=} s_\omega^{-1/7}\omega.$$

Note that the function F and the field of endomorphisms

$$A \stackrel{\text{def}}{=} s_\omega^{-1/7} A_\omega$$

are invariants.

It is easy to see that

$$\begin{aligned} \theta_1 &= -X_2 \rfloor \Omega, & \theta_3 &= -X_4 \rfloor \Omega, \\ \theta_2 &= X_1 \rfloor \Omega, & \theta_4 &= X_3 \rfloor \Omega \end{aligned} \tag{20.17}$$

and

$$AX_2 = X_4, \qquad AX_3 = X_1.$$

In addition, the bases $X_1, \ldots, X_4$ and $\theta_1, \ldots, \theta_4$ are dual.

Summarizing we get the following result.

Theorem 20.4.1 *Suppose that $\omega_a \neq 0$, $s_\omega(a) \neq 0$ and $g_\omega(a) \neq 0$. Then in a neighborhood of the point a we have the following representation of the forms Ω and $\widehat{\omega}$:*

$$\begin{aligned} \Omega &= \theta_1 \wedge \theta_2 + \theta_3 \wedge \theta_4, \\ \widehat{\omega} &= \theta_3 \wedge \theta_2 + F\theta_4 \wedge \theta_1. \end{aligned} \tag{20.18}$$

This theorem shows that two Monge–Ampère equations $E_{\widehat{\omega}_1}$ and $E_{\widehat{\omega}_2}$ are locally equivalent if there exists a local diffeomorphism φ and such that $\varphi^*(F_2) = F_1, \varphi^*\left(\theta_j^2\right) = \theta_j^1$, $j = 1, 2, 3, 4$, and $\varphi(a) = a$.

Theorem 20.4.2 *Symmetries of a Monge–Ampère equation are symmetries of the e-structure preserving the function F.*

20.4.2 Normal forms of mixed-type equations

We use canonical representation (20.18) to construct normal forms of mixed-type Monge–Ampère equations.

We shall suppose that for the form ω the following conditions hold:

1. $F(a) = 0$ and the differential of the function F does not vanish at the point a: $dF_a \neq 0$;
2. the first covariant derivatives of the function F are constants, i.e., $X_j(F) = \text{constant}(j = 1, \ldots, 4)$;

3. the vector fields $X_1, \ldots, X_4$ form the four-dimensional Lie algebra $\mathfrak{g} = \bigoplus_{i=1}^4 \mathbb{R}X_i$.

Denote by c_{ij}^k the structure constants of the Lie algebra $\mathfrak{g}$:

$$[X_i, X_j] = \sum_{k=1}^{4} c_{ij}^k X_k.$$

The condition $d\Omega = 0$ gives the following extra conditions for the structure constants:

$$\begin{aligned} c_{13}^1 + c_{23}^2 + c_{12}^4 = 0, \qquad & c_{14}^1 + c_{24}^2 - c_{12}^3 = 0, \\ c_{34}^2 - c_{13}^3 - c_{14}^4 = 0, \qquad & c_{34}^1 + c_{23}^3 - c_{24}^4 = 0. \end{aligned} \tag{20.19}$$

For the 2-form $\widehat{\omega}$ one has

$$\sigma_{1,\widehat{\omega}} = \theta_1, \quad \sigma_{2,\widehat{\omega}} = \theta_3, \quad \sigma_{3,\widehat{\omega}} = \theta_4 + r_{\widehat{\omega}}\theta_2, \quad \sigma_{4,\widehat{\omega}} = \theta_2 - Fr_{\widehat{\omega}}\theta_4,$$

and from the third property of Lemma 20.4.2 it follows that

$$c_{23}^1 = c_{34}^1 = 0, \quad c_{24}^1 = -1$$

and $r_{\widehat{\omega}} = 0$.

Note that

$$\begin{aligned} X_{\widehat{\omega}} &= -f^2 X_1 + f^1 X_2 - f^4 X_3 + f^3 X_4, \\ V_{\widehat{\omega}} &= \alpha_3 X_1 - \alpha_4 F X_2 - \alpha_1 F X_3 + \alpha_2 X_4, \end{aligned}$$

where $f^i \stackrel{\text{def}}{=} X_i(F)$, and

$$\begin{aligned} \alpha_1 &= -c_{34}^3 - c_{24}^2, \\ \alpha_2 &= c_{14}^2 + (c_{34}^4 - c_{13}^1)F - f^3, \\ \alpha_3 &= c_{14}^3 + (c_{12}^1 - c_{24}^4)F + f^2, \\ \alpha_4 &= -c_{12}^2 - c_{13}^3 + c_{23}^4 F. \end{aligned}$$

Then,

$$Y_1 = (2\alpha_3 - f^2)X_1 + (f^1 - 2\alpha_4 F)X_2 - (f^4 + 2\alpha_1 F)X_3 + (f^3 + 2\alpha_2)X_4.$$

Remark that $s_{\widehat{\omega}} = 1$ and, therefore, $Y_1 = -X_2$, and

$$2\alpha_3 - f^2 = f^4 + 2\alpha_1 F = f^3 + 2\alpha_2 = 0, \qquad f^1 - 2\alpha_4 F = -1.$$

Finally,

$$\begin{aligned} f^1 &= -1 - 2(c_{12}^2 + c_{13}^3)F + 2c_{23}^4 F^2, \\ f^2 &= -2c_{14}^3 + 2(c_{24}^4 - c_{12}^1)F, \\ f^3 &= 2c_{14}^2 + 2(c_{34}^4 - c_{13}^1)F, \\ f^4 &= 2(c_{24}^2 + c_{34}^3)F. \end{aligned}$$

Since all derivatives are constants and $dF \neq 0$, we obtain

$$\begin{aligned} &c_{23}^4 = c_{12}^2 + c_{13}^3 = c_{34}^4 - c_{13}^1 = c_{24}^4 - c_{12}^1 = c_{24}^2 + c_{34}^3 = 0 \\ &f^1 = -1, \quad f^2 = -2c_{14}^3, \quad f^3 = 2c_{14}^2, \quad f^4 = 0. \end{aligned} \tag{20.20}$$

From the Jacobi identity for the Lie algebra $\mathfrak{g}$ we find, that

$$c_{13}^1 - c_{23}^2 + c_{34}^4 = 0.$$

Note that

$$f^{ji} - f^{ij} = \left[X_i, X_j\right](F) = \sum_{k=1}^{4} c_{ij}^k f^k,$$

where $f^{ji} \overset{\text{def}}{=} X_i(f^j)$. But from condition 2 we see that $f^{ij} = 0$ and $\sum_{k=1}^{4} c_{ij}^k f^k = 0$.

Summarizing, we obtain the following relations for the structure constants c_{ij}^k:

$$\begin{aligned} &c_{14}^1 = 0, && 2(c_{24}^2 c_{14}^3 - c_{24}^3 c_{14}^2) = 1, \\ &c_{14}^2 c_{23}^3 - c_{23}^2 c_{14}^3 = 0, && c_{34}^2 c_{14}^3 - c_{34}^3 c_{14}^2 = 0, \\ &c_{12}^1 + 2(c_{12}^2 c_{14}^3 - c_{12}^3 c_{14}^2) = 0, && c_{13}^1 + 2(c_{13}^2 c_{14}^3 - c_{13}^3 c_{14}^2) = 0. \end{aligned} \tag{20.21}$$

There exists a unique Lie algebra $\mathfrak{g}$ satisfying conditions (20.19)–(20.21) and its structure is given by the following multiplication table.

$[\uparrow,\rightarrow]$	X_1	X_2	X_3	X_4
X_1	0	0	0	$bX_2 + aX_3$
X_2	0	0	0	$-X_1 - \frac{1}{2b}X_3$
X_3	0	0	0	0
X_4	$-bX_2 - aX_3$	$X_1 + \frac{1}{2b}X_3$	0	0

Here $a \neq 0$ and $b \neq 0$ are arbitrary constants. This algebra has the following representation by vector fields:

$$\begin{aligned}
X_1 &= \frac{\partial}{\partial q_1} - bq_1\frac{\partial}{\partial q_2},\\
X_2 &= -p_2\frac{\partial}{\partial q_2} + \frac{\partial}{\partial p_1},\\
X_3 &= \frac{\partial}{\partial q_2},\\
X_4 &= -p_1\frac{\partial}{\partial q_1} + \left(aq_1 - \frac{1}{2b}p_1\right)\frac{\partial}{\partial q_2} + bq_1\frac{\partial}{\partial p_1} + \frac{\partial}{\partial p_2}.
\end{aligned}$$

Therefore,

$$\begin{aligned}
\theta_1 &= dq_1 + p_1\,dp_2,\\
\theta_2 &= dp_1 - bq_1\,dp_2,\\
\theta_3 &= dq_2 + bq_1\,dq_1 + p_1 dp_1 + \left(\frac{1}{2b}p_1 - aq_1\right)dp_2,\\
\theta_4 &= dp_2.
\end{aligned}$$

Assume that $a = (0, 0, 0, 0)$. Then

$$F = -q_1 + b^2q_1^2 + 2bq_2 - 2ap_1 + bp_1^2$$

and we obtain the following normal effective 2-form:

$$\begin{aligned}
\widehat{\omega} = &-(F + b^2q_1^2)\,dq_1\wedge dp_2 + dq_2\wedge dp_1\\
&+ bq_1(dq_1\wedge dp_1 - dq_2\wedge dp_2) + \left(aq_1 - \frac{1}{2b}p_1 - bp_1q_1\right)dp_1\wedge dp_2.
\end{aligned}$$

The corresponding Monge–Ampère equation E_{ϖ} is

$$v_{xx} - 2bxv_{xy} + (2by - x - 2av_x - bv_x^2)v_{yy} + \left(\left(\frac{1}{2b} + bx\right) v_x - ax\right) \operatorname{Hess} v = 0, \tag{20.22}$$

where $x = q_1$and $y = q_2$.

Theorem 20.4.3 *Assume that for equation E_ω derivatives of the function F along $X_1, \ldots, X_4$ are constant and these vector fields form a four-dimensional Lie algebra. Then this equation at the point $a \in \{F_\omega = 0\}$ is locally equivalent to (20.22).*

20.5 Applications

In this section we apply the obtained results to some popular differential equations.

20.5.1 The Born–Infeld equation

The effective differential 2-form

$$\widetilde{\omega} = (1 - p_1^2)\, dq_1 \wedge dp_2 + p_1 p_2 (dq_1 \wedge dp_1 - dq_2 \wedge dp_2) + (1 + p_2^2)\, dq_2 \wedge dp_1$$

corresponds to the Born–Infeld equation [11, 111]

$$(1 - \varphi_t^2)\varphi_{xx} + 2\varphi_t\varphi_x\varphi_{tx} - (1 + \varphi_x^2)\varphi_{tt} = 0, \tag{20.23}$$

where $q_1 = t, q_2 = x$. Its Pfaffian is $\operatorname{Pf}(\widetilde{\omega}) = p_1^2 - p_2^2 - 1$. We consider a domain where this equation is hyperbolic, i.e. $p_1^2 - p_2^2 - 1 < 0$. The normed form is

$$\omega = \frac{\widetilde{\omega}}{\sqrt{1 - p_1^2 + p_2^2}}.$$

For this form

$$\mu_+ = \frac{-p_1 + p_2\sqrt{1 - p_1^2 + p_2^2}}{-1 + p_1^2 - p_2^2}\, dp_1 + \frac{p_2 - p_1\sqrt{1 - p_1^2 + p_2^2}}{-1 + p_1^2 - p_2^2}\, dp_2,$$

$$\mu_- = \frac{p_1 + p_2\sqrt{1 - p_1^2 + p_2^2}}{-1 + p_1^2 - p_2^2}\, dp_1 - \frac{p_2 + p_1\sqrt{1 - p_1^2 + p_2^2}}{-1 + p_1^2 - p_2^2}\, dp_2$$

and the distributions $\ker \mu_+$ and $\ker \mu_-$ are completely integrable and

$$g_0 = g_+ = g_- = 0.$$

Due to Theorem 19.2.3, the form $\widetilde{\omega}$ is equivalent to the form (19.8). Integrals of the distributions $\ker \mu_+$ and $\ker \mu_-$ are

$$H_+ = \frac{p_1 + p_2}{1 + \sqrt{1 - p_1^2 + p_2^2}}$$

and

$$H_- = \frac{-p_1 + p_2}{1 + \sqrt{1 - p_1^2 + p_2^2}}$$

respectively. Follow the proof of Theorem 19.2.2, we can introduce new coordinates

$$\begin{aligned}
Q_1 &= H_+, \\
Q_2 &= H_-, \\
P_1 &= -\frac{1}{2}\Big[q_1\Big(1 + p_1p_2 - p_1^2 + \sqrt{1 - p_1^2 + p_2^2}\Big) \\
&\quad + q_2(1 - p_1p_2 + p_2^2 + \sqrt{1 - p_1^2 + p_2^2}\Big)\Big], \\
P_2 &= \frac{1}{2}\Big[q_1\Big(1 - p_1^2 - p_1p_2 + \sqrt{1 - p_1^2 + p_2^2}\Big) \\
&\quad - q_2\Big(1 + p_1p_2 + p_2^2 + \sqrt{1 - p_1^2 + p_2^2}\Big)\Big].
\end{aligned}$$

The inverse symplectic transformation is

$$\begin{aligned}
q_1 &= \frac{(Q_1Q_2 - 1)((1 + Q_1^2)P_1 - (1 + Q_2^2)P_2)}{2(1 + Q_1Q_2)}, \\
q_2 &= \frac{(Q_1Q_2 - 1)((1 - Q_1^2)P_1 + (1 - Q_2^2)P_2)}{2(1 + Q_1Q_2)}, \\
p_1 &= \frac{-Q_1 + Q_2}{-1 + Q_1Q_2}, \\
p_2 &= \frac{Q_1 + Q_2}{1 - Q_1Q_2}
\end{aligned}$$

and we get the following 2-form:

$$\omega = \frac{4(Q_1P_1 + Q_2P_2)}{(Q_1Q_2)^2 - 1}\, dQ_1 \wedge dQ_2 - dP_1 \wedge dQ_1 + dP_2 \wedge dQ_2.$$

This means that in the domain of hyperbolicity Born–Infeld's equation (20.23) is equivalent to the following linear wave equation:

$$v_{Q_1Q_2} = \frac{2(Q_1 v_{Q_1} + Q_2 v_{Q_2})}{(Q_1Q_2)^2 - 1}. \tag{20.24}$$

20.5.2 Gas-dynamic equations

The following equation describes one-dimensional non-stationary gas flow:

$$\varphi_{tt} + 2\varphi_x\varphi_{tx} + \left((k-1)\varphi_t + \frac{k+1}{2}\varphi_x^2\right)\varphi_{xx} = 0. \tag{20.25}$$

Here φ is a velocity potential and k is a constant (an adiabatic index), $k > 1$. This equation is generated by the following differential 2-form:

$$\begin{aligned}\widetilde{\omega} &= dq_1 \wedge dp_1 + (dq_1 \wedge dp_1 - dq_2 \wedge dp_2)p_1 \\ &\quad - \left(\frac{k+1}{2}p_1^2 + (k-1)p_2\right) dq_2 \wedge dp_1.\end{aligned}$$

Here $q_1 = x$ and $q_2 = t$.

Suppose that the Pfaffian of ω is negative:

$$\mathrm{Pf}(\omega) = \frac{1}{2}(k-1)(p_1^2 + 2p_2) < 0.$$

Then the corresponding normed differential 2-form is

$$\omega = \sqrt{\frac{2}{(1-k)(p_1^2 + 2p_2)}}\,\widetilde{\omega}.$$

For this 2-form the conditions of Theorem 19.2.3 are satisfied and

$$\mu_+ = \frac{k-3}{4(k-1)(p_1^2 + 2p_2)}\left(\left(2p_1 + \sqrt{2}\sqrt{(1-k)(p_1^2 + 2p_2)}\right) dp_1 + 2\, dp_2\right),$$

$$\mu_- = \frac{k-3}{4(k-1)(p_1^2 + 2p_2)}\left(\left(2p_1 - \sqrt{2}\sqrt{(1-k)(p_1^2 + 2p_2)}\right) dp_1 + 2\, dp_2\right).$$

Integrals of the distributions $\ker \mu_+$ and $\ker \mu_-$ are

$$H_+ = \frac{p_1}{\sqrt{2}} + \frac{\sqrt{(1-k)(p_1^2+2p_2)}}{1-k},$$

$$H_- = -\frac{p_1}{\sqrt{2}} + \frac{\sqrt{(1-k)(p_1^2+2p_2)}}{1-k}$$

respectively.

The following symplectic transformation

$$q_1 \longmapsto \frac{P_2((k-3)Q_2+(1+k)Q_1 - P_1((k-3)Q_1+(1+k)Q_2))}{\sqrt{2}(k-1)(Q_1+Q_2)},$$

$$q_2 \longmapsto \frac{2(P_1+P_1)}{(k-1)(Q_1+Q_2)},$$

$$p_1 \longmapsto \frac{1}{\sqrt{2}}(Q_1-Q_2),$$

$$p_2 \longmapsto -\frac{1}{8}(2(k-3)Q_1Q_2+(k+1)(Q_1^2+Q_2^2))$$

transforms integrals H_+ and H_- into Q_1 and Q_2 and the differential 1-forms μ_+ and μ_- into

$$\frac{3-k}{(k-1)(Q_1+Q_2)}dQ_1 \quad \text{and} \quad \frac{k-3}{(k-1)(Q_1+Q_2)}dQ_2$$

respectively.

In these coordinates the differential 2-form ω has the form

$$\omega = \frac{(k-3)(P_1+P_2)}{(1-k)(Q_1+Q_2)}dQ_1 \wedge dQ_2 + dQ_1 \wedge dP_1 - dQ_2 \wedge dP_2.$$

So, in a hyperbolicity domain equation (20.25) is equivalent to the following linear wave equation:

$$v_{Q_1Q_2} = \frac{(k-3)(v_{Q_1}+v_{Q_2})}{2(1-k)(Q_1+Q_2)}. \tag{20.26}$$

For (20.26) one can construct a class of partial solutions $v(Q_1, Q_2) = \Phi(Q_1 + Q_2)$. For the function Φ we have an ordinary differential equation of second order:

$$\Phi'' = \frac{k-3}{(1-k)z}\Phi',$$

where $z = Q_1 + Q_2$.

Its general solution is

$$\Phi(z) = C_1 z^{2/(k-1)} + C_2$$

(C_1 and C_2 are arbitrary constants) and we obtain the following two-parameter family of solutions of (20.26):

$$v(Q_1, Q_2) = C_1(Q_1 + Q_2)^{2/(k-1)} + C_2. \tag{20.27}$$

To construct multivalued solutions of the non-linear equation (20.25) corresponding to (20.27), we write the multivalued solution of (20.26):

$$P_1 = P_2 = \frac{2C_1}{k-1}(Q_1 + Q_2)^{(3-k)/(k-1)}.$$

But

$$Q_1 = \frac{p_1}{\sqrt{2}} + \frac{\sqrt{(1-k)(p_1^2 + 2p_2)}}{1-k},$$

$$Q_2 = -\frac{p_1}{\sqrt{2}} + \frac{\sqrt{(1-k)(p_1^2 + 2p_2)}}{1-k},$$

$$P_1 = -\frac{1}{2\sqrt{2}}\Big(2q_1 + q_2\Big(-2p_1 + \sqrt{2}\sqrt{(1-k)(p_1^2 + 2p_2)}\Big)\Big),$$

$$P_2 = -\frac{1}{2\sqrt{2}}\Big(-2q_1 + q_2\Big(2p_1 + \sqrt{2}\sqrt{(1-k)(p_1^2 + 2p_2)}\Big)\Big),$$

and we obtain the following family of multivalued solutions of (20.25):

$$q_1 - q_2 p_1 = 0, \qquad 2p_2 + p_1^2 = z(q_2, C)$$

where a function $z(q_2)$ is a solution of the following functional equation:

$$C(-z)^{1/(k-1)} + (k-1)^{k/(k-1)} q_2 z = 0.$$

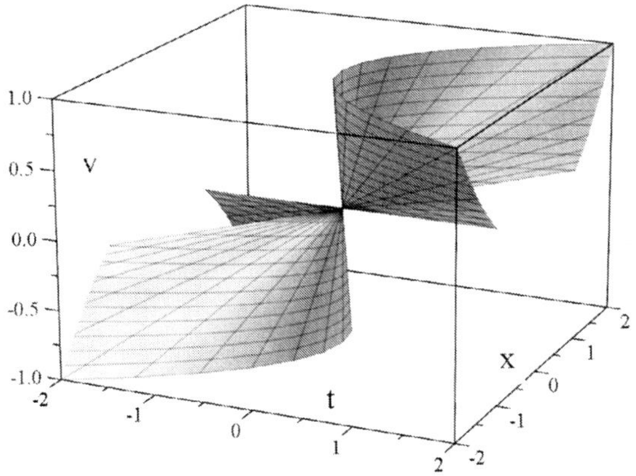

Figure 20.1. The solution $\varphi(x,t) = x^2/2t$.

For example, we can put $C = 0$ and then for any $k > 1, k \neq 3$ we have the multivalued solution

$$L = \{q_1 - q_2 p_1 = 0, 2p_2 + p_1^2 = 0\}.$$

The corresponding classical solution is (see Figure 20.1)

$$\varphi(x,t) = \frac{x^2}{2t}.$$

In the case when $k = 3$ we have $d\omega = 0$ and (20.25) is equivalent to the wave equation with constant coefficients

$$v_{Q_1 Q_2} = 0.$$

20.5.3 Two-dimensional stationary irrotational isentropic flow of a gas

The following equation

$$(\varphi_x^2 - c^2)\varphi_{xx} + 2\varphi_x\varphi_y\varphi_{xy} + (\varphi_y^2 - c^2)\varphi_{yy} = 0 \qquad (20.28)$$

describes two-dimensional stationary irrotational isentropic flow of a gas [7]. Here φ is the velocity potential and c^2 is the square of the velocity of sound. It is easy to see that the corresponding form ω is a wave type and $g_0 = g_+ = g_- = 0$. Due to Theorem 19.2.3, this equation is equivalent to the equation

$$v_{q_1 q_2} = A(q)v_{q_1} + B(q)v_{q_2}$$

for some functions $A(q)$ and $B(q)$. Indeed, it is easy to see that the godograph transformation

$$(q_1, q_2, p_1, p_2) \mapsto (p_1, p_2, -q_1, -q_2)$$

sends (20.28) into a linear equation.

21

Contact classification of MAEs on two-dimensional manifolds

Let $\omega, \tilde{\omega} \in \Omega^2_\varepsilon(J^1M)$. We say that two Monge–Ampère equations E_ω and $E_{\tilde{\omega}}$ are *equivalent* if there exists a contact transformation φ of J^1M and a non-vanishing function $h_\varphi \in C^\infty(J^1M)$, such that $\varphi^*(\omega)_\varepsilon = h\tilde{\omega}$.

We say that two Monge–Ampère equations E_ω and $E_{\tilde{\omega}}$ are *locally equivalent at a point* $a \in J^1M$ if there exists a contact transformation φ of a neighborhood $\mathcal{O}_a$ of a and a function $h_\varphi \in C^\infty(\mathcal{O}_a)$, such that $\varphi^*(\omega)_\varepsilon = h_\varphi\tilde{\omega}$, $\varphi(a) = a$, where $h_\varphi(a) \neq 0$.

21.1 Classes $H_{k,l}$

In this chapter we consider Monge–Ampère equations of constant types (hyperbolic, elliptic or parabolic). In order to unify all types, we consider a complex tangent bundle $T^{\mathbb{C}}(J^1M)$. Let E_ω be a Monge–Ampère equation and let C_j ($j = 1, 2$) be a subbundle of $T^{\mathbb{C}}(J^1M)$ whose fibre $C_j(a)$ at each point $a \in J^1M$ is a complex eigenspace of the operator A_ω, i.e.,

$$C_j(a) = \{X \in T^{\mathbb{C}}_a(J^1M) |\ A_\omega X = \lambda_j X\},$$

where λ_j is an eigenvalue of A_ω. Thus we have the complex distributions C_j on J^1M:

$$C_j : a \in J^1M \mapsto C_j(a) \subset T^{\mathbb{C}}_a(J^1M).$$

Let $C_j^{(k)}$ be the kth derivative of the distribution C_j: $C_j^{(0)} \stackrel{\text{def}}{=} C_j$.

We assume that all $C_j^{(k)}$ are distributions. Then for these distributions we expect one of the following five cases:

(0) $C_j^{(1)} = C_j$;

(1) $C_j \neq C_j^{(1)} = C_j^{(2)}$ and $\dim_{\mathbb{C}} C_j^{(1)} = 3$;

(2) $C_j \neq C_j^{(1)} \neq C_j^{(2)} = C_j^{(3)}$ and $\dim_{\mathbb{C}} C_j^{(1)} = 3$, $\dim_{\mathbb{C}} C_j^{(2)} = 4$;
(3) $C_j \neq C_j^{(1)} \neq C_j^{(2)} \neq C_j^{(3)} = T^{\mathbb{C}}(J^1M)$ and $\dim_{\mathbb{C}} C_j^{(1)} = 3$, $\dim_{\mathbb{C}} C_j^{(2)} = 4$;
(4) $C_j \neq C_j^{(1)} \neq C_j^{(2)} = T^{\mathbb{C}}(J^1M)$ and $\dim_{\mathbb{C}} C_j^{(1)} = 3$.

We say that the Monge–Ampère equation *belongs to the class* $H_{k,l}$ at a point $a \in J^1M$ $(k,l = 0,\ldots,4; k \leq l)$ if case (k) holds for one of C_j and case (l) holds for the other. Classes $H_{k,l}$ are invariant under contact transformations.

Note that the class $H_{0,0}$ contains only parabolic equations. The classes $H_{k,k}$ $(1 \leq k \leq 4)$ contain only parabolic or elliptic equations. For hyperbolic equations eight classes $H_{k,l}$ $(k,l = 1,\ldots,4; k \leq l)$ are possible.

Lemma 21.1.1 *Let functions $F_1, F_2 \in C^\infty(J^1\mathbb{R}^2)$ be such that $(dF_1 \wedge dF_2 \wedge \omega_0)_a \neq 0$ and $dF_1 \wedge dF_2 \wedge \omega_0 \wedge d\omega_0 = 0$. Then there exists a contact diffeomorphism φ such that $\varphi^*(F_1) = q_1$, $\varphi^*(F_2) = q_2$.*

Proof It follows from the conditions of the lemma that the two-dimensional distribution $P = \mathcal{F}\langle dF_1, dF_2, \omega_0\rangle$ is completely integrable. Let F_0 be an integral of P such that $dF_0 \wedge dF_1 \wedge dF_2 \neq 0$. Then $\omega_0 = \sum_{i=0}^{2} a_i\, dF_i$ for some smooth functions a_i $(i = 0,1,2)$. Since $(dF_1 \wedge dF_2 \wedge \omega_0)_a \neq 0$, we see that $a_0 \neq 0$. Therefore $\omega_0 = a_0(dF_0 - F_3\, dF_1 - F_4\, dF_2)$, where $F_3 \stackrel{\text{def}}{=} a_1/a_0$, $F_4 \stackrel{\text{def}}{=} -a_2/a_0$. Moreover, the condition $\omega_0 \wedge d\omega_0 \wedge d\omega_0 \neq 0$ implies that the functions $F_0,\ldots,F_4$ are functionaly independent. Then $\psi_1 : (q_1, q_2, u, p_1, p_2) \to (F_1, F_2, F_0, F_3, F_4)$ is the contact diffeomorphism. ■

Let us introduce the following modules of differential 1-forms:

$$\Omega^1(C_j) \stackrel{\text{def}}{=} \{\alpha \in \Omega^1(J^1M)^{\mathbb{C}} | X_1 \rfloor \alpha = 0 \text{ and } X \rfloor \alpha = 0 \quad \forall X \in D(C_{3-j})\}.$$

Since the form $\Omega_a = d\omega_0|_{C(a)}$ is non-degenerate on $C(a)$, we see that the map

$$D(C_j) \ni X \mapsto X \rfloor \Omega \in \Omega^1(C_j)$$

is an isomorphism of the modules.

Moreover, since C_1 and C_2 are skew-orthogonal

$$\alpha_1 \wedge \alpha_2 \wedge d\omega_0 = 0 \tag{21.1}$$

for $\alpha_1 \in \Omega^1(C_1)$ and $\alpha_2 \in \Omega^1(C_2)$.

Theorem 21.1.1 (D. Tunitskii) *Assume that a Monge–Ampère equation belongs to one of the classes $H_{k,l}$ $(0 \leq k \leq l \leq 2)$. Then this equation is* *locally contact equivalent to a Monge–Ampère equation of type*

$$v_{xx} + \varepsilon v_{yy} = f(x, y, v, v_x, v_y), \tag{21.2}$$

where $\varepsilon = \operatorname{sgn}(\operatorname{Pf}(\omega))$ *and* f *is a smooth function.*

Proof **Hyperbolic case.** In this case the distributions $C_1 = C_+$ and $C_2 = C_-$ are real and

$$C_\pm(a) = \{X \in C(a) | A_{\omega,a}X = \pm\lambda X\}$$

where $\lambda = \sqrt{\operatorname{Pf}(\omega)}$.

Denote by F_+ and F_- first integrals of the distributions $C_+^{(2)}$ and $C_-^{(2)}$ respectively. Then $d_pF_+ \in \Omega^1(C_-)$ and $d_pF_- \in \Omega^1(C_+)$. By (21.1), we see that $d_pF_+ \wedge d_pF_- \wedge d\omega_0 = 0$. Therefore $dF_+ \wedge dF_- \wedge d\omega_0 \wedge \omega_0 = 0$. Moreover, $dF_+ \wedge dF_- \wedge \omega_0 \neq 0$. Due to Lemma 21.1.1, suppose that $F_+ = q_1$ and $F_- = q_2$ in local canonical coordinates. We will still denote the corresponding form $\varphi^*(\omega)_\varepsilon$ by ω.

Then

$$\omega \wedge \omega - \operatorname{Pf}(\omega)\Omega \wedge \Omega = (\omega + \lambda\Omega) \wedge (\omega - \lambda\Omega) = 0.$$

Therefore, the 2-forms $\omega + \lambda\Omega$ and $\omega - \lambda\Omega$ are decomposable and $\omega \pm \lambda\Omega = \alpha^\pm \wedge \beta^\pm$ for some differential 1-forms $\alpha^+, \beta^+, \alpha^-, \beta^-$. Note that $\alpha^+, \beta^+ \in \Omega^1(C_+)$ and $\alpha^-, \beta^- \in \Omega^1(C_-)$. Indeed for a vector field $X \in D(C_-)$ we have $X_1 \rfloor (\omega + \lambda\Omega) = 0$ and

$$X \rfloor (\omega + \lambda\Omega) = X \rfloor \omega + \lambda X \rfloor \Omega = (A_\omega X + \lambda X) \rfloor \Omega = (-\lambda X + \lambda X) \rfloor \Omega = 0.$$

We can choose $\alpha^+ = dq_1$ and $\alpha^- = dq_2$.

Then

$$\omega + \lambda\Omega = dq_1 \wedge \beta^+,$$
$$\omega - \lambda\Omega = dq_2 \wedge \beta^-$$

and

$$\Omega = \frac{1}{2\lambda}(dq_1 \wedge \beta^+ - dq_2 \wedge \beta^-).$$

On the other hand, $\Omega = dq_1 \wedge dp_1 + dq_2 \wedge dp_2$, therefore

$$\beta^+ = a_1\, dq_1 + a_2\, dq_2 + 2\,\lambda\, dp_1$$

and

$$\beta^- = b_1\, dq_1 + b_2\, dq_2 - 2\,\lambda\, dp_2,$$

where a_i, b_i $(i = 1, 2)$ are smooth functions, such that $a_2 + b_1 = 0$.

Then

$$\omega = \tfrac{1}{2}(dq_1 \wedge \beta^+ + dq_2 \wedge \beta^-) = s\, dq_1 \wedge dq_2 + dq_1 \wedge dp_1 - dq_2 \wedge dp_2,$$

with $s = a_2 - b_1$.

Finally the following contact transformation

$$\chi : \begin{cases} q_1 \to \tfrac{1}{2}(q_1 - q_2 + q_1^0 + q_2^0), \;\; q_2 \to \tfrac{1}{2}(q_1 + q_2 - q_1^0 + q_2^0), \\ u \to u - \tfrac{1}{2}((q_1 - q_1^0)(p_1^0 - p_2^0) + (q_2 - q_2^0)(p_1^0 + p_2^0)), \\ p_1 \to p_1 - p_2 + p_2^0, \;\; p_2 \to p_1 + p_2 - p_1^0 \end{cases}$$

takes ω to the form

$$\tilde{\omega} = dq_1 \wedge dp_2 + dq_2 \wedge dp_1 - f\, dq_1 \wedge dq_2$$

for some smooth function f.

Elliptic case

In this case the distributions $C_1 = C_+$ and $C_2 = C_-$ are complex distributions, and

$$C_\pm(a) = \{X \in C(a) | A_{\omega,a} X = \pm\iota\lambda X\},$$

where $\lambda = \sqrt{-\operatorname{Pf}(\omega)}$. Note that $\dim_{\mathbb{C}} C_\pm^{(2)} \le 4$, $C_1 = \overline{C_2}$, and the distributions $C_j^{(2)} + \overline{C}_j^{(2)}$ $(j = 1, 2)$ are involutive. Due to the Frobenius–Nirenberg theorem, the complex distributions $C_1^{(2)}$ and $C_2^{(2)}$ are completely integrable. Let G_j be a first integral of the distribution $C_j^{(2)}$. Note that $\overline{G}_1 = G_2$, $d_p G_1 \in \Omega^1(C_2)$ and $d_p G_2 \in \Omega^1(C_1)$. By (21.1), we see that $d_p G_1 \wedge d_p G_2 \wedge d\omega_0^{\mathbb{C}} = 0$. Therefore $dG_1 \wedge dG_2 \wedge d\omega_0^{\mathbb{C}} \wedge \omega_0^{\mathbb{C}} = 0$. We set $F_1 = \operatorname{Re} G_1$, $F_2 = \operatorname{Im} G_1$. From the inequalities $dG_1 \wedge dG_2 \wedge \omega_0^{\mathbb{C}} \neq 0$ and $dG_1 \wedge dG_2 \wedge d\omega_0^{\mathbb{C}} \wedge \omega_0^{\mathbb{C}} = 0$ it follows that $dF_1 \wedge dF_2 \wedge \omega_0 \neq 0$ and $dF_1 \wedge dF_2 \wedge d\omega_0 \wedge \omega_0 = 0$.

Due to Lemma 21.1.1, we suppose that $F_1 = q_1$ and $F_2 = q_2$. Similar to the hyperbolic case, we have: $(\omega^{\mathbb{C}} \pm \iota\lambda\Omega^{\mathbb{C}}) \wedge (\omega^{\mathbb{C}} \pm \iota\lambda\Omega^{\mathbb{C}}) = 0$. Therefore, the 2-forms $\omega^{\mathbb{C}} + \iota\lambda\Omega^{\mathbb{C}}$ and $\omega^{\mathbb{C}} - \iota\lambda\Omega^{\mathbb{C}}$ are decomposable: $\omega^{\mathbb{C}} + \iota\lambda\Omega^{\mathbb{C}} = \alpha \wedge \beta$

and $\omega^{\mathbb{C}} - \iota\lambda\Omega^{\mathbb{C}} = \overline{\alpha}\wedge\overline{\beta}$, where $\alpha,\beta\in\Omega^1(C_1)$ and $\overline{\alpha},\overline{\beta}\in\Omega^1(C_2)=\Omega^1(\overline{C}_1)$. Let us put $\alpha = dG_1$and $\beta = \gamma_1 + \iota\gamma_2$.

Then

$$\omega^{\mathbb{C}} + \iota\lambda\Omega^{\mathbb{C}} = (dq_1 + \iota\, dq_2)\wedge(\gamma_1 + \iota\gamma_2),$$

$$\omega^{\mathbb{C}} - \iota\lambda\Omega^{\mathbb{C}} = (dq_1 - \iota\, dq_2)\wedge(\gamma_1 - \iota\gamma_2)$$

and

$$\Omega^{\mathbb{C}} = \frac{1}{\lambda}(dq_1\wedge\gamma_2 + dq_2\wedge\gamma_1).$$

On the other hand $\Omega^{\mathbb{C}} = dq_1\wedge dp_1 + dq_2\wedge dp_2$, and we obtain $\gamma_1 = a\,dq_2 + \lambda\,dp_2$ and $\gamma_2 = b\,dq_1 + \lambda\,dp_1$. Then

$$\omega = s\,dq_1\wedge dq_2 + dq_1\wedge dp_2 - dq_2\wedge dp_1,$$

for a smooth real function s.

Parabolic case

In this case the real distributions C_1 and C_2 coincide: $P \overset{\text{def}}{=} C_1 = C_2$. The equation E_ω can be in one of the classes $H_{0,0}, H_{1,1}$ or $H_{2,2}$ only.

Suppose that E_ω belongs to one of the classes $H_{0,0}$ or $H_{1,1}$. In this case $\dim P^{(2)} \leq 3$. Let F_1 and F_2 be functional independent first integrals of the distribution $P^{(2)}$. The conditions of Lemma 21.1.1 hold, therefore we can put $F_1 = q_1$ and $F_2 = q_2$. Suppose now that E_ω belongs to one of the classes $H_{2,2}$. In this case $\dim P^{(2)} = 4$ and we have only one first integral F_1 of the distribution $P^{(2)}$. The missing function F_2 can be found from the relation $dF_1\wedge dF_2\wedge\omega_0\wedge d\omega_0 = 0$. This relation is a first-order scalar linear differential equation with respect to function F_2, and therefore has a solution. By choosing initial values in a suitable way the solution can be chosen in such a way that $dF_1\wedge dF_2\wedge\omega_0 \neq 0$.

For the parabolic equation E_ω the form ω is decomposable: $\omega = \alpha\wedge\beta$, and the 1-forms ω_0, α, β generate the distribution P: $P = \mathcal{F}(\omega_0,\alpha,\beta)$. Due to Lemma 21.1.1, we have $\omega = dq_2\wedge\gamma$, where γ is some 1-form. Since ω is effective, $\omega\wedge\Omega = dq_2\wedge\gamma\wedge(dq_1\wedge dp_1 + dq_2\wedge dp_2) = 0$, we obtain $\gamma = a\,dq_1 + b\,dq_2 + c\,dp$, where a, b and c are functions. Then

$$\omega = s\,dq_1\wedge dq_2 + t\,dq_2\wedge dp_1.$$

■

Corollary 21.1.1 *Let a parabolic Monge–Ampère equation belong to the classe $H_{2,2}$. Then it is locally contact equivalent to an equation of type*

$$v_y = g(x, y, v, v_x, v_{xx}),$$

where g is a smooth function.

Let us consider the following complex distributions:

$$l \stackrel{\text{def}}{=} C_1^{(1)} \cap C_2^{(1)}$$

and

$$l_1 = C_1 \cap C_2^{(2)},$$
$$l_2 = C_2 \cap C_1^{(2)}.$$

Assume that in some domain of J^1M the Monge–Ampère equation E_ω belongs to the classes $H_{2,2}$ and $\mathrm{Pf}(\omega) \neq 0$. Then $l \oplus l_j$ $(j = 1, 2)$ are two-dimensional complex distributions.

The following lemma gives a conditions of linearization with respect to the first derivatives of non-degenerate equations of class $H_{2,2}$.

Theorem 21.1.2 (D. Tunitsky) *Assume that a non-degenerate Monge–Ampère equation E_ω belongs to the class $H_{2,2}$ and $\mathrm{Pf}(\omega) \neq 0$. Then this equation is locally contact equivalent to a Monge–Ampère equation of the type*

$$v_{xx} + \varepsilon v_{yy} = r(x, y, v)v_x + s(x, y, v)v_y + g(x, y, v)$$

if and only if the distributions $l \oplus l_1$ and $l \oplus l_2$ are completely integrable.

Proof Due to Theorem 21.1.1, E_ω is equivalent to (21.2). For this equation the corresponding effective differential 2-form is

$$\omega = -f(q, u, p)\, dq_1 \wedge dq_2 + dq_1 \wedge dp_2 + \varepsilon\, dq_2 \wedge dp_1$$

and the distribution l is generated by the vector field

$$2\frac{\partial}{\partial u} + \varepsilon f_{p_1}\frac{\partial}{\partial p_1} + f_{p_2}\frac{\partial}{\partial p_2}.$$

The distributions $l_{1,0}$ and $l_{0,1}$ are generated by the vector fields $\partial/\partial p_1$ and $\partial/\partial p_2$ respectively. Since the distributions $l \oplus l_1$ and $l \oplus l_2$ are completely integrable, we see that $f_{p_1p_1} = f_{p_2p_2} = 0$. Therefore the function f has the following form:

$$f(q,u,p) = \hat{r}(q,u)p_1 + \hat{s}(q,u)p_2 + \hat{w}(q,u)(p_1^2 + \varepsilon p_2^2) + \hat{g}(q,u)$$

for some functions $\hat{r}, \hat{s}, \hat{w}$ and $\hat{g}$. To prove that (21.2) with this function f is locally contact equivalent to the same equation with the right part:

$$r(q,u)p_1 + s(q,u)p_2 + g(q,u)$$

for some r, s, g we apply the following contact transformation

$$\varphi : (q_1, q_2, p_1, p_2, u) \mapsto (q_1, q_2, Z_{q_1} + p_1 Z_u, Z_{q_2} + p_2 Z_u, Z)$$

that preserves the point a and where $Z = Z(q,u)$ is a smooth function. Then the effective part of $\varphi^*(\omega)$ is

$$\begin{aligned}
\varphi^*(\omega)_\varepsilon &= Z_u(dq_2 \wedge dp_1 - \varepsilon\, dq_1 \wedge dp_2) \\
&\quad + (\tilde{r}p_1 Z_u + \tilde{s}p_2 Z_u + \tilde{g} - (Z_{uu} - \tilde{w}Z_u^2)(p_1^2 + \varepsilon p_2^2) \\
&\quad + \tilde{s}Z_{q_2} + 2\varepsilon\tilde{w}p_2 Z_u Z_{q_2} + \varepsilon\tilde{w}Z_{q_2}^2 - 2\varepsilon p_2 Z_{q_2 u} \\
&\quad - \varepsilon Z_{q_2 q_2} + rZ_{q_1} + 2\tilde{w}p_1 Z_u Z_{q_1} + \tilde{w}Z_{q_1}^2 \\
&\quad - 2p_1 Z_{q_1 u} - Z_{q_1 q_1})\, dq_1 \wedge dq_2,
\end{aligned}$$

where $\tilde{r}, \tilde{s}, \tilde{g}, \tilde{w}$ are functions in q, u. Let (q^0, u^0, p^0) be local coordinates of the point a. Take Z as the solution of the ordinary differential equation $Z_{uu} - \tilde{w}Z_u^2 = 0$ with the Cauchy data $Z(q, u^0) = u^0$, $Z_u(q, u^0) = 1$, and we obtain the required function Z. ■

The following theorem is a contact analog of Theorem 20.1.1 for symplectic Monge–Ampère equations.

Theorem 21.1.3 *A Monge–Ampère equation is locally contact equivalent to a Monge–Ampère equation of type*

$$v_{xx} + \varepsilon v_{yy} = 0$$

if and only if it belongs either to the class $H_{1,1}$ (in the case of a non-zero Pfaffian), or to the class $H_{0,0}$ (in the case where the Pfaffian is zero).

21.2 Invariants of non-degenerate Monge–Ampère equations

21.2.1 Tensor invariants

As we have seen, any non-degenerate Monge–Ampère equation determines a triplet of the real (for hyperbolic case) and complex (for elliptic case) distributions $\mathcal{P} = (\{C_{1,0}, C_{0,1}\}, l,)$ on J^1M.

Moreover, Monge–Ampère equations E and $\tilde{E}$ are contact equivalent if and only if there exists a diffeomorphism φ such that $\varphi_*(\mathcal{P}) = \tilde{\mathcal{P}}$. Here $\tilde{\mathcal{P}}$ is the triplet corresponding to the equation $\tilde{E}$.

In Part I we have constructed the tensors invariants $d_{-1,1,1}$, $d_{1,1,-1}$, $d_{1,-1,1}$, $d_{0,-1,2}$, $d_{2,-1,0}$, $d_{2,1,-2}$ and $d_{-2,1,2}$ of the corresponding $\mathcal{AP}$-structure. In this special case, we have $d_{1,-1,1} = 0$ because the distributions $C_{1,0}$ and $C_{0,1}$ are skew-orthogonal. Moreover, tensors $d_{2,1,-2}$,$d_{2,0,-1}$,$d_{-1,0,2}$ and $d_{-2,1,2}$ are also equal to zero.

Let us prove, for example, that tensors $d_{2,0,-1}$ and $d_{-1,0,2}$ are trivial.

Since

$$d_{2,0,-1} : \Omega^{p,r,q}(J^1M) \to \Omega^{p+2,r,q-1}(J^1M),$$

we see that $d_{2,0,-1} = \alpha_1 \wedge \alpha_2 \otimes Q$, where $\alpha_i \in \Omega^{1,0,0}(J^1M)$ and $Q \in D(C_{0,1})$. Let $\beta \in \Omega^{0,0,1}(J^1M)$ be an arbitrary 1-form and P_1, P_2 be arbitrary vector fields from the distribution $C_{1,0}$. Then since $\beta \in \Omega^{0,0,1}(J^1M)$ and $[P_1, P_2] \in D(C_{1,0}^{(1)})$, we have

$$d\beta(P_1, P_2) = -\beta([P_1, P_2]) = 0.$$

But $d\beta(P_1, P_2) = d_{2,0,-1}\beta(P_1, P_2)$. Therefore, $d_{2,0,-1}\beta(P_1, P_2) = 0$ for any $\beta \in \Omega^{0,0,1}(J^1M)$ and $P_1, P_2 \in D(C_{1,0})$.

Thus, we have four 2-covariant and 1-contravariant tensors $d_{-1,1,1}$, $d_{1,1,-1}$, $d_{0,-1,2}$ and $d_{2,-1,0}$:

$$\begin{aligned}
d_{1,1,-1} &= \omega_0^{\mathbb{C}} \wedge \alpha_1 \otimes Q_1 + \omega_0^{\mathbb{C}} \wedge \alpha_2 \otimes Q_2, \\
d_{-1,1,1} &= \omega_0^{\mathbb{C}} \wedge \beta_1 \otimes P_1 + \omega_0^{\mathbb{C}} \wedge \beta_2 \otimes P_2, \\
d_{2,-1,0} &= \alpha_3 \wedge \alpha_4 \otimes Z, \\
d_{0,-1,2} &= \beta_3 \wedge \beta_4 \otimes Z.
\end{aligned} \tag{21.3}$$

Here $\alpha_i \in \Omega^1(C_{1,0})$, $\beta_i \in \Omega^1(C_{0,1})$, $P_j \in D(C_{1,0})$, $Q_j \in D(C_{0,1})$ and $Z \in D(l)$ $(i = 1, \ldots 3; j = 1, 2)$.

These tensors can be considered as maps:

$$d_{1,1,-1} : C^{(1)}_{1,0} \times C^{(1)}_{1,0} \to C_{0,1},$$

$$d_{-1,1,1} : C^{(1)}_{0,1} \times C^{(1)}_{0,1} \to C_{1,0},$$

$$d_{2,-1,0} : C_{1,0} \times C_{1,0} \to l,$$

$$d_{0,-1,2} : C_{0,1} \times C_{0,1} \to l.$$

Recall that a tensor $\theta \otimes X$, where θ is a 2-form and X is a vector field, acts on a differential form α as

$$(\theta \otimes X)(\alpha) = \theta \wedge (X \rfloor \alpha),$$

and acts on a pair of vector fields Y_1, Y_2 as

$$(\theta \otimes X)(Y_1, Y_2) = \theta(Y_1, Y_2)X.$$

Proposition 21.2.1 *We have* $C^{(2)}_{1,0} = C^{(1)}_{1,0} \oplus \operatorname{Im} d_{1,1,-1}$ *and* $C^{(2)}_{0,1} = C^{(1)}_{0,1} \oplus \operatorname{Im} d_{-1,1,1}$.

Proof We prove the first statement only. The proof of the second one is similar. Let P and Z be vector fields from the distributions $C_{1,0}$ and l respectively, and

$$V_{P,Z} = d_{1,1,-1}(Z, P) \in C_{0,1}.$$

Using the representation (21.3), we obtain

$$V_{P,Z} = \omega_0(Z)(\alpha_1(P)Q_1 + \alpha_2(P)Q_2).$$

Let $\hat{\beta}_1, \hat{\beta}_2$ be a basis in $\Omega^{1,0,0}(J^1M)$. Then

$$\hat{\beta}_i(V_{P,Z}) = \omega_0(Z)(\hat{\beta}_i(Q_1)\alpha_1(P) + \hat{\beta}_i(Q_2)\alpha_2(P)).$$

On the other hand, the commutator $[Z, P]$ is uniquely represented in the form $[Z, P] = X + hZ + Q$, where $X \in D(C_{1,0})$, $Q \in D(C_{0,1})$, $h \in C^\infty(J^1M)$. We see that

$$\begin{aligned}\hat{\beta}_i([Z, P]) &= -d\hat{\beta}_i(Z, P) = -d_{1,1,-1}\hat{\beta}_i(Z, P)\\ &= \omega_0(Z)(\hat{\beta}_i(Q_1)\alpha_1(P) + \hat{\beta}_i(Q_2)\alpha_2(P)) = \hat{\beta}_i(V_{P,Z})\end{aligned}$$

for $i = 1, 2$. Therefore $V_{P,Z} = Q$ and

$$[Z, P] = X + hZ + d_{1,1,-1}(Z, P).$$

Since the vector fields P and Z are arbitrary and $C^{(2)}_{1,0} = \langle P_1, P_2, Z, [Z, P_1], [Z, P_2]\rangle$, we have $C^{(2)}_{1,0} = C^{(1)}_{1,0} \oplus \operatorname{Im} d_{1,1,-1}$. ■

Corollary 21.2.1

1. *One has:*

$$\dim C^{(2)}_{1,0} = \dim C^{(1)}_{1,0} + \dim \operatorname{Im} d_{1,1,-1}$$

and

$$\dim C^{(2)}_{0,1} = \dim C^{(1)}_{0,1} + \dim \operatorname{Im} d_{-1,1,1};$$

2. *the distribution $C^{(1)}_{1,0}$ is completely integrable if and only if $d_{1,1,-1} = 0$ and the distribution $C^{(1)}_{0,1}$ is completely integrable if and only if $d_{-1,1,1} = 0$;*
3. *a non-degenerate Monge–Ampère equation E is locally contact equivalent to the equation $v_{xx} + \varepsilon v_{yy} = 0$ if and only if $d_{1,1,-1} = d_{-1,1,1} = 0$.*

Remark 21.2.1 *Tensors $d_{1,1,-1}$ and $d_{-1,1,1}$ are curvatures of the distributions $C^{(1)}_{1,0}$ and $C^{(1)}_{0,1}$ respectively.*

21.2.2 Absolute and relative invariants

Assume that the distributions $C^{(1)}_{1,0}$ and $C^{(1)}_{0,1}$ are not completely integrable, and a non-degenerate Monge–Ampère equation E_ω belongs to one of the classes $H_{k,l} (0 \le k \le l \le 2)$.

Note that the tensors $d_{1,1,-1}$ and $d_{-1,1,1}$ are decomposable. Indeed E_ω is equivalent to (21.2) and for this case the tensors are decomposable.

Let

$$d_{1,1,-1} = \omega_0^{\mathbb{C}} \wedge \alpha \otimes Q,$$

$$d_{-1,1,1} = \omega_0^{\mathbb{C}} \wedge \beta \otimes P,$$

and define 2-forms $\xi_1, \xi_2 \in \Omega^{1,0,1}(J^1M)$ as follows:

$$\xi_1 \overset{\text{def}}{=} P \rfloor d_{2,-1,0}(\omega_0^{\mathbb{C}} \wedge \beta),$$

$$\xi_2 \overset{\text{def}}{=} Q \rfloor d_{0,-1,2}(\omega_0^{\mathbb{C}} \wedge \alpha).$$

Since $d_{0,-1,2}$ and $d_{2,-1,0}$ are $C^\infty(J^1M)$-homomorphisms, we see that the forms ξ_1 and ξ_2 are invariants.

Using representations (21.3) for $d_{2,-1,0}$ and $d_{0,-1,2}$, we get the following representation of ξ_1 and ξ_2:

$$\xi_1 = P\rfloor(\alpha_3 \wedge \alpha_4 \wedge \beta) = (\alpha_3(P)\alpha_4 - \alpha_4(P)\alpha_3) \wedge \beta,$$

$$\xi_2 = Q\rfloor(\beta_3 \wedge \beta_4 \wedge \alpha) = (\beta_3(Q)\beta_4 - \beta_4(Q)\beta_3) \wedge \alpha.$$

Note that for elliptic equations these forms are complex conjugate: $\xi_1 = \overline{\xi_2}$.

Differential 1-forms

$$\nu_1 \stackrel{\text{def}}{=} Q\rfloor\xi_1,$$

$$\nu_2 \stackrel{\text{def}}{=} P\rfloor\xi_2,$$

are relative invariants, because vector fields Q and P are defined up to the multiplication by a function.

Example 21.2.1 (Non-linear wave equation) *Consider the following non-linear wave equation:*

$$v_{xy} = f(x, y, v, v_x, v_y).$$

Vector fields

$$P_1 = \frac{d}{dq_1} + f\frac{\partial}{\partial p_2},$$

$$P_2 = \frac{\partial}{\partial p_1}$$

constitute a basis for the distribution $C_{1,0}$ and

$$Q_1 = \frac{d}{dq_2} + f\frac{\partial}{\partial p_1},$$

$$Q_2 = \frac{\partial}{\partial p_2}$$

for the distribution $C_{0,1}$.

The distribution l is generated by the following vector field:

$$Z = \frac{\partial}{\partial u} + f_{p_2}\frac{\partial}{\partial p_1} + f_{p_1}\frac{\partial}{\partial p_2}.$$

Thus the vector fields P_1, P_2, Z, Q_1, Q_2 form a basis in vector fields on J^1M. The dual basis is

$$\begin{aligned}
\alpha_1 &= dq_1,\\
\alpha_2 &= dp_1 + p_1 f_{p_2}\, dq_1 + (p_2 f_{p_2} - f)\, dq_2 - f_{p_2}\, du,\\
\omega_0 &= du - p_1\, dq_1 - p_2\, dq_2,\\
\beta_1 &= dq_2,\\
\beta_2 &= dp_2 + (p_1 f_{p_1} - f)\, dq_1 + p_2 f_{p_1}\, dq_2 - f_{p_1}\, du.
\end{aligned}$$

For this case we have the following representation of the four constructed tensor invariants:

$$\begin{aligned}
d_{-1,1,1} = {} & (f f_{p_2p_2}\, dq_1 \wedge du - f_{p_2p_2}\, dp_2 \wedge du - p_1 f_{p_2p_2}\, dq_1 \wedge dp_2 - p_2 f_{p_2p_2}\, dq_2\\
& \wedge dp_2 + (f_u - p_2 f_{p_2u} + f_{p_1} f_{p_2} - f f_{p_1p_2} - f_{q_2p_2})\, dq_2 \wedge du\\
& + (p_1 f_u - p_1 p_2 f_{p_2u} - p_2 f f_{p_2p_2} + p_1 f_{p_1} f_{p_2} - p_1 f f_{p_1p_2} - p_1 f_{q_2p_2})\, dq_1\\
& \wedge dq_2) \otimes \frac{\partial}{\partial p_1},\\
d_{1,1,-1} = {} & (f f_{p_1p_1}\, dq_2 \wedge du - f_{p_1p_1}\, dp_1 \wedge du - p_1 f_{p_1p_1}\, dq_1 \wedge dp_1\\
& - p_2 f_{p_1p_1} dq_2 \wedge dp_1 + (f_u + f_{p_1} f_{p_2} - p_1 f_{p_1u} - f f_{p_1p_2} - f_{q_1p_1})\, dq_1\\
& \wedge du + (-p_2 f_u - p_2 f_{p_1} f_{p_2} + p_1 p_2 f_{p_1u} + p_2 f f_{p_1p_2} + p_1 f f_{p_1p_1}\\
& + p_2 f_{q_1p_1})\, dq_1 \wedge dq_2) \otimes \frac{\partial}{\partial p_2},\\
d_{2,-1,0} = {} & (dq_1 \wedge dp_1 - f_{p_2}\, dq_1 \wedge du + (p_2 f_{p_2} - f)\, dq_1 \wedge dq_2)\\
& \otimes \left(\frac{\partial}{\partial u} + f_{p_2} \frac{\partial}{\partial p_1} + f_{p_1} \frac{\partial}{\partial p_2}\right),\\
d_{0,-1,2} = {} & (dq_2 \wedge dp_2 - f_{p_1}\, dq_2 \wedge du - (p_1 f_{p_1} - f)\, dq_1 \wedge dq_2)\\
& \otimes \left(\frac{\partial}{\partial u} + f_{p_2} \frac{\partial}{\partial p_1} + f_{p_1} \frac{\partial}{\partial p_2}\right).
\end{aligned}$$

The differential 2-forms ξ_1 and ξ_2 are:

$$\begin{aligned}
\xi_1 = {} & f_{p_2p_2}(f_{p_1}\, dq_1 \wedge du - dq_1 \wedge dp_2)\\
& + (f_u - p_2 f_{p_2u} + f_{p_1} f_{p_2} - p_2 f_{p_1} f_{p_2p_2} - f f_{p_1p_2} - f_{q_2p_2})\, dq_1 \wedge dq_2,\\
\xi_2 = {} & f_{p_1p_1}(f_{p_2}\, dq_2 \wedge du - dq_2 \wedge dp_1)\\
& + (-f_u + p_1 f_{p_1u} - f_{p_1} f_{p_2} + p_1 f_{p_2} f_{p_1p_1} + f f_{p_1p_2} + f_{q_1p_1})\, dq_1 \wedge dq_2
\end{aligned}$$

and

$$\nu_1 = f_{p_2p_2}\, dq_1,$$

$$\nu_2 = f_{p_1p_1}\, dq_2.$$

Example 21.2.2 *For (21.2) we have:*

$$\nu_1 = h_1((2\sqrt{\varepsilon}f - \iota p_1 f_{p_2} - \sqrt{\varepsilon}f_{p_1})\, dq_1 - p_2(\iota f_{p_2} + \sqrt{\varepsilon}f_{p_1})\, dq_2$$

$$+ (\iota f_{p_2} + \sqrt{\varepsilon}f_{p_1})\, du - 2\sqrt{\varepsilon}\, dp_1 - 2\iota\varepsilon\, dp_2),$$

$$\nu_2 = h_2(\varepsilon p_1(\sqrt{\varepsilon}f_{p_1} - \iota f_{p_2})\, dq_1 + \varepsilon(2\iota f + \sqrt{\varepsilon}p_2 f_{p_1} - \iota p_2 f_{p_2})\, dq_2$$

$$+ \varepsilon(\iota f_{p_2} - \sqrt{\varepsilon}f_{p_1})\, du + 2\sqrt{\varepsilon}\, dp_1 - 2\,\iota\, dp_2),$$

where $h_1 = \varepsilon f_{p_1p_1} + 2\iota f_{p_1p_2} - f_{p_2p_2}$ *and* $h_2 = \varepsilon f_{p_1p_1} - 2\iota f_{p_1p_2} - f_{p_2p_2}$.

Using forms ν_1 and ν_2 we get the following generalization of Theorem 21.1.2.

Theorem 21.2.1 *Assume that a non-degenerate Monge–Ampère equation* E_ω *belongs to one of the classes* $H_{k,l}$*, where* $1 \le k \le l \le 2$*. Then* E_ω *is locally equivalent to a Monge–Ampère equation of the type*

$$v_{xx} + \varepsilon v_{yy} = r(x, y, v)v_x + s(x, y, v)v_y + g(x, y, v) \tag{21.4}$$

if and only if $\nu_1 = \nu_2 = 0$.

Proof Due to Theorem 21.1.1 the equation E_ω is locally contact equivalent to (21.2). Due to Example 21.2.2 for this equation $\nu_1 = \nu_2 = 0$ if and only if $h_1 = h_2 = 0$. This means that

$$f_{p_1p_2} = 0,$$

$$\varepsilon f_{p_1p_1} = f_{p_2p_2}.$$

Therefore $f(q, u, p) = r(q, u)p_1 + s(q, u)p_2 + g(q, u)$ for some functions r, s, g. ∎

21.3 The problem of contact linearization

In this section we use the constructed tensors to solve the problem of contact linearization of non-degenerate Monge–Ampère equations.

That is, find a class of Monge–Ampère equations that are locally contact equivalent to non-homogeneous linear equations

$$v_{xx} + \varepsilon v_{yy} = r(x,y)v_x + s(x,y)v_y + c(x,y)v + d(x,y). \tag{21.5}$$

The following result gives a solution of the problem.

Theorem 21.3.1 (Linearizaton of non-degenerate MAEs) *A non-degenerate Monge–Ampère equation E_ω of the class $H_{2,2}$ is locally contact equivalent to the linear equation (21.5) if and only if the following conditions hold:*

1. $\nu_1 = \nu_2 = 0$;
2. $d\xi_1 = d\xi_2 = 0$.

Proof The proof of the necessity it trivial. Indeed, is easy to check that (21.5) belongs to one of the classes $H_{2,2}$ and for this equation we have $\nu_1 = \nu_2 = 0$ and $d\xi_1 = d\xi_2 = 0$. Let us prove the sufficiency.

Hyperbolic case

Due to Theorem 21.2.1, the equation E_ω is contact equivalent to (21.4)

$$v_{xx} - v_{yy} = r(x,y,v)v_x + s(x,y,v)v_y + g(x,y,v).$$

The corresponding effective differential 2-form is

$$\omega = dq_1 \wedge dp_2 + dq_2 \wedge dp_1 + (rp_1 + sp_2 + g)\, dq_1 \wedge dq_2.$$

The contact transformation

$$\varphi : q_1 \to \frac{q_1 - q_2}{\sqrt{2}}, \quad q_2 \to \frac{q_1 + q_2}{\sqrt{2}}, \quad u \to u,$$
$$p_1 \to \frac{p_1 - p_2}{\sqrt{2}}, \quad p_2 \to \frac{p_1 + p_2}{\sqrt{2}}$$

takes it to the form

$$\omega_1 \stackrel{\text{def}}{=} \varphi^*(\omega) = dq_1 \wedge dp_1 - dq_2 \wedge dp_2 + (Rp_1 + Sp_2 + G)\, dq_1 \wedge dq_2, \tag{21.6}$$

where

$$R(q_1,q_2,u) = \frac{1}{\sqrt{2}}\left(s\left(\frac{q_1-q_2}{\sqrt{2}}, \frac{q_1+q_2}{\sqrt{2}}, u\right) + r\left(\frac{q_1-q_2}{\sqrt{2}}, \frac{q_1+q_2}{\sqrt{2}}, u\right)\right),$$

$$S(q_1,q_2,u) = \frac{1}{\sqrt{2}}\left(s\left(\frac{q_1-q_2}{\sqrt{2}}, \frac{q_1+q_2}{\sqrt{2}}, u\right) - r\left(\frac{q_1-q_2}{\sqrt{2}}, \frac{q_1+q_2}{\sqrt{2}}, u\right)\right),$$

$$G(q_1,q_2,u) = g\left(\frac{q_1-q_2}{\sqrt{2}}, \frac{q_1+q_2}{\sqrt{2}}, u\right).$$

We have the following coordinate representation of the tensors:

$$d_{1,1,-1} = (Rs + G_u + p_2 S_u - R_{q_1})(dq_1 \wedge du - p_2\, dq_1 \wedge dq_2) \otimes \frac{\partial}{\partial p_2},$$

$$d_{-1,1,1} = (RS + G_u + p_1 R_u - S_{q_2})(dq_2 \wedge du + p_1\, dq_1 \wedge dq_2) \otimes \frac{\partial}{\partial p_1},$$

$$d_{2,-1,0} = (dq_1 \wedge dp_1 - S\, dq_1 \wedge du - (G + p_1 R)\, dq_1 \wedge dq_2)$$
$$\otimes \left(\frac{\partial}{\partial u} + S\frac{\partial}{\partial p_1} + R\frac{\partial}{\partial p_2}\right),$$

$$d_{0,-1,2} = (dq_2 \wedge dp_2 - R\, dq_2 \wedge du + (G + p_2 S)\, dq_1 \wedge dq_2)$$
$$\otimes \left(\frac{\partial}{\partial u} + S\frac{\partial}{\partial p_1} + R\frac{\partial}{\partial p_2}\right).$$

and the invariants:

$$\xi_1 = (RS + G_u + p_1 R_u - S_{q_2})\, dq_1 \wedge dq_2,$$

$$\xi_2 = -(RS + G_u + p_2 S_u - R_{q_1})\, dq_1 \wedge dq_2.$$

Then,

$$d\xi_1 = R_u\, dq_1 \wedge dq_2 \wedge dp_1 + (SR_u + RS_u + G_{uu}$$
$$+ p_1 R_{uu} - S_{uq_2})\, dq_1 \wedge dq_2 \wedge du,$$

$$d\xi_2 = -S_u\, dq_1 \wedge dq_2 \wedge dp_2 - (SR_u + RS_u + G_{uu}$$
$$+ p_2 S_{uu} - R_{uq_1})\, dq_1 \wedge dq_2 \wedge du.$$

We see that $R_u = S_u = G_{uu} = 0$ if and only if $d\xi_1 = d\xi_2 = 0$. In this case

$$G(q_1,q_2,u) = c(q_1,q_2)u + z(q_1,q_2)$$

and we obtain

$$\omega_1 = 2(R(q)p_1 + S(q)p_2 + c(q)u + z(q))\,dq_1 \wedge dq_2 + dq_1 \wedge dp_1 - dq_2 \wedge dp_2.$$

Elliptic case

Due to Theorem 21.2.1, the equation E_ω is equivalent to the equation

$$v_{xx} + v_{yy} = r(x,y,v)v_x + s(x,y,v)v_y + g(x,y,v).$$

The invariant 2-form ξ_1 is

$$\xi_1 = \left(\frac{1}{4}\left(\frac{ds}{dq_1} - \frac{dr}{dq_2}\right) + \frac{\iota}{8}(r^2 + s^2 + 4g_u \right.$$
$$\left. +2(p_1 r_u - r_{q_1} + p_2 s_u - s_{q_2}))\right) dq_1 \wedge dq_2$$

and $\xi_2 = \overline{\xi_1}$. Then

$$d\xi_1 = \frac{1}{4}(-r_u + \iota s_u)\,dq_1 \wedge dq_2 \wedge dp_2 + \frac{1}{4}(\iota r_u + s_u)\,dq_1 \wedge dq_2 \wedge dp_1$$
$$+ \frac{1}{4}((p_1 s_{uu} - p_2 r_{uu} - r_{uq_2} + s_{uq_1})$$
$$+ \iota(r r_u + s s_u + 2g_{uu} + p_1 r_{uu} + p_2 s_{uu} - s_{q_2 u} - r_{q_1 u}))\,dq_1 \wedge dq_2 \wedge du.$$

We see that $r_u = s_u = g_{uu} = 0$ if and only if $d\xi_1 = 0$. In this case

$$g(q_1, q_2, u) = c(q_1, q_2)u + z(q_1, q_2)$$

and we obtain the following effective 2-form:

$$\omega = -(r(q)p_1 + s(q)p_2 + c(q)u + z(q))\,dq_1 \wedge dq_2 + dq_1 \wedge dp_2 - dq_2 \wedge dp_1.$$

■

Example 21.3.1 (The generalized Hunter–Saxton equation) *Let us consider the generalized Hunter–Saxton equation*

$$v_{tx} = v v_{xx} + \kappa u_x^2.$$

This equation has applications in the theory of liquid crystals [39] and in the geometry of Einstein–Weil spaces. This equation is hyperbolic,

$$\omega = 2u\,dq_2 \wedge dp_1 + dq_1 \wedge dp_1 - dq_2 \wedge dp_2 - 2\kappa p_1^2\,dq_1 \wedge dq_2$$

and

$$A_\omega = \left\| \begin{array}{cccc} 1 & 2u & 0 & 0 \\ 0 & -1 & 0 & 0 \\ 0 & -2\kappa p_1^2 & 1 & 0 \\ 2\kappa p_1^2 & 0 & 2u & -1 \end{array} \right\|$$

in the basis

$$\frac{d}{dq_1}, \frac{d}{dq_2}, \frac{\partial}{\partial p_1}, \frac{\partial}{\partial p_2}.$$

Let us consider the following basis:

$$P_1 = \frac{\partial}{\partial q_1} + p_1 \frac{\partial}{\partial u} + \kappa p_1^2 \frac{\partial}{\partial p_2},$$

$$P_2 = \frac{\partial}{\partial p_1} + u \frac{\partial}{\partial p_2},$$

$$Q_1 = \frac{\partial}{\partial q_2} + \kappa p_1^2 \frac{\partial}{\partial p_1} - u \frac{\partial}{\partial q_1} + (p_2 - u p_1) \frac{\partial}{\partial u}$$

$$Q_2 = \frac{\partial}{\partial p_2},$$

$$Z = \frac{\partial}{\partial u} + (2\kappa - 1) p_1 \frac{\partial}{\partial p_2},$$

and the dual one

$$\alpha_1 = dq_1 + u\, dq_2,$$

$$\alpha_2 = dp_1 - \kappa p_1^2\, dq_2,$$

$$\beta_1 = dq_2,$$

$$\beta_2 = dp_2 + (1 - 2\kappa) p_1\, du + (\kappa - 1) p_1^2\, dq_1 + (2\kappa - 1) p_1 p_2\, dq_2 - u\, dp_1,$$

$$\omega_0 = du - p_1\, dq_1 - p_2\, dq_2.$$

Then $P_1, P_2 \in D(C_{1,0})$, $Q_1, Q_2 \in D(C_{0,1})$, $Z \in D(l)$, $\alpha_1, \alpha_2 \in \Omega^{1,0,0}(J^1M)$, $\beta_1, \beta_2 \in \Omega^{0,0,1}(J^1M)$ *and we have the following representation of the tensors:*

$$
\begin{aligned}
d_{-1,1,1} &= -(p_1\,dq_1\wedge dq_2 + dq_2\wedge du)\otimes\left(\frac{\partial}{\partial q_1} + p_1\frac{\partial}{\partial u} + \kappa p_1^2\frac{\partial}{\partial p_2}\right),\\
d_{1,1,-1} &= 2(\kappa-1)(\kappa p_1^3\,dq_1\wedge dq_2 + \kappa p_1^2\,dq_2\wedge du - dp_1\wedge du - p_1\,dq_1\\
&\qquad \wedge dp_1 - p_2\,dq_2\wedge dp_1)\otimes\frac{\partial}{\partial p_2},\\
d_{2,-1,0} &= (dq_1\wedge dp_1 - \kappa p_1^2\,dq_1\wedge dq_2 + u\,dq_2\wedge dp_1)\\
&\qquad \otimes\left(\frac{\partial}{\partial u} + (2\kappa-1)p_1\frac{\partial}{\partial p_2}\right),\\
d_{0,-1,2} &= (dq_2\wedge dp_2 + (1-2\kappa)p_1\,dq_2\wedge du + (1-\kappa)p_1^2\,dq_1\wedge dq_2 - u\,dq_2\wedge dp_1)\\
&\qquad \otimes\left(\frac{\partial}{\partial u} + (2\kappa-1)p_1\frac{\partial}{\partial p_2}\right)
\end{aligned}
$$

and

$$
\begin{aligned}
\xi_1 &= -dq_2\wedge dp_1,\\
\xi_2 &= 2(1-\kappa)\,dq_2\wedge dp_1.
\end{aligned}
$$

Due to theorem 21.3.1, the generalized Hunter–Saxton equation is linearized. The corresponding linear equation is the Euler–Poisson equation

$$
v_{tx} = \frac{1}{\kappa(t+x)}v_t + \frac{2(1-\kappa)}{\kappa(t+x)}v_x - \frac{2(1-\kappa)}{(\kappa(t+x))^2}u
$$

(cf. [86]).

21.4 The problem of equivalence for non-degenerate equations

21.4.1 e-Structure for non-degenerate equations

In this section we construct an e-structure for generic non-degenerate Monge–Ampère equations.

Let $E \overset{\text{def}}{=} E_\omega$ be a non-degenerate Monge–Ampère equations with triplet $\mathcal{P} = (\{C_{1,0}, C_{0,1}\}, l)$. Let Z be a real vector field which generates the complex distribution l and that is normed by the condition $\omega_0(Z) = 1$.

Let us introduce submodules $\underline{\Omega}^s(J^1M) \subset \Omega^s(J^1M)$ of all differential s-forms vanishing on Z:

$$
\underline{\Omega}^s(J^1M) \overset{\text{def}}{=} \{\alpha \in \Omega^s(J^1M) \mid Z\rfloor\alpha = 0\}.
$$

Elements of the submodule $\underline{\Omega}^s(J^1M)$ we will call *l-horizontal* forms. The set of all l-horizontal forms is an algebra $\underline{\Omega}^*(J^1M)$ with respect to the operation of exterior multiplication.

Let us define a projection $\Pi : \Omega^s(J^1M) \to \underline{\Omega}^s(J^1M)$ by

$$\Pi(\alpha) \stackrel{\text{def}}{=} \alpha - \omega_0 \wedge (Z \rfloor \alpha)$$

and an operator

$$\partial : \Omega^s(J^1M) \to \underline{\Omega}^{s+1}(J^1M) \text{ by } \partial \stackrel{\text{def}}{=} \Pi \circ d.$$

The kernel of the operator Π is

$$\ker \Pi = \{\omega_0 \wedge \alpha | \alpha \in \Omega^*(J^1M)\}.$$

Let $\tilde{E} \stackrel{\text{def}}{=} E_{\tilde{\omega}}$ be another non-degenerate Monge–Ampère equation with triplet $\tilde{\mathcal{P}}=(\{\tilde{C}_{1,0}, \tilde{C}_{0,1}\}, \tilde{l})$. If $\varphi : J^1M \to J^1M$, $\varphi^*(\omega_0) = \lambda\omega_0$ is a contact transformation such that $\varphi^*(\omega)_\varepsilon = h\tilde{\omega}$, then $\varphi_*(\tilde{\mathcal{P}}) = \mathcal{P}$ and

$$\varphi_*^{-1}(Z) = \frac{1}{\lambda}\tilde{Z}.$$

Lemma 21.4.1 *Operators* Π *and* ∂ *are invariant with respect to contact diffeomorphisms, i.e.,*

$$\varphi^* \circ \Pi = \tilde{\Pi} \circ \varphi^* \tag{21.7}$$

and

$$\varphi^* \circ \partial = \tilde{\partial} \circ \varphi^*. \tag{21.8}$$

Proof For an s-form $\alpha \in \Omega^s(J^1M)$ we have

$$\begin{aligned}\varphi^*(\Pi(\alpha)) &= \varphi^*(\alpha) - \varphi^*(\omega_0) \wedge \varphi^*(Z \rfloor \alpha)\\ &= \varphi^*(\alpha) - \omega_0 \wedge (\varphi_*^{-1}(Z) \rfloor \varphi^*(\alpha))\\ &= \varphi^*(\alpha) - \lambda\omega_0 \wedge \left(\frac{1}{\lambda}\tilde{Z} \rfloor \varphi^*(\alpha)\right)\\ &= \tilde{\Pi}(\varphi^*(\alpha)),\end{aligned}$$

and for an arbitrary differential l-horizontal form α we have

$$\begin{aligned}\varphi^*(\partial\alpha) &= \varphi^* \circ \Pi \circ d(\alpha) = \tilde{\Pi} \circ \varphi^* \circ d(\alpha)\\ &= \tilde{\Pi} \circ d \circ \varphi^*(\alpha) = \tilde{\partial}(\varphi^*(\alpha)).\end{aligned}$$

■

The restriction of ∂ to the algebra $\underline{\Omega}^*(J^1M)$ we denote by δ and will call *an operator of non-holonomic differentiation*

$$\delta : \underline{\Omega}^s(J^1M) \to \underline{\Omega}^{s+1}(J^1M).$$

Note that the sequence

$$0 \xrightarrow{\delta} \underline{\Omega}^0(J^1M) \xrightarrow{\delta} \underline{\Omega}^1(J^1M) \xrightarrow{\delta} \underline{\Omega}^2(J^1M) \xrightarrow{\delta} \underline{\Omega}^3(J^1M) \xrightarrow{\delta} \underline{\Omega}^4(J^1M) \xrightarrow{\delta} 0 \tag{21.9}$$

is a complex.

This complex we call a *non-holonomic de Rham complex*. Note that

$$\delta(\alpha \wedge \beta) = \delta\alpha \wedge \beta + (-1)^{\deg \alpha} \alpha \wedge \delta\beta$$

for any differential forms $\alpha, \beta \in \Omega^s(J^1M)$.

A differential l-horizontal 2-form ω is called *l-effective* if $\omega \wedge \partial\omega_0 = 0$.

Note also that

$$\partial(g\omega_0) = g\partial\omega_0$$

for any function $g \in C^\infty(J^1M)$.

Let ω be an l-effective differential 2-form and let $\varphi : J^1M \to J^1M$ be a contact diffeomorphism such that $\varphi^*(\omega) = \hat{\omega}$. Then the form $\hat{\omega}$ is $\hat{l}$-effective also.

For any differential l-horizontal 2-form α we can construct its l-effective part

$$\alpha_l \stackrel{\text{def}}{=} \alpha - s_\alpha \partial\omega_0 \wedge \partial\omega_0,$$

where the function s_α is defined by

$$\alpha \wedge \partial\omega_0 = s_\alpha \partial\omega_0 \wedge \partial\omega_0.$$

Obviously, the effective 2-form ω and its l-effective part ω_l generates the same Monge–Ampère equation.

Thus we can reformulate the condition of contact equivalence of Monge–Ampère equations E_ω and $E_{\tilde{\omega}}$ in terms of l-effective forms:

$$\varphi^*(\omega_l) = h\tilde{\omega}_{\tilde{l}}.$$

From now on we assume that ω and $\tilde{\omega}$ are l-effective and $\tilde{l}$-effective differential 2-forms respectively and $\varphi^*(\omega_0) = \lambda\omega_0$, $\varphi^*(\omega) = h\tilde{\omega}$.

If we introduce an *l-Pfaffian* F of ω by

$$F\partial\omega_0 \wedge \partial\omega_0 = \omega \wedge \omega,$$

then

$$\varphi^*(F) = \frac{h^2}{\lambda^2}\tilde{F}.$$

Here $\tilde{F}$ is the $\tilde{l}$-Pfaffian of $\tilde{\omega}$.

Define an operator $A : D(C) \to D(C)$ by the formula:

$$AX \rfloor \partial\omega_0 = X \rfloor \omega, \tag{21.10}$$

where the vector field $X \in D(C)$.

As above the square of the operator A is scalar and $A^2 + F = 0$.

Lemma 21.4.2 *We have*

$$\tilde{A} = \frac{\lambda}{h}\varphi_*^{-1} \circ A \circ \varphi_*.$$

Proof From (21.10), for any vector field $X \in D(C)$ we have: $\lambda\varphi_*^{-1}(AX) \rfloor \tilde{\partial}\omega_0 = h\varphi_*^{-1}(X) \rfloor \tilde{\omega}$. Moreover, $\varphi_*^{-1}(X) \rfloor \tilde{\omega} = \tilde{A}\varphi_*^{-1}(X) \rfloor \tilde{\partial}\omega_0$. Since $\partial\omega_0$ is non-degenerate, we obtain $\lambda\varphi_*^{-1} \circ A = h\tilde{A} \circ \varphi_*^{-1}$. Therefore $\tilde{A} = (\lambda/h)\varphi_*^{-1} \circ A \circ \varphi_*$. ∎

The differential 2-form $\partial\omega_0$ is non-degenerate on the module of vector fields from the Cartan distribution. This means that if $X \rfloor \partial\omega_0 = 0$ for $X \in D(C)$, then $X = 0$.

For a function $H \in C^\infty(J^1M)$ the formula

$$X_H \rfloor \partial\omega_0 = -\partial H \tag{21.11}$$

uniquely defines a vector field $X_H \in D(C)$, and

$$\varphi_*^{-1}(X_H) = \frac{1}{\lambda}\tilde{X}_{\varphi^*(H)}.$$

Similarly with Lemma 20.4.1 one can prove that

$$\partial h \wedge \omega = \frac{1}{2}AX_h \rfloor (\partial\omega_0 \wedge \partial\omega_0).$$

for any function $h \in C^\infty(J^1M)$.

The relations

$$W \rfloor (\partial\omega_0 \wedge \partial\omega_0) = 2\partial\omega,$$
$$\omega_0(W) = 0.$$

uniquely determine a vector field W in the Cartan distribution.

Let $\langle Z, d_{1,1,-1}\rangle$ and $\langle Z, d_{-1,1,1}\rangle$ be contractions of the vector field Z and the tensors $d_{1,1,-1}$ and $d_{-1,1,1}$ respectively. Define a function k as a contraction of the $(1,1)$-tensors $\langle Z, d_{1,1,-1}\rangle$ and $\langle Z, d_{-1,1,1}\rangle$:

$$k \stackrel{\text{def}}{=} \langle\langle Z, d_{1,1,-1}\rangle, \langle Z, d_{-1,1,1}\rangle\rangle.$$

Then

$$\varphi^*(k) = \frac{1}{\lambda^2}\tilde{k}$$

and therefore k is a relative invariant.

Suppose that $k \neq 0$. Let us introduce the function

$$F_0 \stackrel{\text{def}}{=} \frac{F}{k}$$

and the vector field

$$V \stackrel{\text{def}}{=} \frac{1}{k} AW.$$

Since Lemma 21.4.2, and the facts that

$$\varphi_*^{-1}(W) = \frac{1}{\lambda^2}(h\tilde{W} + \tilde{A}\tilde{X}_h),$$
$$\varphi^*(F_0) = h^2 \tilde{F}_0,$$

we get

$$\varphi_*^{-1}(V) = \frac{1}{\varphi^*(k)}(\varphi_*^{-1} \circ A \circ \varphi_*) \circ \varphi_*^{-1}(W)$$
$$= \frac{h^2}{\lambda}\tilde{V} - \frac{h}{\lambda}\tilde{F}_0\tilde{X}_h.$$

For the vector fields X_{F_0} we have

$$\varphi_*^{-1}(X_{F_0}) = \frac{h^2}{\lambda}\tilde{X}_{\tilde{F}_0} + \frac{2h}{\lambda}\tilde{F}_0\tilde{X}_h.$$

Then the vector field

$$\mathcal{Y} \stackrel{\text{def}}{=} \frac{\sqrt{|k|}}{F_0}(X_{F_0} + 2V)$$

is an invariant (up to the sign):

$$\varphi_*^{-1}(\mathcal{Y}) = \tilde{\mathcal{Y}}.$$

This vector field falls into the sum

$$\mathcal{Y} = \mathcal{Y}_{1,0} + \mathcal{Y}_{0,1},$$

where $Y_{1,0} \in D(C_{1,0})$ and $Y_{0,1} \in D(C_{0,1})$ are also invariants.

Define the vector field $\mathcal{Z} \in D(l)$ by

$$\omega_0(\mathcal{Z}) = \frac{1}{\sqrt{|k|}}.$$

This vector field is an invariant (up to the sign).

Applying tensors $d_{-1,1,1}$ and $d_{1,1,-1}$ to these invariants we obtain two invariant vector fields from the distributions $C_{1,0}$ and $C_{0,1}$:

$$\mathcal{X}_{1,0} \stackrel{\text{def}}{=} d_{-1,1,1}(\mathcal{Z}, \mathcal{Y}_{0,1}),$$

$$\mathcal{X}_{0,1} \stackrel{\text{def}}{=} d_{1,1,-1}(\mathcal{Z}, \mathcal{Y}_{1,0}).$$

For the case of the general Monge–Ampère equation the vector fields $\mathcal{Z}, \mathcal{X}_{1,0}, \mathcal{Y}_{1,0}, \mathcal{X}_{0,1}, \mathcal{Y}_{0,1}$ form an e-structure on J^1M. Denote the constructed e-structure by

$$e_E = (\mathcal{Z}, \{\mathcal{X}_{1,0}, \mathcal{X}_{0,1}\}, \{\mathcal{Y}_{1,0}, \mathcal{Y}_{0,1}\}).$$

If the equation E is hyperbolic, then this e-structure is real. If E is elliptic, then this e-structure is complex. In the last case we can construct a real e-structure using the conjugate operation.

Theorem 21.4.1 *Two non-degenerate Monge–Ampère equations E and $\tilde{E}$ are contact equivalent if and only if the e-structures e_E and $e_{\tilde{E}}$ are equivalent.*

Example 21.4.1 *Let us consider the following non-linear wave equation:*

$$v_{xy} = (v_x + v_y)^3.$$

In order to simplify formulae we assume that $p_1 + p_2 > 0$*. We have* $k = 36(p_1+p_2)^2$*. The l-effective part of the form* ω *and the form* $\partial\omega_0$ *are*

$$\begin{aligned}
\tilde{\omega} &= dq_1 \wedge dp_1 - dq_2 \wedge dp_2 + 3(p_1+p_2)^2(dq_2 \wedge du - dq_1 \wedge du+) \\
&\quad + (p_1+p_2)^3 dq_1 \wedge dq_2, \\
\partial\omega_0 &= dq_1 \wedge dp_1 + dq_2 \wedge dp_2 - 3(p_1+p_2)^2(dq_1 \wedge du + dq_2 \wedge du) \\
&\quad + (3(p_1+p_2)^2(p_2-p_1))\, dq_1 \wedge dq_2
\end{aligned}$$

respectively. In the free basis $X_{1,0}, Y_{1,0}, X_{0,1}, Y_{1,0}$ *the operator* A *has a diagonal form:*

$$A = \begin{Vmatrix} 1 & 0 & 0 & 0 \\ 0 & 1 & 0 & 0 \\ 0 & 0 & -1 & 0 \\ 0 & 0 & 0 & -1 \end{Vmatrix}.$$

The following vector fields define the e-structure:

$$\begin{aligned}
\mathcal{Z} &= \frac{1}{6(p_1+p_2)}\left(\frac{\partial}{\partial u} + 3(p_1+p_2)^2\left(\frac{\partial}{\partial p_1} + \frac{\partial}{\partial p_2}\right)\right), \\
\mathcal{Y}_{1,0} &= 12\left((p_1+p_2)^3\left(\frac{\partial}{\partial p_2} - 4\frac{\partial}{\partial p_1}\right) + \frac{d}{dq_1}\right), \\
\mathcal{X}_{1,0} &= -54(p_1+p_2)^3\frac{\partial}{\partial p_1}, \\
\mathcal{Y}_{0,1} &= 12\left((p_1+p_2)^3\left(\frac{\partial}{\partial p_1} - 4\frac{\partial}{\partial p_2}\right) + \frac{d}{dq_2}\right), \\
\mathcal{X}_{0,1} &= -54(p_1+p_2)^3\frac{\partial}{\partial p_2}.
\end{aligned}$$

21.4.2 Functional invariants

Let us construct two functional invariants for non-degenerate Monge–Ampère equations.

First of all we define the contraction operations

$$\rfloor : \Omega^i(N) \otimes D(N) \times \Omega^j(N) \otimes D(N) \to \Omega^{i+j-1}(N) \otimes D(N),$$

(see [49]) and

$$\langle\, ,\, \rangle : \Omega^i(N) \otimes D(N) \times \Omega^j(N) \otimes D(N) \to \Omega^{i+j-2}(N)$$

on a smooth manifold N.

For decomposable tensors $\alpha \otimes X$ and $\beta \otimes Y$ we put

$$(\alpha \otimes X) \rfloor (\beta \otimes Y) \stackrel{\text{def}}{=} \alpha \wedge (X \rfloor \beta) \otimes Y$$

and

$$\langle \alpha \otimes X, \beta \otimes Y \rangle \stackrel{\text{def}}{=} (Y \rfloor \alpha) \wedge (X \rfloor \beta).$$

Suppose that $\Pi\langle d_{-1,1,1} \rfloor d_{2,-1,0}, d_{1,1,-1} \rfloor d_{0,-1,2}\rangle \neq 0$ and by definition, let

$$I_{1,0} \stackrel{\text{def}}{=} \frac{\Pi\langle d_{-1,1,1} \rfloor d_{2,-1,0}, d_{-1,1,1} \rfloor d_{2,-1,0}\rangle}{\Pi\langle d_{-1,1,1} \rfloor d_{2,-1,0}, d_{1,1,-1} \rfloor d_{0,-1,2}\rangle}$$

and

$$I_{0,1} \stackrel{\text{def}}{=} \frac{\Pi\langle d_{1,1,-1} \rfloor d_{0,-1,2}, d_{1,1,-1} \rfloor d_{0,-1,2}\rangle}{\Pi\langle d_{-1,1,1} \rfloor d_{2,-1,0}, d_{1,1,-1} \rfloor d_{0,-1,2}\rangle}.$$

Since the numerators and denominators in these fractions are differential forms from $\underline{\Omega}^4(J^1M)$, these definitions are correct. Therefore $I_{1,0}$ and $I_{0,1}$ are functional invariants of Monge–Ampère equations.

For elliptic equations its real and imaginary parts are real functional invariants.

Note that for (21.2) $I_{1,0}$ and $I_{0,1}$ are zero.

22
Symplectic classification of MAEs on three-dimensional manifolds

In this chapter we consider local classification of general symplectic Monge–Ampère equations. We will discuss the conditions which we need to impose on the effective forms ω_i, $i = 1, 2$ such that there exist a local symplectic diffeomorphism $F : T^*M \to T^*M$ transforming one form into another: $F^*(\omega_1) = \omega_2$.

The problem is purely local so we can suppose without loss of generality that $M = \mathbb{R}^n$. We will also restrict ourselves to the case when one of these forms has *constant* coefficients. Then we will reduce the equivalence problem to integration of a system of non-linear differential equations in the space of

1-jets of Lagrangian sabmanifolds in $\mathbb{R}^{4n}$. The local equivalence problem has a solution in a neighborhood of $0 \in \mathbb{R}^{2n}$ if and only if this system has a local solution.

The main sources of the results in this chapter are the original papers [74] and [77] but extended by results of Banos [5] who had adapted and simplified our original proofs. In particular, he had used the original idea of [77] to replace the Cartan–Kähler integrability theorem by the Frobenius theorem, which is much easies but can be applied only under some conditions on the higher prolongations of the $Sp(n)$-orbit stabilizers.

We refer interested readers to the original paper [77], where we treat more general cases but in much less detail.

22.1 Jets of submanifolds and differential equations on submanifolds

In this section we briefly recall main structures on the manifolds of jets (a more detailed exposition can be found in [1]).

Let M be a manifold. We consider submanifolds $N \subset M$ of fixed codimension m. Let N be such a submanifold, then we denote by $[N]_a^k$ the k-jet of N at the point $a \in M$.

Let $J_a^k(M,m)$ be the set of all k-jets of submanifolds at the point $a \in M$, and

$$J^k(M,m) = \bigcup_{a \in M} J_a^k(M,m)$$

which is the set of all k-jets.

Then $J^k(M,m)$ are smooth manifolds and the natural projections $\pi_{k,l}$: $J^k(M,m) \to J^l(M,m), k > l$, are smooth bundles. Here $\pi_{k,l} : [N]_a^k \longmapsto [N]_a^l$.

Moreover, the bundles $\pi_{k,k-1}$ are affine bundles when $k \geq 2$. Namely, fibres $F(a_{k-1}) = \pi_{k,k-1}^{-1}(a_{k-1})$, where $a_{k-1} = [N]_a^{k-1}$, are affine spaces. The vector spaces associated with the fibres are $S^k(T_a^*N) \otimes \nu_a(N)$ where $\nu_a(N) = T_aM/T_aN$ is the normal space to the submanifold N at the point $a \in N$. The affine structure is invariant with respect to diffeomorphisms of M.

The bundle $\pi_{1,0}$ can be described as follows. First of all $J^0(M,m) = M$, and $J_a^1(M,m)$ can be identified with the Grassmanian $\mathrm{Gr}_m(T_aM)$ of vector subspaces in T_aM having codimension m.

Therefore, fibres $\pi_{1,0}^{-1}(a) = \mathrm{Gr}_m(T_aM)$ are Grassmanians and the bundle has Grassmanian structure.

Any submanifold $N \subset M$ of the codimension m defines submanifolds $j_k(N) \subset J^k(M,m)$ by

$$j_k(N) = \{[N]_a^k \mid a \in N\}.$$

The submanifold $j_k(N)$ can be viewed as a "derivative" of $j_{k-1}(N)$. Geometrically, a point $a_k = [N]_a^k$ can be identified with the tangent space to submanifold $j_{k-1}(N)$ at the point $a_{k-1} = [N]_a^{k-1}$. The tangent space obviously depends on a_k only, and we will denote it by $L(a_k)$.

The linear span of all subspaces $L(a_k)$ at the point a_{k-1} forms a *Cartan subspace* $C(a_{k-1})$ of the tangent space $T_{a_{k-1}}(J^{k-1}(M,m))$. This subspace consists of all vectors $v \in T_{a_{k-1}}(J^{k-1}(M,m))$ such that $(\pi_{k-1,k-2})_* v \in L(a_{k-1})$, or

$$C(a_{k-1}) = (\pi_{k-1,k-2})_*^{-1}(L(a_{k-1})).$$

The distribution $C : a_{k-1} \longmapsto C(a_{k-1})$ of Cartan subspaces on $J^{k-1}(M,m)$ is called the *Cartan distribution.*

From the construction of the Cartan distribution we see that the submanifolds $j_{k-1}(N)$ are integral for C. Moreover, one can easily prove that any integral manifold $L \subset J^{k-1}(M,m)$ of the Cartan distribution has the form $j_{k-1}(N)$ if and only if the map $\pi_{k-1,0} : L \to M$ is an embedding.

By a system of differential equations of order k we mean a submanifold $E \subset J^k(M,m)$. A solution of this system is an integral manifold L of the Cartan distribution on $J^k(M,m)$ such that $L \subset E$ and $\dim L = \dim M - m$. If the map $\pi_{k,0} : L \to M$ is an embedding, then $L = j_k(N)$ for a submanifold $N \subset M$.

We call the last submanifold a *smooth solution of* E.

With a system $E \subset J^k(M,m)$ of order k we associate a *derived* or *prolongated system* $E^{(1)} \subset J^{k+1}(M,m)$ of order $k+1$ as follows: a point $a_{k+1} \in J^{k+1}(M,m)$ belongs to $E^{(1)}$ if and only if $L(a_{k+1}) \subset T_{a_k}(E)$.

The set $E^{(1)}$ we call first *prolongation of* E. If $E^{(1)}$ is smooth, then one can repeat the procedure and get the second *prolongation* $E^{(2)} = (E^{(1)})^{(1)} \subset J^{k+2}(M,m)$, and so on.

A system $E \subset J^k(M,m)$ is called *formally integrable* if all prolongations $E^{(i)} \subset J^{k+i}(M,m)$ are smooth submanifolds and all projections $\pi_{k+i,k+i-1} : E^{(i)} \to E^{(i-1)}$ are smooth bundles.

For any system $E \subset J^k(M,m)$ fibres of the map $\pi_{k+1,k} : E^{(1)} \to E$, if they are not empty, are affine subspaces of the affine bundle $\pi_{k+1,k}$. Therefore, in order to check the smoothness of $E^{(1)}$ one should estimate dimensions of the fibres. This can be done in the following way.

A *symbol of* E at point $a_k \in E$ is the subspace $g_k(a_k)$ of the vertical tangent vectors, that is

$$g_k(a_k) = T_{a_k}(E) \cap T_{a_k}(F(a_{k-1}))$$

where $a_{k-1} = \pi_{k,k-1}(a_k)$.

Because $T_{a_k}(F(a_{k-1})) = S^k(T_a^* N) \otimes \nu_a(N)$ the symbol is a vector subspace of the tensor space

$$g_k(a_k) \subset S^k(T_a^* N) \otimes \nu_a(N).$$

Let $g_k^{(l)}(a_{k+l})$ be the symbol of the lth prolongation at $a_{k+l} \in E^{(l)}$, then

$$g_k^{(l)}(a_{k+l}) \subset S^{k+l}(T_a^* N) \otimes \nu_a(N)$$

and all of these spaces can be defined by the symbol $g_k(a_k)$ only.

Namely, the first prolongation symbol $g_k^{(1)}(a_{k+1})$ consists of tensors $\theta \in S^{k+1}(T_a^* N) \otimes \nu_a(N)$ such that $v(\theta) \in g_k(a_k)$ for all vectors $v \in T_a N$, and $g_k^{(l+1)}(a_{k+l+1}) = (g_k^{(l)}(a_{k+l}))^{(1)}$. Here we denote the derivative of θ along v by $v(\theta)$.

We say, that a system E has a *finite type* if $g_k^{(l)}(a_k) = 0$ for all $a_k \in E$ and all $l \geq l_0$.

Let E be such a system. Assume that $E^{(l_0)}$ is a smooth manifold and $\pi_{k+l_0,k+l_0-1} : E^{(l_0)} \to E^{(l_0-1)}$ is a diffeomorphism.

Note that $\pi_{k+l_0,k+l_0-1} \; : \; E^{(l_0)} \;\; \to \;\; E^{(l_0-1)}$ is a local diffeomorphism if $g_k^{(l_0)} = 0$.

Denote by $a'_{k+l_0-1} \in E^{(l_0)}$ the preimage of $a_{k+l_0-1} \in E^{(l_0-1)}$, $\pi_{k+l_0,k+l_0-1}$ $(a'_{k+l_0-1}) = a_{k+l_0-1}$.

Then the correspondence

$$\mathcal{L} : a_{k+l_0-1} \in E^{(l_0-1)} \longmapsto L(a'_{k+l_0-1}) \subset T_{a_{k+l_0-1}}(E^{(l_0-1)})$$

defines a distribution on $E^{(l_0-1)}$.

Note that this distribution is completely integrable if and only if $E^{(l_0+1)}$ is not empty and $\pi_{k+l_0+1,k+l_0} : E^{(l_0+1)} \to E^{(l_0)}$ is a diffeomorphism.

Indeed, if the distribution $\mathcal{L}$ is completely integrable then its integral manifold has the form $j_{k+l_0-1}(N)$ and $[N]_a^{k+l_0+1}$ are the only points over $[N]_a^{k+l_0} \in E^{(l_0)}$.

On the other hand, let $\pi_{k+l_0+1,k+l_0} : E^{(l_0+1)} \to E^{(l_0)}$ be a diffeomorphism. To find value of the curvature form of $\mathcal{L}$ at a point $b = a_{k+l_0-1} \in E^{(l_0-1)}$ we should find a value of the commutator $[X, Y]_b$ for any pair of vector fields belonging to $\mathcal{L}$. Note that this value depends on 1-jets of vector fields X and Y at the point b.

Let $a''_{k+l_0-1} \in E^{(l_0+1)}, a'_{k+l_0-1} \in E^{(l_0)}$ be the preimages of the point b, and let $a''_{k+l_0-1} = [N]_a^{k+l_0+1}$. Then for given 1-jets $[X]_b^1$ and $[Y]_b^1$ the vector fields X and Y can be chosen to be tangent to $j_{k+l_0-1}(N)$. Therefore, $[X,Y]_b \in T_b(j_{k+l_0-1}(N)) = \mathcal{L}(b)$, and $\mathcal{L}$ is completely integrable.

Summrizing we obtain the following result.

Theorem 22.1.1 *Let $E \subset J^k(M,m)$ be a finite type system such that*

1. $g_k^{(l_0)} = 0$,
2. $\pi_{k+l_0+1,k+l_0} : E^{(l_0+1)} \to E^{(l_0)}$ *and* $\pi_{k+l_0,k+l_0-1} : E^{(l_0)} \to E^{(l_0-1)}$ *are diffeomorphisms.*

Then

- *the distribution $\mathcal{L}$ on $E^{(l_0-1)}$ defined by $E^{(l_0)}$ is completly integrable, and*
- *for any point $a_{k+l_0-1} \in E^{(l_0-1)}$ there is a solution $N \subset M$, such that $[N]_a^{k+l_0-1} = a_{k+l_0-1}$.*

22.2 Prolongations of contact and symplectic manifolds and overdetermined Monge–Ampère equations

In this section we shall apply the previous constructions to Lagrangian and Legendrean submanifolds of symplectic and contact manifolds, and to overdetermined Monge–Ampère equations.

22.2.1 Prolongations of symplectic manifolds

Let (M,Ω) be a symplectic manifold. We consider Lagrangian submanifolds $N \subset M$. If N is such a submanifold, then, as above, we denote by $[N]_a^k$ the k-jet of N at the point $a \in M$.

Let $J_a^k(M,\Omega)$ be the set of all k-jets of Lagrangian submanifolds at the point $a \in M$, and

$$J^k(M,\Omega) = \bigcup_{a \in M} J_a^k(M,\Omega)$$

the set of all k-jets.

Then $J^k(M,\Omega)$ are smooth manifolds and the natural projections $\pi_{k,l} : J^k(M,\Omega) \to J^l(M,\Omega), k > l$, are smooth bundles.

Moreover, the bundles $\pi_{k,k-1}$ are affine bundles, when $k \geq 2$.

Namely, fibres $F(a_{k-1}) = \pi_{k,k-1}^{-1}(a_{k-1})$, where $a_{k-1} = [N]_a^{k-1}$, are affine spaces, and the vector spaces associated with the fibres are $S^{k+1}(T_a^*N)$. The affine structure is invariant with respect to symplectic diffeomorphisms of M.

The bundle $\pi_{1,0}$ can be described as follows. First of all $J^0(M,\Omega) = M$, and $J_a^1(M,\Omega)$ can be identified with the Lagrangian Grassmanian $\mathrm{LGr}(T_aM,\Omega)$ of Lagrangian subspaces in T_aM.

Therefore, fibres $\pi_{1,0}^{-1}(a) = \mathrm{LGr}(T_aM,\Omega)$ are Lagrangian Grassmanians and the bundle has the Grassmanian structure.

Any Lagrangian submanifold $N \subset M$ defines submanifolds $j_k(N) \subset J^k(M,\Omega)$ by

$$j_k(N) = \{[N]_a^k \mid a \in N\}.$$

The submanifold $j_k(N)$ can be viewed as a "derivative" of $j_{k-1}(N)$.

Geometrically, a point $a_k = [N]_a^k$ can be identified with the tangent space to the submanifold $j_{k-1}(N)$ at the point $a_{k-1} = [N]_a^{k-1}$. The tangent space obviously depends on a_k only, and we will denote it by $L(a_k)$.

The linear span of all subspaces $L(a_k)$ at the point a_{k-1} forms a *Cartan subspace* $C(a_{k-1})$ of the tangent space $T_{a_{k-1}}(J^{k-1}(M,\Omega))$. This subspace consists of all vectors $v \in T_{a_{k-1}}(J^{k-1}(M,\Omega))$ such that $(\pi_{k-1,k-2})_* v \in L(a_{k-1})$, or

$$C(a_{k-1}) = (\pi_{k-1,k-2})_*^{-1}(L(a_{k-1})).$$

The distribution $C : a_{k-1} \longmapsto C(a_{k-1})$ of Cartan subspaces on $J^{k-1}(M,\Omega)$ is called the *Cartan distribution.*

From the construction of the Cartan distribution we see that the submanifolds $j_{k-1}(N)$ are integral for C. Moreover, one can easily prove that any integral manifold $L \subset J^{k-1}(M,\Omega)$ of the Cartan distribution has the form $j_{k-1}(N)$ if and only if the map $\pi_{k-1,0} : L \to M$ is an embedding, and L is Lagrangian.

By a system of differential equations of order k on Lagrangian submanifolds we mean a submanifold $E \subset J^k(M,\Omega)$. A solution of this system is an integral manifold L of the Cartan distribution on $J^k(M,\Omega)$ such that $L \subset E$ and $\dim L = \frac{1}{2}\dim M$.

If the map $\pi_{k,0} : L \to M$ is an embedding, then $L = j_k(N)$ for some Lagrangian submanifold $N \subset M$. We call the last submanifold a *smooth solution of* E.

With a system $E \subset J^k(M,\Omega)$ of order k we associate a *derived,* or *prolongated system* $E^{(1)} \subset J^{k+1}(M,\Omega)$ of order $k+1$ as follows: a point $a_{k+1} \in J^{k+1}(M,\Omega)$ belongs to $E^{(1)}$ if and only if $L(a_{k+1}) \subset T_{a_k}(E)$.

The set $E^{(1)}$ we call first *prolongation of* E.

If $E^{(1)}$ is smooth, then one can repeat the procedure and get the second *prolongation* $E^{(2)} = (E^{(1)})^{(1)} \subset J^{k+2}(M, \Omega)$, and so on.

A system $E \subset J^k(M, \Omega)$ is called *formally integrable* if all prolongations $E^{(i)} \subset J^{k+i}(M, \Omega)$ are smooth submanifolds and all projections $\pi_{k+i,k+i-1} : E^{(i)} \to E^{(i-1)}$ are smooth bundles.

For any system $E \subset J^k(M, \Omega)$ fibres of the map $\pi_{k+1,k} : E^{(1)} \to E$, if they are not empty, are affine subspaces of the affine bundle $\pi_{k+1,k}$. Therefore, in order to check the smoothness of $E^{(1)}$ one should estimate dimensions of the fibres. This can be done in the same way as for general case.

A *symbol of E* at point $a_k \in E$ is the following subspace $g_k(a_k)$ of the vertical tangent space, that is,

$$g_k(a_k) = T_{a_k}(E) \cap T_{a_k}(F(a_{k-1})),$$

where $a_{k-1} = \pi_{k,k-1}(a_k)$.

Because $T_{a_k}(F(a_{k-1})) = S^{k+1}(T_a^* N)$ the symbol is a vector subspace of the tensor space

$$g_k(a_k) \subset S^{k+1}(T_a^* N).$$

Let $g_k^{(l)}(a_{k+l})$ be the symbol of the lth prolongation at $a_{k+l} \in E^{(l)}$, then

$$g_k^{(l)}(a_{k+l}) \subset S^{k+l+1}(T_a^* N)$$

and all of these spaces can be defined by the symbol $g_k(a_k)$ only.

Namely, the first prolongation symbol $g_k^{(1)}(a_{k+1})$ consist of tensors $\theta \in S^{k+2}(T_a^* N)$ such that $v(\theta) \in g_k(a_k)$ for all vectors $v \in T_a N$, and $g_k^{(l+1)}(a_{k+l+1}) = (g_k^{(l)}(a_{k+l}))^{(1)}$.

We say, that a system E has a *finite type* if $g_k^{(l)}(a_k) = 0$ for all $a_k \in E$ and $l \geq l_0$.

Let E be such a system. Assume that $E^{(l_0)}$ is a smooth manifold and $\pi_{k+l_0,k+l_0-1} : E^{(l_0)} \to E^{(l_0-1)}$ is a diffeomorphism.

Denote by $a'_{k+l_0-1} \in E^{(l_0)}$ the preimage of $a_{k+l_0-1} \in E^{(l_0-1)}$, $\pi_{k+l_0,k+l_0-1}(a'_{k+l_0-1}) = a_{k+l_0-1}$.

Then the correspondence

$$\mathcal{L} : a_{k+l_0-1} \in E^{(l_0-1)} \longmapsto L(a'_{k+l_0-1}) \subset T_{a_{k+l_0-1}}(E^{(l_0-1)})$$

defines a distribution on $E^{(l_0-1)}$.

As above one can prove that this distribution is completely integrable if and only if $E^{(l_0+1)}$ is not empty and $\pi_{k+l_0+1,k+l_0} : E^{(l_0+1)} \to E^{(l_0)}$ is a diffeomorphism.

Summrizing we obtain the following result.

Theorem 22.2.1 *Let $E \subset J^k(M,\Omega)$ be a finite type system of differential equations on Lagrangian submanifols such that:*

1. $g_k^{(l_0)} = 0$,
2. $\pi_{k+l_0+1,k+l_0} : E^{(l_0+1)} \to E^{(l_0)}$ *and* $\pi_{k+l_0,k+l_0-1} : E^{(l_0)} \to E^{(l_0-1)}$ *are diffeomorphisms.*

Then,

- *the distribution $\mathcal{L}$ on $E^{(l_0-1)}$ defined by $E^{(l_0)}$ is completly integrable, and*
- *for any point $a_{k+l_0-1} \in E^{(l_0-1)}$ there is a solution, that is, a Lagrangian submanifold $N \subset M$, such that $j_{k+l_0-1}(N) \subset E^{(l_0\;1)}$, and $[N]_a^{k+l_0-1} = a_{k+l_0-1}$.*

Note that any effective r-form $\omega \in \Omega_\varepsilon^r(M)$ defines a first-order system of differential equations on Lagrangian submanifolds $E_\omega \subset J^1(M,\Omega)$, where

$$E_\omega = \{[N]_a^1 \in \mathrm{LGr}(T_aM,\Omega) \mid \omega|_{T_aN} = 0\}.$$

We call this system the *overedermined Monge–Ampère equation.*

If $r < \frac{1}{2}\dim M$ then E_ω has the finite type for ωs "in the general position" and one can use the above theorem in order to get smooth solvability of the system.

Later on we will apply this remark to overedermined Monge–Ampère equations arising in the classification problems.

22.2.2 Prolongations of contact manifolds

Let $(M,\mathcal{P})$ be a contact manifold. We consider Legendrean submanifolds $N \subset M$.

Let $J_a^k(M,\mathcal{P})$ be the set of all k-jets of Legendrean submanifolds at the point $a \in M$, and

$$J^k(M,\mathcal{P}) = \bigcup_{a\in M} J_a^k(M,\mathcal{P})$$

the set of all k-jets.

Then $J^k(M,\mathcal{P})$ are smooth manifolds and the natural projections $\pi_{k,l}$: $J^k(M,\mathcal{P}) \to J^l(M,\mathcal{P}), k > l$, are smooth bundles.

Moreover, the bundles $\pi_{k,k-1}$ are affine bundles when $k \geq 2$. Namely, fibres $F(a_{k-1}) = \pi_{k,k-1}^{-1}(a_{k-1})$, where $a_{k-1} = [N]_a^{k-1}$, are affine spaces. The vector spaces associated with the fibres are $S^{k+1}(T_a^*N) \otimes \nu_a$, where $\nu_a = T_aM/\mathcal{P}_a$. The affine structure is invariant with respect to contact diffeomorphisms of M.

The bundle $\pi_{1,0}$ can be described as follows. First of all $J^0(M,\mathcal{P}) = M$, and $J_a^1(M,\mathcal{P})$ can be identified with the Lagrangian Grassmanian $\mathrm{LGr}(\mathcal{P}_a)$ of Lagrangian subspaces in $\mathcal{P}_a$.

Therefore fibres $\pi_{1,0}^{-1}(a) = \mathrm{LGr}(\mathcal{P}_a)$ are Lagrangian Grassmanians and the bundle has Grassmanian structure.

Any Legendrean submanifold $N \subset M$ defines submanifolds $j_k(N) \subset J^k(M,\mathcal{P})$ by

$$j_k(N) = \{[N]_a^k \mid a \in N\}.$$

Geometrically, a point $a_k = [N]_a^k$ can be identified with the tangent space to the submanifold $j_{k-1}(N)$ at the point $a_{k-1} = [N]_a^{k-1}$. The tangent space obviously depends on a_k only, and we will denote it by $L(a_k)$.

The linear span of all subspaces $L(a_k)$ at the point a_{k-1} forms a *Cartan subspace* $C(a_{k-1})$ of the tangent space $T_{a_{k-1}}(J^{k-1}(M,\mathcal{P}))$.

This subspace consist of all vectors $v \in T_{a_{k-1}}(J^{k-1}(M,\mathcal{P}))$ such that $(\pi_{k-1,k-2})_*v \in L(a_{k-1})$, or

$$C(a_{k-1}) = (\pi_{k-1,k-2})_*^{-1}(L(a_{k-1})).$$

The distribution $C : a_{k-1} \longmapsto C(a_{k-1})$ of Cartan subspaces on $J^{k-1}(M,\mathcal{P})$ is called the *Cartan distribution.*

From the construction of the Cartan distribution we see that the submanifolds $j_{k-1}(N)$ are integral for C. Moreover, one can easily prove that any integral manifold $L \subset J^{k-1}(M,\mathcal{P})$ of the Cartan distribution has the form $j_{k-1}(N)$ if and only if the map $\pi_{k-1,0} : L \to M$ is an embedding, and L is Legendrean.

By a system of differential equations of order k on Legendrean submanifolds we mean a submanifold $E \subset J^k(M,\mathcal{P})$. A solution of this system is an integral manifold L of the Cartan distribution on $J^k(M,\mathcal{P})$ such that $L \subset E$ and $\dim L = \frac{1}{2}(\dim M - 1)$. If the map $\pi_{k,0} : L \to M$ is an embedding, then $L = j_k(N)$ for some Legendrean submanifold $N \subset M$. We call the last submanifold a *smooth solution of* E.

With a system $E \subset J^k(M,\mathcal{P})$ of order k we associate a *derived*, or *prolongated system* $E^{(1)} \subset J^{k+1}(M,\mathcal{P})$ of order $k+1$ as follows: a point $a_{k+1} \in J^{k+1}(M,\mathcal{P})$ belongs to $E^{(1)}$ if and only if $L(a_{k+1}) \subset T_{a_k}(E)$.

The set $E^{(1)}$ we call first *prolongation of* E. If $E^{(1)}$ is smooth, then we can repeat the procedure and get the second *prolongation* $E^{(2)} = (E^{(1)})^{(1)} \subset J^{k+2}(M, \Omega)$, and so on.

A system $E \subset J^k(M, \mathcal{P})$ is called *formally integrable* if all prolongations $E^{(i)} \subset J^{k+i}(M, \mathcal{P})$ are smooth submanifolds and all projections $\pi_{k+i,k+i-1} : E^{(i)} \to E^{(i-1)}$ are smooth bundles.

For any system $E \subset J^k(M, \mathcal{P})$ fibres of the map $\pi_{k+1,k} : E^{(1)} \to E$, if they are not empty, are affine subspaces of the affine bundle $\pi_{k+1,k}$. Therefore, in order to check smoothness of $E^{(1)}$ one should estimate dimensions of the fibres. This can be done in the same way as for general and symplectic cases.

A *symbol of* E at point $a_k \in E$ is the following subspace $g_k(a_k)$ of the vertical tangent space, that is,

$$g_k(a_k) = T_{a_k}(E) \cap T_{a_k}(F(a_{k-1}))$$

where $a_{k-1} = \pi_{k,k-1}(a_k)$.

Because $T_{a_k}(F(a_{k-1})) = S^{k+1}(T_a^* N)$ the symbol is a vector subspace of the tensor space

$$g_k(a_k) \subset S^{k+1}(T_a^* N) \otimes \nu_a.$$

Let $g_k^{(l)}(a_{k+l})$ be the symbol of the lth prolongation at $a_{k+l} \in E^{(l)}$, then

$$g_k^{(l)}(a_{k+l}) \subset S^{k+l+1}(T_a^* N) \otimes \nu_a$$

and all of these spaces can be defined by the symbol $g_k(a_k)$ only.

Namely, the first prolongation symbol $g_k^{(1)}(a_{k+1})$ consists of tensors $\theta \in S^{k+2}(T_a^* N) \otimes \nu_a$ such that $v(\theta) \in g_k(a_k)$ for all vectors $v \in T_a N$, and $g_k^{(l+1)}(a_{k+l+1}) = (g_k^{(l)}(a_{k+l}))^{(1)}$.

We say that a system E has a *finite type* if $g_k^{(l)}(a_k) = 0$ for all $a_k \in E$ and $l \geq l_0$.

Let E be such a system. Assume that $E^{(l_0)}$ is a smooth manifold and $\pi_{k+l_0,k+l_0-1} : E^{(l_0)} \to E^{(l_0-1)}$ is a diffeomorphism.

Denote by $a'_{k+l_0-1} \in E^{(l_0)}$ the preimage of $a_{k+l_0-1} \in E^{(l_0-1)}$, $\pi_{k+l_0,k+l_0-1}(a'_{k+l_0-1}) = a_{k+l_0-1}$.

Then the correspondence

$$\mathcal{L} : a_{k+l_0-1} \in E^{(l_0-1)} \longmapsto L(a'_{k+l_0-1}) \subset T_{a_{k+l_0-1}}(E^{(l_0-1)})$$

defines a distribution on $E^{(l_0-1)}$.

As above one can prove that this distribution is completely integrable if and only if $E^{(l_0+1)}$ is not empty and $\pi_{k+l_0+1,k+l_0} : E^{(l_0+1)} \to E^{(l_0)}$ is a diffeomorphism.

Summarizing we obtain the following result.

Theorem 22.2.2 *Let $E \subset J^k(M,\Omega)$ be a finite type system of differential equations on Legendrean submanifolds such that*

1. $g_k^{(l_0)} = 0$,
2. $\pi_{k+l_0+1,k+l_0} : E^{(l_0+1)} \to E^{(l_0)}$ *and* $\pi_{k+l_0,k+l_0-1} : E^{(l_0)} \to E^{(l_0-1)}$ *are diffeomorphisms.*

Then,

- *the distribution $\mathcal{L}$ on $E^{(l_0-1)}$ defined by $E^{(l_0)}$ is completly integrable, and*
- *for any point $a_{k+l_0-1} \in E^{(l_0-1)}$ there is a solution, that is, a Legendrean submanifold $N \subset M$, such that $j_{k+l_0-1}(N) \subset E^{(l_0-1)}$, and $[N]_a^{k+l_0-1} = a_{k+l_0-1}$.*

Note that any effective r-form $\omega \in \Omega_\varepsilon^r(M)$ defines a first-order system of differential equations on Legendrean submanifolds $E_\omega \subset J^1(M,\Omega)$ where

$$E_\omega = \{[N]_a^1 \in \mathrm{LGr}(T_aM) \mid \omega|_{T_aN} = 0\}.$$

We call this system the *(overdetermined) Monge–Ampère equation.*

If $r < \frac{1}{2}(\dim M - 1)$ then E_ω has finite type for ωs "in the general position" and one can use the above theorem in order to obtain smooth solvability of the system.

22.3 Differential equations for symplectic equivalence

Let ω_1 and ω_2 be effective n-forms on T^*M, $n = \dim M$. We reformulate the problem of finding a symplectic diffeomorphism ϕ such that $\phi^*(\omega_2) = \omega_1$ as follows. Denote the symplectic manifold (T^*M,Ω) by (Φ,Ω), and let (Φ',Ω') be a copy. The graph $L_\phi \subset \Phi\times\Phi'$ of a symplectic diffeomorphism $\phi : \Phi \to \Phi'$ is a Lagrangian submanifold of the symplectic manifold $(\Phi\times\Phi', \pi^*\Omega - \pi'^*\Omega')$, where $\pi : \Phi\times\Phi' \to \Phi$, and $\pi' : \Phi\times\Phi' \to \Phi'$ are the natural projections. Moreover, $\phi^*(\omega_2) = \omega_1$ if and only if L_ϕ is a solution of the Monge–Ampère equations corresponding to the effective form $\omega = \pi^*\omega_1 - \pi'^*\omega_2$.

In other words, local symplectic equivalence of effective forms ω_1 and ω_2 follows from the local solvability of the Monge–Ampère equation

$$E_\omega \subset J^1(\Phi\times\Phi', \pi^*\Omega - \pi'^*\Omega').$$

Note that $[\phi]^1_{(a,b)} \in E_\omega$ if and only if the differential $\phi_{*,a} : T_a\Phi \to T_b\Phi$ sends $\omega_{2,b}$ to $\omega_{1,a}$. Therefore, $\pi_{1,0} : E_\omega \to \Phi \times \Phi'$ is onto if and only if effective n-forms $\omega_{1,a}$ and $\omega_{2,b}$ are symplectically equivalent for any points $a, b \in \Phi$. In this case, symbols of E_ω at different points are isomorphic.

Assume, for example, that ω_1 and ω_2 are equal at a point a_0. Then the symbol at $[\mathrm{id}]^1_{(a_0,a_0)} \in E_\omega$, which we denote by $g \subset S^2T^*_{a_0}$, is isomorphic to the stabilizer of ω_{1,a_0} in the Lie algebra of Hamiltonian operators. It is easy to check that symbols of the prolongations of E_ω are exactly the prolongations of g.

Assume now that the first prolongation $g^{(1)}$ is trivial. This means that the fibres $\pi_{2,1} : E^{(1)}_\omega \to E_\omega$ are discrete or empty. Let $[\phi]^1_{(a,b)} \in E_\omega$ or $\phi^*_a(\omega_{2,b}) = \omega_{1,a}$. Then a 1-jet of the n-form $\theta = \phi^*(\omega_2) - \omega_1$ is equal to zero at point a. Consider the 1-jet of θ

$$\varepsilon = [\theta]^1_a \in \Lambda^n_{\mathrm{ef},a} \otimes T^*_a,$$

and deformations $\theta_t = (A_t \circ \phi)^*(\omega_2) - \omega_1$ by symplectic diffeomorphisms which preserve the point a. Then the fibre of $\pi_{2,1} : E^{(1)}_\omega \to E_\omega$ at $[\phi]^1_{(a,b)}$ is not empty if one can find a deformation such that $[\theta_t]^1_a = 0$ for some value of t. Remark that $A^*_t = 1 + tL_H + o(t)$ for a Hamiltonian H with $d_bH = 0$, and

$$\theta_t = \theta + t\phi^* L_H(\omega_2) + o(t).$$

Therefore,

$$[\theta_t]^1_a = [\theta]^1_a + t[\phi^* L_H \omega_2)]^1_a.$$

Let

$$K_a = \{[\phi^* L_H \omega_2)]^1_a, d_a H = 0\} \subset \Lambda^n_{ef,a}$$

be the subspace of the space of effective forms.

Then $[\phi]^1_{(a,b)} \in E_\omega$ has a prolongation if and only if $[\theta]^1_a \in K_a$. Consider the following function:

$$\sigma_1([\phi]^1_{(a,b)}) = [\theta]^1_a \bmod K_a.$$

Then $[\phi]^1_{(a,b)} \in E_\omega$ admits a prolongation if and only if

$$\sigma_1([\phi]^1_{(a,b)}) = 0.$$

Assume now that σ_1 vanishes on the equation E_ω, then bundles of the projection $\pi_{2,1} : E^{(1)}_\omega \to E_\omega$ isomorphic to $g^{(1)} = 0$, and, as above, we obtain a distribution

$\mathcal{L}$ on E_ω. The curvature σ_2 of the distribution defines a g-valued function on E_ω, and if $\sigma_2 = 0$ on E_ω then the classification problem has a (local) solution.

Summarizing, we obtain the following theorem.

Theorem 22.3.1 *Let ω_1 and ω_2 be effective n-forms on T^*M, $n = \dim M$, such that*

1. *exterior forms $\omega_{1,x}$ and $\omega_{2,y}$ are simplectically equivalent for all points x, y from a neighborhood of a point $a \in T^*M$,*
2. *the first prolongation of the stabilizer of $\omega_{1,a}$ in $\mathrm{sp}(T_a, \Omega_a)$ is trivial, and*
3. *the obstractions σ_1 and σ_2 vanish.*

Then the n-form ω_1 is symplectically equivalent to ω_2 in a neighborhood of the point $a \in T^*M$.

Let apply this result for $M = \mathbb{R}^3$. Denote by ω_{const} an effective form with constant coefficients on $T^*\mathbb{R}^3$. We say that an effective 3-form ω is *regular* if the Sp-orbit of ω_a is regular for all points a in the domain of consideration. In our case this means that it belongs to one of the five first orbits listed in the classification table. Then the first prolongation of the stabilizer is trivial and we can apply the previous result.

Theorem 22.3.2 *Consider a Monge–Ampère equation in $\mathbb{R}^3$, corresponding to a regular effective 3-form ω, such that* Sp*-orbits of ω_q do not depend on q. Let ω_{const} be an effective form with constant coefficients with the same* Sp*-orbit. Assume that the obstractions σ_1 and σ_2, constructed for ω and ω_{const}, vanish. Then this differential equation is locally equivalent to one of the following equations:*

$$\lambda + \det \operatorname{Hess} v = 0, \qquad \lambda \neq 0;$$

$$\frac{\partial^2 v}{\partial q_1^2} - \frac{\partial^2 v}{\partial q_2^2} + \frac{\partial^2 v}{\partial q_3^2} + \lambda^2 \det \operatorname{Hess} v = 0, \quad \lambda \neq 0;$$

$$\frac{\partial^2 v}{\partial q_1^2} + \frac{\partial^2 v}{\partial q_2^2} + \frac{\partial^2 v}{\partial q_3^2} - \eta^2 \det \operatorname{Hess} v = 0, \quad \eta \neq 0;$$

$$\frac{\partial^2 v}{\partial q_1^2} - \frac{\partial^2 v}{\partial q_2^2} + \frac{\partial^2 v}{\partial q_3^2} = 0;$$

$$\frac{\partial^2 v}{\partial q_1^2} + \frac{\partial^2 v}{\partial q_2^2} + \frac{\partial^2 v}{\partial q_3^2} = 0.$$

References

[1] D. V. Alekseevskij, V. V. Lychagin and A. M. Vinogradov, *Basic Ideas and Concepts of Differential Geometry, Geometry* vol. I. (Berlin: Springer–Verlag, 1991), pp. 1–264.

[2] J. S. Allen, J. A. Barth and P. A. Newberger, On intermediate models for barotropic continental shelf and slope flowfields. Part I: formulation and comparison of exact solutions. *Phys. Oceanogr.* **20** (1990), 1017–1042.

[3] E. Artin, *Geometric Algebra.* (New York: John Wiley & Sons, Inc., 1988).

[4] B. Banos, Nondegenerate Monge–Ampère structures in dimension 3. *Lett. Math. Phys.* **62**:1 (2002), 1–15.

[5] B. Banos, On symplectic classification of effective 3-forms and Monge–Ampère equations. *Differ. Geom. Appl.* **19**:2 (2003), 147–166.

[6] N. Belotelov and A. Lobanov, Migration and demographic processes. In *Nonlinear Diffusion, Proceeding of Mathematics, Computer, Education, International Conference.* (Moscow: Dubna, 1999), pp. 434–443.

[7] L. Bers, *Mathematical Aspects of Subsonic and Transonic Gas Dynamics.* (New York: John Wiley & Sons, Inc., 1958).

[8] A. V. Bitsadze, *Some Classes of Partial Differential Equations.* (Moscow: Nauka, 1981).

[9] E. Bonan, Decomposition of the exterior algebra of hyper-Kähler manifolds. *C. R. Acad. Sci. Paris, Math.* **320**:4. (1995), 457–462.

[10] W. M. Boothby and H. C. Wang, On contact manifolds. *Ann. Math.* **2**:68 (1958), 721–734.

[11] M. Born and L. Infeld, Foundation of a new field theory. *Proc. R. Soc.* **144** (1934), 425–451.

[12] N. E. Byers, Noether's discovery of the deep connection between symmetries and conservation laws. In *The Heritage of Emmy Noether.* (RamatGan, 1996), pp. 67–81.

[13] G. R. Cavalcanti and M. Gualtieri Generalized complex structures on nilmanifolds. *J. Symplectic Geom.* **2**:3 (2004), 393–410.

[14] S. A. Chaplygin, Gas jets. *Uch. Zap. Mosk. Univ., Otdel. Fiz.-Mat. Nauk* **21** (1904).

[15] S. Chynoweth and M. J. Sewell, Dual variables in semi-geostrophic theory. *Proc. R. Soc. Lond.* A **424** (1989), 155–186.

[16] S. Chynoweth and M. J. Sewel, A concise derivation of the semi-geostrophic equations. *Q. J. R. Meteorol. Soc.* **117** (1991), 1109–1128.

[17] M. J. P. Cullen, J. Norbury and R. J. Purser, Generalised Lagrangian solutions for atmospheric and oceanic flows. *Siam J. Appl. Math.* **51** (1991), 20–31.

[18] M. J. P. Cullen and R. J. Purser, An extended Lagrangian theory of semi-geostrophic frontogenesis. *J. Atmos. Sci.* **41** (1984), 1477–1497.

[19] G. A. Desroziers, Coordinate change for data assimilation in spherical geometry of frontal structures. *Mon. Wea. Rev.* **125** (1997), 3030–3038.

[20] V. A. Dorodnitsin, Group properties and invariant solutions of non-linear thermal conductivity equations with source or drain, Preprint No 57 of Applied Mathematics Institute of the Academy of Sciences of the USSR (1979).

[21] B. Doubrov and A. Kushner, *The Morimoto Problem, Geometry in Partial Differential Equations.* (River Edge, NJ: World Scientific Publishing, 1994), pp. 91–99.

[22] S. V. Duzhin and V. V. Lychagin, Symmetries of distributions and quadrature of ordinary differential equations. *Acta Appl. Math.* **24**:1 (1991), 29–57.

[23] F. J. Ernst, Black holes in a magnetic universe. *J. Math. Phys.* **17** (1976), 54–56.

[24] D. S. Freed, Special Kähler manifolds. *Commun. Math. Phys.* **203** (1999), 31–52.

[25] O.-K. Fossum, On classification of a system of non-linear first-order PDEs of Jacobi type, Preprint Tromsö University (2002).

[26] A. Frölicher and A. Nijenhuis, Theory of vector valued differential forms. Part 1: derivations in the graded ring of differential forms. *Indag. Math.* **18** (1956), 338–359.

[27] Gibbons J., Tsarev S. P., Reduction of the Benney equations, *Phys. Lett. A* **211** (1996), 19–24.

[28] H. Goldschmidt, Formal integrability criteria for systems of non-linear partial differential equations *J. Differ. Geom.* **1** (1967), 269–307.

[29] E. Goursat, *Leçon sur l'intégration des Équations aux Dérivée Partielles du Second Ordre a Deux Variables Indépendantes*, vol. 1 (Paris, 1896).

[30] E. Goursat, Sur le problème de Monge. *Bull. Soc. Math. France* **33** (1905), 201–210.

[31] E. Goursat, Sur les èquations du second ordre à n variables analogues à l'equation de Monge–Ampère. *Bull. Soc. Math. France* **27** (1899), 1–34.

[32] M. Gross, Special Lagrangian fibrations II: geometry. *Surv. Differ. Geom.* **5** (1999), 341–403.

[33] D. I. Gurevich, V. V. Lychagin and V. N. Roubtsov, Cohomologie de "grand crochet" et filtration "non-holonome." *C. R. Acad. Sci. Paris Sr. I Math.* **313**:12 (1991), 855–858.

[34] D. I. Gurevich, V. V. Lychagin and Roubtsov V. N., Nonholonomic filtration of the cohomology of Lie algebras and "large brackets." *Mat. Zametki* **52**:1 (1992), 36–41.

[35] G. Hart, *Geometric Quantization in Action: Applications of Harmonic Analysis in Quantum Statistical Mechanics and Quantum Field Theory.* (Dordrecht: Reidel Publishing, 1983).

[36] N. J. Hitchin, *Monopoles, Minimal Surfaces and Algebraic Curves, NATO Advanced Study Institute* vol. 105 (1987).

[37] N. J. Hitchin, The moduli space of special Lagrangian submanifolds. *Ann. Scuola Norm. Sup. Pisa Cl. Sci.* **25** (1997), 503–515.

[38] N. J. Hitchin, The geometry of three-forms in six dimensions. *Differ. Geom.* **55**:3 (2000), 547–576.

[39] J. K. Hunter and R. Saxton, Dynamics of director fields. *SIAM J. Appl. Math.* **51**:6 (1991), 1498–1521.

[40] Igusa, Jun-ichi, A classification of spinors up to dimension twelve. *Amer. J. Math.* **92** (1970), 997–1028.

[41] P. Jakobsen, V. Lychagin and Y. Romanovsky, Symmetries and non-linear phenomena, I. Preprint, Tromsö University (1997).

[42] P. Jakobsen, V. Lychagin and Y. Romanovsky, Symmetries and non linear phenomena, II, applications to non-linear acoustics. Preprint, Tromsö University (1998).

[43] K. Jörgens, Differentialgleichung $rt - s^2 = 1$. *Math. Ann.* **127** (1954), 130–134.

[44] T. Karman, The similarity law of transonic flow. *Math. Phys.* **26** (1947), 182–190.

[45] Kazuhiko Aomoto and Toru Tsujishita, Open problems in structure theory of non-linear integrable differential and difference systems. (Nagoya University, 1984), p. 44.

[46] M. V. Keldysh, Cases of degeneracy of equations of elliptic type on the boundary of a domain. *Dokl. Akad. Nauk SSSR* **77**:2 (1951), 181–183.

[47] A. A. Kirillov, *Geometric Quantization*, vol. 4. (2001), 139–176.

[48] A. N. Kolmogorov, I. G. Petrovsky and I. S Piskunov. Examining of diffusion equation with increasing of substance amount, and its application to one biological problem. *MSU Bull.* A **6** (1937), 1–26.

[49] I. B. Kovalenko and A. G. Kushner, The non-linear diffusion and thermal conductivity equation: group classification and exact solutions. *Regular Chaotic Dynam.* **8**:2 (2003), 8–31.

[50] I. S. Krasilshchik, V. V. Lychagin and A. M. Vinogradov, *Geometry of Jet Spaces and Non-linear Partial Differential Equations.* (New York: Gordon and Breach, 1986).

[51] I. S. Krasilshchik, Some new cohomological invariants for non-linear differential equations. *Differ. Geom. Appl.* **2**:4 (1992), 307–350.

[52] B. S. Kruglikov, Nijenhuis tensors and obstructions to the construction of pseudoholomorphic mappings. *Mat. Zametki* **63**:4 (1998), 541–561.

[53] B. S. Kruglikov, On some classification problems in four-dimensional geometry: distributions, almost complex structures, and the generalized Monge–Ampère equations. *Mat. Sb.* **189**:11 (1998), 61–74.

[54] B. S. Kruglikov, Symplectic and contact Lie algebras with application to the Monge–Ampère equations. *Tr. Mat. Inst. Steklova* **221** (1998), 232–246.

[55] B. S. Kruglikov, Classification of Monge–Ampère equations with two variables. *CAUSTICS '98, Warsaw.* (Warsaw: Polish Academy of Sciences, 1999), pp. 179–194.

[56] B. S. Kruglikov and V. V. Lychagin, On equivalence of differential equations. *Acta Comment. Univ. Tartu. Math.* (1999), 7–29.

[57] B. S. Kruglikov and V. V. Lychagin, Mayer brackets and solvability of PDEs, I. *Differ. Geom. Appl.* **17**:2–3 (2002), 251–272.

[58] A. Kushner, Chaplygin and Keldysh normal forms of Monge–Ampère equations of variable type. *Math. Zametki* **52**:5 (1992), 63–67.

[59] A. Kushner, Classification of mixed type Monge–Ampère equations. *Geometry in Partial Differential Equations*. (1993) pp. 173–188.

[60] A. Kushner, Symplectic geometry of mixed type equations. *Amer. Math. Soc. Transl. Ser.* **2** (1995), 131–142.

[61] A. Kushner, Monge–Ampère equations and e-structures. *Dokl. Akad. Nauk* **361**:5 (1998), 595–596.

[62] A. Kushner, Contact linearization of nondegenerate Monge–Ampère equations. *Natural Sci.*

[63] I. B. Kovalenko and A. G. Kushner, Symmetries and exact solutions of non-linear diffusion equation. *Math. Model. Comput. Biol. Medicine.* July (2002), 239.

[64] N. Larkin, B. Novikov and N. Yanenko, *Non-Linear Mixed Type Equations.* (Novosibirsk: Nauka, 1983).

[65] S. Lie, Classification und integration von gewöhnlichen Differentialgleichungen zwischen x, y, die eine Gruppe von Transformationen gestatten. *Math. Ann.* **32** (1888), 213–281.

[66] S. Lie, Begrundung einer Invarianten-Theorie der Beruhrungs-Transformationen. *Math. Ann.* **8** (1874), 215–303.

[67] S. Lie Über einige partielle Differential-Gleichungen zweiter Orduung. *Math. Ann.* **5** (1872), 209–256.

[68] V. Lychagin and O. Lychagina, Finite dimensional dynamics for evolutionary equations. *Nonlin. Dynam.* (2006), to appear.

[69] V. V. Lychagin, Contact geometry and second-order non-linear differential equations. *Russian Math.* **34** (1979), 137–165.

[70] V. V. Lychagin, Singularities of multivalued solutions of non-linear differential equations and non-linear phenomena. *Acta Appl. Math.* **3** (1985), 135–173.

[71] V. V. Lychagin, Differential equations on two-dimensional manifolds. *Izv. Vyssh. Uchebn. Zaved. Mat.* **5** (1992), 43–57.

[72] V. Lychagin, *Lectures on Geometry of Differential Equations*, vols 1, 2. (Rome: La Sapienza, 1993).

[73] V. V. Lychagin and V. N. Roubtsov, On Sophus Lie theorems for Monge–Ampère equations. *Dokl. Akad. Nauk BSSR* **27**:5 (1983), 396–398.

[74] V. V. Lychagin and V. N. Roubtsov, Local classification of Monge–Ampère differential equations. *Dokl. Akad. Nauk SSSR* **272**:1 (1983), 34–38.

[75] V. V. Lychagin and V. N. Roubtsov, Non-holonomic filtration: algebraic and geometric aspects of non-integrability. In *Geometry in Partial Differential Equations.* (River Edge, NJ: World Scientific Publishing, 1994).

[76] V. V. Lychagin and R. Romanovsky Yu., Nonholonomic intermediate integrals of partial differential equations. In *Differential Geometry and its Applications* (Brno, 1989). (Teaneck, NJ: World Scientific Publishing, 1990).

[77] V. V. Lychagin, V. N. Roubtsov and I. V. Chekalov, A classification of Monge–Ampère equations. *Ann. Sci. Ecole Norm. Sup.(4)* **26**:3 (1993), 281–308.

[78] L. J. Mason and N. M. J. Woodhouse, *Integrability, Self-duality, and Twistor Theory, London Mathematical Society Monographs, New Series*, vol. 15. (New York: Oxford University Press, 1996).

[79] M. Matsuda, Two methods of integrating Monge–Ampère equations I. *Trans. Amer. Math. Soc.* **150**:1 (1970), 327–343.

[80] J. C. McWilliams and P. R. Gent, Intermediate models of planetary circulations in the atmosphere and ocean. *J. Atmos. Sci.* **37** (1980), 1657–1678.

[81] M. E. McIntyre and I. Roulstone, Hamiltonian balanced models: constraints, slow manifolds and velocity splitting. *Forecasting Research Scientific Paper* 41 (1996), http://www.atm.damtp.cam.ac.uk/people/mem/

[82] M. E. McIntyre and I. Roulstone, Are there higher-accuracy analogues of semi-geostrophic theory? *Large Scale Atmosphere–Ocean Dynamics*, vol. II, *Geometric Methods and Models*, ed. I. Roulstone and J. Norbury. (Cambridge: Cambridge University Press), to appear.

[83] T. Morimoto, La géométrie des équations de Monge–Ampére. *C. R. Acad. Sci. Paris* A-B **289**:1 (1979), A25–A28.

[84] T. Morimoto, Open problem in str. theory of non-linear integrable differential and difference systems. *The 15 International Symposium, Katata.* (1984), pp. 27–29.

[85] O. I. Morozov, Contact equivalence problem for non-linear wave equations. Preprint arXiv: math-ph/0306007 v.1 (2003), pp. 1–13.

[86] O. I. Morozov, Contact equivalence of the generalized Hunter–Saxton equation and the Euler–Poisson equation. Preprint arXiv: math-ph/0406016 (2004), pp. 1–3.

[87] J. D. Murray, *Mathematical Biology*. (Berlin: Springer-Verag, 1993).

[88] A. Newlender and L. Nirenberg, Complex analytic coordinates in almost complex manifolds. *Ann. Math.* **65** (1954), 391–404.

[89] J. Nielsen, On the intermediate integral for Monge–Ampère equations. *Proc. Amer. Math. Soc.* **128**:2 (2000), 527–531.

[90] V. L. Popov, A classification of spinors of dimension fourteen. *Trudy Moskov. Mat. Obshch.* **37** (1978), 173–217.

[91] R. J. Purser, Contact transformations and Hamiltonian dynamics in generalized semi-geostrophic theories. *J. Atmos. Sci.* **50** (1993), 1449–1468.

[92] R. J. Purser and M. J. P. Cullen, A duality principle in semi-geostrophic theory. *J. Atmos. Sci.* **44** (1987), 3449–3468.

[93] B. L. Rojdestvensky and N. N. Yanenko, *Systems of Quasilinear Equations and their Applications to Gas Dynamics*. (Moscow: Nauka, 1968).

[94] V. N. Roubtsov and I. Roulstone, Examples of quaternionic and Kähler structures in Hamiltonian models of nearly geostrophic flow. *J. Phys.* A **30**:4 (1997), L63–L68.

[95] Roubtsov V. N. and I. Roulstone, Holomorphic structures in hydrodynamical models of nearly geostrophic flow. *R. Soc. Lond. Proc.* A **457** (2001), 1519–1531.

[96] I. Roulstone and J. Norbury, A Hamiltonian structure with contact geometry for the semi-geostrophic equations. *J. Fluid Mech.* **272** (1994), 211–233.

[97] I. Roulstone and M. J. Sewell, The mathematical structure of theories of semi-geostrophic type. *Phil. Trans. R. Soc. Lond.* **355** (1997), 2489–2517.

[98] O. V. Rudenko and S. I. Soluyan, *Theoretical Foundations of Non-linear Acoustics*, transl. from Russian by R. T. Beyer, *Studies in Soviet Science*. (New York: Consultants Bureau, 1997).

[99] A. A. Samarsky, S. P. Kurdyumov, V. A. Galaktionov and Michailov A. P., *Peaking Regimes in Problem for Quasilinear Parabolic Equations* (Moscow: Nauka, 1987).

[100] W. Slebodzinski, *Exterior Forms and their Applications*. (Warsaw: PWN-Polish Scientific Publishers, 1970).

[101] R. Salmon, Practical use of Hamilton's principle. *J. Fluid Mech.* **132** (1983), 431–444.

[102] R. Salmon, New equations for nearly geostrophic flow. *J. Fluid Mech.* **153** (1985), 461–477.

[103] R. Salmon, Semi-geostrophic theory as a Dirac bracket projection. *J. Fluid Mech.* **196** (1988), 345–358.

[104] M. J. Sewell and I. Roulstone, Families of lift and contact transformations. *Proc. R. Soc. Lond.* **447** (1994), 493–512.

[105] D. V. Tunitskii, Contact equivalence of Monge–Ampère equations with transitive symmetries. In *Differential Geometry and Applications*, Brno. (1995), pp. 479–485.

[106] D. V. Tunitskii, On the contact linearization of Monge–Ampère equations. *Izv. Ross. Akad. Nauk Ser. Mat.* **60**:2 (1996), 195–220.

[107] D. V. Tunitskii, Monge–Ampère equations and characteristic connection functors. *Izv. Ross. Akad. Nauk Ser. Mat.* **65**:6 (2001), 173–222.

[108] F. Tricomi, Sulle equazioni lineari alle derivate parziali di secondo ordine di tipo misto. *Rend. Atti Acad. Nazion. Lincei* **5**:14 (1923), 134–247.

[109] B. Trubnikov and S. Zhdanov, *Quasi-Gaseous Unstable Media*. (Moscow: Nauka, 1991).

[110] C. Udriste and N. Bila, Symmetry group of Titeica surfaces, PDE. *Balkan J. Geom. Applic.* **4**:2 (1999), 123–140.

[111] G. B. Whitham, *Linear and Non-linear Waves*. (New York: Wiley-Interscience, 1974).

[112] V. F. Zaitsev and A. D. Polyanin, *Non-Linear Partial Differential Equations, Handbook: Exact Solutions*. (Moscow: Nauka, 1996).

[113] L. V. Zilbergleit, Second-order differential equation associated with contact geometry: symmetries, conservation laws and shock waves. *Acta Appl. Math.* **45**:1 (1996), 51–71.

[114] L. V. Zilbergleit, Geometry of the Hugoniot–Rankine conditions. *Russ. J. Math. Phys.* **7**:2 (2000), 234–249.

Index